T0189494

Women in Engineering and Science

Series Editor
Jill S. Tietjen
Greenwood Village, Colorado, USA

The Springer Women in Engineering and Science series highlights women's accomplishments in these critical fields. The foundational volume in the series provides a broad overview of women's multi-faceted contributions to engineering over the last century. Each subsequent volume is dedicated to illuminating women's research and achievements in key, targeted areas of contemporary engineering and science endeavors. The goal for the series is to raise awareness of the pivotal work women are undertaking in areas of keen importance to our global community.

More information about this series at https://link.springer.com/bookseries/15424

Alice E Smith
Editor

Women in Computational Intelligence

Key Advances and Perspectives on Emerging Topics

Editor
Alice E Smith
Department of Industrial
and Systems Engineering
and Department of Computer Science
and Software Engineering
Auburn University
Auburn, Al, USA

ISSN 2509-6427 ISSN 2509-6435 (electronic)
Women in Engineering and Science
ISBN 978-3-030-79094-3 ISBN 978-3-030-79092-9 (eBook)
https://doi.org/10.1007/978-3-030-79092-9

This Springer imprint is published by the registered company Springer Nature Switzerland AG
The registered company address is: Gewerbestrasse 11, 6330 Cham, Switzerland

This book is dedicated to my immediate family – my husband Randy Michael Smith, and our children Elaine Michelle Smith Prince, Alexander David Smith, and Nicholas Charles Smith. The children are all engineers and happily progressing in their chosen professions of logistics, manufacturing, and healthcare. Currently I have four grandchildren, and I anticipate and look forward to multiple engineers/computer scientists among them.

Preface

This book is the first of its kind – showcasing the many diverse research breakthroughs achieved by women-led investigative teams in computational intelligence. Computational intelligence (also sometimes termed artificial intelligence or computing inspired by natural systems) is multi-disciplinary and embraces contributions by engineers, computer scientists, mathematicians, and even social and natural scientists. The field of computational intelligence stretches back to the mid-twentieth century but has seen an explosion of activity over the past several decades with the advent of powerful, plentiful, and inexpensive computing. It is a rapidly developing science that is applicable to almost any sector including healthcare, education, logistics, transportation, finance, and energy.

This book is structured into four main sections of Intelligence, Learning, Modeling, and Optimization. The primary technical methods include artificial neural networks, evolutionary and swarm computation, and fuzzy logic and systems. The wealth of applications can be seen throughout the nineteen chapters within this volume. These include natural language processing, intelligent tutoring, autonomous systems, digital pathology, intrusion detection, and energy management. The 34 authors of this book are nearly all women and represent thirteen countries in five continents. A unique part of this book is the biographies of the authors which include information concerning their beginnings and advancement in computational intelligence research. Many biographies also give advice for those considering this field and its possibilities.

As a final note, the biographical chapter on a founding luminary of computing, Admiral Grace Murray Hopper, is especially meaningful to me. As a young engineer, I attended a Society of Women Engineers national convention and had the privilege of listening firsthand to Admiral Hopper as she delivered the keynote address. It is said that we stand on giants' shoulders – certainly, Grace Hopper is one of those giants. It is hoped that future researchers in computational intelligence can remember the chapters within this volume as inspirations for them to choose this exciting field for their lifework.

Auburn, Alabama, USA
December 2021

Alice E Smith

Contents

Amazing Grace – Computer Pioneer Admiral Grace Murray Hopper

Jill S. Tietjen

1 Introduction

No one thought of having a computer translate the letters of the alphabet into machine code before Admiral Grace Murray Hopper because "they weren't as lazy as I was," she said. This ability – to tell the computer how to translate English (actually any human language) into machine code – laid the foundation for the computer age as we know it today. Admiral Hopper's development of the computer compiler was a giant step toward the widespread usage of computers that now affect almost every aspect of our lives. She received many awards and honors in recognition of her efforts during her lifetime, including the 1991 National Medal of Technology, presented by President H.W. Bush. She was the first individual woman to receive this Medal. The citation for the National Medal of Technology reads "For her pioneering accomplishments in the development of computer programming languages that simplified computer technology and opened the door to a significantly larger universe of users."

"Amazing Grace" successfully combined careers in academia, business, and the US Navy while making history in the computer field. Admiral Hopper is referred to as the mother of computerized data automation in the naval service. Her pioneering computer compilers enabled the computer user to communicate with the computer using human languages, not just mathematical symbols. Furthermore, she loved taking credit for finding the first computer bug – it was a moth stuck in the relay of a computer (Fig. 1).

When asked about the many accomplishments in her life, Admiral Hopper said, "If you ask me what accomplishment I'm most proud of, the answer would be all of

J. S. Tietjen (✉)
Technically Speaking, Inc., Greenwood Village, CO, USA

A. E. Smith (ed.), *Women in Computational Intelligence*, Women in Engineering and Science, https://doi.org/10.1007/978-3-030-79092-9_1

Fig. 1 Admiral Grace Murray Hopper. (Source: United States Navy)

the young people that I've trained over the years, that's more important than writing the first compiler."

2 Early Years

The oldest of three children, Grace Murray Hopper was born in 1906 in New York City to a father (Walter Murray) who was an insurance broker and a mother (Mary Campbell Van Horne Murray) who was the daughter of a civil engineer. Their father's legs were amputated by the time Admiral Hopper was in high school – the standard treatment for hardening of the arteries at the time. He told his children that if he could walk with two wooden legs and a cane, then they could do anything. He would say, "You can overcome things if you want to." His handicap was extra incentive for the children to excel – and to bring home A's from school – which they did. And, his handicap was also a reason for their mother to take care of the family's financial matters – so that Admiral Hopper learned a fascination with numbers – and with getting answers! [1, 12]

Her father's handicap and his response was not the only life philosophy that formed Admiral Hopper's life and career. Her father also believed that his daughters should have the same educational opportunities as his son. He wanted all of his children to go to college, and he wanted his daughters to work for at least one year after they finished college to ensure they knew how to economically fend for themselves [1].

As the oldest and being very adventuresome, Hopper was often fingered as the culprit when trouble was afoot. During the summers, her family vacationed at their summer home in Wolfeboro, New Hampshire: "On one occasion a bunch of cousins and I were caught scrambling in a pine tree. Since I was at the top, it was obvious who started it." [10]

She also loved gadgets from the beginning, a trait that would serve her well in the computer industry: *When I was about seven years old, each room in our Wolfeboro home had an alarm clock. Those round clocks had a bell on top and two feet that shake apart when the alarm goes off. When we were going on a summer hike or a trip, mother would go around and set the clocks for her children and all of the cousins who were visiting. She went around one night and found all the clocks taken apart. What had happened was that I had taken the first one apart and couldn't get it together .... and then I'd opened the next one and the next one, and ... But I'd always loved a good gadget. I'm afraid that in my heart, when I saw the first Mark I, my first thought was "Gee, that's the prettiest gadget I ever saw." Which probably would not have pleased Howard Aiken, had he heard it* [10].

She was a child whose obvious inquisitiveness stimulated a genius and whose excitement was still generated by the thrill and joy of the learning process [3]. Summing up her upbringing and the influence of her family on her life and career, Admiral Hopper said, "My mother's very great interest in mathematics and my father's, a house full of books, a constant interest in learning, an early interest in reading, and insatiable curiosity ... these were a primary influence all the way along." [12]

3 Education and Early Career

Admiral Hopper first attended private schools in New York City, a common practice at the time. The schools primarily educated their female students on how to become ladies. Later, she attended a boarding school, Hartridge School in New Jersey, which focused on college preparation. She took the required entrance exams for Vassar College in 1923 when she was 17 years old but failed the Latin portion. After studying languages diligently for the next year, she passed the Latin portion and was able to enroll in Vassar in 1924 [1, 12].

Reflecting on her high school years, she said: *We had to pass tests to prove we could read, write plain English, and spell. Each summer we had to read twenty books and write reports on them. You were educated and had some background then, not like today. It didn't give us any inhibitions. It gave us an interest in reading and history* [10].

Admiral Hopper studied mathematics and physics while she was at Vassar. When a professor asked her to help another student who was having trouble with physics, Admiral Hopper discovered her love of teaching. She also liked to bring real-world applications to her tutoring sessions – even having students learn about the theory of displacement by dropping objects – including themselves – into a bathtub! She was elected to Phi Beta Kappa, graduated with a bachelor of arts (1928), and won a Vassar College fellowship, which provided funding for further education. She used the money to enroll at Yale University and earned her master's degree in mathematics (1930) [1, 12].

After marrying and returning from her honeymoon to Europe, Admiral Hopper did something that not many women did in the 1930s – she went to work. Vassar College offered her a position teaching mathematics and she jumped at the opportunity. In addition, she decided she wanted further education and completed her PhD in mathematics from Yale University in 1934. During her time at Yale, she was awarded two Sterling Scholarships. Her dissertation topic was "The Irreducibility of Algebraic Equations." Admiral Hopper stayed in the Mathematics Department at Vassar from 1931 to 1943 and rose through the academic ranks from instructor to associate professor. She and her husband divorced in 1945 – no children had been born of the marriage [1, 12] (Fig. 2).

Fig. 2 Grace Murray Hopper. (Source: Library of Congress)

4 World War II and Harvard

After the US entered World War II, Admiral Hopper wondered what her role should be in helping her country. There was Navy blood in her family as her great-grandfather had been a rear admiral. In the summer of 1942, when she was teaching at Barnard College in New York City, she would see marching sailors go by her dormitory. Later, she said, "The more they went by, the more I wanted to be in the Navy also." The Navy didn't accept women at that time [1, 12].

But that would shortly change. The US Congress passed Public Law 869 authorizing the women's reserve as part of the US Navy, and it was signed into law by President Franklin D. Roosevelt on July 30, 1942. The Women Accepted for Voluntary Emergency Service – the WAVES – provided Admiral Hopper with the opportunity that she was looking for, although at first, the Navy didn't think so [9]. When she tried to enlist, the Navy rejected her – too old (36 years of age), too light (she only weighed 105 pounds and the minimum for her height of 5′6″ was 121 pounds), and she was a math teacher – whom the Navy believed needed to stay at their institutions to teach future soldiers and sailors [1, 12].

But Admiral Hopper had learned to be persistent from her childhood. She got a six-month leave of absence and associated permissions from Vassar College and waivers for her age and weight. In December 1943, Grace Murray Hopper was sworn into the US Navy. To her relief, as she would later say, "There was a war on! It was not unusual for a woman at that time to join the Navy; there were 30,000 to 40,000 women there at the time." Besides, she said, "It was the only thing to do." Off she went to Midshipman's School for Women at Smith College (Northampton, Massachusetts), where almost everyone else was the age of the students she had recently taught in her classes at Vassar. She loved Midshipman's School! [1, 2, 12]

Upon graduating at the top of her class from Midshipman's School, she was commissioned a lieutenant junior grade and assigned to work at the Bureau of Ordnance Computation Project at Harvard University. She said, "I'd never thought of going into computer work because there weren't any computers to go into – except for the Mark I, which I didn't know anything about. In the 1940s, you know, you could have put all the computer people in the country into one small room." She reported to work on July 2, 1944 and finally found the office of her new boss – Commander Howard Aiken [1, 12].

First I went to Com I headquarters; they said, 'It's out at Harvard'; so I went to Harvard, and after several hours, about two o'clock in the afternoon, I found the Bureau in the basement of Cruft Laboratory. I walked in the door, and Commander Aiken looked up at me and said, 'Where the hell have you been?' Well, heck, I'd just gotten out of Midshipman school, they said I could have two days off, and I had spent the whole day trying to find the place, so I tried to explain all that. And he said 'I mean for the last two months!' I said I was at Midshipman school. He said, 'I told them you didn't need that; we've got to get to work.' He then waved his hand at Mark I, all 51 feet of her, and he said, 'That's a computing engine.' He then informed me that he would be delighted to have the coefficients for the interpolation of the arc tangent by next Thursday. Fortunately, I was rescued by two young ensigns [Robert V.D. Campbell and Richard Bloch] [10].

I found that they'd been very thoughtful I preparing for me. They'd heard that this white-haired old schoolteacher was coming, so they bribed each other as to who would have to sit next to her. However, they were most kind to me, and they did help me get my first program onto the computer, which computed the coefficients of the arc tangent interpolation function [10].

And, thus, Admiral Hopper became, as she called herself, the "third programmer" on the Mark I, the world's first large-scale, automatically sequenced digital computer. Also known as the Automatic Sequence Controlled Calculator (ASCC), this new computing machine was a joint project between IBM and the Harvard Research Facility. It was officially dedicated in August of 1944 and leased by the US Navy for the remainder of World War II [1].

The Mark I weighed five tons, had about 800,000 parts and contained over 500 miles of wire. The 3300 relays could handle numbers up to 23 digits in length plus an algebraic sign for positive or negative. It could perform addition, multiplication and division and compute mathematical functions including logarithms and sines. Its capacity was three additions every second. Heralded as a modern miracle, it could do calculations in one day that previously had required six months to do by hand. It stayed in service until 1959. Said Admiral Hopper about the Mark I, "It was man's first attempt to build a machine that would assist the power of his brain rather than the strength of his arm." [1]

The Mark I provided the calculations that the Navy gunners needed during the war. New technologies that the fighting troops used included self-propelled rockets, large and complex guns, and new powerful bombs. The Ballistics Research Laboratory needed to compute more than 40 firing tables each week; they were capable of completing fifteen. Thus, the Mark I was desperately needed to support the war effort [1, 12].

In order to aim the new Navy anti-aircraft guns, gunners needed to know the angle necessary to elevate the gun from the horizon of the earth in order to hit a target at a known distance. The calculation needed to incorporate crosswinds, air density and temperature, and the weight of the shell. In addition to rocket trajectories and range tables, Admiral Hopper and her colleagues programmed the Mark I to calculate the area covered by a mine-sweeping detector towed behind a ship and the top secret mathematical simulation of the shock waves that were expected from the explosion of the first atomic bomb [1, 2, 12].

The only woman among the eight-member crew tasked with keeping the Mark I computer running 24 hours a day, seven days a week, Admiral Hopper and the rest of the crew sometimes slept on their desks in order to be available if the computer ran into problems. Social time was out of the question. As Admiral Hopper related, "You didn't go out in the evening and drink beer and compare notes on problems. You were dead tired and you went home and you went to bed because you were going to be there at the crack of dawn the next morning." [1, 12]

In order to program the Mark I, instructions had to be written in machine code telling the computer what operations it needed to perform and in what order to perform those calculations. These calculations told the computer which switches were to be set in the ON position and which switches were to be set in the OFF position. Each task or problem required a new set of code [1]. Thus, Admiral Hopper broke new ground every time she wrote code for a new problem. And she acquired detailed expertise in the design and use of early computers and, in particular, an understanding of the difficulty of programming problems directly in machine language – the only approach then available [3].

Admiral Hopper explained the coding process for the Mark I as thus: "The coding sheets we used had three columns on the left (for code numbers) and we wrote comments on the right which didn't go into the computer." Holes were then punched into paper tape standing for the code numbers on the coding sheets. Each horizontal line of holes on the paper tape represented a single command. Mistakes were very easy to make. The programming had to be done by mathematics experts [1].

At Harvard, she also helped develop the Mark II, the computer that came after Mark I, leading to the discovery of the first computer bug [3]. As Admiral Hopper tells the story,

In the summer of 1945 we were building Mark II; we had to build it in an awful rush – it was wartime – out of components we could get our hands on. We were working in a World War I temporary building. It was a hot summer and there was no air-conditioning, so all the windows were open. Mark II stopped, and we were trying to get her going. We finally found the relay that had failed. Inside the relay – and these were large relays – was a moth that had been beaten to death by the relay. We got a pair of tweezers. Very carefully we took the moth out of the relay, put it in the logbook, and put scotch tape over it.

Now, Commander Howard Aiken had a habit of coming into the room and saying "Are you making any numbers?" We had to have an excuse when we weren't making any numbers. From then on if we weren't making any numbers, we told him that we were debugging the computer. To the best of my knowledge that's where it started. I'm delighted to report that the first bug still exists [10] (Fig. 3).

The Mark II was designed, built, and tested within the span of three years. Five times faster than the Mark I, it was built for the Navy Bureau of Ordnance and installed at the Naval Proving Ground in Dahlgren, Virginia. Admiral Hopper described the Mark II as the first multiprocessor. She said, "you could split it into two computers, side by side. It worked in parallel and exchanged data through the transfer registers. And then under program control you could throw it back into being one computer again. So it was the beginning of multiprocessors." [1]

The first subroutines also originated at Harvard. Admiral Hopper and her colleagues collected pieces of programming that they knew worked. She relates, "When we got a correct subroutine, a piece of a program that had been checked out and debugged – one that we knew worked – we put it in a notebook." [1]

Admiral Hopper viewed Commander Aiken as good teacher for whom you could make a mistake once but not twice. She credits him with leading her and others to go into the field of computing [1]. She relates a story from her time at Harvard: *He [Howard Aiken] was a tough taskmaster. I was sitting at my desk one day, and he came up behind me, and I got to my feet real fast, and he said, 'You're going to*

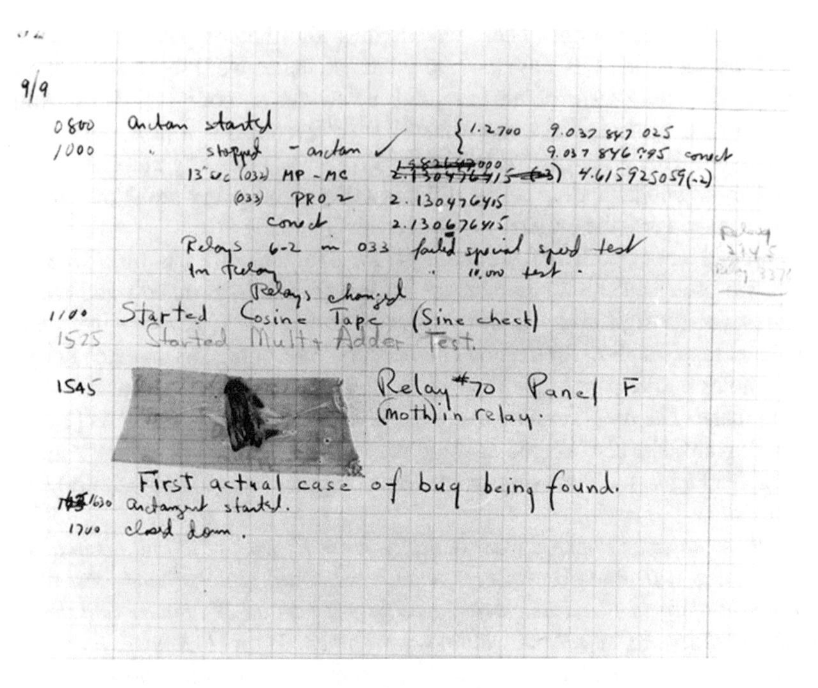

Fig. 3 First computer bug. (Source: Smithsonian Institution)

write a book." I said, "I can't write a book!" He said, "You're in the Navy now." So I wrote a book [The Manual of Operation for the Automatic Sequence Controlled Calculator] [10].

This manual was published in the *Annals of the Harvard Computation Laboratory*, Volume I, Harvard University Press, 1946. The book's foreword was written by James Bryant Conant, who served as President of Harvard University from 1933 to 1953. He wrote: *No combination of printed words can ever do justice to the real story of an undertaking in which cooperation between men of capacity and genius is of the essence. The development of the IBM Automatic Sequence Controlled Calculator is such a story, with many fascinating chapters . . . On August 7, 1944, Mr. Thomas J. Watson, on behalf of the International Business Machines Corporation, presented Harvard University with the IBM Automatic Sequence Controlled Calculator. Since that date, the machine has been in constant use by the Navy Department on confidential work. Therefore, Mr. Watson's gift came at a time when the new instrument his company had created was able to serve the country in time of war . . .* [1]

At the end of 1946, Admiral Hopper had reached age 40. Although she wanted to stay in the Navy, since WAVES could now transfer to the regular Navy, she had surpassed the maximum allowable age of 38. Admiral Hopper says, "I always explain to everybody it's better to be told you're too old when you are forty because then you go through the experience and it doesn't bother you again." Instead of returning to Vassar to teach (she turned down a full professorship), she allowed her leave of absence to expire. She stayed in the Naval Reserve and became a research fellow in Engineering and Applied Sciences at Harvard in the Computation Laboratory because, she said, "Computers are more fun." She got to participate in the building of the next computer – the Mark III. It was 50 times faster than the Mark I and used vacuum tubes and magnetic tape [1, 2, 12].

5 Moving into Industry

Admiral Hopper left Harvard in 1949, as there were no permanent positions for women, to join the Eckert-Mauchly Computer Corporation in Philadelphia as a senior mathematician. She said, "I always had a queer feeling that you shouldn't take a job unless you can learn and grow in that job." She saw that type of a future at Eckert-Mauchly – a company whose goal was to create a revolution in the computer industry [1, 2, 12].

When Admiral Hopper joined Eckert-Mauchly in 1949, many people thought there wouldn't be a need for computers in large numbers. Admiral Hopper states, "Back in those days, everybody was using punched cards, and they thought they'd use punched cards forever." But Admiral Hopper and others foresaw many business uses for the machines and a time when nonresearchers, including factory workers, businesspeople, and salespeople, would work on computers. It was one of many times in her career when she was told that "That can't be done." But Admiral Hopper would not be deterred [1, 12].

J. Presper Eckert and John Mauchly had built the first general-purpose computer at the University of Pennsylvania during World War II. This behemoth, the Electronic Numerical Integrator and Calculator (ENIAC), was over one hundred feet long and ten feet high, weighed over thirty tons, and contained more than 18,000 glass electronic vacuum tubes. ENIAC was a decimal machine based on the 0–9 system, which the general public uses every day. It required more than two and a half years to build [1]. But this first all-electronic digital computer led to the first mass-produced commercial computers.

They left the University of Pennsylvania in 1946 to establish their own company. Their first client was Northrop Aircraft Corporation, for whom they built a computer called BINAC or Binary Automatic Computer for Northrop's secret Snark Missile project. In 1949, when Admiral Hopper joined Eckert-Mauchly, BINAC was just being completed. One of her first jobs was to help install the computer and then teach Northrop employees how to use it [1].

The next computer was a mass-produced model aimed for the business market. It was called the Universal Automatic Computer (UNIVAC I) and used storage tubes and magnetic cores for memory and high-speed magnetic tape instead of punched cards for data. It was 1000 times faster than the Mark I and could process 3000 additions or subtractions each second. It was also much smaller than its predecessors – fourteen and one-half feet long, nine feet wide, and seven and one-half feet high. It was also revolutionary and Admiral Hopper relates that they weren't sure it was going to be successful, "We used to say that if UNIVAC I didn't work, we were going to throw it out one side of the factory, which was a junkyard, and we were going to jump out the other side, which was a cemetery!" Successful, it was [1].

In order to program the UNIVAC, Admiral Hopper used code in octal (base eight) that John Mauchly had developed. She hired people to be computer programmers by determining who was curious and liked to solve problems; there weren't computer programmers in the work force to hire – such a job category didn't exist. Admiral Hopper's staff consisted of four women and four men [1, 12].

She believed that a computer could be used to write its own programs. And she understood that these first computers were not at all user friendly. She further believed that in order for computers to be used widely, they would need to be accessible to a much wider audience than just computer specialists. As she later related, "What I was after in beginning English language [programming] was to bring another whole group of people able to use the computer easily… I kept calling for more user-friendly languages. Most of the stuff we get from academicians, computer science people, is in no way adapted to people." [1, 2, 12]

The Sort-Merge Generator, a program developed by Betty Holberton, was a start in this direction. Betty Holberton had been one of the six women who were the original programmers for ENIAC. She joined Eckert-Mauchly at the company's founding. Holberton's program sorted files of data and arranged them in a desired order, such as by date or code number. Admiral Hopper remembers that Holberton also taught Admiral Hopper how to write flowcharts for computer programs [1].

Admiral Hopper relates how her thinking about computers and compilers evolved from the Sort-Merge Generator: *I think the first step to tell us that we could actually use a computer to write programs was Betty Holberton's "Sort-Merge Generator" … Of course, at that time the Establishment promptly told us – at least they told me quite frequently – that a computer could not write a program… Sometime along the middle of 1952 I flatly made the statement that I could make a computer do anything which I could completely define. I'm still involved proving that, because I'm not sure if anyone believes me yet* [10].

6 The First Compiler

The first completed compiler was Admiral Hopper's A-0 system, developed at Remington Rand[1] in 1952. A-0 (where the A stood for algebraic) was a set of instructions for the computer that translated symbolic mathematical code into machine code.

[1]Remington Rand bought the Eckert-Mauchly Corporation in 1950. In 1955, Remington Rand was merged into the Sperry Corporation. Admiral Hopper was an employee until her retirement in 1971 from Sperry, although she was on military leave from 1967 until her 1971 retirement [6, 10].

From the machine code, the computer could then perform calculations. Admiral Hopper then took all of the subroutines she had collected over the years, put them on magnetic tape, and gave them each a unique call number. Instead of writing each subroutine out in machine code, she would refer to the call number in her programming instructions. The computer could then find the call number on the magnetic tape and complete the calculations [1].

The development that made the A-0 a single-pass compiler was explained by Hopper this way: *It so happened that when I was an undergraduate at college, I played basketball under the old women's rules which divided the court into two halves, and there were six on a team; we had both a center and a side center, and I was the side center. Under the rules, you could dribble only once and you couldn't take a step while you had the ball in your hands. Therefore, if you got the ball and you wanted to get down there under the basket, you used what we called a "forward pass." You looked for a member of your team, threw the ball over, ran like the dickens up ahead, and she threw the ball back to you. So it seemed to me that this was an appropriate way of solving the problem I was facing of the forward jumps! I tucked a little section down at the end of the memory which I called the "neutral corner." At the time I wanted to jump forward from the routine I was working on, I jumped to a spot in the "neutral corner." I then set up a flag for an Operation which said, "I've got a message for you." This meant that each routine, as I processed it, had to look and see if it had a flag; if it did, it put a second jump from the neutral corner to the beginning of the routine, and it was possible to make a single-pass compiler – and the concept did come from playing basketball!* [10]

Her original paper, "The Education of a Computer," in the Proceedings of the Association of Computing Machinery Conference, Pittsburgh, May 1952, is regarded by programming language compiler developers and writers as the primordial presentation on this topic. It was reprinted in the *Annals of the History of Computing* in 1988, with a preface by Dr. David Gries, Cornell University. He stated, "Hopper anticipates well what will happen – even, perhaps, artificial intelligence – when she says 'it is the current aim to replace, as far as possible, the human brain by an electronic digital computer.' She is one of the first to recognize that the software not the hardware will turn out to be the most expensive." [3]

The presentation of this paper led to her appointment as Systems Engineer, Director of Automatic Programming Development for Remington Rand. It was also the first of more than 50 papers that she published on programming languages and computer software [1].

However, not everyone was excited by her groundbreaking development. Admiral Hopper recalls, "The usual response was 'You can't do that.'" It took two years before others in the computer industry accepted her idea: "I had a running compiler and nobody would touch it because, they carefully told me, computers could only do arithmetic; they could not do programs. It was a selling job to get people to try it. I think with any new idea, because people are allergic to change, you have to get out and sell the idea." [1]

A-0 was followed by A-1 and A-2, which were both improved versions. A-2 was completed in 1955 and is said to be the first compiler to be used extensively. Computers were starting to be sold to businesses both in the US and abroad, and they needed a language for businesses that was easy to use – that did not require mathematicians [3]. Admiral Hopper understood that in order for computers to be successful, programming languages were going to need to be developed [1].

Here as well Admiral Hopper had the key insight. She believed that computer programs could be written in English (or any other human language). She understood that the letters of the alphabet were symbols. Alphabetic symbols could be translated into machine code just like mathematical symbols. Admiral Hopper explains, "No one thought of that earlier because they weren't as lazy as I was. A lot of our programmers like to play with the bits. I wanted to get jobs done. That's what the computer was there for." [1]

Not everyone believed that programming languages were necessary either. Admiral Hopper, though, was convinced of their need. She went ahead with the B-0 compiler (where B stands for business) following her philosophy of "go ahead and do it. You can always apologize later." Compiler B-0 would later be known as Flow-matic, the first English-based programming language. By the end of 1956, UNIVAC I and II were able to understand 20 statements in English [1].

By 1957, there were three computer languages being used in the US on various computers – Automatically Programmed Tools (APT), IBM's FORTRAN, and Admiral Hopper's Flow-matic. Hers was the only one using English commands, and each only worked on a certain type of computer. Recognizing that a Tower of Babel was developing, Admiral Hopper put together 500 typical programs and identified 30 verbs that should be used in computer languages going forward. Discussions began that progressed toward eliminating the Tower of Babel [1].

The first meeting to discuss movement toward a common business programming language was held in 1959 with Admiral Hopper in attendance [1]. At this time, there were two major inputs to its consideration. The only relevant language in actual use at that time was Flow-matic, and the only other significant input was a set of paper specifications from IBM. Flow-matic was a major factor in the COBOL (Common Business Oriented Language) committee's deliberations. As one of the members of the committee stated, "Flow-matic was the only business-oriented programming language in use at the time COBOL development started ... Without Flow-matic we probably never would have had a COBOL." [3, 6]

The committee that was established set itself very aggressive goals. The committee membership included Betty Holberton. Admiral Hopper served as one of two technical advisors. In 1960, the Government Printing Office issued the COBOL 60 Report. Although IBM didn't officially accept COBOL until 1962, many other companies announced in 1960 that they would be developing COBOL compilers for their computers and users. On August 26, 1960, *The New York Times* contained an article stating "COBOL, for Common Business Oriented Language, substitutes simple English key words for present complicated numerical jargon understood only by electronic computer specialists to 'instruct' a computer in its functions."

Admiral Hopper was quoted in the article as stating that Sperry Corporation's Remington Rand Division would have a COBOL programming system available on their UNIVAC II by October 31, 1960 [1, 6]. COBOL then became the standard language used in a wide variety of computer applications [3]. By the 1970s, COBOL was the "most extensively used computer language" in the world [2].

The standardization of languages and selling that concept to management were both strengths of Admiral Hopper. Betty Holberton said that Admiral Hopper was able to convince management of the importance of high-level languages for computers. And more and more people were now calling her "Amazing Grace." [1, 12]

7 The US Navy

Admiral Hopper had been affiliated with the US Naval Reserve since 1946, promoted to lieutenant commander in 1952, and promoted to commander in 1966. However, also in 1966, she received a letter pointing out that she had been on reserve duty for 23 years (when the limit was 20 years). "I knew that," Admiral Hopper said. It further pointed out that she was 60 years old. "I knew that too," she said. And it requested that she retire. She complied with the request and on December 31, 1966, which she recalls as "the saddest day of my life," Admiral Hopper retired. That retirement lasted for seven months [1, 12].

Admiral Hopper was recalled by the Navy to standardize COBOL for the Navy and to persuade everyone across the Navy to use this higher-level language. She returned to "temporary" active duty on August 1, 1967, taking a leave of absence from Remington Rand. She recalls, "I came running – I always do when the Navy sends for me." Later, "temporary" was amended to "indefinite." Indefinite lasted for almost twenty years – "The longest six months I've ever spent." [1] (Fig. 4)

As an active duty officer in the US Navy, from 1967 through her retirement in 1986 she served in a variety of capacities. Admiral Hopper led the Navy effort to develop a validation procedure for COBOL. Many of the technical and administrative concepts involved were adopted and used by governmental and standards organizations for other languages. She managed a computer applications department, served as major Navy spokesperson on computers, and later served as Special Advisor to the Director, Naval Automation Command. And she continuously battled the attitude "But we've always done it this way." She even said, "I'm going to shoot somebody for saying that some day. In the computer industry, with changes coming as fast as they do, you just can't afford to have people saying that." [3, 12]

In 1973, Admiral Hopper was promoted to captain by President Richard M. Nixon. In 1983, she became a Commodore by a special act of Congress (because she was too old to otherwise be promoted). In 1985, when she was elevated to rear admiral, she became the first female rear admiral in the history of the US Navy [1] (Fig. 5).

Fig. 4 Admiral Hopper teaching a COBOL class. (Source: Walter P. Reuther Library, Wayne State University)

In 1986, when she was 79, Admiral Hopper retired again from the US Navy. She asked that the retirement ceremony be held on the deck of the *USS Constitution* in Boston Harbor and the Navy honored her request. This ship is the oldest commissioned ship in the Navy and Admiral Hopper said, "We belong together. After all, I'm the oldest sailor in the Navy." [12]

Former Secretary of the Navy John Lehman sums up her contributions this way: *Military success is in no small measure because of the efficiencies and innovation provided by pioneering introduction of signal and data processing led by Grace Hopper. These accomplishments are of heroic proportions* [3].

Admiral Hopper closed each of her public appearances with these words: *I've been grateful for all of the help I've been given and all of the wonderful things that have happened to me. I've also received a large number of the honors that are given to anyone in the computer industry. But I've already received the highest award I will ever receive – no matter how long I live, no matter how many different jobs I may have – and that has been the privilege and the honor of serving proudly in the United States Navy* [10].

She was buried with full military honors at Arlington National Cemetery [12].

Fig. 5 Grace Murray Hopper being promoted to Commodore. (Source: United States Department of Defense)

8 Teaching Career

Admiral Hopper, who had discovered her love of teaching while in college, continued to instruct throughout her career and life. She was a visiting lecturer, visiting assistant professor, visiting associate professor, and then adjunct professor of engineering at the Moore School of Engineering at the University of Pennsylvania. Later she served as a professional lecturer in management science at George Washington University [1].

Admiral Hopper said it was very important to her to open the minds of young people and encourage them to probe the unknown and not be afraid of going against the conventional wisdom. For more than 20 years, she traveled the world to pass on what she knew and to inspire young people [3]. During those years, she spoke hundreds of time each year to many different kinds of audiences across the country. Any time she was paid for her talk, Admiral Hopper donated the fees to the Navy Relief Fund, which was money set aside for sailors and their families in need [1, 12].

She famously handed out "nanoseconds," wire cut to 11.8 inches long to represent how far light traveled in a billionth of a second, at these speaking engagements. She wanted people to understand the value of efficiency to computer programming and how fast computers operated.

She related how as computers became faster and faster, she wanted to understand that speed: *I didn't know what a billion is. I don't think most of those men in Washington do, either. So how could I know what a billionth is? I'm an extremely annoying employee because I won't do anything till I understand it. I could see a second go by, but darned if I could see a billionth of a second. I asked them to please cut off a nanosecond and send one over. This* [referring to a coil 984 feet long] *is a microsecond, a millionth of a second. We should hang one over every programmer's desk – so he'd know what he's throwing away all the time* [4]. As for picoseconds (one trillionth of a second), Admiral Hopper told her audiences that they can be found wrapped in small paper packets at fast-food restaurants: these packets are labeled "pepper," but those black specs inside are really picoseconds [1].

She had a special place in her heart for young people. As she said, "If you ask me what accomplishment I'm most proud of, the answer would be all of the young people that I've trained over the years, that's more important than writing the first compiler." She encouraged young people to go for it and to contribute their own genius for the advancement of mankind [3]. *I do have a maxim – I teach it to all youngsters: "A ship in port is safe, but that's not what ships are built for." I like the world of today much better than that a half-century ago. I am very sorry for the people who, at thirty, forty, or fifty, retire mentally and stop learning. Today, the challenges are greater. I like our young people: they know more, they question more, and they learn more. They are the greatest asset this country has* [3, 10].

And although she had retired from the Navy in 1986, she became a senior consultant for Digital Equipment Corporation a month after the ceremony aboard the *USS Constitution*. She said, "I don't think I will ever be able to really retire." Her job was to represent the company in contacts with other businesses as well as schools and colleges. She was teaching in this capacity as she always did, "People in a way are very much wanting for someone to express confidence in them. Once you'll do it, they'll take off." [12]

9 Awards and Honors

Admiral Hopper received many awards and honors in recognition of her efforts during her lifetime, including more than 40 honorary doctorates, as well as some honors that have been awarded posthumously. She said that the Naval Ordnance Development Award in 1946 was particularly meaning to her as it indicated "You're on the right track. Keep up the good work." [1, 2]

In 1964, the Society of Women Engineers presented her its Achievement Award: "In recognition of her significant contributions to the burgeoning computer industry as an engineering manager and originator of automatic programming systems." [3]

In 1969, she received the first computer science "Man of the Year" award from the Data Processing Management Association. In 1973, she was elected to membership in the National Academy of Engineering. Also in that year she received the Legion of Merit from the US, which was created by Congress in 1942 and is given for the performance of outstanding service. Admiral Hopper was also named a distinguished fellow of the British Computer Society in 1973, the first woman and first person from the US to receive that distinction [1].

In 1984, her induction citation into the Engineering and Science Hall of Fame read: *In tribute to your superior technical competence and mathematical genius; in tribute to your creative leadership, vision, and commitment as a computer pioneer; in tribute to your setting a foremost example as an author and inventor; in tribute to your dedication to human and humane elements of teaching, learning, and scholarship; and in tribute to your insights and innovations in meeting the challenges of rapidly changing times, we induct you – Commodore Grace Murray Hopper – into the Engineering and Science Hall of Fame* [1].

In 1988, she received The Emanuel E. Piore Award from the Institute of Electrical and Electronics Engineers, which is awarded for outstanding achievement in the field of information processing in relation to computer science, deemed to have contributed significantly to the advancement of science and to the betterment of society.

In 1991, she received the National Medal of Technology, presented by President H.W. Bush. Admiral Hopper was the first individual woman to receive this Medal.[2] The citation for the National Medal of Technology reads "For her pioneering accomplishments in the development of computer programming languages that simplified computer technology and opened the door to a significantly larger universe of users."

Admiral Hopper was inducted posthumously into the National Women's Hall of Fame (1994). In 1996, the *USS Hopper* (DDG 70), an Arleigh Burke-class guided missile destroyer, was launched.[3] It was commissioned in 1997. This destroyer was only the second to be named for a woman who had served in the Navy.

In 2016, President Barack Obama posthumously awarded her the Presidential Medal of Freedom, the nation's highest civilian honor in recognition of her "lifelong leadership in the field of computer science." [2, 5]

10 Lasting Influences

Admiral Hopper had a long and distinguished career in the computer field with major contributions to the development and use of software, to the encouragement and leadership of workers in the field, and to the overall stimulation and promotion of the field. Without a doubt Dr. Hopper has been one of the most influential

[2]Physicist Helen Edwards was on a four-person team for the Fermi National Accelerator Laboratory that received the 1989 National Medal of Technology for contributions to the TEVATRON particle accelerator [7, 11].

[3]DDG is the NATO code for Guided Missile Destroyer and 70 is the position that the Hopper occupies within the ships commissioned [8].

persons in the field of computer science in the promotion of the development of programming languages and the expansion of our knowledge and expertise in language translation [3].

She early recognized the need for better programming languages and techniques, understood also the close interaction between hardware design and software design and used a firm grounding in computer operation and principles of computation, to take the innovative technical steps necessary for new language development. Without Dr. Hopper's pioneering efforts there would have been little software to run on computers, their basic efficiency of operating would have been severely limited and their ease of use would have been at a standstill. She contributed significantly to the more effective use of computers through the development of major new software tools – tools that were essential to the realization of the potential of the emerging technology for computer hardware [3].

Her role in the emergence of the computer field was an extremely broad one and one that extended over a period of more than 40 years. From the beginning, Admiral Hopper challenged the industry to accept new technological advances and to avoid the "but we've always done it that way" mentality. She confirmed computers could be used for more than just mathematics, and computers users owe her a debt of gratitude for proving computers could be programmed in English, not just be using symbols. In addition to the development of essential new technology, she had the vision and persistence to exert overall leadership. She was an effective and ubiquitous spokesperson for the Navy, for the computer field, and for its career opportunities for young people. Jean Sammet says in *Programming Languages* (Prentice Hall, 1969), "In my opinion, Dr. Grace Hopper did as much as any other single person to sell many of these concepts from an administrative and management, as well as a technical point of view." [3]

Admiral Hopper distinguished herself as a Naval officer in the service of her country, as a computer pioneer in the evolution of the high technology industry, and as an educator who inspired generations of young people. She was not only a pioneer in the development of data processing and computer languages, she was also a significant pioneer and role model for women in the Navy and for the whole nation. For many decades she inspired generation after generation of engineers and naval officers and enlisted to take ideas and turn them into real accomplishments. Grace Hopper not only achieved great things herself but inspired literally hundreds of others to great accomplishments of their own [3].

References

1. C.W. Billings, *Grace Hopper: Navy Admiral and Computer Pioneer* (Enslow Publishers, Inc, Hillside, 1989)
2. Biography of Grace Murray Hopper, Yale University – Office of the President. https://president.yale.edu/biography-grace-murray-hopper. Accessed 28 Aug 2019
3. From the author's files, Nomination and Supporting Letters for the National Medal of Technology, Submitted December 4, 1989

4. D.A. Fryxell, *Outfiguring the Navy* (American Way, 1984, May), pp. 38–43
5. Fuller, Rear Admiral John, U.S. *Navy Toughness: 'Amazing Grace and Namesake USS Hopper*, Navy Live. https://navylive.dodlive.mil/2017/03/17/u-s-navy-toughness-amazing-grace-and-namesake-uss-hopper/. Accessed 1 Sept 2019
6. Grace Murray Hopper. http://www.cs.yale.edu/homes/tap/Files/hopper-story.html. Accessed 28 Aug 2019
7. National Science and Technology Medals Foundation. Helen T. Edwards. https://www.nationalmedals.org/laureates/helen-t-edwards#. Accessed 31 Aug 2020
8. Naval Vessel Register – DDG: Guided Missile Destroyer. https://www.nvr.navy.mil/NVRSHIPS/HULL_SHIPS_BY_CATEGORY_DDG_99.HTML. Accessed 31 Aug 2020
9. The WAVES 75th Birthday, Naval History and Heritage Command. https://www.history.navy.mil/browse-by-topic/wars-conflicts-and-operations/world-war-ii/1942/manning-the-us-navy/waves_75th.html. Accessed 29 Aug 2019
10. H.S. Tropp, Grace hopper: the youthful teacher of us all. Abacus **2**(1, Fall), 7–18 (1984)
11. United States Patent and Trademark Office, 1989 Laureates – National Medal of Technology and Innovation. https://www.uspto.gov/learning-and-resources/ip-programs-and-awards/national-medal-technology-and-innovation/recipients/1989. Accessed 31 Aug 2020
12. N. Whitelaw, *Grace Hopper: Programming Pioneer* (W. H. Freeman and Company, New York, 1995)

Jill S. Tietjen P.E., entered the University of Virginia in the fall of 1972 (the third year that women were admitted as undergraduates – after suit was filed in the US District Court) intending to be a mathematics major. But midway through her first semester, she found engineering and made all of the arrangements necessary to transfer. In 1976, she graduated with a BS in Applied Mathematics (minor in Electrical Engineering) (Tau Beta Pi, Virginia Alpha) and went to work in the electric utility industry.

Galvanized by the fact that no one, not even her PhD engineer father, had encouraged her to pursue an engineering education and that only after her graduation did she discover that her degree was not ABET-accredited, she joined the Society of Women Engineers (SWE) and for more than 40 years worked to encourage young women to pursue science, technology, engineering, and mathematics (STEM) careers. In 1982, she became licensed as a professional engineer in Colorado.

Tietjen starting working jigsaw puzzles at age two and has always loved to solve problems. She derives tremendous satisfaction seeing the result of her work – the electricity product that is so reliable that most Americans just take its provision for granted. Flying at night and seeing the lights below, she knows that she had a hand in this infrastructure miracle. An expert witness, she works to plan new power plants.

Her efforts to nominate women for awards began in SWE and have progressed to her acknowledgement as one of the top nominators of women in the country. She nominated Admiral Hopper for the National Medal of Technology and received the Medal for her in the

White House Rose Garden from President George H.W. Bush. Admiral Hopper was Tietjen's first successful nomination into the National Women's Hall of Fame and Tietjen accepted that medallion posthumously at Admiral Hopper's family's request. A second nominee later received the National Medal of Technology and Innovation, as it is now called. Other of her nominees have received the Kate Gleason Medal; they have been inducted into the National Women's Hall of Fame and state Halls including Colorado, Maryland and Delaware; and have received university and professional society recognition. Tietjen believes that it is imperative to nominate women for awards – for the role modeling and knowledge of women's accomplishments that it provides for the youth of our country.

Tietjen received her MBA from the University of North Carolina at Charlotte. She has been the recipient of many awards, including the Distinguished Service Award from SWE (of which she has been named a fellow and is a National Past President), the Distinguished Alumna Award from the University of Virginia and the University of North Carolina at Charlotte, and she has been inducted into the Colorado Women's Hall of Fame and the Colorado Authors' Hall of Fame. Tietjen sits on the boards of Georgia Transmission Corporation and Merrick & Company. She has published 10 books, including the bestselling and award-winning book *Her Story: A Timeline of the Women Who Changed America*, for which she received the Daughters of the American Revolution History Award Medal. Her 2019 book, *Hollywood: Her Story, An Illustrated History of Women and the Movies,* is also a bestseller and has received numerous awards.

Message:

When I arrived at the University of Virginia in the Fall of 1972, I was a mathematics major in the College of Arts and Sciences. As I learned about what my friends in engineering were studying, I decided that those were the classes I wanted to take and I transferred to the School of Engineering and Applied Sciences. I'm not sure now that I had any idea about what I would do with an engineering education – but my PhD engineer father and his colleagues all loved their jobs – so I wasn't very worried.

Engineers make the world work. We solve problems, generally with very creative solutions. Engineers apply science.

Going to work in the electric utility industry, which really happened by serendipity and not intentionally, was a perfect outcome for me. My first employer – Duke Power Company – espoused community and service. Our primary job, in the words of one of my electric utility CEO friends, is to KLO – Keep the Lights On!

Our economy and all of the economies around the world rely on a safe, economic, and reliable source of electricity to power everything in our lives. What a tremendous sense of satisfaction I derive from knowing that I help ensure that safe, economic, and reliable product. My sense of pride about the work that electric utilities do has only increased over the years – and was enhanced even more during the pandemic. It is hard to deliver electricity, which most of us take for granted, safely, economically, and reliably 24 hours a day, 7 days a week. But, I am grateful to be part of that cadre of people.

Part I
Intelligence

XAI: A Natural Application Domain for Fuzzy Set Theory

Bernadette Bouchon-Meunier, Anne Laurent, and Marie-Jeanne Lesot

1 Introduction

Explainable Artificial Intelligence (XAI) has been at the core of many developments in Artificial Intelligence in the recent years, following the DARPA incentives[1] to not only produce more explainable artificial intelligence models while maintaining good learning performances, but also to enable the user to easily interact with intelligent systems. Several related concepts, such as interpretability, accountability, understandability, transparency or expressiveness are inherent in the capacity of intelligent systems to be explainable.

Along the same lines, the guidelines for Artificial Intelligence presented by the European Commission in April 2019[2] point out the importance of transparency of data, systems and Artificial Intelligence business models, as well as the necessity for AI systems and their decisions to be explained in a manner adapted to the stakeholder concerned. This has become a crucial issue as AI can be considered as a pervasive methodology, present in most digital systems, covering almost all personal and professional activities. To cite a single example, in 2019, the journal ComputerWeekly[3] established that "*Almost one-third of UK bosses surveyed in the*

[1]https://www.darpa.mil/program/explainable-artificial-intelligence.

[2]https://ec.europa.eu/digital-single-market/en/news/ethics-guidelines-trustworthy-ai.

[3]https://www.computerweekly.com (30 Apr 2019).

B. Bouchon-Meunier (✉) · M.-J. Lesot
LIP6, Sorbonne Université, CNRS, Paris, France
e-mail: bernadette.bouchon-meunier@lip6.fr; Marie-Jeanne.Lesot@lip6.fr

A. Laurent
LIRMM, Univ. Montpellier, CNRS, Montpellier, France
e-mail: anne.laurent@umontpellier.fr

A. E. Smith (ed.), *Women in Computational Intelligence*, Women in Engineering and Science, https://doi.org/10.1007/978-3-030-79092-9_2

latest PwC global CEO survey plan to use artificial intelligence in their businesses – but explainable AI is critical".

Indeed, among others, the understandability of decisions implies an increase in the decision trustworthiness, as is for instance the explicit aim of one of the first and now best-known explanation systems, LIME [61]. The understandability of decisions also implies a decrease in ethical problems, e.g. those associated with diversity (gender, culture, origin, age, etc.), as it opens the way for a deep analysis of the considered model: As such, it relates to the notion of *fair AI* (see for instance [7]). Besides, another beneficial outcome of such models is the robustness of decisions. The interpretability of decisions requests better interactions with the user, raising the question of the design of interfaces dedicated to explainable systems [51].

The capacity of an AI system to meet these interpretability requirements lies in all its components. At the origin, data must be easily apprehended through their description or their visualisation, and with the help of an explanation of the characteristics used. An AI model is then applied to these data in order to extract knowledge. Its description must be clear and explanations on how the results are obtained must be available. The expressiveness of the resulting information and its legibility by the user are also part of the explainability of the AI. As a consequence, the concepts of explainability, understandability, expressiveness and interpretability must be considered as closely intertwined; among others the explainability of the model is based on its transparency and includes its expressiveness. Numerous surveys describing the state-of-the-art of this wide field have been recently published, see for instance [7, 10, 39, 56].

This chapter studies the issue from the point of view of fuzzy set theory; it argues that the latter, as elaborated as early as the 1970s by Lotfi Zadeh [73] and its numerous developments and applications since then, can be seen as having taken into account explainability considerations from its very beginning, in its motivations and principles: Fuzzy set-based methods are natural solutions to the construction of interpretable AI systems, because they provide natural language-like knowledge representations, with a capacity to manage subjective and gradual information which is familiar to human beings. The easiness of interactions with the user and the transparency of features are qualities of fuzzy systems involved in their explainability; expressiveness and interpretability have been extensively investigated in the construction and analysis of fuzzy intelligent systems. In addition, fuzzy set theory and its developments consider human cognition to be the source of inspiration for the proposed methods and tools, naturally leading to legible outputs, that follow the same way of thoughts.

This chapter proposes to review some of the aspects that make the domain of fuzzy set theory and its developments appropriate tools to deal with the issues of XAI, mainly dealing with the capacity of an intelligent system to explain how it obtains results, and to provide the user with easily understandable outcomes, in light of the fuzzy paradigm.

The chapter is organised as follows: Sect. 2 describes the key concept called *Computing With Words*. It relies on user interactions based on using natural language, both regarding the inputs and the outputs, which obviously makes systems easily legible. Section 3 discusses the application of fuzzy logic to model natural human ways of reasoning, providing outputs that are interpretable for the users. Section 4 is dedicated to the most visible part of Artificial Intelligence nowadays, namely machine learning, and discusses several fuzzy approaches that make it possible to increase the expressiveness and thus the explainability capacities of machine learning. Section 5 discusses some conclusions.

2 Computing with Words

Computing With Words is a key concept proposed by Zadeh [78–80] that aims at allowing to compute and reason using words instead of numbers. It allows to cope with imprecision, when for instance we know that a person is *tall* without knowing precisely her size or when precision is not necessary. For instance, even if we know precisely the heights of the people, it can be the case that we are looking for *tall* people without designing such a characteristic in a crisp manner. This thus provides flexibility and/or robustness to the systems. Beyond this flexible knowledge representation discussed in this section, Computing With Words integrates a reasoning component that mimics common sense reasoning familiar to human beings, as discussed in Sect. 3.

We introduce below the main concepts and formal tools before presenting some of the main applications, namely fuzzy databases and fuzzy summaries, including gradual patterns.

2.1 Formal Tools

The fuzzy set theory underlies Computing With Words by providing a formal framework: The key idea is that most of the objects that are manipulated in natural language and computation are not just members of a set or not, but they rather *gradually belong* to sets.

For instance, firms can be categorised according to their size, but the definition of categories depends on the considered country rules or domain of activity. As a result, a firm with 2 employees could then be considered as a micro-company (or Small Office Home Office a.k.a. SOHO) with no doubt, while one with 10 employees could still be considered as fairly micro but may rather be considered a small-size company, depending on the environment.

In this chapter, we claim that when considering explainable systems, this point of *gradual belonging* is crucial so that decisions can take into account the fact that every object or situation cannot always be categorised in a binary manner, as *belongs/does not belong*. This is the case in particular when complex systems are considered where there are very little simplistic black-and-white visions. The same idea holds for all domains. For instance, a system could not easily justify that it does not recommend a medicine to an infant of weight 19.8kg because it is for children over the weight of 20kg.

This points out a critical question between explainability and confidence in a system that would mimic human intelligence.

2.1.1 Fuzzy Sets

Formally, a fuzzy set A is defined over the universe U by its membership function

$$\mu_A : U \to [0, 1]$$

that can be considered as extending the classical binary characteristic function of a set: Whereas the latter allows only values 0 or 1, the membership function allows for the whole gradual range between 0 and 1.

For instance, the fuzzy set *micro* can be defined over the universe of company sizes $U = \mathbb{N}^+$ by the membership function

$$\mu_{micro}(u) = \begin{cases} 1 & \text{if } 0 < u < 8 \\ \dfrac{-u+15}{7} & \text{if } 8 \leq u < 15 \\ 0 & \text{if } u \geq 15 \end{cases}$$

Such a fuzzy set can be represented graphically, as illustrated by Fig. 1.

In this example, a firm of size 10 is considered as being fairly micro, with a degree $\mu_{micro}(10) = \frac{5}{7}$. By considering such fuzzy sets, it is then possible to cope with the ambiguity of the term over the world, as micro companies are not officially

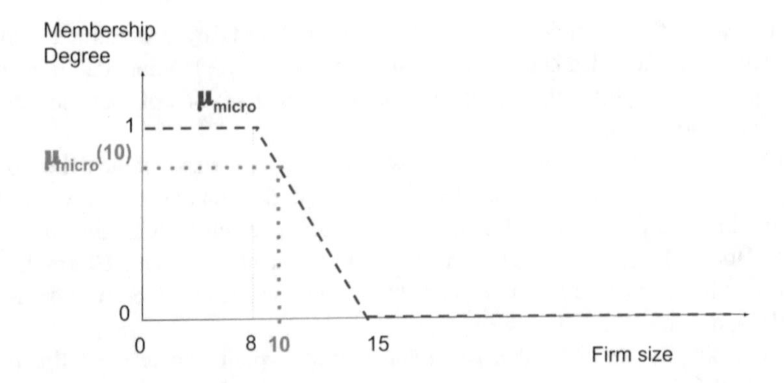

Fig. 1 Illustration of the fuzzy set *micro*

considered of the same size in the USA compared to Europe. Using such gradual membership also allows to mitigate a difference of one employee in the number of employees: Classic binary set definitions are such that a firm of size 199 and a firm of size 200 belong to two different categories (medium and large, respectively). The fuzzy set theory allows to make such crisp, and therefore somehow arbitrary, boundaries more gradual, and thus easier to understand.

Fuzzy sets defined on numbers can represent various types of information, as for instance:

- Fuzzy intervals: A profession is said to be gender-balanced if the gender rate ranges between 40% and 60%,[4] or
- Fuzzy numbers: Almost 10% of employees working in AI in French companies are women.[5]

The definition of the fuzzy sets to be used in the system is equivalent to the definition of their membership functions. Three types of methods to design them can be mentioned: First, this design can rely on user expertise, so that for instance the formal definition of *micro firms* correspond to the user point of view. This makes it possible to personalise the whole system, adjusting its components (here the fuzzy sets) to her own representation. A second method to design fuzzy sets relies on cognitive considerations, see for instance [49]: It can be shown that fuzzy sets representing the imprecise notion *around* x, where x is a numerical value, actually do not vary so much with each individual user, and that they can be defined based on cognitive studies. It has been shown that they mainly depend on x magnitude, last significant digit, granularity and a notion of cognitive saliency. A third method to design fuzzy set consists in extracting them automatically from possibly available data, so that the fuzzy sets match the underlying distribution of the data. Such an approach can be applied in a supervised paradigm (see e.g. [54]) or a non-supervised one (see e.g. [66]).

As for classical sets, fuzzy sets are provided with operators such as union, intersection, difference, etc. that are extended from crisp membership in $\{0,1\}$ to gradual membership ranging in $[0,1]$, based on operations on their membership functions [46].

2.1.2 Linguistic Variables and Fuzzy Partitions

One of the central concepts for Computing With Words is the *linguistic variable* [74]. A linguistic variable is defined by a name, a universe and a set of fuzzy sets. For instance *firm size* can be defined over the universe $\mathbb{N}^+$ with the elements {micro, small, medium, large}. These fuzzy elements are meant to cover

[4] https://cutt.ly/8t3Kt3m.

[5] https://cutt.ly/tt3KoM3.

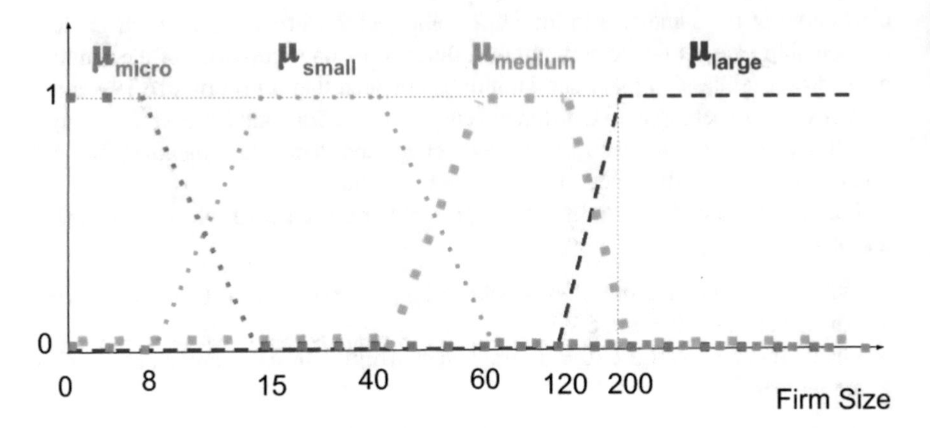

Fig. 2 A fuzzy partition

all values of the universe as a partition would do. In order to convey the possibility of imprecision, fuzzy sets can overlap in the thus so-called *fuzzy partition*.

Fuzzy partitions can be defined in two main manners. First, a partition P over the universe U can be defined as a set of fuzzy sets $\{p_1, \ldots, p_n\}$ with membership functions $\{\mu_{p_1}, \ldots, \mu_{p_n}\}$ such that

$$\forall u \in U, \quad \sum_{i=1}^{n} \mu_{p_i}(u) = 1$$

In such a definition, all fuzzy sets cross so that the sum of the membership degrees is 1 for any element that thus has total membership. For instance, we may consider $P = \{micro, small, medium, large\}$ with membership functions represented by Fig. 2.

Another way to define a partition P over the universe U is to consider that:

$$\forall u \in U, \quad \exists i \in [\![1, n]\!] \text{ such that } \mu_{p_i}(u) > 0$$

This definition leaves more flexibility for setting the membership functions that build the partition. Membership functions can be built by an expert of the domain, on the basis of a subjective assessment, or automatically derived from real data, by means of supervised learning or optimisation techniques (see e.g. [16]). In the first case, the partitions are generally of the first type, while they can be of the second type when obtained automatically.

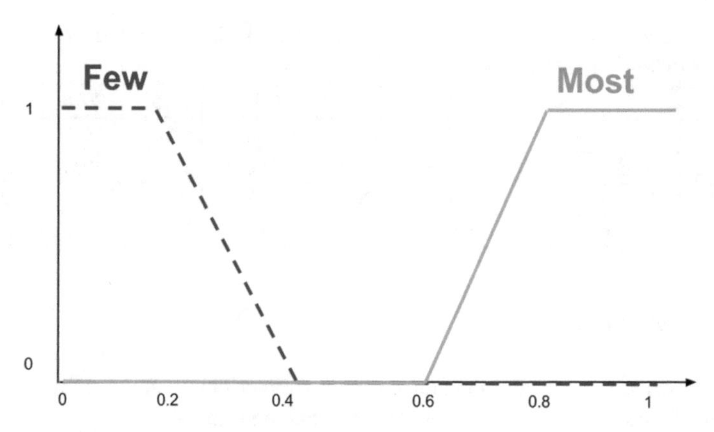

Fig. 3 Example of fuzzy quantifier membership functions

2.1.3 Fuzzy Quantifiers

Fuzzy set theory also enables to represent another type of imprecise concepts, using so-called *fuzzy quantifiers*, that for instance allow to model the linguistic quantifiers *a few* or *most*, beyond the classical mathematical quantifiers ∀ and ∃. To that aim, the fuzzy quantifiers are modelled through their membership functions. Absolute fuzzy quantifiers, like *at least 20* or *almost 5*, are distinguished from relative ones, like *most*: The first ones are defined over the universe of natural numbers, the second ones over the [0, 1] universe of proportions. For instance, the two quantifiers *few* and *most* can be defined by the membership functions shown in Fig. 3.

These fuzzy quantifiers allow to describe situations with words instead of crisp numbers whose details can for instance make difficult to cope with the given message, or even undermine the message. We can for instance feel uncomfortable with a sentence providing information about *more than 96.3%* of the data, and rather replace it with *almost all data.*

2.2 *Examples of Methodological Utilisation*

In this section, we present some applications based on fuzzy sets presented above. These applications provide users with tools that help them to better understand and deal with their data. As many data are stored in databases, the first part is thus devoted to fuzzy databases, before presenting methods for data summarisation.

Table 1 Gender pay gap, measured in %, for all employees by business size & year UK, ASHE, 2011–2017 (https://cutt.ly/4t3KqP8)

	2011	2012	2013	2014	2015	2016	2017
1 to 9 employees	16.5	15.8	15.4	14.8	14.1	14.2	12.6
10 to 49 employees	20.2	20.5	20.5	19.6	21.5	20.0	20.4
50 to 249 employees	21.3	19.3	20.5	20.1	20.6	21.0	19.3
250+ employees	21.5	20.7	20.6	19.7	19.7	19.2	19.3

2.2.1 Fuzzy Databases

As databases are a key component in many numerical systems, they are a key element to consider in order to provide explainable systems.

The literature provides many works on fuzzy databases [57], especially for the relational model which has been extensively considered. In such systems, imprecision is considered for both data and/or queries. Flexible queries can thus be of different forms [72]: They can allow users to propose either "crisp" or "fuzzy" terms in the queries themselves and can rely on "crisp" or "fuzzy" data. In flexible queries, fuzzy terms (words) are integrated within the various clauses of extended query languages, as for instance SQLf for relational databases [15, 65].

Such extensions provide ways to better get information from raw data. For instance, Table 1 reports the gender pay gap depending on the business size in 2017 in UK, as published within the 2017 Annual Survey for Hours and Earnings from the Office for National Statistics.[6] From such data, if the user wants to extract the years and business sizes when the pay gap has been *lower than 14%*, then she will not come out with "2015 for businesses of size 1 to 9 employees" because it was 14.1. This result is likely to appear as non-explainable, as 14.1 is so close to the requested threshold. With fuzzy queries, the condition *lower than 14%* can be made flexible so as to gradually consider the situations as described above with fuzzy membership functions.

As relational databases have been extended to multidimensional databases and NoSQL engines, works have proposed models and query languages for these frameworks. [47] proposes fuzzy operators for OLAP databases which are meant to help decision makers to navigate through OLAP cubes. Such cubes provide aggregated data structured by measures analysed over dimensions, as for instance the *number of sales* as a measure and *category, time, location* as dimensions. Fuzzy queries can be expressed, like navigating through the *number of products of category Outdoor being sold in Winter near Paris* where *Winter* can be defined in a fuzzy manner, so as the *Outdoor category* and *near Paris*.

More recently, NoSQL databases have emerged to cope with very large databases. Some of them specifically address the need to represent and query graph data. In this case, data and links are represented by two families of objects: the so-called

[6]https://cutt.ly/4t3KqP8.

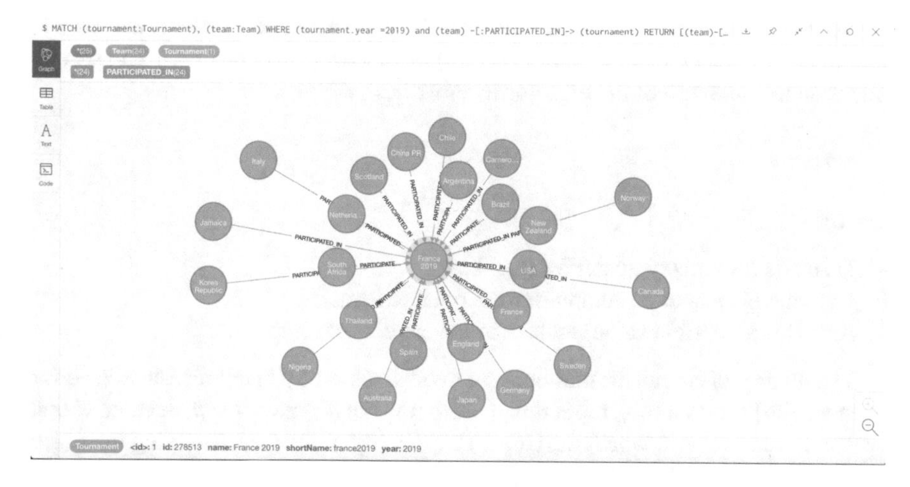

Fig. 4 Nodes, properties, and relationships labels in a property graph database [58]

nodes and *relationships* which have labels and types. These two families of objects are similar to any graph structure. But in property graphs, so-called properties can be defined over nodes and relationships. For instance, persons may be described by their name, birth date, etc.

Consider the example of the data set describing 2019 women's world cup [58], in this database, we have:[7] Persons, Teams participating in Matches from Tournaments. Figure 4 represents the teams participating in the 2019 tournament in France.

It has been shown that such data engines allow to process complex queries that would exceed the performance limitations of the relational databases. This is especially the case when queries require too many "join" operations (as for instance to retrieve persons playing in teams involved in matches from specific tournaments).

These engines are provided with their own query language that has been extended to allow fuzzy queries such as *retrieve the teams participating in most of tournaments* [25, 60].

2.2.2 Fuzzy Linguistic Summaries

Pursuing the same goal of making information as understandable as possible, fuzzy summaries have been proposed in the 1980s [43, 71] (see [18, 44] for surveys). They summarise data using so-called *protoforms*, defined as sentence patterns to be filled with appropriate values depending on the considered data, based on fuzzy sets for expressing concepts and quantifiers.

[7] Data can be retrieved from https://neo4j.com/sandbox/.

Such fuzzy summaries can be illustrated by sentences such as *Most women are less-payed than men* or *Few large firms display a low gender pay gap*. The second one is an instantiation of the protoform defined as

$$Qy \ are \ P$$

where

- Q stands for a fuzzy quantifier (e.g. Few),
- y are the objects to be summarised (e.g. large firms),
- And P is a possible value, such as in *low gender pay gap*.

The quality of linguistic summaries can be assessed by many measures (see for instance [26] for a survey), the seminal one being the *degree of truth*, denoted T that computes the extent to which the (fuzzy) cardinality of the data satisfying the fuzzy set P satisfies the fuzzy quantifier Q. Formally

$$T(Qy's \ are \ P) = \mu_Q\left(\frac{1}{n}\sum_{i=1}^{n}\mu_P(y_i)\right)$$

where n is the number of objects (y_i) that are summarised and μ_P and μ_Q are the membership functions of the summariser P and quantifier Q, respectively.

In order to extract such summaries from large data sets, efficient methods have been proposed [32, 67].

2.2.3 Extended Linguistic Summaries

Other types of patterns can help to capture and extract knowledge, as for instance gradual patterns [31, 42, 48] that are of the form *the more/less X, the more Y*. For instance, sadly, the following gradual pattern holds:[8] *The higher the level of responsibility, the less the proportion of women*. Such patterns can take a fuzzy form, when X and Y represent membership degrees to fuzzy linguistic variables [4], as in "the more the age is *middle-aged*, the more the gender pay gap is *large*".

Such gradual patterns can be extended to more complex protoforms or more complex data, thus helping to better describe, understand and explain phenomena. For instance, protoforms can be extended to take into account the fact that some contexts can strengthen a pattern [22]: *The younger, the lower the salary*, **all the more as** *employees are women*.

They can also help to understand and explain how the spatial dimension impacts some phenomena: Spatial gradual patterns have been proposed to recommend decisions in public health policies by explaining the spatial dynamic of potentially

[8] https://www.insee.fr/fr/statistiques/3363525.

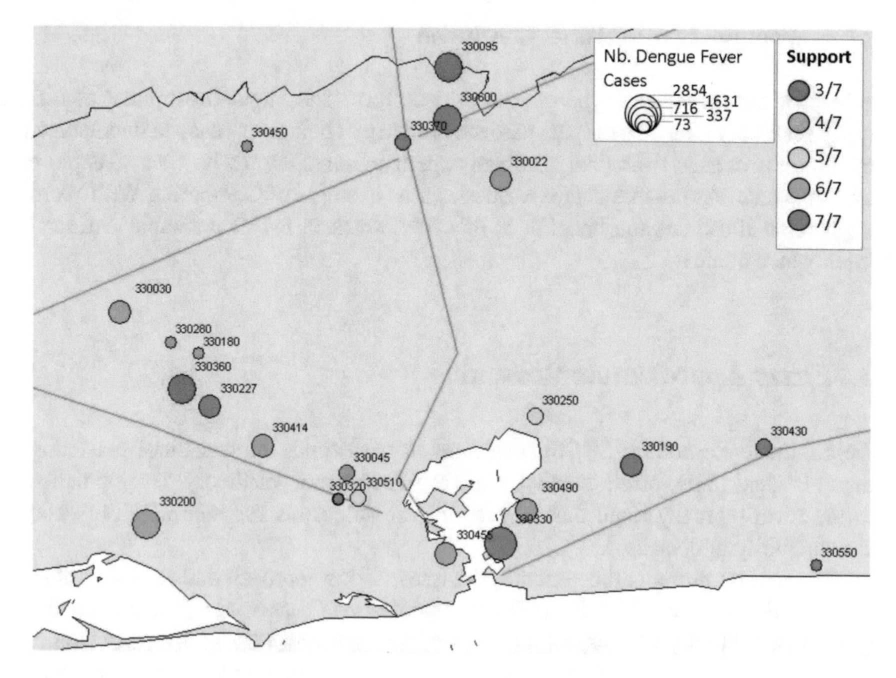

Fig. 5 Gradual Map (Temperature,↑), (Mobility,↓), (NbCases,↓) [3]

avoidable hospitalisations [59] or to describe epidemiological dengue fever data from Brazil, by mining for gradual patterns in a geographical information system [3]. In the latter work, patterns like "*the higher the temperature, the lower the mobility, the lower the number of cases*" can be extracted and information can be plotted on so-called gradual maps, as shown in Fig. 5. Here, it can be seen that emerging epidemic dengue fever mostly appeared in cities located around the coast (displayed in green on the map).

Protoforms can also be extended from complex data, such as time series [1, 45] or property graphs (introduced above), as in [64].

From such property graph data sets, various summaries and gradual patterns can be extracted. For instance[9] "*the higher the number of matches played, the higher the number of won tournaments*". Exceptions may be retrieved, such as "*the higher the number of matches played, the higher the number of won tournaments, unless the tournament is not in final draw*".

[9]These summaries are given as examples and have not been extracted from the data set.

2.2.4 Computing with Words Conclusion

Whatever the considered type of data or protoform, the linguistic summaries that fuzzy set theory allows to output share the property of being very easy to understand, by all kinds of user, from data or model experts to naive users who want to explore available data. As flexible database queries, this example of Computing With Word application illustrates the benefits of this framework in the eXplainable Artificial Intelligence context.

3 Fuzzy Approximate Reasoning

Approximate Reasoning [9, 76], that aims at performing more natural reasoning than classical logic, offers another domain where fuzzy set theory developments, in the form of fuzzy logic theory, provide valuable tools for interpretability and explainability of decisions.

This section presents the general principles of this approach and then describes in more details several tools for fuzzy approximate reasoning: It discusses the Generalised Modus Ponens, which is the inference rule offered by fuzzy logic, and gives a short presentation of other inference rules that allow to model gradual, analogical and interpolative reasoning and, therefore, to increase the legibility of the reasoning as well. It finally discusses some examples of utilisation of these formal tools.

3.1 General Principles

Formal logic provides a theoretical framework to model inference and reasoning; among others, it aims at guaranteeing that all reasoning steps are correct, free from any error and bias. Yet classical logic is binary, insofar as any assertion can only be *true* of *false*, without any intermediate or other truth value. Now limiting the possibilities to these two exclusive values actually remains far from the natural way of thinking for human beings, who intuitively may use degrees,[10] such as *rather false*, *more or less true*, or who may answer *I do not know*. Fuzzy logic, as a generalisation of multi-valued logic [37], aims at providing a theoretical framework to model reasoning with such extended *truth degrees*. The latter are of course related to the notions of membership degrees and fuzzy sets presented in the previous section.

[10]Another increase of expressiveness, beside the use of degrees, is to consider modalities, such as *I believe it is true* or *it ought to be true*. Formalising reasoning with these modalities is the aim of modal logic [11], which is out of scope of this chapter.

As such, fuzzy logic offers a formal framework that is closer to the natural way of thinking of human beings and, therefore, more easily legible and understandable: From its very introduction by Zadeh [76], fuzzy logic can be seen as a tool for explainable artificial intelligence.

3.2 *Generalised Modus Ponens*

3.2.1 Modus Ponens

The classic Modus Ponens inference rule, also named implication elimination in the natural deduction or sequent calculus inference systems, can be sketched as follows:

$$\begin{array}{lcl} A & \Longrightarrow & B \\ A & & \\ \hline & & B \end{array}$$

It means that from the pieces of knowledge that the rule *if* A, *then* B holds and that A holds as well, the piece of knowledge that B holds can be inferred.

Let us for instance consider the rule that characterises the French demographic balance statistics[11] in 2019: *if age is between 50 and 60 years old, the proportion of women is between 53.4% and 56.8%*. When observing people aged 53, the rule premise A = *age is between 50 and 60 years old* holds and thus allows to infer that the rule conclusion B holds, i.e. that the proportion of women among people of that age is between 53.4% and 56.8%.

3.2.2 Fuzzy Extension

The Modus Ponens rule can be considered as strict, both in terms of inference and representation. Indeed, it first requires that the implication premise is exactly satisfied and does not allow for any tolerance: In the previous example, if the considered age is 49, then condition A is considered not satisfied and the inference process does take place. Thus no new piece of knowledge can be derived. This can be considered as difficult to interpret for a human being, who may intuitively expect that the implication rule applies to some extent and that the proportion is not far from this interval.[12] Approximate reasoning aims at performing more natural reasoning and among others at allowing to perform inference even in such a case where the observation A' does not perfectly match the rule premise A.

[11] https://www.insee.fr/en/statistiques/2382597?sommaire=2382613.

[12] Actually, the proportion of women among people aged 49 is 50.5%, according to the same source.

Another difficulty with classical logic is that it only allows for precise, binary, concepts to be used, as for instance the mathematical condition "between 50 and 60" or exact value, e.g. "equals 53.4". It does not allow to take into account imprecise concepts, such as "young people" or "slightly above balance". Such fuzzy concepts are however easier to understand and to process for human beings, as discussed in Sect. 2.

Fuzzy logic proposes to generalise binary logic on both points, regarding representation and inference tools. Regarding the representation issue, the notion of fuzzy set and fuzzy linguistic, as well as fuzzy quantifiers, as presented in Sect. 2.1, can be used to model imprecise concepts. Regarding the inference process, the Generalised Modus Ponens [75] implements a solution to approximate reasoning and can be sketched as:

$$\begin{array}{ccc} A & \Longrightarrow & B \\ A' & & \\ \hline & & B' \end{array}$$

The main difference with classical Modus Ponens is that A' may be different from A and still allow to infer a piece of knowledge B'. However, B' is then usually different from B, to take into account the difference between A and A'.

Formally, the membership function of the output is defined from that of the rule premise, rule conclusion and observation as

$$\forall y \in \mathcal{Y},\ \mu_{B'}(y) = \sup_{x \in \mathcal{X}} \top(\mu_{A'}(x), \mathcal{I}(\mu_A(x), \mu_B(y)))$$

where $\top$ is a so-called t-norm, i.e. a conjunctive operator, and $\mathcal{I}$ a fuzzy implication operator. This equation means that a value y has a high membership degree to B' if there exists (sup operator) at least one value x that both ($\top$ conjunctive operator) belongs to the observation A' and can imply ($\mathcal{I}$ implication) y when considering the rule $A \Longrightarrow B$.

Figure 6 shows an example of result that can be obtained when considering fuzzy subsets defined on real numbers, i.e. $\mathcal{X} = \mathcal{Y} = \mathbb{R}$, with the Łukasiewicz operators $\top(u, v) = \max(u + v - 1, 0)$ and $\mathcal{I}(u, v) = \min(1 - u + v, 1)$. The left part of the figure shows the rule premise A in blue and the observation A' in red, that only partially matches A; the right part shows the rule conclusion B in blue and the obtained output B' in red: Although A' is not equal to A, an informative inference can be performed. B' can be observed to be a fuzzy subset similar to B, although with a larger kernel (set of values with membership degrees equal 1). The main difference is that, for B', all values in the universe have a non-zero membership degree: the minimal value is 0.2. This is interpreted as an uncertainty level, expressing that, because the observation only partially matches the rule premise, the rule does not completely apply. Therefore, some values not implied by the rule may not be excluded, due to the fact that the rule premise only partially

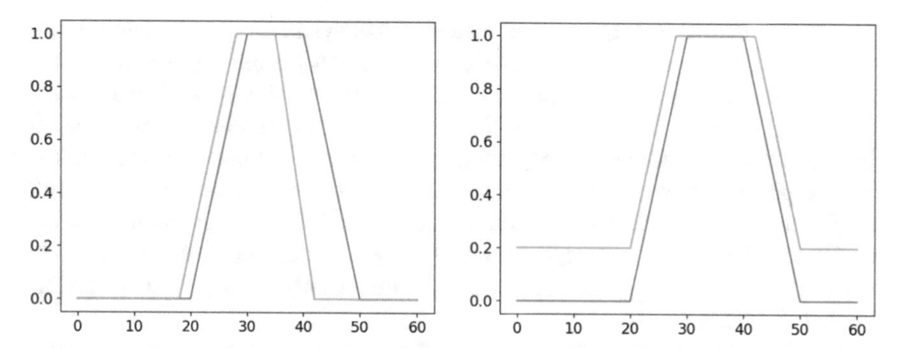

Fig. 6 Illustration of the Generalised Modus Ponens with Łukasiewicz operators: (left) rule premise A in blue, observation A' in red; (right) rule conclusion B in blue, output B' in red

accounts for the observation. However, the rule still provides some information, whereas a classical logic inference process would not allow to draw any conclusion from the observation.

3.3 *Other Reasoning Forms*

Generalised Modus Ponens makes it possible to perform more natural reasoning than the strict classical logic Modus Ponens; however, it can be considered that it also suffers from some limitations: This section briefly presents other types of approximate reasoning variants, namely gradual, analogical and interpolative reasoning.

3.3.1 Gradual Reasoning

In the case where there is no match at all between the observation and the rule premise, the Generalised Modus Ponens usually outputs a conclusion with full uncertainty: It thus expresses that the rule does not allow to perform any informative inference, which can be considered as justified insofar as the rule does not actually apply to the observation. Gradual reasoning has been proposed as a method to take into account other principles to guide inference, and in particular monotonicity constraints:[13] It considers the case where the rule *if* A, *then* B is actually not to be understood as a pure implication, but in the form *the more* A, *the more* B. As such, it is a reasoning, logical, counterpart to the gradual patterns discussed in the previous section.

[13]This notion of monotonicity is here understood as a global constraint across the universes, and not in terms of truth values, as is sometimes the case [34].

Gradual reasoning requires a monotonous behaviour of the inference result with respect to the rule premise: It takes into account the relative position of the observation with respect to the premise. Considering, e.g. fuzzy sets defined on the universe $\mathcal{X} = \mathbb{R}$, it allows to get different results when the observation is on the left or on the right of the premise. Roughly speaking, if the observation is shifted with respect to the rule premise, then the inferred result should also be shifted with respect to the rule conclusion, which is not the case with Generalised Modus Ponens, as illustrated on Fig. 6. Indeed, the result B' is symmetrical with respect to the rule conclusion B, not taking into account the fact that the observation A' is on the left of the rule premise A.

For the example given previously, regarding gender balance, one can consider that the actual rule knowledge is of the form *the higher the age, the higher the proportion of women*, with the reference proportion given in the previous rule. Considering gradual reasoning would then lead to different results in the case where the observations A' is *age is around 48* or *age is around 52*.

There exist several formal expressions to model this type of natural reasoning, which are not developed in this chapter. The interested reader is for instance referred to [12, 69].

3.3.2 Analogical Reasoning

Another natural way of thinking, called analogical reasoning (see for instance [8]), stems from the philosophical concept of analogy, regarded as a process to transfer information from a domain to another one on the basis of similarities. Analogy is widely used by human beings to make decisions, use metaphors or understand complex systems. It became popular in artificial intelligence in the form of the so-called case-based reasoning. A first fuzzy version of case-based reasoning was proposed in [13], followed by many derived versions. More generally, fuzzy analogical reasoning relies on a comparison between the observation and the rule premise. Schematically, from a "rule" that links A and B and from the observation, it outputs B' such that B' *is to* B *what* A' *is to* A. As such, it does not explicitly consider an implication relation between A and B and it does not perform logical inference: Rather, it exploits comparison measures, to assess the similarity between the observation and the rule premise and to build the output with an equivalent similarity relation to the rule conclusion.

This approach has also been developed as a method of approximate reasoning, with many variants, see [17, 19, 28, 29] to name a few. Another form of reasoning based on similarities was introduced as *similarity-based reasoning* by [63] and developed in [35]. It can be observed that, to some extent, gradual reasoning is also related to analogical reasoning, because of its integration of a relative position comparison. Other approaches [12] combine even more explicitly the analogical similarity principle with the Generalised Modus Ponens inference process.

3.3.3 Interpolative Reasoning

A derived case of analogical reasoning concerns an extension of classic interpolation, very useful in the case of sparse data, to make decisions in the interval between two existing cases: It also makes it possible to perform inference even when no rule applies to the considered observation. It relies on the comparison of the observation with the premises of other rules that differ.

When the available information is imprecise, fuzzy interpolative reasoning enables the user to make a decision in a region of the universe where there is no explicit data, considering both a graduality in decisions with respect to inputs and an analogy with existing data [6, 20, 21].

3.3.4 Fuzzy Reasoning Conclusion

In all cases, the underlying motivation for fuzzy reasoning approaches is to model natural ways of thinking, so that a user can easily understand the performed inference and thus the obtained result, while still offering guarantees about the reasoning validity. The latter come from the proposed formalisation, using a reference formula for building the output and predefined parameters (e.g. conjunctive or implication operators or comparison measures).

3.4 Examples of Methodological Utilisation

The formal tools described in the previous subsection lead to many different types of applications. In the framework of machine learning, they open the way for the successful fuzzy rule-based systems, presented in Sect. 4.1. Two other examples of methodological utilisation are presented here.

3.4.1 Fuzzy Ontologies

Reasoning tools are of particular use when knowledge is represented using ontologies, for instance in the framework of description logic [5]. Fuzzy logic has been integrated into ontologies [23], so as to benefit from its advantages regarding imprecise or vague knowledge representation, suitable to model and process concepts as they are handled by human beings and they thus improve the interactions with users. Fuzzy description logic [68] provides solutions to instance checking, relation checking, subsumption, consistency, in an imprecise environment. It has in particular proved to be highly useful in the semantic web.

Another development of fuzzy set theory and fuzzy logic, not detailed in this chapter, is the notion of possibility theory and possibilistic logic [77]: They make it possible to deal with uncertain knowledge, whereas fuzzy logic is more dedicated to the case of imprecise knowledge. Possibilistic description logic [27, 50] is based on possibilistic logic and, in particular, allows to represent and manage certainty about instances and relations. It is based on gradual and subjective uncertainties and evaluates the non-statistical confidence assigned to a piece of information. It can be used in multisource and heterogeneous information environments.

3.4.2 Fuzzy Control

The theoretical principles for extended formal logic also founded the practical domain of fuzzy control [38]: Similar principles are applied to control systems, using rules that define the desired behaviour and fuzzy inference definitions to derive the applied command for a given observed state of the system. Fuzzy control has numerous real world applications in many domains for instance including electrical household appliances (washing machines, fridges, air conditioning, vacuum cleaner), autonomous vehicles (smart metro, helicopter, train inclination in curves) or industrial applications (steel industry, cement works, paper mills) to name a few. Among these, smart transportation, internet of things and robotics are examples of domains where the need of explainable AI is the most important because the users request security and explanations about the functioning of the systems in which they are personally involved.

4 Fuzzy Machine Learning

Machine learning is nowadays the most visible part of Artificial Intelligence. Statistical machine learning provides very good results in classification or decision-making, in particular in the case of big data. Nevertheless the reasons why an object is classified in a given class or why a decision is made are not obvious to the user. On the opposite, fuzzy inductive learning based on fuzzy decision trees provides the explicit list of attributes involved in the result, which enables the user to understand the reasons of the decision. A fuzzy set-based knowledge representation offers the advantage of taking into account imprecise or linguistically expressed descriptions of the cases, which facilitates interactions with end users in allowing them to use their natural manner to describe situations.

These considerations have nevertheless their limits. It has been pointed out that a trade-off is necessary between explainability and accuracy [30] where it is often considered that graphical models or decision trees are more explainable than deep learning or random forests, although sometimes at the expense of a lower prediction accuracy. This principle also applies to the case of fuzzy models [24].

Experimental approaches to the interpretability of fuzzy systems have been investigated [2] in an attempt to cope with the subjective perception of explainability. Multi-objective optimisation strategies can be used to maximise the interpretability of fuzzy AI systems while looking for high accuracy [52].

This section discusses two examples of classification systems, namely fuzzy rule-based systems and fuzzy decision trees, as the unsupervised learning case of clustering.

4.1 *Fuzzy Rule-Based Systems*

Fuzzy rule-based systems are historically the first fuzzy AI systems, developed in the 1970s after their inception by L.A. Zadeh [73]. They are recognised to provide an easy interaction with the user.

Rules have the form "if V_1 is A_1 and V_2 is A_2, ..., then W is B", for instance "If the temperature is low and the wind is speedy the risk is high". V_1, V_2, ..., W are linguistic variables associated with fuzzy modalities A_1, A_2, ..., B belonging to fuzzy partitions of their respective universes, as introduced in Sect. 2.1. They can be processed to infer decision for a given observation using tools such as the Generalised Modus Ponens, as described in Sect. 3.2.

Such rules were originally elicited from experts, as it was the case for expert systems. They were later obtained automatically through machine learning. Modalities are tuned, either by a consensus among experts made possible thanks to aggregation methods available in fuzzy logic, or automatically by various methods such as genetic algorithms or evolutionary computation [36]. Techniques for rule pruning or rule merging participate in the reduction of complexity [40].

Various types of properties are involved in the explainability of fuzzy rule-based systems. They have in particular been decomposed into high level properties, such as compactness, completeness, consistency and transparency of the set of rules, as opposed to low level properties for the set of modalities associated with each attribute, namely the coverage of the universe, convexity of membership functions or distinguishability between modalities [81]. Measures of focus and specificity of information have also been proposed to evaluate the informativeness of information in such systems [70].

4.2 *Fuzzy Decision Trees*

Fuzzy decision trees constitute classifiers that are built by inductive learning from a set of training data. As their crisp counterpart, they are constructed from the root to the leaves, by successively selecting the most discriminant attribute regarding classes. However, each edge from a vertex is associated with one of the *fuzzy* modalities describing the attribute selected for the vertex [53, 55]. As it was the case for fuzzy rules, modalities can be defined by experts or automatically constructed, for instance by means of mathematical morphology [54].

A path from the root to a leaf of the fuzzy decision tree, that represents the decision for a given observation data point, is then associated with a series of linguistic variables described by a fuzzy modality: It provides a linguistic rule which can be easy for the user to understand. Therefore, fuzzy decision trees are deemed to be explicit. It must still be noted that their complexity, required to achieve better accuracy, may make difficult the understanding of reasons of a decision or a classification, in the case when the number of involved features is too big to be grasped by the user.

4.3 Fuzzy Clustering

Unsupervised machine learning also has been considered from the point of view of fuzzy tools, in particular for the topic of clustering. Indeed, in classic clustering, each object is assigned to a cluster, in such a way that the elements of each cluster are similar to each other. It can be observed that in many cases, the boundaries between clusters do not look natural, but somehow arbitrary and are difficult to justify.

Probabilistic approaches to clustering, among others using the reference Gaussian Mixture Model, have been proposed to integrate more flexibility. However, they consider a different cost function for the clustering task: They do not aim at optimising an objective function expressing a balance between intra-cluster homogeneity and inter-cluster separability, but aim at modelling the data underlying probability distribution. Even if the two aims are related, they are not identical.

Fuzzy clustering, as first introduced by [62] and widely developed since then (see e.g. [14, 33, 41]), preserves the initial cluster compactness/separability cost function while also enabling an object to belong to several clusters, at least to a certain extent. The degree of membership of objects to clusters defines fuzzy classes, with imprecise boundaries. A fuzzy clustering is therefore more explainable to the user than a crisp clustering.

5 Conclusion

This chapter illustrated that fuzzy models effectively participate in solutions to achieve Explainable AI, due to their underlying principles: the concepts, methods and tools of fuzzy logic on which they are based, among which in particular imprecision management and Computing With Words, are key components to design interpretable and legible systems with high expressiveness. The latter then naturally implement human-in-the-loop requirements and are currently used in a very large spectrum of applications: They for instance include instrumentation, transportation, sentiment analysis, fuzzy question answering systems, information quality and many more.

There are still many research avenues in fuzzy logic, and it seems clear that, as XAI is developing, the role of fuzzy logic will be growing. Even when it is not called fuzzy logic, the concepts of graduality or management of imprecision, are indeed of crucial importance to comply with the needs. Although currently too much reduced to deep learning, artificial intelligence is a very large domain that most likely needs to take into account complex cognitive paradigms so as to achieve its explainability aim and its friendly acceptance by users. The general framework of fuzzy set theory and fuzzy logic can be seen as a key concept that offers many tools to progress in this direction.

The authors do not pretend to present an exhaustive list of fuzzy logic-based solutions to the difficult paradigm of explainable AI, but they rather describe various ways to integrate expressiveness, transparency and interpretability in smart devices or artificial intelligence-based decision support systems, to comply with the requirements of explainable artificial intelligence. Fuzzy artificial intelligence should not be limited to fuzzy rule bases, as it is often the case. Likewise, explainable AI must not be restricted to the interpretability of statistical machine learning or statistical data science.

The purpose of this chapter is to open the door to a multiplicity of solutions linking Computing With Words and, more generally fuzzy logic, with XAI and to show that many more developments can be expected.

Let us point out that the human component of all AI systems, be it the user, the expert, the patient or the customer, is carefully taken into account in all fuzzy models, whose main mission is to mimic natural language and common sense reasoning, with a capacity to deal with subjective information and qualitative uncertainty. This is a reason why many women feel interested in the field and the gender balance is very good in the fuzzy community.

References

1. R.J. Almeida, M. Lesot, B. Bouchon-Meunier, U. Kaymak, G. Moyse, Linguistic summaries of categorical series for septic shock patient data, in *Proceedings of the IEEE International Conference on Fuzzy Systems, FUZZ-IEEE 2013*, Hyderabad, India, 7–10 July, 2013 (IEEE, 2013), pp. 1–8. https://doi.org/10.1109/FUZZ-IEEE.2013.6622581
2. J. Alonso, L. Magdalena, G. González-Rodríguez, Looking for a good fuzzy system interpretability index: An experimental approach. Int. J. Approx. Reason. **51**(1), 115–134 (2009)
3. Y.S. Aryadinata, Y. Lin, C. Barcellos, A. Laurent, T. Libourel, Mining epidemiological dengue fever data from Brazil: A gradual pattern based geographical information system, in *Proc. of the 15th Int. Conf. on Information Processing and Management of Uncertainty in Knowledge-Based Systems IPMU 2014*, vol. 443 (Springer, 2014), pp. 414–423
4. S. Ayouni, S.B. Yahia, A. Laurent, P. Poncelet, Genetic programming for optimizing fuzzy gradual pattern discovery, in *Proceedings of the 7th Conference of the European Society for Fuzzy Logic and Technology, EUSFLAT 2011*, ed. by S. Galichet, J. Montero, G. Mauris (Atlantis Press, 2011), pp. 305–310. https://doi.org/10.2991/eusflat.2011.41
5. F. Baader, I. Horrocks, U.: Sattler, Description logics, in *Handbook on Ontologies*, ed. by S. Staab, R. Studer (Springer, 2004), pp. 3–28
6. P. Baranyi, L.T. Koczy, T.D. Gedeon, A generalized concept for fuzzy rule interpolation. IEEE Trans. Fuzzy Syst. **12**(6), 820–837 (2004)
7. A. Barredo Arrieta, N. Díaz-Rodríguez, J. Del Ser, A. Bennetot, S. Tabik, A. Barbado, S. García, S. Gil-López, D. Molina, R. Benjamins, R. Chatila, F. Herrera, Explainable artificial

intelligence (XAI): concepts, taxonomies, opportunities and challenges toward responsible AI. Information Fusion **58**, 82–115 (2019)
8. P. Bartha, Analogy and analogical reasoning, in *The Stanford Encyclopedia of Philosophy*, ed. by E.N. Zalta (Spring 2019 Edition) (2019)
9. J. Bezdek, D. Dubois, H. Prade (eds.), *Fuzzy Set in Approximate Reasoning and Information Systems* (Springer, 1999)
10. O. Biran, C. Cotton, Explanation and justification in machine learning: A survey, in *Proc. of the IJCAI Workshop on eXplainable AI* (2017)
11. P. Blackburn, M. De Rijke, Y. Venema, *Modal Logic* (Cambridge University Press, 2001)
12. M. Blot, M.J. Lesot, M. Detyniecki, Transformation-based constraint-guided generalised modus ponens, in *Proc. of the IEEE Int. Conf. on Fuzzy Systems, FuzzIEEE'16* (IEEE, 2016)
13. P. Bonissone, W. Cheetham, Fuzzy case-based reasoning for decision making, in *Proc. of the 10th IEEE Int. Conf. on Fuzzy Systems*, vol. 3, Melbourne, Australia, pp. 995–998 (2001)
14. C. Borgelt, Objective functions for fuzzy clustering, in *Computational Intelligence in Intelligent Data Analysis*, ed. by C. Moewes, A. Nürnberger (Springer, 2012), pp. 3–16
15. P. Bosc, O. Pivert, SQLf: a relational database language for fuzzy querying. IEEE Trans. Fuzzy Syst. **3**(1), 1–17 (1995)
16. B. Bouchon-Meunier, N. Aladenise, Acquisition de connaissances imparfaites : mise en évidence d'une fonction d'appartenance. Revue Internationale de Systémique **11**(1), 109–127 (1997)
17. B. Bouchon-Meunier, L. Valverde, A fuzzy approach to analogical reasoning. Soft Computing **3**, 141–147 (1999)
18. B. Bouchon-Meunier, G. Moyse, Fuzzy linguistic summaries: where are we, where can we go? in *IEEE Conf. on Computational Intelligence for Financial Engineering & Economics (CIFEr), CIFEr 2012* (IEEE, 2012), pp. 317–324
19. B. Bouchon-Meunier, J. Delechamp, C. Marsala, M. Rifqi, Several forms of analogical reasoning, in *Proc. of the IEEE Int. Conf. on Fuzzy Systems, FuzzIEEE'97* (IEEE, 1997)
20. B. Bouchon-Meunier, C. Marsala, M. Rifqi, Interpolative reasoning based on graduality, in *Proc. of the IEEE Int. Conf. on Fuzzy Systems, FuzzIEEE'00* (IEEE, 2000), pp. 483–487
21. B. Bouchon-Meunier, F. Esteva, L. Godo, M. Rifqi, S. Sandri, A principled approach to fuzzy rule-based interpolation using similarity relations, in *Proc. of EUSFLAT-LFA 2005*, Barcelona, Spain, pp. 757–763 (2005)
22. B. Bouchon-Meunier, A. Laurent, M. Lesot, M. Rifqi, Strengthening fuzzy gradual rules through "all the more" clauses, in *Proceedings of the IEEE International Conference on Fuzzy Systems, FUZZ-IEEE 2010*, Barcelona, Spain, 18–23 July, 2010 (IEEE, 2010), pp. 1–7. https://doi.org/10.1109/FUZZY.2010.5584858
23. S. Calegari, D. Ciucc, Integrating fuzzy logic in ontologies, in *Proc. of the 8th Int. Conf. on Enterprise Information Systems, ICEIS 2006*, pp. 66–73 (2006)
24. J. Casillas, O. Cordon, F. Herrera, L. Magdalena, Interpretability improvements to find the balance interpretability-accuracy in fuzzy modeling: An overview, in *Interpretability issues in fuzzy modeling, Studies in Fuzziness and Soft Computing*, vol. 128, ed. by J. Casillas, O. Cordon, F. Herrera, L. Magdalena (Springer, 2003), pp. 3–22
25. A. Castelltort, A. Laurent, Fuzzy queries over NoSQL graph databases: Perspectives for extending the cypher language, in *Information Processing and Management of Uncertainty in Knowledge-Based Systems - 15th International Conference, IPMU 2014, Proceedings, Part III*, ed. by A. Laurent, O. Strauss, B. Bouchon-Meunier, R.R. Yager, Communications in Computer and Information Science, vol. 444 (Springer, 2014), pp. 384–395. https://doi.org/10.1007/978-3-319-08852-5_40
26. R. Castillo-Ortega, N. Marin, D. Sanchez, A. Tettamanzi, Quality assessment in linguistic summaries of data, in *Proc. of the Int. Conf. Information Processing and Management of Uncertainty in Knowledge-Based Systems, IPMU2012* (Springer, 2012), pp. 285–294
27. O. Coucharière, M. Lesot, B. Bouchon-Meunier, Consistency checking for extended description logics, in *Proc. of the 21st International Workshop on Description Logics (DL 2008)*, vol. 353 (2008)

28. V. Cross, Patterns of fuzzy rule based inference. Int. J. Approx. Reason. **11**, 235–255 (1994)
29. V. Cross, M.J. Lesot, Fuzzy inferences using geometric compatibility or using graduality and ambiguity constraints, in *Proc. of the IEEE Int. Conf. on Fuzzy Systems, FuzzIEEE'17* (IEEE, 2017), pp. 483–487
30. H. Dam, T.T. Ghose, Explainable software analytics, in *Proc. of ICSE'18* (2018)
31. L. Di Jorio, A. Laurent, M. Teisseire, Mining frequent gradual itemsets from large databases, in *Proc. of the Symposium on Intelligent Data Analysis, IDA09*, pp. 297–308 (2009)
32. T.D.T. Do, A. Termier, A. Laurent, B. Négrevergne, B. Omidvar-Tehrani, S. Amer-Yahia, PGLCM: efficient parallel mining of closed frequent gradual itemsets. Knowl. Inf. Syst. **43**(3), 497–527 (2015). https://doi.org/10.1007/s10115-014-0749-8
33. C. Döring, M.J. Lesot, R. Kruse, Data analysis with fuzzy clustering methods. Comput. Stat. Data Anal. **51**(1), 192–214 (2006)
34. D. Dubois, H. Prade, Gradual inference rules in approximate reasoning. Information sciences **61**, 103–122 (1992)
35. F. Esteva, P. Garcia, L. Godo, R. Rodríguez, A modal account of similarity-based reasoning. Int. J. Approx. Reason. **16**(3), 235–260 (1997)
36. A. Fernandez, F. Herrera, O. Cordon, M. del Jesus, F. Marcelloni, Evolutionary fuzzy systems for explainable artificial intelligence: Why, when, what for, and where to? IEEE Comput. Intell. Mag. **14**(1), 69–81 (2019)
37. S. Gottwald, *A Treatise on Many-Valued Logics, Studies in Logic and Computation*, vol. 9 (Baldock, 2001)
38. T.M. Guerra, K. Tanaka, A. Sala, Fuzzy control turns 50: 10 years later. Fuzzy Sets Syst. **281**, 168–182 (2015)
39. R. Guidotti, A. Monreale, S. Ruggieri, F. Turini, F. Giannotti, D. Pedreschi, A survey of methods for explaining black box models. ACM Comput. Surv. **51**(5), 93:1–93:42 (2018)
40. S. Guillaume, Designing fuzzy inference systems from data: An interpretability-oriented review. IEEE Trans. Fuzzy Syst. **9**(3), 426–443 (2001)
41. F. Höppner, F. Klawonn, R. Kruse, T. Runkler, *Fuzzy Cluster Analysis: Methods for Classification, Data Analysis and Image Recognition* (Wiley-Blackwell, 1999)
42. E. Hüllermeier, Association rules for expressing gradual dependencies, in *Proc. of the 6th European Conference on Principles of Data Mining and Knowledge Discovery* (Springer, 2002), pp. 200–211
43. J. Kacprzyk, S. Zadrozny, Linguistic database summaries and their protoforms: towards natural language based knowledge discovery tools. Inf. Sci. **173**(4), 281–304 (2005). https://doi.org/10.1016/j.ins.2005.03.002
44. J. Kacprzyk, S. Zadrozny, Comprehensiveness of linguistic data summaries: A crucial role of protoforms, in *Computational Intelligence in Intelligent Data Analysis*, ed. by C. Moewes, A. Nürnberger (Springer, 2012), pp. 207–221
45. J. Kacprzyk, A. Wilbik, S. Zadrozny, An approach to the linguistic summarization of time series using a fuzzy quantifier driven aggregation. Int. J. Intell. Syst. **25**(5), 411–439 (2010). https://doi.org/10.1002/int.20405
46. A. Kaufmann, L. Zadeh, D.L. Swanson, *Introduction to the Theory of Fuzzy Subsets*, vol. 1 (Academic Press, 1975)
47. A. Laurent, Querying fuzzy multidimensional databases: Unary operators and their properties. Int. J. Uncertainty Fuzziness Knowledge Based Syst. **11**(Supplement-1), 31–46 (2003). https://doi.org/10.1142/S0218488503002259
48. A. Laurent, M.J. Lesot, M. Rifqi, Graank: Exploiting rank correlations for extracting gradual itemsets, in *Proc. Of the Int. Conf. on Flexible Query Answering Systems, FQAS09*, pp. 382–393 (2009)
49. S. Lefort, M.J. Lesot, E. Zibetti, C. Tijus, M. Detyniecki, Interpretation of approximate numerical expressions: Computational model and empirical study. Int. J. Approx. Reason. **82**, 193–209 (2017)

50. M. Lesot, O. Coucharière, B. Bouchon-Meunier, J.-L. Rogier, Inconsistency degree computation for possibilistic description logic: an extension of the tableau algorithm, in *Proc. of the 27th North American Fuzzy Information Processing Society Annual Conference (NAFIPS 2008)* (IEEE, 2008)
51. V. Liao, D. Gruen, S. Miller, Questioning the ai: Informing design practices for explainable ai user experiences, in *Proc. of the CHI Conf. on Human Factors in Computing Systems (CHI'20)* (2020)
52. L. Magdalena, Do hierarchical fuzzy systems really improve interpretability? in *Information Processing and Management of Uncertainty in Knowledge-Based Systems. Theory and Foundations, IPMU 2018*, ed. by J. Medina, et al., Communications in Computer and Information Science, vol. 854 (Sprinter, 2018), pp. 16–26
53. C. Marsala, Apprentissage inductif en présence de données imprécises : construction et utilisation d'arbres de décision flous. Ph.D. thesis, Université Paris VI (1998)
54. C. Marsala, B. Bouchon-Meunier, Fuzzy partitioning using mathematical morphology in a learning scheme, in *Proc. of the IEEE 5th Int. Conf. on Fuzzy Systems*, pp. 1512–1517 (1996)
55. C. Marsala, B. Bouchon-Meunier, Choice of a method for the construction of fuzzy decision trees, in *Proc. of the 12th IEEE International Conference on Fuzzy Systems*, vol. 1, St Louis, MO, USA, pp. 584–589 (2003)
56. T. Miller, Explanation in artificial intelligence: Insights from the social sciences. Artificial Intelligence **267**, 1–38 (2019)
57. A. Motro, Imprecision and incompleteness in relational databases: survey. Inf. Softw. Technol. **32**(9), 579–588 (1990). https://www.sciencedirect.com/science/article/pii/0950584990902045
58. M. Needham, Explore the data behind the women's world cup with our world cup graph (2019). https://medium.com/neo4j/now-available-womens-world-cup-2019-graph-cf3bd9e44e22
59. T. Ngo, V. Georgescu, A. Laurent, T. Libourel, G. Mercier, Mining spatial gradual patterns: Application to measurement of potentially avoidable hospitalizations, in *Proc. of the 44th Int. Conf. on Current Trends in Theory and Practice of Computer Science. Lecture Notes in Computer Science*, vol. 10706 (Springer, 2018), pp. 596–608
60. O. Pivert, O. Slama, G. Smits, V. Thion, SUGAR: A graph database fuzzy querying system, in *Tenth IEEE International Conference on Research Challenges in Information Science, RCIS 2016* (IEEE, 2016), pp. 1–2. https://doi.org/10.1109/RCIS.2016.7549366
61. M. Ribeiro, S. Singh, C. Guestrin, Why should i trust you?: Explaining the predictions of any classifier, in *Proc. of the ACM SIGKDD Int. Conf. on Knowledge Discovery and Data Mining*, pp. 1135–1144 (2016)
62. E. Ruspini, A new approach to clustering. Inf. Control **15**(1), 22–32 (1969)
63. E. Ruspini, On the semantics of fuzzy logic. Int. J. Approx. Reason. **5**(1), 45–88 (1991)
64. F. Shah, A. Castelltort, A. Laurent, Extracting fuzzy gradual patterns from property graphs, in *2019 IEEE International Conference on Fuzzy Systems, FUZZ-IEEE 2019*, New Orleans, LA, USA, June 23–26, 2019 (IEEE, 2019), pp. 1–6. https://doi.org/10.1109/FUZZ-IEEE.2019.8858936
65. G. Smits, O. Pivert, T. Girault, Reqflex: Fuzzy queries for everyone. Proc. VLDB Endow. **6**(12), 1206–1209 (2013). https://www.vldb.org/pvldb/vol6/p1206-smits.pdf
66. G. Smits, O. Pivert, M.J. Lesot, A vocabulary revision method based on modality splitting, in *Proc. of IPMU'14*, vol. CCIS442 (Springer, 2014), pp. 376–385
67. G. Smits, O. Pivert, R.R. Yager, P. Nerzic, A soft computing approach to big data summarization. Fuzzy Sets Syst. **348**, 4–20 (2018). https://doi.org/10.1016/j.fss.2018.02.017
68. U. Straccia, A fuzzy description logic for the semantic web, in *Capturing Intelligence*, chap. 4, ed. by E. Sanchez (Elsevier, 2006), pp. 73–90
69. P.N. Vo, M. Detyniecki, B. Bouchon-Meunier, Gradual generalized modus ponens, in *Proc. of the IEEE Int. Conf. on Fuzzy Systems, FuzzIEEE'13* (IEEE, 2013)
70. A. Wilbik, J. Kacprzyk, On the evaluation of the linguistic summarisation of temporally focused time series using a measure of informativeness, in *Proc. of the Int. Multiconf. on Computer Science and Information Technology, IMCSIT 2010*, pp. 155–162 (2010)

71. R.R. Yager, A new approach to the summarization of data. *Information Sciences* **28**(1), 69–86 (1982)
72. R.R. Yager, *Social Network Database Querying Based on Computing with Words* (Springer International Publishing, Cham, 2014), pp. 241–257. https://doi.org/10.1007/978-3-319-00954-4_11
73. L.A. Zadeh, Outline of a new approach to the analysis of complex systems and decision processes. IEEE Trans. Syst. Man Cybern. **SMC-3**(1), 28–44 (1973)
74. L.A. Zadeh, The concept of a linguistic variable and its application to approximate reasoning - i. Inf. Sci. **8**, 199–249 (1975)
75. L.A. Zadeh, The concept of a linguistic variable and its application to approximate reasoning—ii. Inf. Sci. **8**(4), 301–357 (1975)
76. L.A. Zadeh, Fuzzy logic and approximate reasoning. Synthese **30**(3–4), 407–428 (1975)
77. L.A. Zadeh, Fuzzy sets as a basis for a theory of possibility. Fuzzy Sets Syst. **1**(1), 3–28 (1978)
78. L.A. Zadeh, Fuzzy logic = computing with words. IEEE Trans. Fuzzy Syst. 4, 103–111 (1996)
79. L.A. Zadeh, From computing with numbers to computing with words—from manipulation of measurements to manipulations of perceptions. IEEE Trans. Circuits Syst. **45**, 105–119 (1999)
80. L.A. Zadeh, Outline of a computational theory of perceptions based on computing with words, in *Soft Computing and Intelligent Systems*, ed. by N. Sinha, M. Gupta, (Academic Press, Boston, 1999), pp. 3–22
81. S. Zhou, J. Gan, Low-level interpretability and high-level interpretability: a unified view of data-driven interpretable fuzzy system modelling. Fuzzy Sets Syst. **159**(23), 3091–3131 (2008)

Bernadette Bouchon-Meunier entered the Ecole Normale Supérieure Cachan in the Mathematics main stream and discovered that a new option on Computer Science had just been created. Finding this option appealing, she chose to follow these courses in addition to courses in Mathematics and she received Master of Science degrees in Computer Science in 1970 and in Mathematics in 1972. She discovered Artificial Intelligence in 1969 under the guidance of the French Artificial Intelligence pioneer, Jacques Pitrat. She then earned a Ph.D. in Applied Mathematics in 1972 and a D.Sc in Computer Science from the University of Paris in 1978. She was hired as a full-time researcher at the National Centre for Scientific Research in 1972 and started to work on information processing and decision support systems. In 1973, she discovered by serendipity the recent concept of fuzzy set created by L.A. Zadeh and it changed the direction of her research. Her professional life would not have been the same without the discovery of this promising paradigm and her immersion in the emerging fuzzy community, led by L.A. Zadeh, who was a true mentor to her.

She began to be the head of a research group in 1979 in the Paris VI-Pierre et Marie Curie University and, in 1990, she was promoted to director of research by the National Centre for Scientific Research. At the end of her career, she was the head of the department of Databases and Machine Learning in the Computer Science Laboratory of the University Paris 6 (LIP6). She is now director of research emeritus. She is the co-executive director of the IPMU International Conference held every other year, that she created with R.R. Yager in 1986. She is the Editor-in-

Chief of the International Journal of Uncertainty, Fuzziness and Knowledge-based Systems that she founded in 1993. She is also the (co)-editor of 27 books and the (co)-author of five. She supervised 52 Ph.D. students, including 15 female students.

She is currently the President of the IEEE Computational Intelligence Society (2020–2021) and the IEEE France Section Computational Intelligence chapter vice-chair. She is an IEEE life fellow, an International Fuzzy Systems Association fellow and an Honorary Member of the EUSFLAT Society. She received the 2012 IEEE Computational Intelligence Society Meritorious Service Award, the 2017 EUSFLAT Scientific Excellence Award and the 2018 IEEE CIS Fuzzy Systems Pioneer Award.

The friendly and dynamic community working on fuzzy sets has certainly been a breeding ground for her research from the early years, as has been later the broader Computational Intelligence community, and more particularly the IEEE Computational Intelligence Society. The latter is very active in terms of scientific leadership, and committed to the principles of diversity and support to all its members, especially women. Bernadette Bouchon-Meunier was the founding chair of the IEEE Women in Computational Intelligence committee, whose purpose is to develop, promote, organise and lead activities to ensure equal opportunities for both genders in the life of the IEEE Computational Intelligence Society and in the arena of computational intelligence.

Anne Laurent is Full Professor at the University of Montpellier, LIRMM lab. As the head of the FADO Research group, she works on open data, semantic web and data mining. She is particularly interested in the study of the use of fuzzy logic to provide more valuable results, while remaining scalable. Anne Laurent has been the head of the Computer Science Department at Polytech Montpellier Engineering School at the University of Montpellier, which prepares a 5-year Masters in computer science and management. She is currently Vice-President at University of Montpellier delegated to open science and research data. She also heads the Montpellier Data Science Institute (ISDM) and the high-performance computing centre (MESO@LR). Anne has (co-)supervised 17 Ph.D. thesis, among which 6 female candidates.

Interested in all subjects taught in high school, entering the field of CIS was quite natural as it was the best way to open multiple studies and career opportunities. It was then quite natural to stay in it for several reasons. Anne has met many high level and committed women who helped her and pushed her. All these women served as examples, which is a key point that Anne tries to reproduce with kids and teenagers (at Coderdojo Montpellier that aims at Enabling young people worldwide to create and explore technology together), students and young colleagues.

Marie-Jeanne Lesot is an associate professor in the department of Computer Science Lab of Paris 6 (LIP6) and a member of the Learning and Fuzzy Intelligent systems (LFI) group. Her research interests focus on fuzzy machine learning with an objective of data interpretation and semantics integration; they include similarity measures, fuzzy clustering, linguistic summaries and information scoring. She is also interested in approximate reasoning and the use of non-classical logics, in particular weighted variants with increased expressiveness that are close to natural human reasoning processes. Marie-Jeanne Lesot has (co-)supervised 15 Ph.D. thesis, among which 3 female candidates.

When encountered during her studies, the domain of Computational Intelligence appeared to her as an optimal combination of theoretical challenges with practical considerations, taking a human-into-the-loop position: users are put in the centre of the solution design process, and considered as source for inspiration so as how to deal with the considered issues, and, as such, as incentive for integrating cognitive points of view.

Adaptive Psychological Profiling from Nonverbal Behavior – Why Are Ethics Just Not Enough to Build Trust?

Keeley Crockett

1 Introduction

Psychological profiling is a term used to describe a person's probable character or behavior and typically stems from criminology. Human deception detection is a vastly studied area that covers strategies to detect low- and high-stake lies through the observation of indicators stemming from nonverbal communication and/or body language such as arm movement. Nonverbal communication can be defined as any form of communication without using words and encompasses visual and auditory channels such as facial expressions, posture, movement of the body, gestures, touch, and nonlinguistic vocal sounds [5]. Human deception detection is a high-trained skill where the human lie catcher must not "misinterpret 'meaningless' movements and commit false positive errors" [57]. There is also an ingrained level of subjectivity between human lie catchers, with different observers potentially reporting different opinions. Therefore, a solution is to perform automated and adaptive physiological profiling using machine learning to assist a human lie catcher in making a decision about whether the person is actually telling a lie. Silent Talker (ST) [6, 14–16] is an automated deception detection system that extracts and analyses multichannels of nonverbal facial behavior using a hierarchy of machine learning classifiers to detect patterns of behavior of facial components (i.e. left eye) over periods of time. The core technology has been successfully evaluated in studies of both deception detection [6, 54, 59] and in human comprehension when engaged in a learning task [9, 10, 38] showing accuracies between 75% and 85% depending on the study. The Silent Talker architecture was designed to be part of a human in the

K. Crockett (✉)
Manchester Metropolitan University, Manchester, UK
e-mail: K.Crockett@mmu.ac.uk

A. E. Smith (ed.), *Women in Computational Intelligence*, Women in Engineering and Science, https://doi.org/10.1007/978-3-030-79092-9_3

loop system, providing an additional dimension of information to a human decision maker.

This chapter first describes two cases studies where the Silent Talker technology has been reengineering and utilized. The first case describes how an automated deception detection system (ADDS) was integrated within a module in a prototype system known as iBorderCtrl (European Union H2020 funded project) and was used to interview travelers when they preregistered for a trip across EU Schengen land borders [12]. The second case study describes how the Silent Talker technology was adapted to detect human comprehension from nonverbal behavior within an educational context to create two systems: FATHOM [10] and COMPASS [38]. For each study, using each system, stringent ethics procedures were followed, including institutional approvals, the use international ethical advisors and panels (where appropriate) for country specific approvals, consultations with partner organizations, advice from legal professionals, and data security experts. However, the granting of ethical approval from all stakeholders did not equate to trust in systems that could be perceived in having an impact on the human rights of an individual. This raises the question: *Are humans ready for machines to automatically profile behavior of complex mental states such as deception and comprehension?* Given the media coverage of the iBorderCtrl system (case study 1), the media perception is clearly not. There are five factors that make a story newsworthy: timing, significance, proximity, prominence, and human interest [46]. The use of automated detection deception within a land border scenario in the context of a significantly larger research and innovation project was a human interest story that met the criteria of provoking emotion and evoking responses, which led to a number of inaccurate news stories appearing across the world. Case study 2, detection of human comprehension in a learning environment using similar technology, sparked no media interest.

Artificial intelligence (AI) is currently very topical and being used in systems that affect the day-to-day lives of people, of which they may not even be aware. While governments and corporate organizations are currently focused on presenting principles of what constitutes Ethical AI, there is little guidance on how to implement, and not the infrastructure to support its implementation, particularly in small businesses. Coupled with GDPR's Article 22 [33] concerning profiling and automated decision-making, case law on companies that violate this Article are in the early stages. This chapter briefly reviews the rise of the ethical charters in terms of the big issues in the use of AI and machine learning, including explainability, transparency, fairness, accountability, trust, bias, sustainability, and data privacy and governance. As ethical processes have been standardized, then perhaps so should the implementation of principles of ethical AI in order to build trust in the use of a system. For example, this will always include the right of an individual to request a human to make any final decisions. Therefore, we should also ask: *Are machines based on AI ready to automatically profile behavior of complex mental states such as deception and comprehension*?

This chapter is structured as follows. The first section defines and reviews the advances in machine-based automated psychological profiling. Next, the rise of ethical charters reviews the current international state of principles and guidelines in ethical AI, which is then followed by a brief overview of the ethical and legal implications of automated profiling. Two case studies based on automated psychological profiling in the areas of deception and comprehension detection are then presented. The impact of the media on the first case study is described, which highlights why ethically aligned design alone is not sufficient in any project or system using AI. Finally, the need for educating the public in ethical AI is discussed in order to empower them to ask questions about AI systems that impact their everyday lives.

2 Machine-Based Automated Psychological Profiling

Automated psychological profiling uses image processing and machine learning techniques to try and detect specific, complex, mental states of a person through establishing patterns indicative of that state from nonverbal behavior. Such a system is complex, not only in its architecture but also in how it can be trained, validated, and tested using laboratory-based experiments designed to capture ground truths. For humans, detecting if a person is deceiving or not is usually determined by a trained psychologist. To the average person, lie detection is often only slightly better than chance. Whereas there are documented signs that a person is deceiving, these are subjective, situation dependent, and challenging for a human to spot. The fact that there is no clear set of rules for a human makes the problem, on one hand, suitable for a machine learning approach; however, detected patterns and results can be extremely hard to verify. This section briefly defines nonverbal behavior (NVB) and provides an overview of automated psychological profiling systems. A brief overview of the ethical and legal landscape of automated profiling are discussed, and the rise of international ethical guidelines and charters is reviewed.

2.1 Nonverbal Behavior

In nonverbal behavior (NVB) research, the term nonverbal channel is used to describe a single isolatable unit of nonverbal behavior, e.g., left eye blink. Traditionally, the measurement of nonverbal behavior of a person is done via manual extraction. A human observer hand codes observed instances of each nonverbal channel under analysis in real-time or from a video recorded after an interview [32]. This method is known to be very subjective depending upon the experience of the observer, time-consuming, and expensive. Miller [48] reported that humans cannot process more than 7 ± 2 pieces of information at the same time in immediate memory. Therefore, in manual extraction, human observers have to adopt a serial approach to coding multiple channels of NVB by continually viewing video-recorded observations. A solution was to use additional human observers, but this contributes to the problem of agreement between coders and the overall reliability of results. The most well-known coding system for human facial expression analysis and classification is known as the Facial Action Coding System (FACS), developed by Ekman and Friesen [21]. FACS, while marketed as a standard tool which offers a consistent approach to encoding, still relies on manually using human observers with the additional overhead of training costs. Therefore, automated extraction of NVB is essential to reduce subjectivity and overheads.

2.2 Automated Psychological Profiling Systems

Silent Talker [59, 60] and AVATAR [36] are two known automated deception detection systems. Avatar, a kiosk-based system, has been trialed at US-Mexico, US-Canada and selected EU borders, with reported deception detection accuracies of between 60% and 80% [17, 36]. Silent Talker is designed to be used in a natural interviewing situation; it classifies multiple visible signals from the upper body to create a comprehensive time-profile of a subject's psychological state [5]. ST is a patented automated psychological profiling system that can be adapted to detect different behavioral states of a person, such as deception or comprehension. ST was used to investigate whether artificial neural networks (ANNs) could learn NVB patterns and provide a more consistent, robust way of classifying deception better than the polygraph, humans, and other semiautomated methods [59]. The system analyzed patterns of very fine-grained micro-gestures to determine psychological states. Determination of the state of an individual is akin to profiling based on their NVB in response to a specific situation or event. ST was developed using hierarchical banks of ANNs which, once trained for a specific mental state and scenario, could operate in real-time in a noninvasive manner. The system uses object locators to detect facial movements and pattern detectors to detect specific behaviors of objects, i.e., right eye blinking and sideways head movements. A further bank of ANN's is then used to classify NVB patterns as behavioral states over different time periods.

ST has been shown in scientific studies to detect both deception and comprehension of an individual while they are engaged in specific tasks. The original study reported in [59], reported that ST could detect deception from NBV from an interview with up to 85% classification accuracy (compared to 54% human level and between 80% and 98% for the polygraph), without introducing human perception bias. More recently, a variant of ST was developed as part of the European Commission H2020 project, known as iBorderCtrl, to interview travelers prior to them travelling across EU Schengen land borders within a traveler preregistration system [39]. The iBorderCtrl Automated Deception Detection System (ADDS) incorporating ST would contribute to an overall estimated deception risk score for individuals that support human border guards in their decision-making. An initial prototype of ADDS, using no ANN optimization, on a small sample of 32 participants of mixed gender and ethnicity, gave an accuracy of 75% on people the system had never seen before [54]. ADDS and its impact are described in detail in case study 1.

It is also possible to detect from a person's NVB their comprehension level within a learning environment. From a teacher/instructor's viewpoint, knowing when a pupil is not comprehending may come from in-class activities, coursework or examinations, or from experience in observing their general NVB. From a learner's perspective, failing to comprehend in a classroom may have a long-term impact on their educational attainment and, consequently, their career path. Early work on automatically detecting comprehension levels from NVB has been achieved by extracting head movements in 1 s slots through use of an Active Appearance Model [69], giving a 71% accuracy. Both Graesser et al. [34] and Rayner et al. [58] conducted experimental studies that used eye tracking to provide insight into human cognition with various success levels. However, the eye tracking technology was expensive and heavy and obtrusive and could potentially hamper a person's natural behavior. In order to create a system which automatically detected comprehension using invasive technology, the ST architecture was adapted to create first FATHOM [10] and then COMPASS [38]. FATHOM involved image-based evaluation of comprehension using mock-informed consent interviews recorded as part of an HIV/AIDS prevention study in North-western Tanzania. Classification accuracy was between 84% and 91% [3]. COMPASS meanwhile used a conversational tutorial system developed in-house with adapted psychological profiling embedded; it achieved 76% classification accuracy for 'comprehension' and 'noncomprehension' states through real-time on-screen learning interactions. The development and application of FATHOM and COMPASS are described in case study 2.

2.3 Ethical and Legal Implications of Automated Profiling

The iBorderCtrl project, introduced in Sect. 2.2, took place in a period of social and legislative turmoil regarding the legal and ethical framework in which artificial intelligence should be operated. The project ran from August 2016 to August 2019 with General Data Protection Regulation (GDPR) (Regulation (EU) 2016/679) coming into effect on May 25, 2018 [33]. During its lifetime, iBorderCtrl met every legal and ethical requirement imposed by the EU on member states as they arose during its lifetime, including the tracking, implementation, and auditing of compliance. Of specific interest within the field of machine-based automated psychological profiling is a specific provision in Article 22 of the GDPR [3] which concerns profiling and automated decision-making. The Article defines an automated decision as a decision based solely on automated processing, including profiling of a person, which produces legal effects concerning him or her or similarly significantly affects him or her [14–16]. In addition, safeguards must be put in place so individuals have choice if they do not wish to be subjected to automated profiling, i.e., the ability to engage instead with a human being. However, when a system such as ADDS is used to generate a deception risk score of a traveler which may (subject to data quality) contribute less than 8% to an overall traveler's risk score, that is presented as additional information to a border guard who makes the final decision – then this is not fully automated profiling. However, if a human who is not trained in the interpretation of the system looks only at this score and makes a decision, then does it by default become automated? This complex problem is very subjective, and the answer may very well depend on the application and its consequence on humans rights [14–16]. In all cases, it is essential that an algorithm does not violate the data subject's fundamental rights. In order to achieve this, the European Union agency for Fundamental Human rights recommends that in addition to a Data Privacy Impact Assessment, projects should also undergo a Fundamental Rights Impact Assessment [28–31].

The impact of a false positive result will be different based on the AI application and context in which it is used. For example, for a law enforcement agency, the impact of a false positive may lead to a wrong person being added to a watch list, or even arrested [30, 31]. Consequently, a false negative may mean a person who should have been identified is missed. A survey by the European Union Agency for Fundamental Rights in 2015 showed that 12% of respondents felt very uncomfortable with facial image recognition being used at border crossings, 18% felt it was intrusive, and 26% said it was humiliating. However, Russian and US citizens were less concerned about its usage. In a 2019 UK public survey conducted by the Ada Lovelace Institute, the main message was that the UK was not ready for the use of face recognition technology, despite 70% of respondents agreeing it was appropriate for use by the police in criminal investigations [1]. However, when asked their opinion on the use of such technology in retail for tracking customers or in human resources for screening job applicants, 76–77% of people reported they would be uncomfortable. A major concern would be the impact on a person if the AI got it wrong.

It is well known that the quality of the data used for machine learning can also impact on the right to nondiscrimination (Article 21, Charter of Fundamental Rights, EU) through the introduction of bias. Bias of any kind can lead to discrimination of a human being. This may be through unrepresentative training data or the sampling technique used. It may also be through the unconscious bias introduced by a human expert in labelling the data. In June 2019, the European Union Agency for Fundamental Rights [31] reported that there were no agreed standards for data quality assessments within the realm of machine learning; however, it recommends a number of key questions that could be asked to determine data quality. The questions cover all aspects of data management and governance and include: Where does the data come from? What information is included in the data? And is it appropriate for the purpose of the algorithm? Who is included/excluded in the data and who is underrepresented?

Adaptive psychological profiling through the application of machine learning algorithms (as described within this chapter) does not use facial recognition technology. Image-processing techniques are used to extract changes in micro gestures over periods of time, across a number of facial channels, which are then fed into a series of machine learning algorithms capable of detecting objects and their states. Each time period, i.e. 1 s, patterns are aggregated over a number of frames and are used to produce numeric vectors which are labelled according to the mental state being classified, derived through a carefully devised experimental scenario which limits confounding factors. Systems such as Silent Talker and ADDS use these numeric, anonymous vectors with which to train and learn patterns of NBV conducive to detecting a specific mental state. After processing the video data, it is no longer required. Recent work investigated if there was a difference between the nonverbal cues to deception generated by males and females [14–16]. On the analysis of the results, the evidence suggested that the nonverbal behavior cues in this small dataset were highly similar in their prominence but were, nonetheless, different. Each gender appeared to be disadvantaged when they were subjected to a combined classifier (trained on males and females) than when they are subjected to classifiers that were trained for each specific gender, thus indicating a gender effect between classifiers trained specifically for each gender. On examination of the top ranking nonverbal cues, it was also shown that the relative power of the cues was also different, even though there was a high degree of similarity. Significantly, more work is required using larger datasets [43] to further investigate the impact of both gender and also cultural differences within adaptive and automated psychological profiling.

2.4 *The Rise of Ethical Charters*

Recently, governments, corporate organizations, and small to medium businesses have proposed principles and charters on what constitutes ethical AI [4, 24, 26, 27, 40, 44, 47, 49, 52, 68]. These principles provide guidelines to different stakeholders in the design, development, operation, and use of AI systems; however, currently they are not embedded in the law. In this work, nine published sets of principles on ethical AI were examined, and as expected there was a high proportion of overlapping concepts. The common themes among all principles were:

- AI should respect human rights not be used to harm or kill any human.
- AI systems must always be fair, unbiased, and transparent in the decision-making process.
- AI systems and solutions should always operate within the law and have human accountability.
- Data governance and data privacy should be incorporated into the AI life cycle.
- Humans should always know when they have interactions with an AI system.
- Human-centered AI design should be undertaken.
- Appropriate levels of explainability should be provided on AI decision-making.
- Humans must always be in the loop when an AI is making a decision that affects other humans.
- Humans responsible for designing, developing, and operating AI systems should be competent in the skills and knowledge required.

- AI systems should be inclusive to all.
- AI systems should be sustainable and work to benefit humans, the society, and the environment.

The most comprehensive guidelines can be found in *Understanding artificial intelligence ethics and safety: A guide for the responsible design and implementation of AI systems in the public sector* by Leslie [44]. Leslie emphasizes the need to ensure that projects are ethically permissible before they even take place and provides practical guidelines for an ethical platform which incorporates data governance throughout the AI system lifecycle. The platform is constructed from three building blocks. The first covers responsible data design which utilizes the SUM principles (Respect, Connect, Care, and Protect), the second comprises of the Fast Track principles (Fairness, Accountability, Sustainability, and Transparency). The last block focuses on constructing the ethical platform for responsible AI using a process-based governance framework which allows the SUM and Fast Track principles to be operationalized. The guide presents in essence a tool kit which can be used to guide the practical implementation of ethical AI, but it is currently up to the moral and ethical values of individuals and companies to follow through on these and other international recommendations. In 2020, the European commission published a white paper entitled "On Artificial Intelligence – A European Approach to excellence and Trust" [25, 27], which reiterates that developers of AI systems should operate within EU legislation on fundamental human rights (e.g., GDPR) but acknowledges that enforcement is difficult. Therefore, current EU legislation will be examined to address the concerns of using AI, the currently legal framework, and whether it needs to be adapted and how it can be enforced. In April 2021, the European Commission published its Proposed Regulation Framework on Artificial Intelligence [29] for legal consideration which takes a risk-based approach to AI. When this becomes law, the challenge will be not only in its implementation but also in managing audit and compliance. As other countries follow suit, it is certain that ethical AI will become a legal requirement for stakeholders to follow and will change the landscape of AI research, innovation, and implementation as we currently see it.

3 Case Study 1: Deception

3.1 Overview of iBorderCtrl

iBorderCtrl (Intelligent Portable Control System) was a 3-year H2020 research and innovation project designed to integrate existing and develop new innovative for novel mobility concepts for land border security across Schengen land borders in the EU [39]. IBorderCtrl-involved collaborative research and development work across 13 European partners, including three end users: the Hungarian National Police, TRAINOSE-Greece and the state Border Guard of the Republic of Latvia.

The main objective of the project was to "*enable faster and thorough border control for third country nationals crossing the land borders of EU Member States (MS), with technologies that adopt the future development of the Schengen Border Management*" [39]. In addition, the iBorderCtrl system aimed to "*speed up the border crossing at the EU external borders and at the same time enhance the security and confidence regarding border control checks by bringing together many state of the art technologies (hardware and software) ranging from biometric verification, automated deception detection, document authentication and risk assessment*" [39]. The system works through a two-stage process. First, travelers undertake a preregistration phase where they will register details about their future trip from their own home or other convenient location. In this phase, relevant data are collected to allow all automated checks to take place prior to travel. It is during this stage that each adult traveler (subject to consent) will undertake a short interview (less than 2 min depending on network speed) with an Avatar Border Guard. The interview is processed live through an automated deception detection system known as ADDS, which is described in Sect. 3.4. When preregistration is completed, the traveler receives a QR code on their device. The second stage of the process takes place at the land border crossing itself where Border Guards are equipped with a portable hardware unit that provides all information about the traveler gathered during the preregistration phase. The analytical component of iBorderCtrl is used to identify risks and produce an overall evaluation of the traveler. Visible on the portable hardware unit, the system provides a wealth of information to support the Border Guard in their decision-making, essentially on whether the traveler passes straight across the border or is referred to a second line of human check [12].

3.2 *Privacy and Security by Design*

Development of the prototype iBorderCtrl system, including the ADDS tool, used both Privacy by Design and Security by Design Approaches, due to inclusions of sensitive personal data being used within this research and innovation project. Privacy by design promotes privacy and data protection compliance throughout the entire project and includes actively adhering to the 7 foundational principles [41]. In order to identify, assess, and mitigate privacy risks within the project, a Data Privacy Impact Assessment was used. Security by design aims to ensure the system is free from software and hardware vulnerabilities and is resistant to attacks. This was achieved through following the 10 principles proposed by OWASP (Open Web Application Security Project) [53].

3.3 Project Ethics

The iBorderCtrl project as a whole, and the individual modules such as ADDS, underwent extensive ethical reviews both at consortium, partner, and EU levels throughout the project. Ethical applications were deemed robust, and approval was granted for all applications and then rigorously monitored. Specifically for ADDS, ethical approval was granted for early experiments to collect training data from within the university, for a public engagement event at a national Museum of Science and Industry for usability testing and for modifications to the project pilots as a result of the extensive media coverage. It is important to emphasize here that a significant amount of time was spent on each ethical application ensuring that participant information sheet, informed consent forms, etc. were appropriate in terms of language and understandability for the participant audience to ensure inclusivity for all. Ethical applications and approvals were checked by iBorderCtrl legal partners and included in EU ethical reviews and monitoring and were accepted by the European Commission. However, despite following all procedures to the highest standard, robust ethics were simply not enough to gain trust of such a system. Ethical procedures were not even mentioned for the vast majority of media portrayals of the project (Sect. 3.5).

3.4 Automated Deception Detection Tool

ADDS is fundamentally based on the Silent Talker technology (Sect. 2.2) and was redesigned for inclusion in the iBorderCtrl system. Further, publically available details can be found in [54]. Figure 1 shows the interview process (part of a traveler's new trip registration) within the iBorderCtrl System where ADDS is deployed. Once the interview commences, a series of instructions on lighting and camera alignment are shown to enable the best possible video quality to be captured. If the video quality is poor, the ADDS system does not calculate a deception risk score and records a null value. The Avatar is selected to be the same as the person's self-identifying gender and asks a series of questions about the traveler and their upcoming trip. These questions were typical of those asked by Border Guards based at Schengen land borders in Europe. The entire interview lasted less than 2 min. A video of the traveler captured from their device was live streamed to the Interview Control Module, where it was processed by the ADDS system and the deception risk score per question was calculated. During the interview, the avatar attitude (body posture and tone of voice) is adapted based on the calculated deception risk score. Research is currently ongoing to assess the effects on Avatar attitudes on a person's own NVB. On completion, all question risk scores and the aggregated interview scores are uploaded to the iBorderCtrl central database, where they are used within the analytical components of iBorderCtrl. If insufficient NVB data is collected at

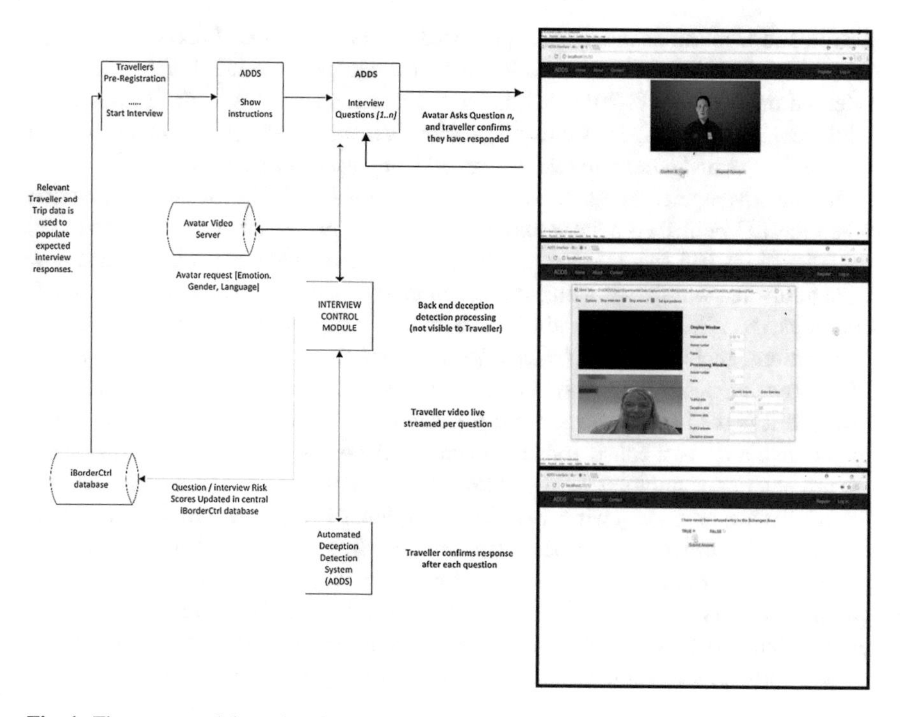

Fig. 1 The automated deception detection tool

any level, a NULL is recorded and the results of ADDS may not be used within iBorderCtrl analytics.

The results of an initial laboratory based experiment of ADDS which involved consenting volunteers to role-play truthful and deceptive scenarios, was peer reviewed and published in [54]. The experiment yielded an 75% accuracy on a small dataset of 32 participants of mixed gender and ethnicity on unseen participants. However, the paper [54] acknowledged the small sample size, and the diversity of the participants would require further experiments to be conducted on large samples.

3.5 *Media and Public Discourse*

In October 2018, the Traveler User Interface (including the ADDS module) to the iBorderCtrl system formed part of the Platform for Investigation entitled "Me Verses the Machine," which featured eight activities designed to introduce families to artificial intelligence, coding, and computer science through offering hands on STEM activities. The Platform was held at the Science and Industry Museum, Manchester, UK, and the aim was to evaluate the usability of the first prototype of the Traveler User Interface. In order to promote the science festival event, MMU

Media and Communications team produced a short YouTube Video entitled "Can you fool an AI lie detector?" which was picked up by the media [19] when it was released on October 18, 2018. While the event was successful, with 51 consenting adults engaging with the system and proving valuable feedback, the main focus of the media was on "*automated lie detectors*" being used "*at borders.*"

Media coverage escalated to levels unprecedented in the European Commission when the EU published a story about iBorderCtrl [39], which also coincided with the science festival event. Since this event, more than 1300 news articles have now been published worldwide with coverage in mainstream national media in most EU countries, the US, Australia, and China, with articles appearing in major mainstream publications, including *New Scientist* [50], *Guardian*, *MIT Review*, *The Telegraph*, *The Independent*, *CNN*, *Newsweek*, and *Euro-news*. iBorderCtrl website analytics revealed it had 11, 242 unique visitors during Nov-2018. Public discourse fueled by the media led to a considerable amount of fake news being widely spread about the project, For example, several articles reported "Passengers to face AI lie detector tests at EU airports" [55], when in fact the iBorderCtrl project was in response to the H2020 call "*Novel mobility concepts for land border security*" [23] and was seeking to "*enable faster and thorough border control for third country nationals crossing the land borders of EU Member States*" [39]. The iBorderCtrl project was focused only on land borders. Members of the consortium engaged with the media [35] to try and correct incorrect facts about the iBorderCtrl system, including:

- iBorderCtrl was a research and innovation project that would require changes in EU law to ever be used.
- The contribution of ADDS to the overall risk profile of an individual was typically <8% due to its technological readiness level being the lowest of all modules, and if the data was not of significantly higher quality, ADDS would not contribute to any traveler risk score.
- The research team emphasized from early experimental work [54] that significant more data were required for training, validation, and testing to ensure it was tested on different genders and ethnicities.
- Misunderstanding of what micro-gestures were (in ADDS), which were constantly referred to in the media as micro-expressions.

From December 2018, the consortium produced a standard agreed response to all media queries and directed interested parties toward the project website where regular updates would be posted. However, this also seemed to fuel further international media interest, with threats being made to consortium members. Interestingly, prior to iBorderCtrl, the prototype technology featured in the 2014 TV Documentary entitled "The Lying Game: Crimes that fooled Britain" and was used to evaluate deception scores of now-convicted criminals in public interviews prior to their arrests, yet it generated very little public discourse and no media attention. In 2014, AI was not such a newsworthy topic. As a further comparison, AVATAR – a border control kiosk based system using deception detection, developed by the University of Arizona, also received similar negative reporting from the media [45], reporting an accuracy of between 60% and 80% [17].

Turner, a Senior Lecturer in media law wrote "We have to assume that most news, while not being phoney is biased to an extent that it is not fully accurate. News tends to be presented, no matter how 'unbiased' the source claims to be from a perspective of the culture in which the news provider is based" [66]. The concept of automated deception detection being conceivably used and contributing less than 8% to a traveler's risk profile in a research and innovation project was clearly enough to be newsworthy, promote discourse and debate and provoke emotion amongst the public. Pinker believes that the media exaggerates negative news, distorting the facts, which leads to consequences for all parties [56]. In general, he believes reporting has become more negative over time. A study [61] conducted in 2019 involving 1000 people in 17 countries found that people paid more attention to negative news than positive. One finding of this study was that personal bias toward negative news was found to drive negative news coverage. With regards to artificial intelligence (AI) communication, a study by The Royal Society found "*Exaggerated expectations and fears about AI, together with an over-emphasis on humanoid representations, can affect public confidence and perceptions. They may contribute to misinformed debate, with potentially significant consequences for AI research, funding, regulation and reception*" [63]. The way in which applications of AI are therefore reported by the media can have long-term consequences.

The iBorderCtrl project was successfully completed at the end of 2019; however, its legacy continues, as it continues to be used as an active case study in university courses in the US and the Netherlands, and be the subject of many Freedom of Information Requests. However, given its technological enhancement level and the future changes to legislation on the use of AI, its actual implementation remains extremely unlikely.

4 Case Study 2: Comprehension

Human comprehension is often assessed through verbal or written language. Nickerson [51] defines comprehension as "*the connecting of facts, the relating of newly acquired information to what is already known, the weaving of bits of knowledge into an integrated and cohesive whole. In short, it requires not only having knowledge but also doing something with it.*" Traditionally, Davis [18] defined "*Comprehension is a thinking process, it is thinking through reading.*" Assessing comprehension in a classroom environment is usually achieved using a rich assortment of assessment tools. Teachers will also be able to determine comprehension of individual pupils during classroom engagement from their behavior in certain contexts. However, this is dependent on classroom experience and can be subjective. In this case study, comprehension is measured only through a person's facial NVB while they are engaged in a task. Two systems known as FATHOM and COMPASS are introduced. In FATHOM, the learning task is in the form of a verbal interview, whilst within COMPASS, the task is visual engagement with an online tutoring system where humans are required to read and answer questions in a short programming tutorial. In both cases, machine learning is used to automatically detect comprehension levels.

4.1 FATHOM

FATHOM is a novel human comprehension detection software application that was built to automatically detect human comprehension and noncomprehension from monitoring multi-channels of nonverbal behavior using hierarchies of artificial neural networks [9]. Similar to ADDS, 38 NVB facial channels were monitored. The core concept was that if comprehension (and noncomprehension) could be detected during a learning event, then appropriate learner intervention strategies could be applied. FATHOM was first evaluated on a mock informed consent process for a sexual and reproductive health clinical trial, which was executed in Tanzania by Family Health International (FHI-360) and the National Institute for Medical Research in Tanzania. The aim was to train, validate, and test FATHOM's Comprehension classifier ANNs on a dataset from the informed consent process of a mock Human Immunodeficiency Virus (HIV) prevention trial, in order to provide a way to measure and give indicators of comprehension/noncomprehension of participants to medical practitioners. By informing the practitioners when a person was not comprehending, the ultimate aim was to communicate information using different approaches. Initial results indicated that FATHOM's Comprehension Classifier ANN was able to detect patterns of human comprehension and noncomprehension from the nonverbal multichannels greater than chance, with an adopted cross-validation strategy which consistently provided grouped classification accuracies above 80% [9] on a sample of 80 African women. However, further work was required to determine if there were general patterns of comprehension-related NVB in the general population.

In a second study, 20 males and 20 females took part in a learning environment study in the UK. Participants watched a factual video on termites (typical to family viewing on an early Sunday evening) and then had their comprehension of the video assessed in a recorded Q&A session. A sketch of the experimental setup can be seen in Fig. 2. Recording equipment was positioned to ensure good lighting with the aim of minimizing shadows. Questions were open and closed and written by experts in entomology to span the spectrum of easy and hard questions. To ensure that there were no confounding factors in the understanding of the video content, the narrative from the termites' video was transcribed and evaluated using the Flesch-Kincaid Grade Level [42] readability metrics and had a rating of 9.8 FKGL, translating to a reading age of 14–15. In the learning environment study, FATHOM achieved normalized classification accuracies above 76% in the testing phase [10].

FATHOM provided evidence that facial nonverbal behaviors do emit human comprehension and noncomprehension patterns and that human comprehension detection knowledge resided across multiple NVB channels. Supporting evidence was therefore found across two different studies within different domains and

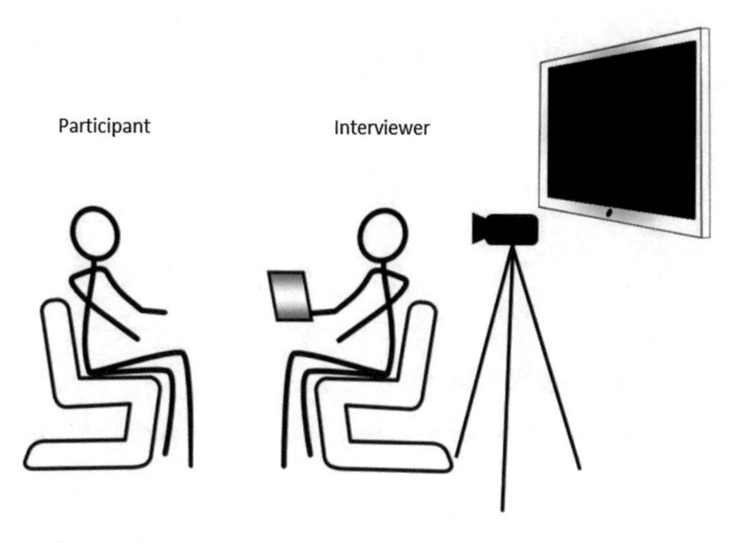

Fig. 2 Participant and interview setup [8]

cultures. However, comprehension-level detection was achieved after the event through the use of prerecorded videos. In order for timely interventions to take place, comprehension detection had to occur in near real time.

4.2 COMPASS

In order to address the shortcomings of FATHOM, work by Holmes [37] involved development of a near real-time comprehension classification system known as COMPASS, which utilized machine learning algorithms to detect patterns of learner nonverbal behavior while they engaged in a java programming tutorial. COMPASS was effectively a comprehension classifier that could learn discriminant behavioral patterns indicative of comprehension states without relying on self-reporting from the student. Utilizing data from a webcam stream, facial features were located from which physiological data were extracted and summarized over arbitrary time periods. An initial study required learners to complete a 21 question, multiple-choice quiz covering topics related to Java programming and logic. Forty-four undergraduate students consented to take part and gave a 75.8% normalized classification accuracy [38]. COMPASS was further evaluated through integration into Hendrix 2.0 – a conversational intelligence tutoring system (CITS) (Fig. 3). The comprehension classification score was used as a feedback channel to provide first adaptation of learning material based on comprehension levels and secondly to inform the automatic intervention provided by the CITS. Hendrix 2.0 demonstrated that it was possible to reliably detect comprehension states within

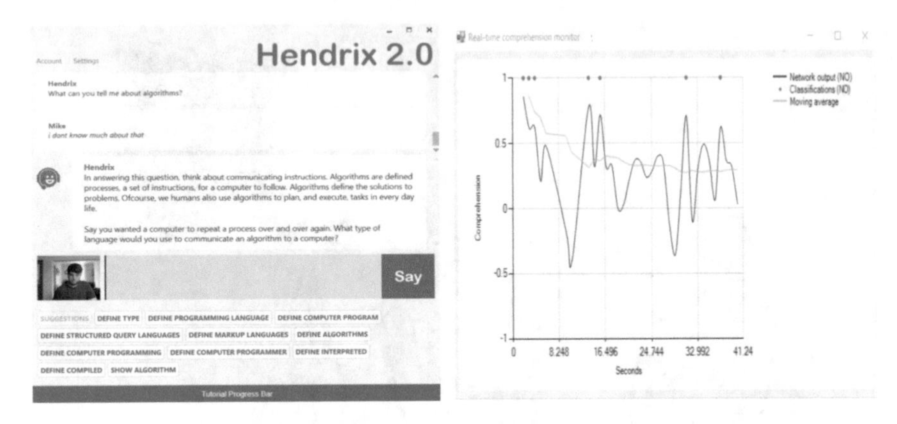

Fig. 3 Example of COMPASS behind the scenes comprehension monitoring within Hendrix 2.0 [37]

a CITS, and the comprehension-based interventions during a tutorial improved the answer scores provided by learners during tuition [37]. Figure 3 shows two elements of the Hendrix 2.0 CITS: (1) the tutorial screen asking a question regarding the knowledge of algorithms and (2) the back end COMPASS system displaying near real-time comprehension levels extracted from NVB. In a 51-participant study, participants answered a total of 1269 questions with comprehension detection having a normalized classification accuracy of 75.44% [37].

Detecting comprehension states from NVB is challenging and dependent on the experimental design where the ground truth is established. The results have shown patterns of NVB can be indicative of comprehension levels across four different scenarios. However, the data samples were not large, and it is noted that FATHOM and COMPASS would both benefit from larger studies to address issues associated with cross-cultural analysis and gender-based differences of NVB. The impact of false positives in detecting comprehension in a system such as Hendrix 2.0 has very low impact on a student, in that they may be offered an intervention, such as an additional explanation, which they can choose to ignore. Strict ethical procedures were followed in the experimental studies conducted using FATHOM and COMPASS, with both systems using similar technology to ADDS. Hendrix and FATHOM were both implemented and tested in the field. There was never any media interest.

5 Empowering the General Public Through Education

The media interpretation of the use of psychological profiling for automated deception detection identified that the public were very engaged in this topic and AI in general (Sect. 3.5). People who engage with AI systems will have very different backgrounds, experiences and educational levels. There have been numerous recent surveys that have asked the general population about their understanding, concerns and aspirations about AI [1, 2, 11, 13, 67]. The question is: Do these surveys represent the view of the general public? The people on the street who may not have the appropriate educational attainment level or knowledge to understand the AI systems that they interact with on a day-to-day basis. For example, a key finding of the 2019 Ada Lovelace survey was that "*Most people do not know enough about facial recognition technology to have an informed opinion on its use*" [1]. The results of a public risk and trust perceptions of AI study [13] indicated that the public is uncomfortable about "*being left behind*" in research and development of AI systems. Both studies conclude that education of the general public, from school children to senior citizens, is essential to enable informed debate and discussion around ethical AI. Public engagement activities are also required to encourage people to feel empowered about AI, to understand the ethical issues involved and the legalities surrounding its use. A positive step in the right direction is the Elements of AI course, a free online course created by Reaktor and the University of Helsinki. The goal of the course is to educate 1% of European citizens by 2021. The prerequisite skills required for the course include basic programming and advanced mathematics is available to anyone; however, it is advertised as a university-level course and therefore may not be suitable for those with lower educational levels [22].

There is a clear need to provide education on many different levels for many different stakeholders (general public, primary and secondary school children, AI procurers, AI designers and developers, public organizations, businesses etc.), and there is no one-size-fits-all solution. Figure 4 proposes a top-level series of questions that are of interest to the general public. These questions have been determined from public engagements activities and from ideas in international surveys. Although AI and ethics have been embedded in some countries in the schools' national curriculum, the level at which it is covered varies [65], and significant investment is needed in order to tackle the myths about AI from an early age.

In terms of automated decision-making, the GDPR states that every person should understand the logic involved, but how should this be presented? Especially if the person does not use a computer, smartphone, or ICT device.[1] Explainable AI is essential to giving users confidence in the system and how it actually works. It can also help safeguard against bias through showing what data were used and how they were interpreted to make a decision. It also supports traceability within the system to allow developers to understand why a system behaves in a particular way. Information on how to use an AI system (as well as explainability of any decisions) should be at an appropriate level to the person requesting it. It is therefore conceivable that explanations need to be on many different levels depending on

[1] In 2019, 45.12% of the world's population owned a smartphone [7], and by 2025, 72% of all users will use only smartphones to access the internet. In 2019, 49.7% of households had access to a PC [62].

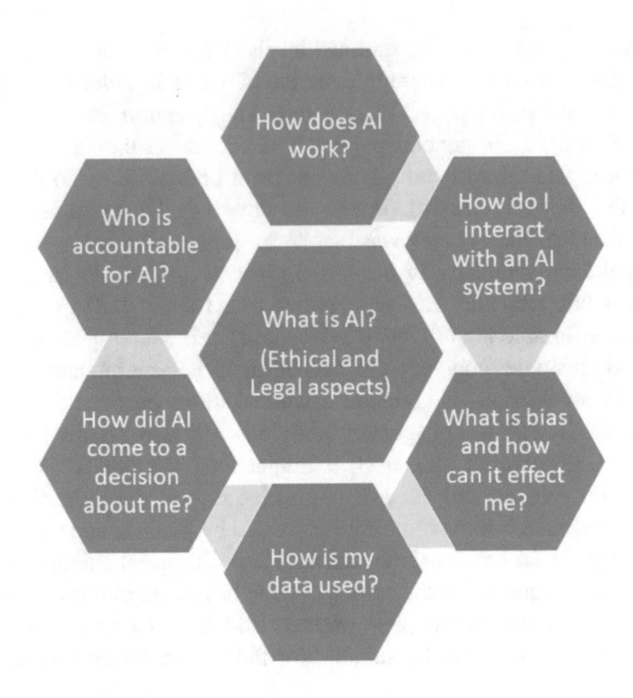

Fig. 4 Initial educational topics

the user. The Royal Society reported that "*Different users require different forms of explanation in different contexts*" [64]. O'Shea [14–16] proposed a Hierarchy of Explainability and Empowerment which aims to ensure transparency through maximizing the readability of text explanations and minimizing the mathematical complexity within explanations to data subjects. However, this hierarchy has yet to be validated and evaluated in the field. The challenge is that in the field of automated and adaptive psychological profiling, detecting patterns of NVB and their relationships in determining a mental state at a specific point in time is hard to understand, explain, and verify. Yet, research has shown that overall behavior is multichannel with AI essentially seeing the overall picture of a person and not just the incongruence between channels. In explaining these complex systems, it is also the psychology behind the AI as well as the AI itself that need to be explained.

Current work is focused on running a series of interactive public engagement workshops where the public can learn, debate, and discuss AI regardless of their background. Contributions by the public at these workshops will help identify the AI knowledge gaps (and interests), which will be used to develop both offline and online AI educational resources for all. This will also help facilitate and engage people in AI upskilling (extending the skills and capabilities of people through increasing their knowledge in AI) that is globally required by organizations and businesses to help to mitigate against social implications of AI systems.

6 Conclusions

Understanding how a machine can learn to detect complex mental states in the realm of psychological profiling is a difficult and challenging topic for human experts due to the subjective nature of determining deception or comprehension. The automatic detection of deception cues from NVB using AI is also challenging, not only from the perceptive of black-box neural networks, but in trying to understand why certain patterns of NVB are indicators of either deception or comprehension. Humans and machine learning algorithms can both get decisions wrong. Whereas humans are subjective and may have unconscious bias, machine-learning algorithms can generate false positives and negatives from poor quality data (from poor lighting and head position in a frame to an underrepresented and bias data sample). In such systems, it is easy for a human to be accountable for a decision (i.e. sending a person to a second line check for an in-depth human interview at a land border, based on border guard assessment of traveler behavior) compared to an AI system providing additional information in the form of an aggregated risk assessment to support the decision of a border guard. This is due to there currently not being any legal frameworks around the accountability of AI systems. Doshi-Velez and Kortz [20] in a Harvard white paper on *Accountability of AI Under the Law* state that it should be "*technically feasible to create AI systems that provide the level of explanation that is currently required of humans. The question, of course, is whether we should?*" They argue that the software development costs of providing such explanations to low resource companies has not yet been quantified, and with the rapid advancements in AI, the landscape is continually changing.

The deception detection case study described in this chapter has shown that ethically aligned design of an AI system and conformance to all ethical protocols required does not equate to trust in a system. Effectively, the findings echo those of a study conducted by a UK Citizens juries in 2018 which found that a person's viewpoint on particular applications of AI was often affected by their perception of who would benefit and who was developing the technology [64]. However, detecting comprehension of an individual in a mock, medical informed consent study or in a learning environment using similar technology received no media attention, discussion, or debate. In February 2020, the EU published a white paper "*On Artificial Intelligence – A European approach to excellence and trust*", which outlines a future regulatory framework for AI in Europe through creation of an "ecosystem of trust" [26]. Within this framework, different regulations will apply based on whether an AI system is deemed to be high-risk or not. Changes in how we develop and use AI ethically will soon be embedded into the law and will impact everyone. Consequently, public awareness of AI and its implications is growing and is often fueled by the media. Therefore, in order to dispel the myths and empower the general public, educational resources on AI need to be developed for all. Knowledge builds understanding, which in turn builds trust allowing for fair, informed debate on when and how AI systems should and should not be used.

References

1. Ada Lovelace Institute, Survey: Beyond face value: public attitudes to facial recognition technology, (2019). Available: https://www.adalovelaceinstitute.org/beyond-face-value-public-attitudes-to-facial-recognition-technology/
2. ARM, Global Artificial Intelligence Survey [Online], (2019). Available: https://www.arm.com/solutions/artificial-intelligence/survey. Accessed 17 Dec 2019
3. Art. 22 GDPR Automated individual decision-making, including profiling [Online], (2018). Available: https://gdpr-info.eu/art-22-gdpr/
4. Australian Government – AI principles, (2019). Available: https://www.industry.gov.au/data-and-publications/building-australias-artificial-intelligence-capability/ai-ethics-framework/ai-ethics-principles
5. E. Babad, Teaching and nonverbal behavior in the classroom, in *International Handbook of Research on Teachers and Teaching*, ed. by L. J. Saha, A. G. Dworkin, (Springer, 2009), pp. 817–827
6. Z. Bandar, D.A. McLean, J.D. O'Shea, J.A. Rothwell, *International Patent Number WO02087443* (World Intellectual Property Organization, Geneva, 2002)
7. Bankmycell, How many people have smartphones in the world? (2020). Available: https://www.bankmycell.com/blog/how-many-phones-are-in-the-world
8. F.J. Buckingham, Detecting human comprehension from nonverbal behaviour using artificial neural networks, Manchester Metropolitan University [Online]. PhD thesis, (2017). Available: https://e-space.mmu.ac.uk/id/eprint/617426
9. F. Buckingham, K. Crockett, Z. Bandar, J. O'Shea, K. MacQueen, M. Chen, *Measuring Human Comprehension from Nonverbal Behaviour Using Artificial Neural Networks* (IEEE World Congress on Computational Intelligence Australia, 2012), pp. 368–375. https://doi.org/10.1109/IJCNN.2012.6252414
10. F.J. Buckingham, K.A. Crockett, Z.A. Bandar, J.D. O'Shea, FATHOM: A neural network-based non-verbal human comprehension detection system for learning environments, in *IEEE Symposium on Computational Intelligence and Data Mining (CIDM)*, (2014), pp. 403–409. https://doi.org/10.1109/CIDM.2014.7008696
11. B. Cheatham, K. Javanmardian, H. Samandari, Confronting the risks of artificial intelligence, McKinsey [Online], (2019). Available: https://www.mckinsey.com/business-functions/mckinsey-analytics/our-insights/confronting-the-risks-of-artificial-intelligence#. Accessed 6 Dec 2019
12. K.A. Crockett, J. O'Shea, Z. Szekely, A. Malamou, G. Boultadakis, S. Zoltan, Do Europe's borders need multi-faceted biometric protection. Biometric Technol. Today **2017**(7), 5–8 (2017) ISSN:0969-4765
13. K. Crockett, S. Goltz, M. Garratt, A. Latham, Trust in computational intelligence systems: A case study in public perceptions, in *IEEE Congress on Evolutionary Computation*, (2019), pp. 3228–3235
14. K. Crockett, J. O'Shea, W. Khan, Automated deception detection of male and females from non-verbal facial micro-gestures, *2020 International Joint Conference on Neural Networks (IJCNN)*, 2020, pp. 1–7, https://doi:10.1109/IJCNN48605.2020.9207684.
15. K. Crockett, M. Garratt, S. Goltz, A. Latham, E. Coyler, Risk and trust perceptions of the public of artificial intelligence applications, *2020 International Joint Conference on Neural Networks (IJCNN)*, 2020, pp. 1–8, https://doi:10.1109/IJCNN48605.2020.9207654
16. K. Crockett, J. Stoklas, J. O'Shea, T. Krügel, W. Khan, Reconciling adapted psychological profiling with the new European data protection legislation, In: C. Sabourin, J. J. Merelo, A. L. Barranco, K. Madani, K. Warwick, (eds) *Computational Intelligence, Studies in Computational Intelligence*, vol 893. Springer, Cham. https://doi.org/10.1007/978-3-030-64731-5_2.
17. J. Daniels, Lie-detecting computer kiosks equipped with artificial intelligence look like the future of border security [Online], (2018). Available: https://www.cnbc.com/2018/05/15/lie-detectors-with-artificial-intelligence-are-future-of-border-security.html. Accessed 04/01/2020

18. F. Davis, Psychometric research on comprehension in reading. Read. Res. Q. **7**(4), 628–678 (1972). https://doi.org/10.2307/747108
19. R. Day, Can you fool a lie detector? Manchester Evening News interview, 27/10/2018 [Online], (2018). Available: https://www.manchestereveningnews.co.uk/news/greater-manchester-news/lie-detector-test-border-control-15319641
20. F. Doshi-Velez, M. Kortz, *Accountability of AI Under the Law: The Role of Explanation* (Berkman Klein Center Working Group on Explanation and the Law, Berkman Klein Center for Internet & Society Working Paper, 2017) Available: https://dash.harvard.edu/bitstream/handle/1/34372584/2017-11_aiexplainability-1.pdf
21. P. Ekman, V.W. Friesen, *The Facial Action Coding System (FACS)* (Consulting Psychologists Press, Palo Alto, 1978)
22. Elements of AI, (2019). Available: https://www.elementsofai.com/eu2019fi
23. European Commission, BES-05-2015 – Border crossing points topic 1: Novel mobility concepts for land border security [Online], (2015). Available: https://cordis.europa.eu/programme/id/H2020_BES-05-2015
24. European Commission, Smart lie-detection system to tighten EU's busy borders, (2018). Available: https://ec.europa.eu/research/infocentre/article_en.cfm?artid=49726
25. European Commission, Shaping Europe's digital future: Commission presents strategies for data and Artificial Intelligence [Online], (2020a). Available: https://ec.europa.eu/commission/presscorner/detail/en/ip_20_273
26. European Commission, White Paper on Artificial Intelligence: A European approach to excellence and trust, (2020b). Available: https://ec.europa.eu/info/sites/info/files/commission-white-paper-artificial-intelligence-feb2020_en.pdf
27. European Commission, The ethics of artificial intelligence: Issues and initiatives [Online], (2020c). Available: https://www.europarl.europa.eu/RegData/etudes/STUD/2020/634452/EPRS_STU(2020)634452_EN.pdf
28. European Commission Ethics guidelines for trustworthy AI, (2019), Available: https://ec.europa.eu/digital-single-market/en/news/ethics-guidelines-trustworthy-ai
29. European Union, "Regulation of the European Parliament and of the Council, Laying down harmonised rules on Artificial Intelligence (Artificial Intelligence Act) and amending certain Union Legislative Acts", (2021). Available: https://eur-lex.europa.eu/legal-content/EN/TXT/?uri=CELEX:52021PC0206
30. European Union Agency for Fundamental Rights, *Preventing Unlawful Profiling Today and in the Future: A Guide* [Online], (2018), p. 12. https://op.europa.eu/en/publication-detail/-/publication/328663bc-f909-11e8-9982-01aa75ed71a1/language-en. Accessed 5/10/2020
31. European Union Agency for Fundamental Rights, Facial recognition technology: Fundamental rights considerations in the context of law enforcement, (2019). Available: https://fra.europa.eu/en/publication/2018/prevent-unlawful-profiling
32. D. Frauendorfer, M.S. Mast, L. Nguyen, D. Gatica-Perez, Nonverbal social sensing in action: Unobtrusive recording and extracting of nonverbal behavior in social interactions illustrated with a research example. J. Nonverbal Behav. **38**(2), 231–245 (2014)
33. GDPR Portal, (2018) [Online]. Available at: https://gdpr-info.eu/. Accessed 27/02/2020
34. A.C. Graesser, S. Lu, B.A. Olde, E. Cooper-Pye, S. Whitten, Question asking and eye tracking during cognitive disequilibrium: Comprehending illustrated texts on devices when the devices break down. Memory Cognit. **33**(7), 1235–1247 (2005)
35. Guardian, The Guardian podcast, "Can we trust AI lie detectors? Chips with Everything podcast", (2018). Available: https://www.theguardian.com/technology/audio/2018/nov/23/can-we-trust-ai-lie-detectors-chips-with-everything-podcast
36. C. Hodgson, AI lie detector developed for airport security, Financial Times [Online], (2019). Available: https://www.ft.com/content/c9997e24-b211-11e9-bec9-fdcab53d6959
37. M. Holmes, Comprehension based adaptive learning systems. PhD thesis, Manchester Metropolitan University, (2017)
38. M. Holmes, A. Latham, K. Crockett, J. O'Shea, Near real-time comprehension classification with artificial neural networks: Decoding e-learner non-verbal behaviour. IEEE Trans. Learn. Technol. **2017**(99) (2017). https://doi.org/10.1109/TLT.2017.2754497

39. iBorderCtrl, (2020). Available: https://www.iborderctrl.eu/
40. IEEE Ethically Aligned Design, Version 2 (EADv2), (2017). Available: https://ethicsinaction.ieee.org/?utm_medium=undefined&utm_source=undefined&utm_campaign=undefined&utm_content=undefined&utm_term=undefined
41. Information Commissioners Office (ICO), Privacy by design, (2020). Available: https://ico.org.uk/for-organisations/guide-to-data-protection/guide-to-the-general-data-protection-regulation-gdpr/accountability-and-governance/data-protection-impact-assessments/
42. J.P. Kincaid, R.P. Fishburne, R.L. Rogers, B.S. Chissom, *Derivation of New Readability Formulas (Automated Readability Index, Fog Count and Flesch Reading Ease Formula) for Navy Enlisted Personnel* (National Technical Information Service. (RBR 8–75), Springfield, 1975)
43. R.V. Krejcie, D.W. Morgan, Determining sample size for research activities. Educ. Psychol. Measur. **30**, 607–610 (1970)
44. D. Leslie, Understanding artificial intelligence and safety: A guide for the responsible design and implementation of AI systems in the public sector. The Alan Turing Institute, (2019). Available: https://zenodo.org/record/3240529#.XjcJhbk3ZaQ
45. A. Marsh, A Brief History of the Lie Detector [Online], IEEE Spectrum, (2019), Available: https://spectrum.ieee.org/tech-history/heroic-failures/a-brief-history-of-the-lie-detector. Accessed 04/01/2020
46. Media College, What Makes a Story Newsworthy, (2020). Available: https://www.mediacollege.com/journalism/news/newsworthy.html
47. Microsoft, AI Principles, (2020). Available: https://www.microsoft.com/en-us/ai/our-approach-to-ai
48. G.A. Miller, The magical number seven, plus or minus two: Some limits on our capacity for processing information. Psychol. Rev. **63**(2), 81–97 (1956). https://doi.org/10.1037/h0043158
49. Mores.Code, Morse.ai, (2020). Available: http://www.morse.ai/
50. New Scientist, AI lie detection at border control should proceed with caution, (2018). Available: https://www.newscientist.com/article/mg24032022-900-ai-lie-detection-at-border-control-should-proceed-with-caution/
51. R.S. Nickerson, Understanding understanding. Am. J. Educ. **93**(2), 201–239 (1985)
52. OECD, Principles on Artificial Intelligence, OCED, (2019). Available: http://www.oecd.org/going-digital/ai/
53. Open Web Application Security Project (OWASP), (2019). Available: https://owasp.org/. Accessed 28/2/2020
54. J. O'Shea, K. Crockett, W. Khan, P. Kindynis, A. Antoniades, Intelligent deception detection through machine based interviewing, in *IEEE International Joint Conference on Artificial Neural Networks (IJCNN)*, (2018). https://doi.org/10.1109/IJCNN.2018.8489392
55. R. Picheta, CNN, Passengers to face AI lie detector tests at EU airports, (2018). Available: https://edition.cnn.com/travel/article/ai-lie-detector-eu-airports-scli-intl/index.html
56. S. Pinker, The media exaggerates negative news. This distortion has consequences, in *Enlightenment Now: The Case for Reason, Science, Humanism, and Progress*, (Penguin Publishing Group, 2018). Available: https://www.theguardian.com/commentisfree/2018/feb/17/steven-pinker-media-negative-news
57. S. Porter, L. ten Brinke, The truth about lies: What works in detecting high-stakes deception? Legal Criminol. Psychol. **15**(1), 57–75 (2010)
58. K. Rayner, K.H. Chace, T.J. Slattery, J. Ashby, Eye movements as reflections of comprehension processes in reading. Sci. Stud. Read. **10**(3), 241–255 (2006)
59. J. Rothwell, Z. Bandar, J. O'Shea, D. McLean, Silent talker: A new computer-based system for the analysis of facial cues to deception. Appl. Cognit. Psychol. **757, 20**(6), –777 (2006). https://doi.org/10.1002/acp.1204
60. Silent Talker, (2020). Available: https://find-and-update.company-information.service.gov.uk/company/09533454/officers

61. S. Soroka, P. Fournier, L. Nir, Cross-national evidence of a negativity bias in psychophysiological reactions to news. Proc. Natl. Acad. Sci. U. S. A. **116**(38), 18888–18892 (2019). https://doi.org/10.1073/pnas.1908369116
62. Statista, Share of households with a computer at home worldwide from 2005 to 2019, (2020). Available: https://www.statista.com/statistics/748551/worldwide-households-with-computer/
63. The Royal Society, Portrayals and perceptions of AI and why they matter [Online], (2018). Available: https://royalsociety.org/-/media/policy/projects/ai-narratives/AI-narratives-workshop-findings.pdf
64. The Royal Society, Explainable AI Policy Briefing, (2019). Available: https://royalsociety.org/-/media/policy/projects/explainable-ai/AI-and-interpretability-policy-briefing.pdf
65. The Times Educational Supplement, AI in UK schools? I'd give us 5 out of 10, (2019). Available: https://www.tes.com/news/ai-uk-schools-id-give-us-5-out-10
66. M. Turner, Research Gate, (2015). Available: https://www.researchgate.net/post/How_do_you_determine_whether_a_news_source_is_reputable_Or_a_news_story_is_reliable
67. UK Gov., Artificial Intelligence: Public awareness survey [Online], (2019). Available https://www.gov.uk/government/publications/artificial-intelligence-public-awareness-survey. Accessed 17 Dec 2019
68. US Government, Guidance for Regulation of Artificial Intelligence Applications, (2020). Available: https://www.whitehouse.gov/wp-content/uploads/2020/01/Draft-OMB-Memo-on-Regulation-of-AI-1-7-19.pdf
69. M. Van Amelsvoort, B. Joosten, E. Krahmer, E. Postma, Using non-verbal cues to (automatically) assess children's performance difficulties with arithmetic problems. Comput. Hum. Behav. **29**(3), 654–664 (2013)

Keeley Crockett is Reader in Computational Intelligence at Manchester Metropolitan University in the UK. She gained a BSc Degree (Hons) in Computation from UMIST in 1993 and a PhD in the field of machine learning from Manchester Metropolitan University (MMU) in 1998. She is senior fellow of the Higher Education Academy and leads the Computational Intelligence Lab in the Centre for Advanced Computational Science. Her main research interests include fuzzy decision trees, semantic text based clustering, conversational agents, fuzzy natural language processing, semantic similarity measures, and artificial Intelligence (AI) for psychological profiling. She was the principal investigator (MMU) on the H2020 funded project iBorderCtrl: Intelligent Smart Border Control, and currently, she is a co-investigator on the H2020 Grant "Populism and Civic Engagement: a fine-grained, dynamic, forward-looking response to the negative impacts of populist movements (PaCE)". She has also worked and is working on a number of Knowledge Transfer Partnerships involving artificial intelligence with Innovate UK and a number of SME's. She is a qualified coach and mentor and a UK STEM Ambassador.

From an early age Keeley was fascinated with programming and logic, studying both computer science and control technology in school. Whilst studying her software engineering and computation degrees at University, Keeley worked part time for Disney, developing confidence, leadership and communication skills receiving a prestigious Disney Spirit Award. Studying artificial intelligence, coupled with the rate of technological innovation inspired her to look at how computational intelligence could be applied to improve people's day to day lives. Keeley studied a PhD in Fuzzy Logic which was funded through teaching many college and university classes. After a time in industry, she returned to academia as a Senior lecturer teaching many topics including database systems, data science and machine learning. On joining the IEEE Computational intelligence society in 2000, Keeley found her lifelong mentor and a host of volunteering opportunities in STEM, focusing on attracting young people to careers in computer science and artificial intelligence. Today, she is passionate in educating young people of all ages on how to code, on the ethical use of artificial intelligence and a coach and mentor for young girls and women who wish for exciting careers Computational Intelligence.

When Keeley Crockett was in school, she wanted to be an astronaut – but she was not the greatest at physics, demonstrating a stronger ability in computer science and control technology. Keeley first learnt to program using the BBC BASIC programming language at the age of 14 in school. During this time, she also studied control technology using simple circuits to build simple traffic lights and small robots. Keeley was inquisitive, liked a challenge, and studied artificial intelligence as part of her first degree. Following a practical Higher National Diploma in Software Engineering, she spent 2 years at the University of Manchester Institute of Science and Technology studying computation, graduating in 1993. Here she gained an appreciation of artificial intelligence and was introduced to fuzzy logic. Whilet studying, Keeley found herself to be in a very small minority of women on the course.

Following graduation, Keeley applied and received a good job offer; however, she choose to carry on in education and pursue a PhD, which involved a teaching role within the University. As well as research, Keeley really enjoyed working with and helping students to understand key computer science concepts and loved seeing them have a Eureka moment when they finally managed to solve a problem. Over the years, she has had the opportunity to work in hospitals with medical professionals on using ICT, through to teaching the more elderly community to use email, and working with young people who have left school with no qualifications on computer science projects to allow them to believe in themselves.

Toward the end of her PhD, Keeley joined the IEEE Computational Intelligence Society and was inspired by other women professors in the field. She attended her first IEEE conference on Fuzzy systems (IEEE-FUZZ) in San Antonio, Texas in 2000 and was inspired and motivated by the quality of the speakers. In 2001, she attended IEEE-FUZZ in Melbourne and was privileged that the founder of fuzzy logic, Professor Lofti Zadeh, attended her paper session and spoke to her briefly afterward about her work – providing motivation. At this conference, she met the most amazing woman – Professor Bernadette Bouchon-Meunier from the Université Pierre et Marie Curie, who had started a group within the IEEE Computational Intelligence Society known as IEEE Women in Computational Intelligence (WCI). Bernadette went on to become her unofficial mentor.

Keeley is an active volunteer within the IEEE Computational Intelligence Society, chairing many sub-committees on travel grants. In 2014, she also became the Chair for Women in Engineering (WIE) in the UK and Ireland until 2019, when she served a year on the IEEE Women in Engineering Leadership Committee. If it was not for the incredible women mentors and role models who provided advice and support throughout her career, she is convinced her path would have been different. Now as a qualified mentor, she has had the privilege to see students grow and follow their goals to achieve their own successful careers (kind of like a proud parent!).

Keeley also has a passion and drive to bring computer science opportunities to rural schools in the UK and can be regularly found in primary schools delivering programming and robotics sessions with children aged between 4 and 10 years old. Running computer science events at National Science festivals and IEEE WIE and WCI events allows young people to have hands on experiences, inspires, and encourages them within their education. Seeing female academics engage in these many activities sends a clear action message, that yes you can be female and work in this field. Despite incredible efforts over the years to encourage females into taking STEM careers from many organizations and people, there is still much work to be done form the grassroots level. Keeley became a national STEM Ambassador in 2018.

Currently, Keeley is Professor in Computational Intelligence at Manchester Metropolitan University. She has supervised 25 successfully completed PhD students to date, 50% whom were women. She believes that the key to successful PhDs is a team partnership where supervisors and students are on a research journey together to try and solve a societal challenge that will have some positive impact on people's lives.

Message:

I was honored to take part in the writing of this book which showcases some of the amazing research undertaken by women in the field of computational intelligence. I will openly admit I suffer from imposter syndrome, I am not very confident, and I end up questioning everything that I do. It took me longer than most to become a full professor in my academic career journey, but I never gave up and instead relished working with some amazing students and people along the way. My advice is to never lose sight of who you are, always be kind, and have the courage to follow your dreams.

Conversational Intelligent Tutoring Systems: The State of the Art

Annabel Latham

1 Introduction

In 2020, the COVID-19 pandemic led to the closure of schools, colleges and universities in many areas of the world, throwing to the fore the challenges of supporting remote learning with currently available technology. A key aspect of remote learning is the need to blend teaching approaches in an attempt to fulfil the individual and social needs of learners, whilst recognizing that in the context of the pandemic, time to reorganize and plan for online learning was not available. A blended approach incorporates moving class and individual teaching/learning online via video- or tele-conferencing, supplemented with computer-supported e-learning and offline independent learning activities; however, in 2020, most available e-learning systems delivered a one-size-fits-all experience.

One of the key challenges for computerized learning systems is to offer automated tutoring that flexibly delivers the benefits of expert human tutoring on a big scale. The research area of intelligent tutoring systems (ITS) applies computational intelligence techniques to model a human one-to-one tutorial. ITS aim to personalize automated tutoring by adapting to each individual learner's knowledge and personal traits, such as emotion and learning style [42]. Additionally, they offer intelligent solution analysis, delivering instant feedback on aspects of a learner's response containing misunderstandings or mistakes. Unlike traditional e-learning systems, ITS also support a learner's problem-solving by giving hints to aid the learner in constructing their own knowledge.

Conversational ITS (CITS) further model human tutorials by incorporating a conversational agent (CA) interface, enabling discussion and direct, non-linear

---

A. Latham (✉)
Computational Intelligence Lab, Department of Computing and Mathematics, Manchester Metropolitan University, Manchester, UK
e-mail: a.latham@mmu.ac.uk

A. E. Smith (ed.), *Women in Computational Intelligence*, Women in Engineering and Science, https://doi.org/10.1007/978-3-030-79092-9_4

access to information through questions. Typically, CITS conversations are mixed-initiative, meaning that they are led by the tutor, but learners are able to take the initiative to explore related topics or ask questions. This approach helps to overcome the issue of learner motivation, which has traditionally been a problem with e-learning systems, which, despite adding adaptations such as reordering of links, typically rely on learner initiative, making them fundamentally not much different to reading a book.

Despite the benefits demonstrated by ITS, and in particular CITS, their implementation into real teaching/learning environments has been limited. The expertise and time involved in the development of ITS/CITS has been a major barrier to their growth. With adaptive systems such as ITS, there is a need to design multiple versions of learning material in order to adapt to individual learner needs. Additionally, to design intelligent feedback on errors, misunderstandings or misconceptions requires considerable time for capturing such expertise and knowledge engineering. With the additional complexity of a natural language interface, these challenges are exacerbated in CITS. CITS require the design of sophisticated human-like tutorial conversations to manage an individual learner's path, intelligently discuss and answer questions relating to the context of the learning, whilst also managing progression towards the learning goal. As with all goal-oriented CAs, the time required for design, development and user testing of tutoring conversations is substantial:

> "One grand challenge for education is to scale up the benefits of expert human tutoring for millions of students individually." [33]

In this chapter, the current state of the art in Conversational Intelligent Tutoring System research is reviewed. The challenges in CITS design and development are explored in order to highlight the barriers to the wide application of CITS in real learning environments. Two novel developments in CITS learner profiling are described:

- Oscar CITS, which dynamically profiles and adapts its conversation to individual learning style using behaviour data captured during the tutoring conversation
- Hendrix CITS, which profiles levels of learner comprehension during a tutoring conversation, using facial micro-expressions captured from a webcam, prompting pedagogical interventions

The chapter then explores the future direction of human-like intelligent tutoring systems and the challenges and scalability issues that need to be addressed before CITS can move into the arena of mass learning.

The rest of the chapter is organized as follows: Section "Conversational Intelligent Tutoring Systems" gives a background on CITS and describes the complex challenges in CITS design. Sections "Oscar CITS" and "Hendrix 2.0 CITS" describe Oscar CITS and Hendrix CITS, respectively. Section "Ethical Use of 'AI' in Education" explores the ethical issues that are particular to the use of computational intelligence in the educational domain. Section "Scalability Challenges for CITS" summarizes the scalability challenges in moving CITS into the mainstream, and Section "Conclusion and Future Directions" concludes and describes future directions for CITS research.

2 Conversational Intelligent Tutoring Systems

Conversational Intelligent Tutoring Systems (CITS) are cutting-edge e-learning systems that use a natural language interface to deliver an adaptive tutoring conversation. They are an extension of Intelligent Tutoring Systems, an area of research that aims to go beyond the one-size-fits-all approach of traditional e-learning systems to provide a personalized learning experience using computational intelligence (CI) techniques. ITS interfaces typically include links to learning material that are selected or reordered according to the individual learner's need, such as their existing level of knowledge of the topic, their personality or learning style [37, 42]. As well as adapting to the learner, ITS often include instant analysis and feedback on solutions and a hinting mechanism to help learners develop their solutions. CITS aim to mimic human tutors by holding a one-to-one tutorial conversation adapted to an individual's need and can deliver instant, human-like feedback on errors, misunderstandings or misconceptions. This scaffolding approach is designed to enable learners to construct their own knowledge, leading to a more effective learning experience. Studies have shown that CITS can rival or outperform novice human tutors on comparable content [36, 43, 44].

What Does the Conversational Agent Interface Add? Despite adding considerable complexity, and thus development time, to CITS, the CA facilitates intuitive communication with learners using natural language and enables the ability to ask questions. Studies have shown that people who are inclined to trust CAs (AKA chatbots), e.g., are more likely to answer personal questions posed by a chatbot than a questionnaire [5, 49]. This trust has been a positive phenomenon in CITS, where studies suggest that learners are more confident in asking questions to an agent, whereas in a classroom of peers they may be worried to do so [15, 28]. Building trust in the CA tutor is critical for engaging the learner and helping to mitigate negative feelings towards the tutor, such as frustration, that may impact the efficacy of the learning experience. Therefore, CITS designers build in traits that engender trust from learners, such as natural language traits (e.g. social and colloquial aspects) and anthropomorphism. There is a fine balance to be struck between developing a human-like online tutor and a CITS becoming too similar to a human, which can lead to the 'uncanny valley' situation whereby the learner may become disengaged and distracted by small differences between a CITS tutor and a real human [39].

CITS work by building a profile of each learner, called a student model, which contains:

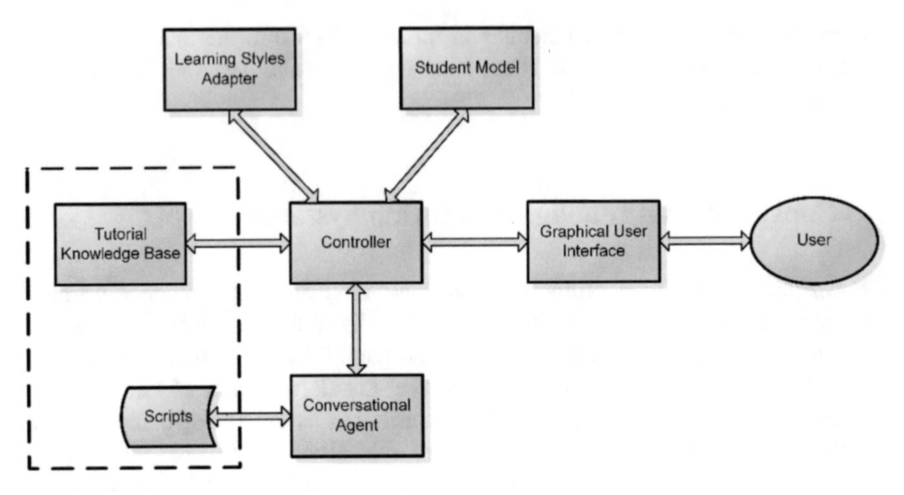

Fig. 1 Oscar CITS Architecture [30]

- Information on the learner's individual traits and preferences that may affect their learning (e.g. learning styles [28] and mood [42])
- Behavioural interaction data from CITS (such as time spent on particular activities, approach to solving problems)
- Learning progress (such as questions attempted, test results, skills and knowledge demonstrated)
- Conversation logs. Note that although they are linked to learner interactions, the transcripts of conversations are often kept separately from the student model for evaluation of CITS (CA or tutorial) or for future knowledge discovery

The information contained in the student model determines the curriculum and the learner's path through the tutorial. The ultimate aim is to populate the student model automatically; however, some CITS use a collaborative approach whereby they ask learners to self-report emotions and mood or complete self-assessment questionnaires [3]. An example of multi-agent CITS architecture is shown in Fig. 1.

In the Oscar CITS architecture shown in Fig. 1, there is a central controller agent that manages the tutorial session and communicates with the other agents to co-ordinate activities. The components grouped together in the dashed-line box are domain-specific (relating to the subject being learned) and can be replaced for new subject domains. The tutorial knowledge base retains knowledge of the learning content and how concepts relate to each other, along with information on the different learning objects available for each concept. The scripts database contains conversation scripts for the CA component, in terms of stimulus-response pairs with additional CA variables that are used to manage conflict and avoid duplication.

2.1 Design Challenges for CITS

CITS are sophisticated expert systems, and as such there are a number of challenges to be overcome and key questions to be answered during their design and development. These are highlighted below:

- **Choice of communication method.** One important design question for CITS is how to communicate with the learner: using a text interface, speech or talking avatar. To some extent, this may be determined by the age of the target learners; however, it is important to note that the method of communication may affect learner behaviour and engagement, either beneficially or detrimentally. For example, the addition of an avatar could add non-verbal behaviour to support communication of tutor feedback, yet the design of the avatar (e.g. object or humanoid, attractiveness or gender) has been shown to affect learners' confidence in the advice given [6]. Studies indicate small or no effect on learning of including an avatar over text or voice interfaces [20, 38]. Text interfaces are prone to user typos (although this can be mitigated by including common mistakes as patterns). Speech interfaces rely on the success of the voice recognition technology and may be affected by background noise as well as other issues. Some CITS (AutoTutor) include a text interface and also allow learners to switch on or off the tutor speech facility; however, this can lead to frustration as the avatar reads slower than the learner. Interface issues such as typos and voice recognition issues may lead to frustration or demotivation, which have been shown to affect learning [10].
- **Conversational agent challenges.** CITS require sophisticated natural language techniques to engage in a convincing human-like tutorial conversation. Unlike more simple Q&A chatbots, CITS must be able to manage the different contexts of a conversation, discussion of the domain concepts linked to the current topic and, more broadly, subject, along with a memory of the conversation, whilst nudging the learner towards the learning goal. There are a number of approaches for developing CAs, including rule-based methods, generative methods (supervised or reinforcement learning from conversation data), pattern matching [35], Latent Semantic Analysis [48] and natural language processing; however, the pattern matching approach currently works best for longer, goal-oriented conversations. The major drawback of the pattern matching technique is the time and expertise needed to develop stimulus-response pairs that can match input text to an output response. Humans use language creatively and can understand when different words and sentences have the same meaning. The pattern-matching technique requires extensive end-user testing to verify that there are sufficient stimulus patterns to cover multiple forms of language with similar meaning. The development of semantic similarity measures aims to overcome this requirement for predefining many different stimulus patterns by using a computational comparison measure to establish the similarity of meaning [4]. However, despite its promise, semantic similarity research is still in its infancy.

- **Representation of knowledge.** As well as deciding how to represent an individual learner's skills and knowledge of the domain concepts, representing common misunderstandings and errors related to domain concepts is vital. Concept networks and maps, ontologies or categories of key concepts and skills are reasonably straightforward to design for a single domain; however, in order to generalize a CITS to multiple domains, a standardized representation is important. Reusable Learning Objects (RLOs) are standardized representations of knowledge that have had limited success in e-learning systems [16], yet no such standard representation exists for tutoring conversations. Nye et al. [34] tried a different approach of overlaying natural language conversation templates onto an existing ITS to provide a CITS-like experience – however, this proved complex and the results were inconclusive, with student reactions being mixed to negative.
- **Profiling method.** How individual traits will be captured (such as use of sensors, webcam, self-report, language, behaviour, etc.) may have an impact on the design of CITS and the tutorial. For example, the location of learning materials on the screen may affect user gaze when using webcam [22], or some tutorial questions may be designed to capture specific behaviours [28].
- **Adaptation decisions.** Pedagogical research has linked many different human traits to the efficacy (or lack of) in teaching/learning. These include mood, emotion, attitude, confidence, comprehension, personality and learning styles amongst others ([23, 28, 36, 42];). CITS designers must decide which factors are most appropriate given the context of learning and design algorithms that determine when and how to adapt the conversation. The granularity of the adaptation is also important – whether the same adaptation is applied to the whole of the tutorial session, or whether the appropriate adaptation is determined by different types of activity and question, or whether adaptation is event driven (e.g. a change in mood or comprehension level [23]). The Oscar CITS adaptation algorithm [30] determines the adaptation at a question level and takes into account both the score of the trait in the student model and the availability of suitable adaptations for each question, thus offering a varied learning experience over a tutorial.
- **Design of learning material.** Careful design of tutoring conversations is critical to the success of a human-like learning experience. This demands expert knowledge of pedagogy and the domain being taught and careful capture and modelling of common learner misconceptions and mistakes. As well as supplemental conversations to facilitate problem-solving support, adaptive systems need multiple versions of learning materials.
- **Learner control of interventions** There has been some research that proposes that learners should be given access to explore their student model and the option to control whether a CITS profiles and adapts to their individual needs. The suggestion is that some learners may feel uncomfortable about an intelligent system deciding on interventions automatically. The ethics and trust issues in the widespread use of computational intelligence in computer systems is currently an active area of debate, discussed further in Section "Ethical Use of 'AI' in Education".

Having explored the context of CITS development and the key debates and challenges of CITS research, the next two sections describe two cutting-edge CITS that take different approaches to learner profiling and adaptation.

3 Oscar CITS

Oscar is an online CITS that models a human tutor by dynamically predicting and adapting to each student's preferred learning style during a tutoring conversation [28, 30]. Learning styles are individual traits, linked to personality, that describe the way the groups of learners prefer to learn. There is much debate in the education community about the usefulness of learning styles, as they are not fixed and thought to change over time and across subjects. If this is the case, a CITS that profiles and adapts dynamically to learning styles during a conversation may be a powerful tool for improving the learning experience and understanding changes in learning style.

Throughout the conversation, Oscar CITS adapts to the learner's knowledge by following a scaffolding approach to change the learning path. Additionally, Oscar CITS aims to mimic a human tutor by leading a mixed initiative two-way tutorial discussion and using cues from the student dialogue and behaviour to dynamically profile and adapt to their preferred learning style. Oscar's intelligent approach includes:

- Automatically profiling the individual's preferred learning styles during the tutorial conversation
- Adapting the conversation to personalise scaffolding and incorporating learning material in a sequence and style most suited to the individual's knowledge and learning style (curriculum sequencing)
- Analysing solutions and giving instant feedback on correct, incomplete erroneous solutions (intelligent solution analysis)
- Providing intelligent hints and discussing learner questions (problem-solving support)

Like human tutors, Oscar CITS promotes a deeper understanding of the topic by using a constructivist style of tutoring, giving intelligent hints and discussing questions with learners rather than presenting the answer straight away. Oscar's natural language interface and classroom tutorial style are intuitive to learners, enabling them to draw on experience of face-to-face tutoring to feel more comfortable and confident in using CITS. Oscar CITS is a personal tutor that can answer questions, provide hints and assistance using natural dialogue and which favours learning material to suit each individual's learning style. The Oscar CITS offers 24-hour personalised learning support at a fixed cost.

Oscar CITS is the first CITS to capture verbal (via text) and interaction behaviour to dynamically profile and adapt to individual learning styles during a tutoring conversation (other systems relied on learners completing self-assessment questionnaires). Latham et al. [28] describes a formal methodology for developing the

Oscar CITS, along with its implementation and initial evaluation, and is summarised in section "Automatic Profiling of Learning Styles". Originally Oscar used a rule-based approach to profiling learning styles from user behaviour with good accuracy, but profiling has been extended by applying several machine learning algorithms and fuzzy techniques to improve accuracy and conflict resolution, as outlined in section "Automatic Profiling of Learning Styles" [1, 9, 29].

The second key contribution of the Oscar CITS research is its generalised adaptation algorithm, described in [30]. The Oscar CITS adaptation algorithm was the first to be published that could dynamically determine the best adaptation at tutorial question level (most CITS determine adaptation in advance at tutorial level). Powerfully, the algorithm takes into account the availability of adaptive material for each question and can incorporate several dimensions of an individual learner's traits (e.g. learning style, mood or personality) as long as they can be assigned a score. By eliminating the requirement for alternative learning material to exist for every question for every learning style, the Oscar CITS adaptation algorithm promotes rapid prototyping of tutorials, and the algorithm adjusts as more styles of learning material are developed. An evaluation of Oscar CITS adaptation algorithm found that participants who used learning material adapted to their individual learning styles performed significantly better (on average 12%) than participants using learning material that was not matched to their learning styles [30].

Oscar CITS was originally developed to deliver a Sequential Query Language (SQL) revision tutorial suitable for undergraduate computing students, and it has been supporting SQL students at Manchester Metropolitan University for a number of years. Rather than being evaluated in a research lab like other CITS, Oscar CITS' empirical evaluation was conducted in a real university learning environment by undergraduate and postgraduate students studying SQL. An example of an interaction where Oscar CITS is helping a student to write an SQL query is shown in Fig. 2. and a full video demonstration highlighting Oscar's intelligent techniques can be seen at https://www.annabellatham.com/my-research/

Evaluations of Oscar CITS in a real university teaching/learning environment showed that 94% of participants found the Oscar CITS SQL tutoring helpful [30]. Of the participants, 92% stated they would use Oscar CITS if it were available, with 78% saying they would prefer Oscar over learning from a book and 46% would choose Oscar over attending face-to-face tutorials [30]. When openly asked for comments about Oscar CITS, participants were positive about the learning experience, with half remarking that Oscar was easy to use and 43% that Oscar CITS was helpful [28]. One participant commented, it "is like having your own friendly tutor", and another said, "it gives instant feedback unlike a traditional test" [28]. Evaluation found that Oscar's conversational style is intuitive to use, helping to improve motivation and build confidence, with one user remarking, "it encouraged me to think rather than simply giving me the answer" [28].

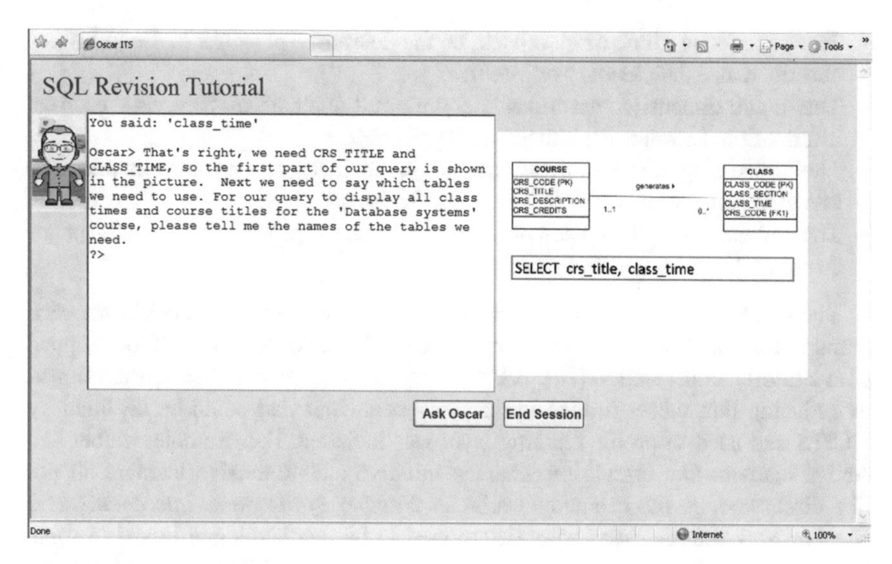

Fig. 2 Oscar CITS tutorial interface

3.1 Automatic Profiling of Learning Styles

There are many formal models of individual learning styles (Coffield et al. [7] identified 71 models) and several contradictory theories about their application in learning. Rather than formally assessing learning styles, human tutors use their knowledge and experience to pick up cues from learners and adapt their teaching style to improve the learning experience, for example, giving an example or drawing a diagram. Oscar CITS mimics this informal approach to automatically profile learning styles by detecting learner behaviours and relating them to a model of learning preferences to select an appropriate adaptation or intervention.

3.1.1 Learning Styles Knowledge Engineering

An established and popular learning styles model is the Felder and Silverman [14] Learning Styles Model (FSLSM), which describes the learning styles in engineering education and proposes different teaching styles to address learner needs. The FSLSM has four dimensions of learning style, relating to steps in the receipt and processing of information. Each individual's learning style preference is described for each dimension, for example, sensory/visual/active/global[1]:

[1]There are 16 (2^4) learning styles in FSLSM

- The processing dimension relates to the learner's perception. Learners are described as either *Sensory* or *Intuitive*.
- The input dimension describes the preferred method of receiving external information. Learners are either *Visual* or *Verbal*.
- The processing dimension describes how learners convert information into knowledge. Learners are either *Active* or *Reflective*.
- The understanding dimension describes individuals' progression towards understanding. Learners are *Sequential* or *Global*.

The FSLSM was analysed to extract typical behaviour characteristics for each learning style and a subset created by selecting behaviours that could be mapped onto a tutorial conversation [28]. A list of important behaviour cues was developed by reducing this subset further so that only behaviour that could be captured by a CITS and used to profile learning style was included. For example, Verbal and Active learners like discussion whereas Intuitive and Reflective learners do not like discussion, so the behaviour variables *Number of discourse interactions* and *Number of questions asked* were determined to be predictors and linked to these four learning styles. The behaviour characteristics extracted from the FSLSM were further analysed to extract language traits that may be used to profile learning styles, and a list of indicative key words and phrases (e.g. *show me* versus *tell me*) was developed, expanded using a thesaurus and linked to relevant learning styles.

The behaviour and language traits to be captured were encoded into 33 logic rules. In the logic rules, some variables could be directly observed from events (e.g. interaction count), whereas others relied on comparisons (e.g. learner has an interaction count greater than the average interaction count).

3.1.2 Capturing CITS Behaviour Dataset

A CITS tutorial was developed to tutor undergraduate computing students in SQL revision. The Oscar CITS SQL Revision tutorial was integrated into two undergraduate units at Manchester Metropolitan University to be evaluated in a real teaching/learning environment. During timetabled classes, 115 undergraduate students were asked to complete the Index of Learning Styles questionnaire[2] to establish their ground truth Learning Style, and then to complete the SQL revision tutorial in Oscar CITS [28]. During the tutorial, 41 behaviour variables were captured for each learner, along with the learning style class. There were 75 usable instances; data from incomplete tutorials were discarded. A description of the dataset can be found in Crockett, Latham and Whitton [9].

[2]The instrument to assess learning styles for FSLSM [40]

Table 1 Oscar CITS best rule-based prediction accuracy [28]

n	Sensory	Intuitive	Visual	Verbal	Active	Reflective	Sequential	Global
75–95	70%	80%	94%	71%	100%	73%	82%	61%

3.1.3 Learning Styles Prediction Approaches

Rule-Based Profiling **[28]** Experiments were conducted to answer the following research questions:

- RQ1: Is it possible to automatically profile learning styles from interaction data captured during a CITS tutorial?
- RQ2: What are the behaviour variables, or set of behaviour variables, that can best predict each learning style in FSLSM?

Knowledge engineering of the FSLSM resulted in 33 logic rules that follow a typical IF..THEN format to increment the learning style scores when a particular behaviour occurs during the tutorial. A simple winner-takes-all strategy was adopted by comparing learning style scores for each dimension, with the higher score deciding the predicted learning style (e.g. for the processing dimension, if the score for Active is higher than that for Reflective, the learner is predicted to be Active).

It was empirically determined that it is possible to predict learning styles (RQ1) and that some individual rules (behaviour variables) are better than the combined logic rules at predicting particular learning styles (RQ2). As shown in Table 1, prediction accuracies ranged from 61–100%, and it was concluded that a combination of best-fit rules per learning style produced the best predictions. However, accuracies reduced when a single best-fit variable *per dimension* was applied.

Profiling with Machine Learning Although it showed that automatic prediction of learning styles was possible, rule-based profiling had a number of limitations:

- There was no combination of behaviour variables to produce a classification for each learning style dimension.
- Conflict was not dealt with adequately.
- There was an assumption that variables are independent of each other.
- There was an assumption that indicative behaviour variables were linked solely to those learning styles identified in FSLSM.

To address these limitations, some exploratory machine learning experiments (summarised below) were done using the same dataset to answer the following research question:

- RQ3: Is it possible to use machine learning to improve predictions of student learning styles from behaviour attributes captured during a tutorial with a CITS?

Four labelled datasets were generated, one for each learning style dimension. Each dataset of 41 behaviour variables captured during an Oscar CITS tutorial

Table 2 Single Variable Prediction (SVP) vs MLP-ANN Classification Accuracy

	ACT	REF	**ACT/REF**	SEQ	GLO	**SEQ/GLO**
n	43	32	**75**	45	30	**75**
Oscar CITS SVP	53%	73%	**63%**	70%	59%	**65%**
MLP-ANN	98%	78%	**89%**	96%	67%	**84%**

(section "Capturing CITS Behaviour Dataset") was associated with the pertinent learning style class for the dimension.

***Multi-layer Perceptron Artificial Neural Network (MLP-ANN) with Attribute Selection* [29]** Initial experiments considered two learning style dimensions with a balance of labelled classes: Active/Reflective (ACT/REF) and Sequential/Global (SEQ/GLO). The open-source machine learning package WEKA was used [19].

First, an attribute selection algorithm was applied to identify a subset of 'best predictor' attributes in order to reduce the number of attributes input to a neural network. The correlation-based feature subset selection algorithm was applied [18] with best-first forward search and ten-fold cross-validation. Two attribute subsets were selected that are highly correlated with the class but with low inter-correlation, each containing three different attributes [29]. Neither subset contained the single best predictor attribute for the learning style class identified in the rule-based profiling experiments.

Two multi-layer perceptron feedforward artificial neural networks were implemented to classify the learning style dimensions. The attribute subsets were input to six input nodes in the ACT/REF MLP-ANN and three input nodes in the SEQ/GLO MLP-ANN, and each classifier had two hidden layers. As can be seen in Table 2, the MLP profiler improved the classification accuracy substantially for both dimensions,[3] with a 26% improvement for ACT/REF and a 19% improvement for SEQ/GLO. These improvements result from much better classification of the Active and Sequential learning styles.

The improvements in accuracy are likely due to the use of a combination of attributes rather than the single best predictor, and the use of a small subset of attributes has avoided adding too much complexity to the model. The results support a positive answer to RQ3.

***Data Mining Algorithms* [1]** Seven data mining algorithms (see Table 3) were applied to the data to further investigate RQ3 across all four learning style dimensions. For each dataset, a number of experiments were run for each of the seven algorithms selected, using ten-fold cross-validation in WEKA. Initially each algorithm was evaluated using the default parameters in WEKA, then the parameters were tuned empirically, which, in some cases, resulted in improved classification accuracy.

[3]Note that the Single Variable Predictor (SVP) in Table 2 is that variable with the best accuracy over the learning style *dimension*, and is thus different to that reported in Table 1

Table 3 Best accuracies for selected data mining algorithms

	ACT/REF	SEQ/GLO	SNS/INT	VIS-VRB
Oscar CITS SVP	63%	65%	65%	61%
Naïve Bayes	78.6%	84%	70.6%	85.3%
Bayes Network	78.6%	84%	72%	84%
J48	81.3%	**85.3**%	**81.3**%	85.3%
Simple CART	82.6%	82.6%	**81.3**%	86.6%
Random Forest	**84**%	81.3%	73.3%	**88**%
IBK	66.6%	50.6%	62.6%	72%
RBF	73.3%	58.6%	69.3%	78.6%

Table 3 shows the best classification accuracies for each algorithm and indicates that accuracies can be improved, answering RQ3. Over the whole dataset, the Simple CART algorithm performed the best, although J48 and Random Forest generated the highest classification accuracy for some dimensions.

Fuzzy Decision Trees **[9]** There is an inherent uncertainty in profiling learning styles from learner behaviour, and small changes in behaviour can lead to incorrect predictions. Fuzzy decision trees are based on classical decision trees (which are robust and interpretable by humans), but are less rigid as they have fuzzy rather than rigid boundaries. Crockett, Latham and Whitton [9] developed a new profiling method that uses fuzzy decision trees to build a series of predictive models. The benefit of using fuzzy decision trees is that when a new case (set of behaviour data for a learner) passes through the tree, all branches in the tree will contribute towards the decision classification.

In Fig. 3, the *Fuzzification Interface* takes captured behaviour variables and maps them to a predefined fuzzy set. The *Knowledge Base* includes fuzzy decision tree models that have been induced from the behaviour variable knowledge base. The knowledge base also contains induced fuzzy rules (obtained from the fuzzy decision tree models) and a series of fuzzy sets representing branches within the tree. The *Fuzzy Inference* mechanism combines grades of membership in all eight classes of the four learning style dimensions. Finally, the *Defuzzification Interface* translates the aggregated fuzzy sets into a fuzzy singleton that predicts the learning style. The generation of the fuzzy decision trees from conversational tutorials is described in full in Crockett, Latham and Whitton [9].

In Table 4, it can be seen that the prediction accuracies have been improved using a traditional decision tree (C4.5), and then further improved using the Fuzzy Learning Styles Predictor.

Conclusion Automatic profiling of learning styles from learner behaviour and conversation during an online tutorial with Oscar CITS is possible. A number of experiments have been described which evaluate the ability of different methods and algorithms to improve the prediction accuracy of learning styles. It seems to be important to combine behaviour variables to produce a better prediction. In conclusion, a hybrid profiling approach would produce the best accuracies, with

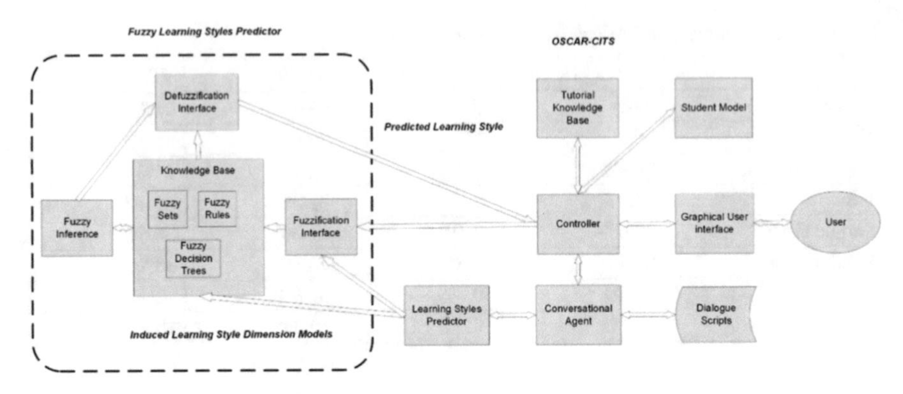

Fig. 3 Oscar CITS with fuzzy learning styles predictor [9]

Table 4 Fuzzy learning styles predictor accuracies

	ACT/REF	SEQ/GLO	SNS/INT	VIS-VRB
Oscar CITS SVP	63%	65%	65%	61%
DT (C4.5)	73%	76%	82%	73%
FLSP	**84%**	**88%**	**86%**	**85%**

different algorithms giving best results for different learning style dimensions.[4] However, for dynamic profiling, there would be a trade-off between acceptable accuracy and performance – any perceived delay in the profile resulting in a slow response from an adaptive system would have negative consequences for the learning experience.

4 Hendrix 2.0 CITS

Hendrix 2.0 CITS is an ontology-based CITS that adapts a tutoring conversation based on a learner's level of understanding, predicted dynamically from webcam images during the tutorial [21–23]. Hendrix CITS incorporates a cutting-edge profiling technique to determine a learner's level of comprehension (an important cognitive state for learning) using a novel image processing and machine learning algorithm called COMPASS [23], described in section "Profiling Comprehension: Comprehension Assessment and Scoring System (COMPASS)". Throughout the tutoring conversation, Hendrix 2.0 uses COMPASS to model the learner's level of comprehension by producing a time-series of comprehension estimates and classifications (see Fig. 4).

[4]MLP-ANN for ACT/REF; FLSP for SEQ/GLO and SNS/INT; Random Forest for VIS/VRB

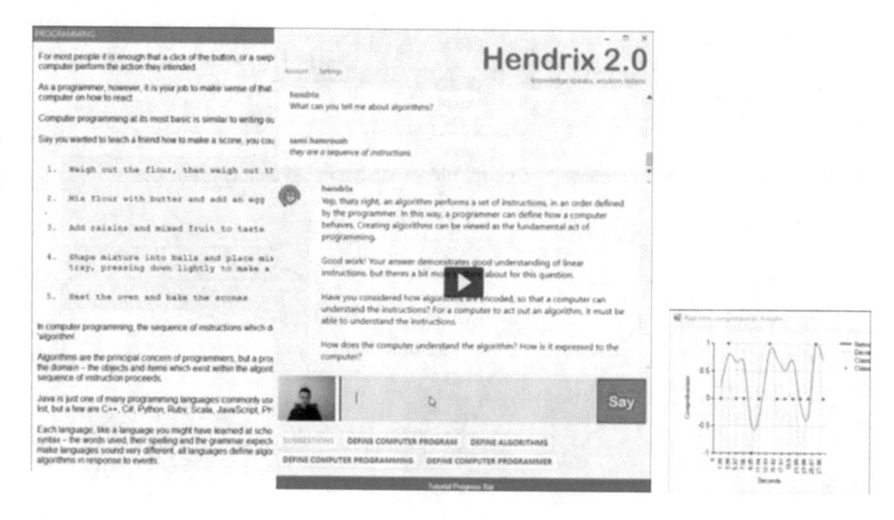

Fig. 4 Hendrix 2.0 tutorial interface (left) and real-time comprehension monitoring window (right)

In Fig. 4 (right), the real-time comprehension monitoring window, which is shown here for demonstration purposes (i.e. learners are not shown this window, which could be distracting), plots levels of learner comprehension against their baseline. Hendrix 2.0 has two levels of micro-adaptation interventions, determined by the intensity of a learner's non-comprehension. When mild non-comprehension is detected, Hendrix adds buttons at the bottom of the chat interface listing suggested topics that a learner may want help with (see Fig. 4 left). When strong non-comprehension is detected, Hendrix 2.0 intervenes directly in the conversation to offer support. The intention is to help the learner overcome non-comprehension and avoid impasse, which leads to loss of motivation over time.

Hendrix CITS uses a graph-based concept map approach to represent knowledge of the Java programming language. Based on the profile of a learner's level of comprehension, Hendrix searches its graph database for semantically relevant information to support learning activities. Hendrix is able to give feedback on both discursive and programming code tutorial answers.

Figure 5 shows how concepts in Hendrix's domain model are connected with directional relations representing conceptual dependency. Hendrix creates the tutorial conversation by creating a route from one concept to another using a shortest path calculation across the graph. Each concept on the graph is a hub node with a cluster of related learning materials, as shown in Fig. 6. Each tutorial question may have multiple ordered remediation questions that Hendrix uses to guide a learner to a solution.

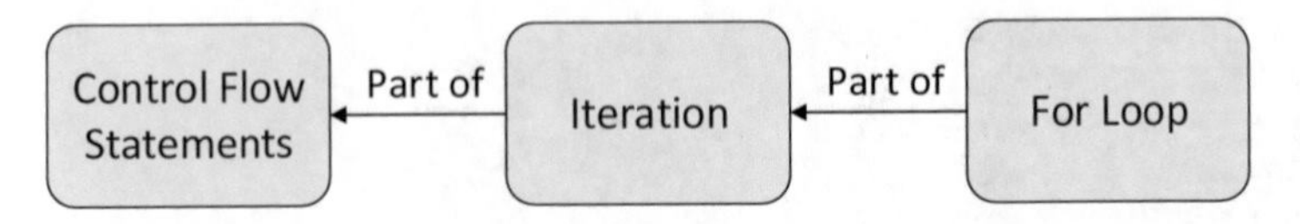

Fig. 5 Ontological path from concept 'Control Flow Statements' to concept 'For Loop'

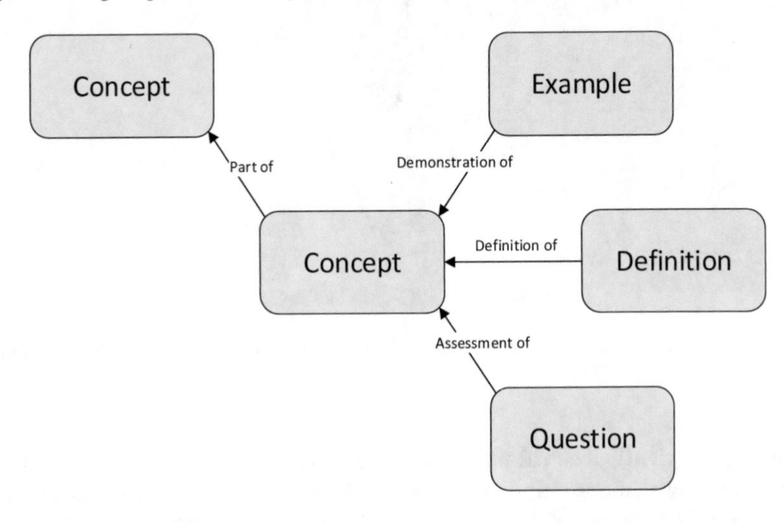

Fig. 6 A basic Java tutorial ontology

4.1 *Profiling Comprehension: Comprehension Assessment and Scoring System (COMPASS)*

Experienced human tutors can accurately estimate a learner's level of subject comprehension using non-verbal behaviour (e.g. gestures, facial expressions, facial actions, physiological, chemical and audible information) [47]. Automating this intelligent human behaviour in a real-world teaching/learning environment in near real-time presents a number of challenges:

- Analysing live video in near real-time using non-specialist equipment to track non-verbal behaviour (NVB) in an uncontrolled environment (in terms of varying lighting, camera and subject position)
- Creating an accurate model of comprehension from NVB during a conversation with a virtual tutor
- Producing an accurate comprehension classification without invasive body-attached sensors or prohibitively expensive specialist equipment

COMPASS is a novel near real-time comprehension assessment and scoring system designed to automatically classify e-learner comprehension of on-screen information as it is being read and appraised [23]. COMPASS uses learner NVB captured from a webcam stream during online learning, which is input to a bank

Table 5 COMPASS behavioural data model [23]

Type	Channels	Examples
Learner	5	Gender, ethnicity, academic level, specialism, experience
Eyes	17	Openness, gaze, blink
Geometries	18	Position, rotation, movement
Physiological	2	Blush, Blanche

of artificial neural networks to profile learner comprehension levels. COMPASS is summarised below, but for full details, see Holmes et al. [23].

Behaviour Modelling The COMPASS learner behaviour model incorporates 42 features, including meta-data about the learner and non-verbal behavioural (NVB) channels, as shown in Table 5. Each NVB channel is a single observed behaviour (e.g. 'head rotated left') that is captured from each image of a webcam image stream.

Over the learning period, the image data is segmented into one-second chunks that comprise 15 time-sequential still images. The behaviour model records both state and change over time for each channel, and the image data is summarised into a single feature vector representing the average behaviour expressed over the time chunk (called a Cumulative Behaviour Feature Vector (CBFV)).

Behaviour Extraction Each still image is firstly reduced in size and converted to greyscale, and then decomposed into regions of interest (ROI) that contain features (such as a left eye). As the position of the learner relative to the camera is not strictly controlled, Haar cascades are used to locate features. ROI pixel data is reduced using principle component analysis, then input to a specific behavioural channel neural network classifier to determine its state. Also extracted is data on state-change, geometries and physiology. The extraction process is repeated for each still image in a time chunk and summarised into a CBFV as described above.

Comprehension Classification A feedforward multi-layer perceptron artificial neural network was developed to classify the level of comprehension from a series of CBFVs. The network has 42 input nodes, a single hidden layer containing 20 fully connected nodes and a single output node. Binary comprehension classification is performed by applying a threshold function to the network output, with better accuracy the higher the threshold. Figure 7 shows COMPASS tracking a learner's comprehension-indicative behaviour and highlights the change as they process information from the screen. At second 10 there is a strong change in behaviour which could be used by Hendrix CITS to trigger an intervention in the learning process.

Evaluation of COMPASS In a study of 44 participants described in Holmes et al. [23], webcam data was collected during interactions with a specially designed quiz application. The bespoke application presented a baseline question, followed by 20 randomly ordered questions on programming and logic, with multiple choice answers and a pass option (to discourage guessing). From 869 questions answered,

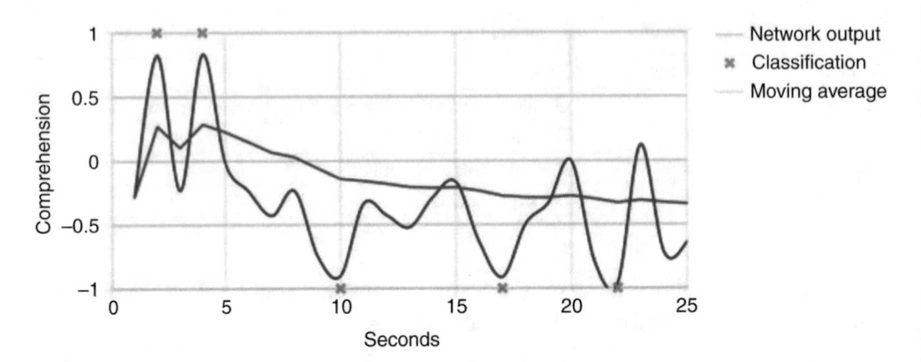

Fig. 7 COMPASS time-series for an incorrect answer period

185,075 webcam video stream images were captured and analysed with the classifier using ten-fold cross-validation. As ethnicity plays an important role in mediating subconscious NVB, the training and evaluation of the classifier were repeated using a subset of the data representing the predominant (59%) white male group of participants.

The COMPASS classifier was able to detect non-comprehension behaviour for the white male group with an accuracy of 75.8% and precision of 75.5% in near real-time, making it a practical, low-cost and non-intrusive method for comprehension classification during human to computer non-verbal interactions [23]. However, comprehension-indicative NVB differs between demographic groups, and the classification accuracy is weaker when the classifier is trained using data for different ethnic groups. Further training data are required to achieve good accuracy for females and other ethnicities.

5 Ethical Use of 'AI'[5] in Education

In order to benefit from the use of AI techniques in educational technology, such as personalization of learning, learners must accept the trade-off of the system gathering personal data and tracking their learning experience [31]. In the wake of recent scandals involving user profiling to further political and marketing gain (such as the Facebook/Cambridge Analytica scandal [17]), there has been much publicity about algorithmic decision making and the extent to which computational intelligence (CI) techniques are integrated into software systems. Concerns raised are whether the decisions made by algorithms can be explained, whether data used

[5]For the purposes of this article, CI and AI are used as having the same meaning because in general parlance the term 'AI' is more commonly used and understood, whereas the term 'CI' is less familiar [8].

in tracking and profiling require explicit consent, and the issue of bias encoded in the training data leading to biased decisions [32]. Recent laws, such as the General Data Protection Regulations (GDPR) [12] and the Californian Consumer Privacy Act (CCPA) [11], have attempted to address some of these concerns. However, the question of whether explanations about how algorithms come to decisions (which are often black box machine learning algorithms) are truly understandable by the general public remains unsolved [45].

The specific ethical concerns of AI in education (AIED) are not new and relate to learner agency, learner awareness of the use of profiling and AI techniques and the ability of learners to make informed decisions. In 2000, Aiken and Epstein proposed two meta-principles as a basic philosophical underpinning for any discussion of AIED systems:

1. "The Negative Meta-Principle for AIED – AIED technology should not diminish the student along any of the fundamental dimensions of human being; and
2. The Positive Meta-Principle for AIED – AIED technology should augment the student along at least one of the fundamental dimensions of human being" [2].

Organizations such as IEEE [26] and European Commission [13] have proposed frameworks for ethical design of AI systems. In 2018, Holmes argued that

> "around the world, virtually no research has been undertaken, no guidelines have been provided, no policies have been developed, and no regulations have been enacted to address the specific ethical issues raised by AIED" [24].

There are two key challenges specific to the use of AI in education that are not sufficiently addressed by the Ethics Guidelines for Trustworthy AI that relate to the ability of stakeholders in education and training systems to ensure that systems meet the guidelines and to make informed decisions on the use of these systems. The Institute for Ethical Artificial Intelligence in Education in the UK is therefore developing a code of practice (due 2021) aiming to protect the vulnerable and disadvantaged and maximize the benefits of AI in education [25].

A survey taken during a public event in the UK explored the question of whether learners are aware of the use of AI in education technology and understand that in order to personalize, learning systems gather user data and profile their personal traits [31]. The survey involved collecting anonymous questionnaires completed voluntarily by general public attendees at a free National Science Festival event held at Manchester Science Museum, called 'Me versus Machine'. The event included a number of activities designed to introduce people of different ages to Computer Science. One exhibit, 'I, Teacher', was dedicated to Artificial Intelligence in Education, where recent research in Conversational Intelligent Tutoring Systems (Oscar CITS and Hendrix CITS) was demonstrated and discussed. Interested attendees were asked to participate in a study of views on AI in Education and completed a questionnaire. It was found that most people knew the term Artificial Intelligence but did not necessarily understand its meaning in detail and were mostly not aware of the extent of the use of profiling and automatic adaptation in common software or that AI had been applied to learning systems. Most participants in the survey were worried in general about the use of their data; however, in the context of learning technology, fewer than 8% of adults were 'not happy' about being tracked, compared to nearly two-thirds (63%) of children surveyed. This seems to suggest that adults feel that the benefits of AI in education outweigh the threats, although a larger study is needed to explore the issue of trust.

The ethical challenges of AI seem to be amplified in an education context due to several factors, for example, subjects are frequently minors (and therefore considered to be vulnerable participants in ethical terms), the sensitive nature of the personal information involved and the importance of this application along with its potential benefit to learners. It remains to be seen how public trust in AI systems in general affects the uptake of CITS.

6 Scalability Challenges for CITS

Widely available e-learning systems are typically non-adaptive, such as video tutorials or websites delivering static content. Despite their significant learning gains over traditional e-learning systems [27], scaling up ITS and CITS into mainstream education remains a significant challenge for researchers. Before CITS can be scaled up and offered in mainstream online education, there are some key challenges to be overcome:

- **Sophistication of available CAs.** Widely available CAs such as Python libraries or ALICE [46] lend themselves to chatbot development, for question and answering, and yet are not sophisticated enough to manage sustained conversations, such as those required for tutoring conversations. Goal-oriented CAs are available [35] but are normally proprietary, thus limiting the wider development of CITS.
- **Generalizability.** The close coupling of CITS learning activities to domain content means that handling a range of subjects is difficult, and as yet there is no evidence to suggest that learning interactions based on universal learning activities add the benefits to justify the cost of development. In Latham et al. [28], a number of templates for conversational learning interactions were presented to speed up development of tutorial conversations, yet widespread use or development of Reusable Learning Objects does not exist.
- **Scripting expertise.** It is a complex task to design a tutoring conversation, requiring pedagogical, domain and scripting knowledge. The Generalized Intelligent Framework for Tutoring (GIFT) [41] is a tool for authoring adaptive tutoring, and its extension or a similar tool for authoring tutoring conversations may go some way to addressing this considerable challenge. Further work is needed to reduce the burden of scripting, either through the production of tools or of more efficient scripting mechanisms, such as the use of semantic conversational agents [35].
- **Development time.** In order to scale up CITS, development time must be reduced. Although implementing universal learning activities, RLOs and author-

ing tools such as GIFT should help speed up development, there is a trade-off in terms of learning gain as some types of learning activity are not readily transferrable between domains (e.g. solving a programming problem vs developing and drawing a system model vs improving essay writing). Particular to CITS is the considerable time required for user evaluation and testing – as variance in use of language affects the correct function of CAs in terms of appropriate responses.
- **Trust and ethical concerns.** As outlined in section "Ethical Use of 'AI' in Education"., public trust in AI and the general acceptance of learner profiling in education requires careful communication of the advantages, limitations and possibility of abuse. This may prove to be a barrier to CITS uptake without more general awareness and education of the public.
- **Processing power.** Considerable processing power is required to deliver multiple conversations without any time lag. Although easily overcome, cost may prove a barrier.

7 Conclusion and Future Directions

This chapter has described current research that aims to improve online educational experiences by employing computational intelligence techniques to make e-learning systems more human-like. Although research into Intelligent Tutoring Systems has been active for several decades, the disruption to education posed by the COVID-19 pandemic has highlighted the need for intelligent tutoring to move into mainstream education. Conversational Intelligent Tutoring Systems are an extension of ITS that adds an intelligent conversational interface to better model a human one-to-one tutorial. CITS offer many advantages over traditional e-learning systems by automatically profiling a learner's knowledge and individual traits which may affect their learning, and then personalizing the tutoring conversation. Like a human tutor, CITS create a scaffold for learners to construct their own knowledge and enable discussion and questions, problem-solving support and intelligent solution analysis to give personalized instant feedback.

The key design challenges and decisions for CITS researchers have been explored and then research on two cutting-edge CITS, Oscar and Hendrix, was described. Oscar CITS uses a non-intrusive profiling approach, modelling preferred learning styles from learner behaviour, and then selecting an adaptation for each tutorial question. A number of profiling methods were summarized, each aiming to improve the profiling predictions and achieving classification accuracies of 86–89% for learning styles dimensions. Learners rated Oscar CITS tutoring positively, and there was a significant (12%) improvement in learning gain for those experiencing a tutorial conversation adapted to their preferences. Hendrix 2.0 CITS uses an algorithm called COMPASS to profile a learner's level of comprehension from webcam images in near real-time. COMPASS was able to detect non-comprehension behaviour using a bank of ANNs with an accuracy of 75.8% for the white-male group; however, classification accuracy is weaker for other demographics. Hendrix

2.0 CITS adapts its tutoring, utilizing two levels of intervention depending on the level of non-comprehension exhibited by the learner.

Public trust in AI is critical to the uptake of CITS. A discussion of research on the ethical use of AI in education concludes that more needs to be done in raising awareness and educating the public. Finally, there is a review of the outstanding challenges and scalability issues that must be overcome by the research community before CITS can move into the mainstream.

As well as overcoming the highlighted scalability challenges, there are a number of future directions for CITS research. There is an ambition of accessing tutorials on the move using mobile devices to talk to an automated tutor. In CA research, the development of more efficient scripting mechanisms through semantic CAs will go a long way to reducing the overheads of time and expertise in designing CITS tutorial conversations. Finally, an investigation of the improvement of motivation and learner confidence when using CITS would be of interest to the educational research community.

References

1. N. Adel, A. Latham, K.A. Crockett, Towards socially intelligent automated tutors: Predicting learning style dimensions from conversational dialogue. In 2016 Intl IEEE Conferences on Ubiquitous Intelligence & Computing, Advanced and Trusted Computing, Scalable Computing and Communications, Cloud and Big Data Computing, Internet of People, and Smart World Congress (UIC/ATC/ScalCom/CBDCom/IoP/SmartWorld) (pp. 315–320). IEEE, July 2016
2. R.M. Aiken, R.G. Epstein, Ethical guidelines for AI in education: starting a conversation. Int. J. Artif. Intell. Educ. **11**, 163–176 (2000)
3. S. Aljameel, J. O'Shea, K. Crockett, A. Latham, M. Kaleem, LANA-I: An Arabic Conversational Intelligent Tutoring System for Children with ASD. In Proceedings of the 2019 Computing Conference, Springer, Volume 1 (vol. 997, pp. 498–516) June 2019
4. N. Alnajran, K. Crockett, D. McLean, A. Latham, An Empirical Performance Evaluation of Semantic-Based Similarity Measures in Microblogging Social Media. In 2018 IEEE/ACM 5th International Conference on Big Data Computing Applications and Technologies (BDCAT) (pp. 126–135). IEEE, December 2018
5. R. Bhakta, M. Savin-Baden, G. Tombs, Sharing Secrets with Robots? In EdMedia: World Conference on Educational Media and Technology (pp. 2295–2301). Association for the Advancement of Computing in Education (AACE), June 2014
6. S.W. Chae, K.C. Lee, Y.W. Seo, Exploring the effect of avatar trust on learners' perceived participation intentions in an e-learning environment. Int. J. Hum. Comput. Int. **32**(5), 373–393 (2016)
7. F. Coffield, D. Moseley, E. Hall, K. Ecclestone, F. Coffield, D. Moseley, E. Hall, K. Ecclestone, *Learning styles and pedagogy in post-16 learning: A systematic and critical review* (The Learning and Skills Research Centre (LRSC), London, 2004)
8. K. Crockett, S. Goltz, M. Garratt, A. Latham, Trust in Computational Intelligence Systems: A Case Study in Public Perceptions. In 2019 IEEE Congress on Evolutionary Computation (CEC) (pp. 3227–3234). IEEE, 2019
9. K. Crockett, A. Latham, N. Whitton, On predicting learning styles in conversational intelligent tutoring systems using fuzzy decision trees. Int. J. Hum.-Comput. Stud **97**, 98–115 (2017)
10. L.A. Díaz, F.B. Entonado, Are the functions of teachers in e-learning and face-to-face learning environments really different? J. Educ. Technol. Soc. **12**(4), 331–343 (2009)

11. Department of Justice, Californian Consumer Privacy Act (CCPA) | State of California – Department of Justice – Office of the Attorney General. (online), (2020), Available: https://oag.ca.gov/privacy/ccpa Accessed 23 Mar 2020
12. European Commission, EU data protection rules | European Commission. (online), (2019) Available: https://oag.ca.gov/privacy/ccpa Accessed 23 Mar 2020
13. European Commission AI High-Level Expert Group, Ethics Guidelines for Trustworthy AI | FUTURIUM | European Commission (online), (2019) Available: https://oag.ca.gov/privacy/ccpa Accessed 23 Mar 2020
14. R.M. Felder, L.K. Silverman, Learning and teaching styles in engineering education. Eng Educ **78**(7), 674–681 (1988)
15. A.C. Graesser, H. Li, C. Forsyth, Learning by communicating in natural language with conversational agents. Curr. Dir. Psychol. Sci. **23**(5), 374–380 (2014)
16. IEEE, IEEE Standard for Learning Object Metadata, in IEEE Std 1484.12.1-2002, pp.1–40, (2002) 6 Sept. 2002, doi: https://doi.org/10.1109/IEEESTD.2002.94128
17. The Guardian, The Cambridge Analytica Files | The Guardian (online), (2019) Available: https://www.theguardian.com/news/series/cambridge-analytica-files Accessed 02 Feb 2019
18. M.A. Hall, *Correlation-Based Feature Subset Selection for Machine Learning* (University of Waikato, Hamilton, 1998)
19. M. Hall, E. Frank, G. Holmes, B. Pfahringer, P. Reutemann, I.H. Witten, The WEKA data mining software: An update. ACM SIGKDD Explorations Newsl. **11**(1), 10–18 (2009)
20. S. Heidig, G. Clarebout, Do pedagogical agents make a difference to student motivation and learning. Educ. Res. Rev. **6**(1), 27–54 (2011)
21. M. Holmes, A. Latham, K. Crockett, C. Lewin, J. O'Shea, Hendrix: A conversational intelligent tutoring system for Java programming. In Proceedings of UKCI 2015 15th UK Workshop on Computational Intelligence (2015)
22. M. Holmes, A. Latham, K. Crockett, J.D. O'Shea, Modelling e-learner comprehension within a conversational intelligent tutoring system, in *Tomorrow's Learning: Involving Everyone. Learning with and about Technologies and Computing*, WCCE 2017. IFIP Advances in Information and Communication Technology, ed. by A. Tatnall, M. Webb, vol. 515, (Springer, Cham, 2017a)
23. M. Holmes, A. Latham, K. Crockett, J.D. O'Shea, Near real-time comprehension classification with artificial neural networks: Decoding e-learner non-verbal behavior. IEEE Trans. Learn. Technol. **11**(1), 5–12 (2017b)
24. W. Holmes, The ethics of Artificial Intelligence in education – University Business (online), (2018) Available: https://universitybusiness.co.uk/Article/the-ethics-of-artificial-intelligence-in-education-who-care/ Accessed 11 Feb 2020
25. Institute for Ethical Artificial Intelligence in Education, Institute for Ethical Artificial Intelligence in Education | Mission (online), (2020) Available: http://instituteforethicalaiineducation.org/#mission Accessed 11 Feb 2020
26. IEEE, Ethically Aligned Design: A Vision for Prioritizing Human Well-being with Autonomous and Intelligent Systems, Version 2, The IEEE Global Initiative on Ethics of Autonomous and Intelligent Systems (2018) (online), (2018) Available: https://ethicsinaction.ieee.org/, Accessed 29 Dec 2018
27. J.A. Kulik, J. Fletcher, Effectiveness of intelligent tutoring systems: A meta-analytic review. Rev. Educ. Res. **86**(1), 42–78 (2016)
28. A. Latham, K. Crockett, D. McLean, B. Edmonds, A conversational intelligent tutoring system to automatically predict learning styles. Comput. Educ. **59**(1), 95–109 (2012)
29. A. Latham, K. Crockett, D. Mclean, Profiling student learning styles with multilayer perceptron neural networks. In 2013 IEEE International Conference on Systems, Man, and Cybernetics (pp. 2510–2515). IEEE, October 2013
30. A. Latham, K. Crockett, D. McLean, An adaptation algorithm for an intelligent natural language tutoring system. Comput. Educ. **71**, 97–110 (2014)
31. A. Latham, S. Goltz, A Survey of the General Public's Views on the Ethics of Using AI in Education. In International Conference on Artificial Intelligence in Education, (Springer, Cham, 2019), pp. 194–206

32. New York Times, Artificial Intelligence's white guy problem. The New York Times (online) (2016). Available: https://www.nytimes.com/2016/06/26/opinion/sunday/artificial-intelligences-white-guy-problem.html Accessed 06 Aug 2018
33. B.D. Nye, A.C. Graesser, X. Hu, AutoTutor and family: A review of 17 years of natural language tutoring. Int. J. Artif. Intell. Educ. **24**(4), 427–469 (2014)
34. B.D. Nye, P.I. Pavlik, A. Windsor, A. Olney, M. Hajeer, X. Hu, SKOPE-IT (shareable knowledge objects as portable intelligent tutors): Overlaying natural language tutoring on an adaptive learning system for mathematics. Int. J. STEM Edu. **5**, 12 (2018)
35. J. O'Shea, Z. Bandar, K. Crockett, Systems engineering and conversational agents, in *Intelligence-based Systems Engineering*, (Springer, Berlin, Heidelberg, 2011), pp. 201–232
36. A. Olney, S. D'Mello, N. Person, W. Cade, P. Hays, C. Williams, B. Lehman, A.C. Graesser, Guru: A computer tutor that models expert human tutors, in *Proceedings of the 11th International Conference on Intelligent Tutoring Systems*, ed. by S. Cerri, W. Clancey, G. Papadourakis, K. Panourgia, (Springer, Berlin, 2012), pp. 256–261
37. E. Sangineto, N. Capuano, M. Gaeta, A. Micarelli, Adaptive course generation through learning styles representation. J. Univers. Access Inf. Soc. **7**(1), 1–23 (2007)
38. N.L. Schroeder, O.O. Adesope, R.B. Gilbert, How effective are pedagogical agents for learning? A meta-analytic review. J. Educ. Comput. Res. **49**(1), 1–39 (2013)
39. A.M. Seeger, A. Heinzl, Human versus machine: Contingency factors of anthropomorphism as a trust-inducing design strategy for conversational agents, in *Information Systems and Neuroscience*, (Springer, Cham, 2018), pp. 129–139
40. B.A. Soloman, R.M. Felder, Index of learning styles questionnaire. NC State University, (2005). Available online at: http://www.engr.ncsu.edu/learningstyles/ilsweb.html (last visited on 14.05. 2010), 70
41. R.A. Sottilare, K.W. Brawner, B.S. Goldberg, H.K. Holden, *The generalized intelligent framework for tutoring (GIFT)* (US Army Research Laboratory–Human Research & Engineering Directorate (ARL-HRED), Orlando, 2012)
42. N. Tsianos, Z. Lekkas, P. Germanakos, C. Mourlas, G. Samaras, User-centered profiling on the basis of cognitive and emotional characteristics: An empirical study. Lect. Notes Comput. Sci **5149**, 214–223 (2008)
43. K. VanLehn, A.C. Graesser, G.T. Jackson, P. Jordan, A. Olney, C.P. Rose, When are tutorial dialogues more effective than Reading? Cogn. Sci. **31**(1), 3–62 (2007)
44. K. VanLehn, The relative effectiveness of human tutoring, intelligent tutoring systems, and other tutoring systems. Educ. Psychol. **46**(4), 197–221 (2011)
45. S. Wachter, B. Mittelstadt, L. Floridi, Why a right to explanation of automated decision-making does not exist in the general data protection regulation. Int. Data Priv. Law **7**(2), 76–99 (2017)
46. R.S. Wallace, The anatomy of a.L.I.C.E, in *Parsing the Turing Test*, ed. by R. Epstein, G. Roberts, G. Beber, (Springer Science + Business Media, London, 2009), pp. 181–210
47. J.M. Webb, E.M. Diana, P. Luft, E.W. Brooks, E.L. Brennan, Influence of pedagogical expertise and feedback on assessing student comprehension from nonverbal behavior. J. Educ. Res. **91**, 89–97 (1997)
48. P. Wiemer-Hastings, D. Allbritton, E. Arnott, RMT: A dialog-based research methods tutor with or without a head, in *International Conference on Intelligent Tutoring Systems*, (Springer, Berlin, Heidelberg, 2004), pp. 614–623
49. L. Yearsley, "We Need to Talk About the Power of AI to Manipulate Humans", MIT Technology Review, (2017), 5 June 2017

Annabel Latham is a Senior Member of IEEE, and chairs IEEE Women in Engineering (UK/Ireland) and the IEEE Computational Intelligence Society Research Student Grants committee.

Annabel was born in the UK, but moved overseas aged 8 to spend the rest of her childhood in the Netherlands and Norway. Moving home and schools so frequently (attending 7 different British and Dutch schools) helped her develop an openness to difference and change, and a sense of adventure. After finishing at the British School in the Netherlands, she moved to Manchester, UK, to read BSc (Hons) Computation at UMIST, graduating in 1991, one of three women in the 120 strong cohort. Following university, Annabel enjoyed travelling and meeting many different people and companies for her work in the financial software industry. Her roles included project management, technical writing, systems implementation and user training. Working with large and small organisations to analyse their business needs and manage the implementation of new hardware and software systems appealed to Annabel's love of problem solving and working with different people in different environments. In 1996, Annabel joined the UK's National Computing Centre to work as a manager for Higher Education courses, responsible for working with universities in course development and marketing, and travelling worldwide for validation and quality assurance of partner colleges in the Far East.

On becoming a mother, Annabel decided to change jobs to reduce her overseas travel, returning to university as Admissions Tutor and Associate Lecturer at Manchester Metropolitan University (MMU). Four years later, with two young children at home, she returned to study at MMU gaining an MSc with Distinction in Computing (2007), followed by a PhD in Computational Intelligence in 2011. This change of career was partly driven by a desire to show her baby daughter that with a little hard work it is possible to follow your dream.

Annabel is a Senior Lecturer in Computer Science in the Department of Computing and Mathematics at MMU, where she enjoys mentoring and inspiring students and delights in watching them grow in confidence to achieve their goals. Her research interests under the Computational Intelligence lab include conversational agents, intelligent tutoring systems, big data, text mining, agent intelligence and knowledge engineering. Her main research focuses on applying computational intelligence (CI) techniques to automated tutoring systems in order to improve access to, and support success in, personalised learning. Annabel is passionate about promoting the power of CI to improve our lives, and enjoys discussing the benefits and ethical risks in her role as STEM ambassador through outreach work including invited talks, panel discussions, media articles, science festivals, school visits and organising events/boot-camps. As Chair of IEEE Women in Engineering for UK and Ireland, her aim is to change the face of computer science by challenging stereotypes and promoting careers in computing and engineering to young girls and women. In 2019 WIE UKI won the IEEE Region 8 (Europe, Africa, Middle East) WIE group of the year award, bringing publicity and recognition to our work.

Design and Validation of a Mini-Game for Player Motive Profiling

Kathryn Kasmarik, Xuejie Liu, and Hussein Abbass

1 Introduction

A mini-game is a game fragment or short scenario that can sit within a more complex game or virtual environment. The game fragment has its own mechanics, characters, mini-plot (storyline) and environment. These are often self-contained and may be isolated from other parts of a larger game. Large games often comprise many mini-games connected by a more complex, common storyline. A mini-game may conclude with a 'win' or a 'loss' for the player, or there may simply be different possible outcomes that influence the next phase of the larger game. We take this latter approach in our mini-game, for detecting player motivation.

Player motivation expresses why players want to participate in gameplay and which part of game players may find most engaging. The diversity of motivation types and underlying cognitive and biological processes, thus, requires that we limit our initial investigation to a specific category of motivation theory. We chose achievement, affiliation and power motivation for a variety of reasons. On the one hand, these three motives can be related to existing player types from the study of player experience in computer games [7]. On the other hand, these motives have been influential in the field of motivation psychology, forming the basis of a number of theories such as the three needs theory [2] and three factor theory [13]. They underlie a wide range of behaviours including social, risk-taking and

K. Kasmarik (✉) · H. Abbass
School of Engineering and Information Technology, University of New South Wales, Canberra, NSW, Australia
e-mail: K.Merrick@adfa.edu.au

X. Liu
School of Physics and Telecommunication Engineering, South China Normal University, Guangzhou, China

A. E. Smith (ed.), *Women in Computational Intelligence*, Women in Engineering and Science, https://doi.org/10.1007/978-3-030-79092-9_5

skill acquisition behaviours. Furthermore, these behaviours, in turn, can be linked to emotion, risk and social attitude.

It has been previously identified that while individuals may express aspects of all three motives, achievement, affiliation and power, they tend to have a dominant motive. The dominant motive has a stronger influence on decision-making than the other two, although the individual will not be conscious of this [2]. Mixed profiles of achievement, affiliation and power motivation have also been identified and associated with distinct individual characteristics. Leadership ability, for example, was found to be associated with mixed profiles of dominant power and achievement motivation.

Likewise, different levels of expression of the fear components of motivation can impact an individual's decision-making. For example, individuals with strong fear of rejection become nervous and insecure in social situations when they meet people they do not know. They may fear that others may not like them and seek to bring the contact to an end. People with strong achievement-related fear of failure are often afraid of failing in situations in which their performance can be compared with that of others. Fear of failure can therefore cause an individual to act with particular thoroughness and care and to strive constantly to make no mistakes. In experiments with goal selection, strong fear of failure coupled with weak hope for success increased the range of risk-taking behaviour exhibited by participants [1]. Finally, individuals with strong fear of loss of control concentrate on avoiding the loss of influence, control or prestige. Where a choice is required, they may prefer to secure their own position rather than consider the welfare of the group.

The space of motive profiles can be conceptualised as in Fig. 1, which shows the three profiles with dominant hope components and three profiles with dominant fear components, on axes of risk and social attitude [6]. The spaces between these named profiles represent the possible hybrids of these profiles. In addition, the relative dominance of each motivation varies from one individual to another. We made the conjecture that this conceptual model of the profiles will be useful for informing investigations into the use of EEG to distinguish motivation [7]. Existing approaches used to distinguish risk and social attitude can be applied and then the results used to place an individual within the space shown in Fig. 1.

Besides our previously published paper [6], the contribution of this chapter is to present the full game design for player motive profiling, which incorporates the non-player characters proposed in [6]. Next, an experimental validation of the game design by means of a human user study is also included in this work.

The challenge addressed in Sect. 2 of this chapter is to design a mini-game with appropriate components to permit players to express different social attitudes and risk-taking behaviour. Section 3 shows the experimental validation of the game elements for assessing player motive profiling accompany with underlying risk-taking and social attitudes.

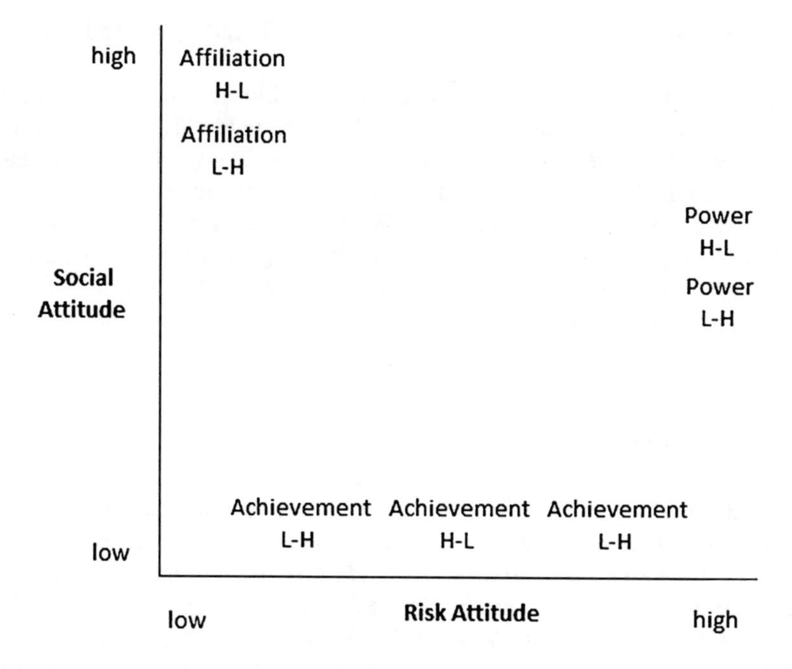

Fig. 1 Comparison of different dominant motivations on axes of risk and social attitude. There are four possible combinations of high and low levels of hope and fear components of each motivation. High hope and low fear (H–L) or low hope and high fear (L–H) are indicators of motive dominance

2 Game Design

This section introduces the storyline, mechanics, characters and gameplay design for our proposed mini-game. We justify each aspect of the game design with reference to the conceptual models in the previous section.

2.1 Storyline

We saw in the previous section that achievement, affiliation and power motivation are linked to risk-taking, social attitude and emotion. These three elements are ubiquitous in our everyday lives and are thus relatively easy to frame into a mini-game. Game theory and the gambling industry offer us many examples of games that abstract real-life decision-making and take into account risk and social attitude [10, 12]. In this chapter, we chose the well-known prisoners' dilemma (PD) [11], which invokes both risk-taking and social interactions, as the basis for the mechanics of our mini-game. As well as being used in EEG studies, precedents for using PD games to study motivation include work by Terhune [14] and Kuhlman et al. [5].

In order to do that, we designed a novel mini-plot, instead of using the traditional prisoner interrogation plot. The modified PD plot illustrates that our abstract scenario could represent multiple concrete scenarios. The storyline of our game revolves around the theme of 'Friends or Fortune'. In the game, players meet four non-player characters (NPCs) who have different play strategies in their virtual lives. First, players get to know the NPCs individually, by playing several rounds of the PD-like game with them. After that, they could team up to make friends or fortune. The player can choose which outcome (friends or fortune) they wish to pursue.

2.2 Game Mechanics

As described in the previous section, the mechanics of our mini-game are based on a PD-like game. Each player chooses cooperation or defection to earn their 'fortune' or build a 'friendship', while friendships and fortunes of each player depend on the choices of both players. The specific 'fortune' (money) payoffs V^{human} and V^{NPC} for each player and their NPC opponent in a single round of the game is shown in Fig 2. The money dimension is named to represent a tangible, valuable item from day-to-day life. It is designed to differentiate achievement and power motivation. We hypothesise that power-motivated individuals may prefer to maximise monetary pay-off. On the other hand, we hypothesise that achievement-motivated individuals will use the money reward to gauge their level of achievement in the game. Both players and NPCs accumulate the money they win in each round of the game.

A player's choice also influences the 'satisfaction' S of an NPC according to [6]:

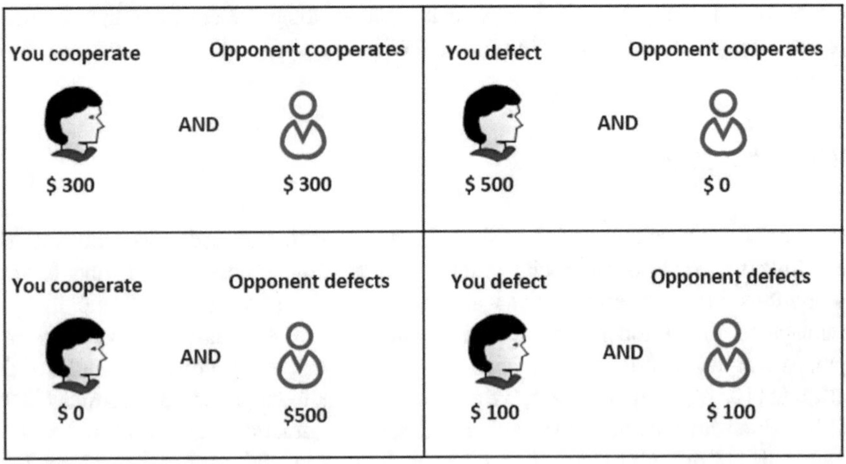

You cooperate	AND	Opponent cooperates	You defect	AND	Opponent cooperates
$ 300		$ 300	$ 500		$ 0
You cooperate	AND	Opponent defects	You defect	AND	Opponent defects
$ 0		$500	$ 100		$ 100

Fig. 2 Game mechanics for accumulating 'fortune'

$$S_t = \frac{V_t^{NPC}}{V_t^{NPC} + V_t^{human}} \tag{1}$$

A dynamic satisfaction value E_t is displayed on the game screen for each NPC. The initial value of E_t is $E_0 = 0.5$. The following update is applied to smooth the change in this value over time:

$$E_{t+1} = (1 - \lambda)S_t + \lambda E_t \tag{2}$$

where $\lambda = 0.5$. The satisfaction dimension is associated with social satisfaction about interactions between players and NPCs, which reflects the need for affiliation we experience in our daily life. Thus, the satisfaction dimension is designed to differentiate between power and affiliation motivations.

2.3 Non-player Characters

In our previous work [6], we have seen two themes emerging from the discussion of achievement, affiliation and power motivation. The first theme concerns social attitude: that is, the number of relationships an individual may choose to initiate and maintain. The second theme concerns risk attitude: that is, the degree of risk that an individual will tolerate when selecting goals. Accordingly, in Fig. 1, we proposed a model of motivation that positions achievement, affiliation and power motivation on two dimensions of risk and social attitude. We proposed that power-motivated players prefer high-risk tasks and have a neutral social attitude, while achievement-motivated players tend to select medium-risk tasks and enjoy working alone. Affiliation-motivated players have a high social tendency and prefer low-risk tasks.

In order to measure motivations during gameplay, we suggested risk-taking behaviour and social attitude should be considered. Thus, a range of non-player characters has been used for detecting cognitive and emotional phenomena as a basis for identifying a player's motive profile [6].

NPC design is a multi-faceted topic, including the design of the visible avatar and the design of the algorithms that control the behaviour of the avatar. A range of different approaches is taken for this latter topic. This includes rule-based crowds and evolutionary approaches to controlling behaviour [9]. In this work, we are primarily interested in aspects of decision-making that force players to reveal their risk attitude and social attitude. We propose to do this through an examination of their behaviour. As such we do not consider the design of the visible avatar but focus on the design of the character's cognitive attributes and decision-making behaviours.

Existing work proposed an abstract character in Fig. 3 with dimensions for money and satisfaction [6]. As described in the previous section, the money dimension is defined as the distribution of in-game money between NPCs and players using the

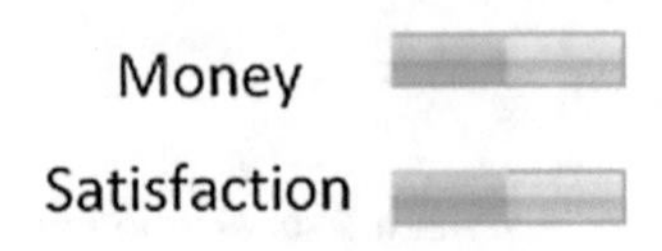

Fig. 3 Our proposed abstract non-player characters have dimensions for money and satisfaction. Image from [6]

mechanism of the prisoner's dilemma. The satisfaction dimension is proposed from a performance approach view, as proportional to the percentage of the winnings pool accumulated by both players. Players are given the option to play with NPCs for maximising their money, maximising NPCs' satisfaction or trading off these objectives. This has the potential to reveal players' risk-taking and social attitudes by examining their in-game behaviour and EEG signals.

Various existing works have proposed artificial intelligence techniques of different complexity for computer-controlled players of PD games [4]. We selected four classic PD techniques [6] to best aid the distinction between the motive profiles. These NPCs are listed as follows. It should be noted that we named the characters according to their play strategies. This also helps players to interact with these virtual characters more naturally.

- **Cooperator Candy.** Candy always cooperates, and she is satisfied when she earns the greatest share of the fortune. Thus, she could be easily exploited if an opponent wishes to do so.
- **Defector Dan.** Dan always defects, and he is satisfied when he earns the greatest share of the fortune. Thus, he can be easily satisfied when an opponent chooses to cooperate with him. However, the opponent will need to sacrifice own monetary gains to do this.
- **Random Ruby.** Ruby chooses her actions at random. Ruby is satisfied when she earns the greatest share of the money. Thus, playing with Ruby presents a risk because, even opponents know the probability with which she will cooperate, they do not know precisely when she will cooperate.
- **Vengeful Vince.** Vince will cooperate at first and continue cooperating as long as his opponent cooperates. However, if his opponent defects, he will take revenge by defecting in the next round. He always chooses the same actions his opponent

played in the previous round. This strategy is commonly known as 'Tit for Tat' (TFT).

Players are instructed that NPCs' satisfaction is an indicator of how likely they are to want to make friends. The NPC strategies above are deliberately simple so that human players can learn quickly how each NPC will behave during the tutorial phase of the mini-game. This phase will be described in detail in the next section. The responses of the human player are more likely to be deliberate rather than exploratory or curiosity motivated in the ensuing parts of the game. This is important for assessing achievement, affiliation and power motivation.

To demonstrate the differences of the four NPCs described above in terms of the way they accumulate money and satisfaction, we present a selection of charts. Each chart shows either the money or satisfaction of each NPC after 20 iterations of the PD game against theoretical human opponents with different probabilities of choosing to cooperate. These probabilities range from $P(C) = 0$ to $P(C) = 1$. The total money and satisfaction are shown in Figs. 4 and 5, respectively.

We can see from the figures that Cooperator Candy will have the highest satisfaction value against an opponent with a similar strategy. This is because both Candy and her opponent will earn a similar amount of money. Candy will have the lowest value of satisfaction against an opponent with high preference for playing defection because the opponent earns the greatest share of the money. An opponent may choose to exploit Candy to earn more money, but they will not maximise her satisfaction if they do this.

Defector Dan will have the highest money value and higher satisfaction value when he plays against an opponent who prefers cooperation, while the lowest money

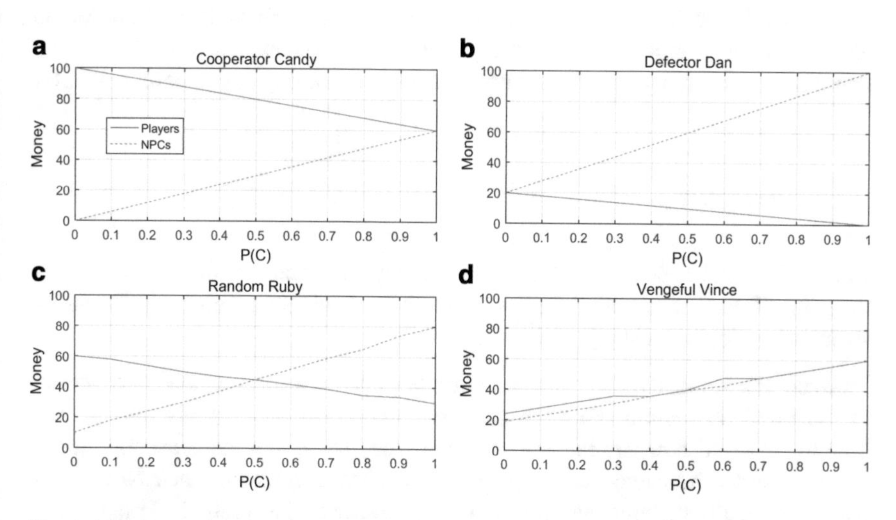

Fig. 4 Differences in money accumulated by each non-player character playing theoretical opponents with different probabilities of choosing to cooperate [6]. The money accumulated by the theoretical human player is also shown. (**a**) Cooperator Candy, (**b**) Defector Dan, (**c**) Random Ruby and (**d**) Vengeful Vince

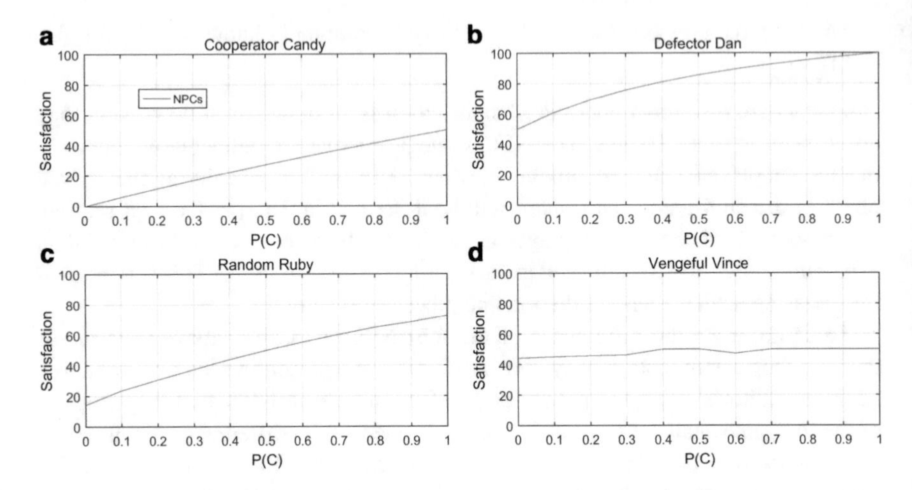

Fig. 5 Differences in final satisfaction of each non-player character playing theoretical opponents with different probabilities of choosing to cooperate [6]. (**a**) Cooperator Candy, (**b**) Defector Dan, (**c**) Random Ruby and (**d**) Vengeful Vince

value if his opponent prefers defection. An opponent may choose to be friend with Dan by playing cooperation; however, they will need to sacrifice their own earnings to do this.

Likewise, Random Ruby's satisfaction will be higher if her opponent chooses cooperation, but the opponent will again need to sacrifice their own earnings. Ruby will be less satisfied when opponents prefer to play defection.

Finally, Vince has generally moderate satisfaction values as a result of his adaptive strategy that permits him to perform well (in monetary terms) against opponents with a range of different preferences for choosing cooperation/defection.

2.4 Gameplay

Our mini-game has several phases in which players interact in different ways with the game. These are the tutorial phase (TP), the individual play phase (IPP) and the social network phase (SNP).

Tutorial Phase In the TP, players learn about each of the four NPCs described in the previous section. First, players need to follow instructions telling them how to play and how to read the user interface. Then, they have an opportunity to demonstrate their understanding of various characteristics of each NPC. Once they successfully demonstrate this, they can move onto the 'individual play' phase (IPP) of the game. A screenshot of the user interface from the tutorial phase is shown in Fig. 6. The TP is important to the game design because it ensures that players understand how the characters work and that they will not be learning

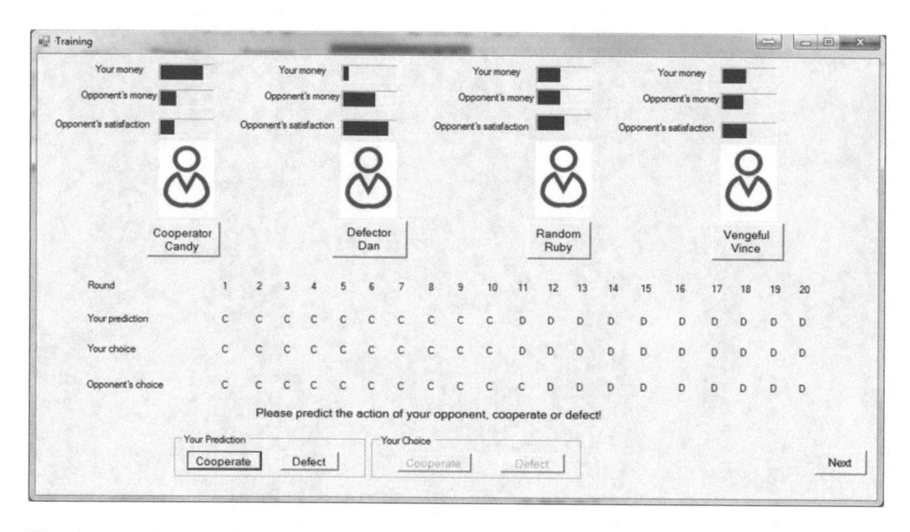

Fig. 6 Tutorial and individual play phase user interface

or experimenting during subsequent phases of the game when measurements are recorded.

Individual Play Phase In the IPP, players interact with each NPC individually. Players are instructed to play 20 rounds with each NPC in whatever way they find fun: 'for friends or fortune'. After they play with each character, they do a short survey to rate their emotion on a three-point scale (positive, neutral or negative). The user interface for the IPP is shown in Fig. 6, with an additional screen to collect player emotion information. The IPP is important to the game design as it gathers behaviour data necessary for understanding achievement motivation in a minimal social setting (one-vs-one play). The responses of NPCs in the IPP were described in the previous section.

Social Network Phase In the SNP, players first build a social network comprising their selection of up to eight instances of the four NPCs they met in the previous phases of the game. They then play 20 rounds of the game with their network. Players score points in a pairwise fashion (that is, this part of the game is not an n-player PD). The user interface for the SNP is shown in Fig. 7. In the SNP, all NPCs' satisfaction is further amplified by the size of the social network they belong to. This is implemented as follows: a factor $N/8$ is incorporated into the equation to represent the influence of network size on network satisfaction. N is the number of selected opponents in the network.

$$E_t^{NPC} = S_t^{NPC}(1 + N/8) \tag{3}$$

where E_t^{NPC} is the satisfaction value displayed on the game interface for each NPC, and S_t^{NPC} is the satisfaction value calculated by the definition for each NPC. The

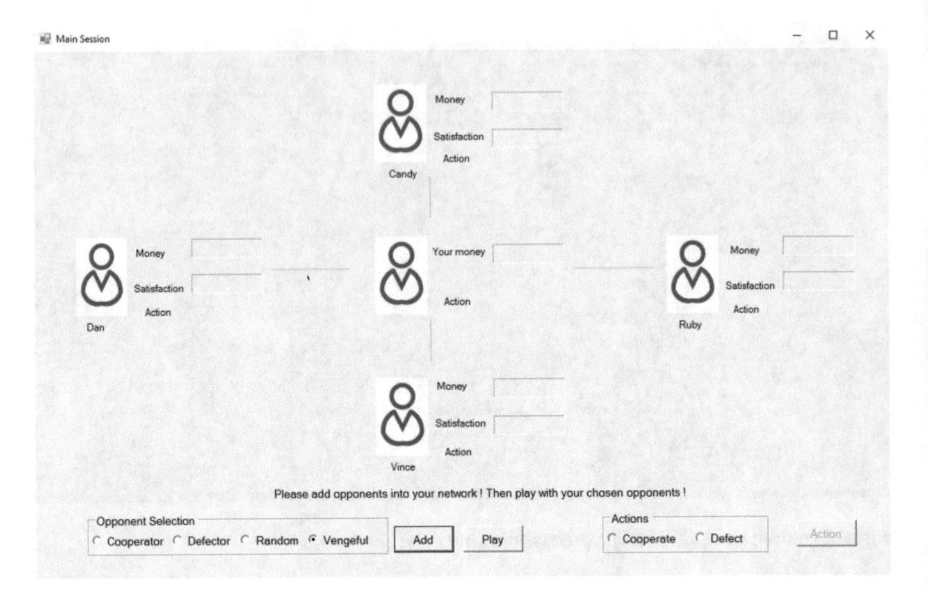

Fig. 7 Social network phase user interface

network satisfaction is also updated using Eq. (2) to smooth the changes of the value throughout the game.

The proposed abstract mini-game has characteristics that make it potentially suitable for use within a commercial game. For example, it has the potential for application in massively multi-player online role-playing games (MMORPGs). In such games, players control avatars to interact with NPCs. Following a storyline and game mechanics, players can control and interact with game elements and receive feedback regularly. The scenario in which players interact with our specially designed NPCs could be appropriately isolated in a larger game through terrain conditions or levelling constraints.

However, currently, our mini-game is designed to be simple, in order to control the variables that may influence player motive profiling. Commercial games generally have clear goals, like winning money/points, and however, the goal of our proposed game is defined by players. They choose how to play the game for earning money, building friendships and trade-offs between these. Moreover, the environment or interface is designed simply without any visual aesthetic, which is also controlled compared to standard games. Furthermore, the dynamics of the mini-game are controlled with a limited level of unpredictable elements (e.g. the Random Ruby NPC). In the next section, a human user study is performed to examine the effectiveness of the proposed game design for assessing player motive profile.

3 Experimental Validation

We proposed non-player characters (NPCs) to suppose player motive profiling by detecting cognitive and emotional phenomena. Our NPCs have different play strategies, money and satisfaction features. The play strategies of NPCs assist in the distinction between different player motive profiles. The money and satisfaction features of NPCs have the potential to evoke risk-taking and social attitudes, respectively. In addition, the interactions between players and NPCs in the IPP and SNP are hypothesised to reveal their risk-taking and social attitudes differently to further help identify player motive profiles. In order to validate the game design, an experiment incorporating the game was performed to collect player behaviour and EEG data from the game. The multi-motive grid (MMG) test was incorporated in the experiment for obtaining the ground truth of player motive profiles. Those data are analysed in this section to examine the design of the NPCs and game scenario for identifying player motive profiles and related risk-taking and social attitudes.

Section 3.1 of this chapter analyses the effectiveness of the play strategies of NPCs for identifying motivation variables using the IPP and SNP. Section 3.2 presents an analysis of the two features of our NPCs (money and satisfaction) for assessing motivation via EEG signals. The effectiveness of the IPP and SNP for assessing motivation is summarised in Sect. 3.3.

3.1 Analysing Play Strategies of Non-player Characters for Assessing Motivation

The aim of this section is to assess whether the play strategies of NPCs in our mini-game are appropriate for differentiating player behaviour that ultimately contributes to player motive profiling. To achieve the goal, we aim to assess whether different play strategies of NPCs in different game phases (individual play and social network play) contribute to differentiation between player motive profiles.

We hypothesise that play strategies of NPCs in the mini-game are appropriate for differentiating between player motive profiles. Specifically, Cooperator Candy is proposed for cooperation by achievement and affiliation motivated players and for exploitation by power motivated players. Vengeful Vince is proposed for cooperation by achievement and affiliation motivated players. Random Ruby's strategy represents a risk-taking challenge potentially preferred by power motivated players. Defector Dan, on the other hand, may be preferred by affiliation-motivated players for cooperation. In addition, we hypothesise that NPCs incorporated in both the IPP and the SNP are appropriate for distinguishing player motive profiles.

Multiple linear regression is used to model motivation from player behaviour with NPCs. Motivation data from the MMG test have six motivation variables, shown in the first column of Table 1, which are regarded as the response. The requirement for using a linear regression model is that the response should be

Table 1 Characteristics of regression models using player behaviour in the individual play phase to predict motivation profile

	P-value	R^2	MSE (%)	Candy	Dan	Ruby	Vince
Hope for success	<0.05	0.3	19.3	✓	✓		
Fear of failure	<0.05	0.6	13.9		✓	✓	✓
Hope for control	>0.1	0.3	17.7				
Fear of loss of control	<0.05	0.2	17.4			✓	
Hope for social acceptance	<0.05	0.5	12.8	✓	✓		✓
Fear of rejection	<0.1	0.2	24.2	✓			

normally distributed. Thus, we used the Kolmogorov–Smirnov test (KS test) to examine the distribution of our six motivational variables from the MMG test. The results showed that these six variables have normal distributions, which means it is appropriate to employ linear regression for motivation modelling. An experiment was performed between player behaviour in the IPP and motivation variables. This helps us to examine whether the design of the IPP is appropriate for player motive profiling.

The mixed stepwise method was used to select relevant predictors. Modelling mainly focused on linear terms, and if these could not represent the relationships, then interaction and quadratic terms were added. All the important variables were included in the model in terms of the principle of regression. The probability of cooperation with Cooperator Candy, Defector Dan, Random Ruby and Vengeful Vince was input into the regression model as the predictors. Due to the different scale between motivational variance and the probability of cooperation, the probability of cooperation was multiplied by 100.

Twenty-three subjects took part in the game, and their gameplay behaviour was analysed. As shown in Table 1, the p values indicate that most of the motivation regression models are statistically significant except for the hope for control variable. This means that the hope for control variable cannot be learned from player behaviour in the IPP. The R^2 show the variability of motivation values that can be explained by the player behaviour in the IPP. We can see from Table 1 that fear of failure and hope for social acceptance have the highest R^2 value with both above 0.5. However, for fear of loss of control and fear of rejection, the R^2 values are relatively small with the value around 0.2. This may indicate that our design of the IPP can reveal fear of failure and hope for social acceptance the best among the six variables, while fear of loss of control and fear of rejection are revealed the least well and the hope for success falls in the middle.

In addition, the MSE shows the random error that is included in the regression model when fitting the data. This is the other indicator that demonstrates how the regression model fits the data and how our game reveals players' motivation values. Table 1 shows that five motivation variables have mean MSE values less than 20%, which indicates the regression models fit the data properly. We indicate the characters that are included in each of the regression models with a tick in Table 1. Player behaviour with Candy and Dan contributed to hope for success, and player

Table 2 Characteristics of regression models using player behaviour in the social network phase to predict motivation profile

	P-value	R^2	MES(%)	Candy	Dan	Ruby	Vince
Hope for success	<0.05	0.2	20.7		✓		
Fear of failure	=0.05	0.5	15.7	✓		✓	✓
Hope for control	<0.005	0.2	20.8		✓		
Fear of loss of control	<0.01	0.3	16.6	✓	✓		
Hope for social acceptance	<0.05	0.4	14.4		✓	✓	
Fear of rejection	<0.05	0.3	25.8				✓

behaviour with Dan, Ruby and Vince related to fear of failure. For fear of loss of control, player behaviour with Ruby was significant. Player behaviour with Candy, Dan and Vince was relevant to hope for social acceptance, whereas fear of rejection depended on player behaviour with Candy.

In the SNP, players can decide how to organise their social networks by choosing to incorporate their preferred characters from Cooperator Candy, Defector Dan, Random Ruby and Vengeful Vince. The regression was performed between the percentage of different characters in subjects' networks and the six MMG motivation variables. Table 2 shows statistical characteristics of the regression models. All of the p values are less than or around 0.05 (except fear of loss of control that is $p < 0.01$), which means the relationships between player behaviour in the SNP and motivation values are statistically significant. The R^2 of fear of failure and hope for social acceptance are higher than the remaining motivation variables, which indicates that the SNP reveals fear of failure and hope for social acceptance more successfully. Hope for success and hope for control have R^2 values around 0.2, which may mean our game cannot assess these two variables very well.

We conclude with the player variables that are included in each of the SNP regression models in Table 2. Fear of failure has player behaviour with three characters included in the regression models, fear of loss of control and hope for social acceptance have two variables, while the hope for success, hope for control and fear of rejection only have one variable.

According to MSE, as shown in Table 2, column 4, hope for success, hope for control and fear of rejection have MSE above 20%. In particular, fear of rejection has 25% MSE, which demonstrates that the SNP may not reveal motivation as well as the IPP. For the rest of the motivation variables, the MSEs are around 15%, suggesting a better model fit.

These results with player behaviour identify the importance of using different NPCs to assess player motive profiles. However, there are still several motivation variables that cannot be learned from the results. One reason for this is perhaps our game design requires further improvements. Another reason may be that the player behaviour provides limited information for assessing player motivation, while the advantage of EEG-based motive measurement is that EEG signals provide continuous and rich information about the human mind. In the next section, we use EEG signals to examine the money and satisfaction features of NPCs in order to assess the mental states of player motive profiles.

3.2 Analysing Features of Non-player Characters for Assessing Motivation

The aim of this experiment is to examine how money and satisfaction features are able to reveal risk-taking and social attitudes, which are key aspects of achievement, affiliation and power motivation. Our NPCs have two features: money and satisfaction. The money feature represents a tangible, valuable item in our daily life. It is treated as monetary rewards in the mini-game, which is hypothesised to assist with the differentiation of risk-taking attitudes of players with different motives. The satisfaction feature, on the other hand, is associated with social satisfaction about interactions between players and NPCs. It is treated as a state of friendship in the mini-game. We hypothesise that it will enable us to distinguish between social attitudes of players with different motives.

In order to understand mental states, in particular actions and responses, we divided the EEG analysis into specific 500 ms time windows for action, money feedback and satisfaction feedback (as the event-related potentials are usually lasting around 300–500 ms). Specifically, the action starts from 0 s to 500 ms, money feedback occurs from 600 ms (including average system delay) to 1.1 s, and satisfaction feedback occurs from 1.2 to 1.7 s in each trial. Features were extracted from these time windows without overlapping for analysing risk-taking and social attitudes when players made actions and received money and satisfaction feedbacks. Features extracted from EEG signals can be categorised into three groups: temporal, time–frequency and asymmetry features.

Time–frequency features were calculated by complex wavelet transformation, and decibel normalisation was utilised. Time–frequency features were divided into five frequency bands: alpha/mu band (8–12 Hz), theta band (4–7 Hz), beta band (13–31 Hz) and gamma band (32–42 Hz). According to the international 10–10 system, the frontal region is electrodes FP1, FP2, FPz, AFP5, AFP6, AF7, AF3, AF1, AFz, AF2, AF4, AF6, AF8, F7, F5, F3, F1, Fz, F2, F4, F6 and F8; the pre-frontal region is electrodes FP1, FP2, FPz, AF7, AF5, AF3, AF1, AFz, AF2, AF4, AF6 and AF8; the ACC is electrodes Fz, FCz and Cz; the parietal region is PO7, PO5, PO3, PO1, POz, PO2, PO4, PO6, PO8, POO7, POO8, O1, O2 and Oz and the temporal region is FT7, T7, TP7, FT8, T8 and TP8.

Asymmetry features were derived from the differences between the left and right hemispheres, and thus a positive value means greater left than right, and a negative value means greater right than left. The electrode pairs for frontal alpha are FP1-FP2, AFP5-AFP6, AF7-AF8, AF5-AF6, AF3-AF4, AF1-AF2, F7-F8, F5-F6, F3-F4, F1-F2, FT7-FT8, FC5-FC6, FC3-FC4, FC1-FC2 and T7-F8 and for prefrontal alpha are FP1-FP2, AFP5-AFP6, AF7-AF8, AF5-AF6, AF3-AF4 and AF1-AF2.

EEG data were epoched in three types of segment depending on the temporal features being analysed. For the analysis of event-related negativity (ERN), the mean amplitude was calculated in segments of 100 ms from 0 to 100 ms time-locked to the onset of money and satisfaction feedbacks. For medial frontal negativity (MFN), we derived mean amplitude from 200 to 300 ms time-locked to the action

Table 3 Features extracted for identifying mental indicators of motive profile

Mental state	Feature	Brain region	Category
Social	Mu band	ACC	Time–frequency feature
	Alpha band	Frontal	Asymmetry
Risk	Alpha band	Pre-frontal	Asymmetry
	Theta–beta ratio	ACC	Time–frequency feature
	ERN	ACC	Temporal
	MFN	ACC	Temporal
	P300	ACC	Temporal

Table 4 Hypothesis about using the conceptual model in Fig. 1 to evaluate risk-taking and social attitudes for each motivation related to the behaviour. The important part of the game is selected from the regression analysis of in-game behaviour developed in Sect. 3.1

Motivation	Important part of game		Attitude	Features
Achievement	*HL*	P_C P_D	Low social	High A_{mu}, greater right than left F_α
	LH	P_D P_R P_V	Medium risk	Medium A_{TBR}, medium PF_α, medium A_{MFN}, medium A_{ERN} and medium A_{P300}
Affiliation	*HL*	P_C P_D P_V	High social	Low A_{mu}, greater left than right F_α
	LH	P_C	Low risk	Low A_{TBR}, right higher PF_α, low A_{MFN}, high A_{ERN} and low A_{P300}
Power	*HL*		Medium social	Medium A_{mu} and medium F_α
	LH	P_R	High risk	High A_{TBR}, right higher PF_α, high A_{MFN}, low A_{ERN} and high A_{P300}

and feedbacks onset, while P300 was also calculated as the mean value of the time segment of 300 and 400 ms to the feedback onset. Following the EEG-based risk-taking literature, all temporal features were computed in the ACC area (Fz, FCz and Cz).

As presented in Table 3, the mu band in the ACC and frontal alpha asymmetry were computed to be indicators of social attitudes. Prefrontal alpha asymmetry, theta–beta ratio, ERN, MFN and P300 in ACC were calculated as EEG features for risk-taking attitudes.

According to the regression analysis of motivation and player behaviour in the IPP, play with certain NPCs was evaluated to be more relevant to the H-L or L-H situation of each motivation. We summarise this in Table 4. As shown in Table 4, play with Candy and Dan was chosen to study H-L achievement, while the study of L-H achievement used the play with Dan, Ruby and Vince. For affiliation, our study

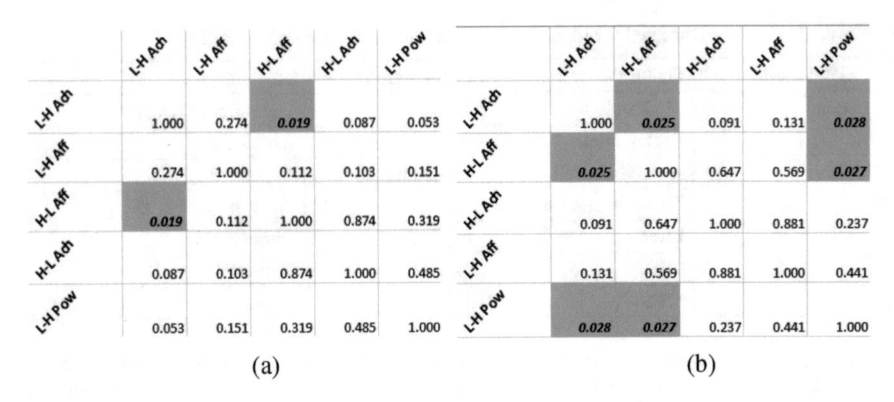

(a)

	L-H Ach	L-H Aff	H-L Aff	H-L Ach	L-H Pow
L-H Ach	1.000	0.274	*0.019*	0.087	0.053
L-H Aff	0.274	1.000	0.112	0.103	0.151
H-L Aff	*0.019*	0.112	1.000	0.874	0.319
H-L Ach	0.087	0.103	0.874	1.000	0.485
L-H Pow	0.053	0.151	0.319	0.485	1.000

(b)

	L-H Ach	H-L Aff	H-L Ach	L-H Aff	L-H Pow
L-H Ach	1.000	*0.025*	0.091	0.131	*0.028*
H-L Aff	*0.025*	1.000	0.647	0.569	*0.027*
H-L Ach	0.091	0.647	1.000	0.881	0.237
L-H Aff	0.131	0.569	0.881	1.000	0.441
L-H Pow	*0.028*	*0.027*	0.237	0.441	1.000

Fig. 8 Results of using ACC mu band in (**a**) money and (**b**) satisfaction feedbacks to reveal social attitude. The p-value of independent t-test between hope and fear components of each motivation is presented. Grey squares indicate significant differences

of the H-L component used play with Candy, Dan and Vince, while only play with Candy contributed to the study of the L-H component. According to the regression results, no player behaviour in the IPP explained H-L power, while play with Ruby contributed to L-H power.

Overall, we evaluated risk-taking and social behaviours using the EEG recordings from play in the IPP. On the basis of the NPCs where player behaviour related to motive profiles, we first evaluated the EEG signals from selected NPCs to see if there was any significant impact on motive profile according to our proposed conceptual model. Later, we synthesised EEG signals to explore the effectiveness of each NPC to study mental states with motive profiles.

To examine the difference between EEG features among different motive profiles, we used the independent t-test to determine the significance of the differences between EEG mental indicators for the H-L and L-H situations of each motivation. A t-test is a statistic that checks if two means (averages) are reliably different from each other, and independent t-test tests the means of two different groups. Only the H-L and L-H situations of achievement, affiliation and power motivation are evaluated pairwise in this chapter. Differences are identified when the p-value of the t-test is less than 0.05 ($p < 0.05$).

We first determined the performance of using selected EEG features to predict social attitudes, when participants play with NPCs chosen from regression models. Figures 8 and 9 show the performance of using EEG features to predict social attitude when players receive money feedback and satisfaction feedback. We found that social attitude is expressed more often by EEG signals collected during receipt of satisfaction feedback than those collected during receipt of money feedback. Furthermore, the ACC mu band and frontal alpha asymmetry performed almost equally to indicate social attitude.

We also explored the use of EEG features to predict risk-taking attitude when players receive money and satisfaction feedbacks. As shown in Figs. 10, 11 and 12,

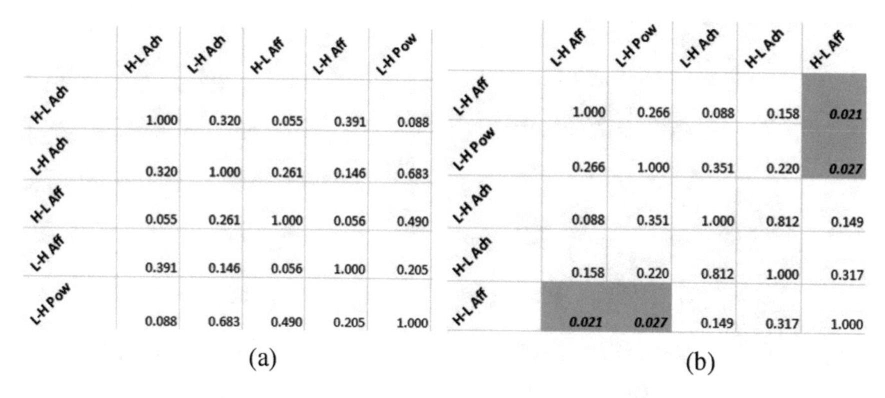

(a)

	H-L Ach	L-H Ach	H-L Aff	L-H Aff	L-H Pow
H-L Ach	1.000	0.320	0.055	0.391	0.088
L-H Ach	0.320	1.000	0.261	0.146	0.683
H-L Aff	0.055	0.261	1.000	0.056	0.490
L-H Aff	0.391	0.146	0.056	1.000	0.205
L-H Pow	0.088	0.683	0.490	0.205	1.000

(b)

	L-H Aff	L-H Pow	L-H Ach	H-L Ach	H-L Aff
L-H Aff	1.000	0.266	0.088	0.158	*0.021*
L-H Pow	0.266	1.000	0.351	0.220	*0.027*
L-H Ach	0.088	0.351	1.000	0.812	0.149
H-L Ach	0.158	0.220	0.812	1.000	0.317
H-L Aff	*0.021*	*0.027*	0.149	0.317	1.000

Fig. 9 Results of using frontal alpha asymmetry in (**a**) money and (**b**) satisfaction feedbacks to reveal social attitude. The p-value of independent t-test between hope and fear components of each motivation is presented. Grey squares indicate significant differences

(a)

	L-H Aff	H-L Ach	H-L Aff	L-H Ach	L-H Pow
L-H Aff	1.000	0.317	*0.039*	*0.010*	*0.009*
H-L Ach	0.317	1.000	0.120	*0.012*	*0.007*
H-L Aff	*0.039*	0.120	1.000	*0.031*	*0.015*
L-H Ach	*0.010*	*0.012*	*0.031*	1.000	0.693
L-H Pow	*0.009*	*0.007*	*0.015*	0.693	1.000

(b)

	H-L Ach	L-H Ach	H-L Aff	L-H Aff	L-H Pow
H-L Ach	1.000	0.751	0.248	0.451	0.815
L-H Ach	0.751	1.000	0.401	0.674	0.619
H-L Aff	0.248	0.401	1.000	0.826	0.274
L-H Aff	0.451	0.674	0.826	1.000	0.529
L-H Pow	0.815	0.619	0.274	0.529	1.000

Fig. 10 Results of using ACC ERN in (**a**) money and (**b**) satisfaction feedbacks to reveal risk-taking attitude. The p-value of independent t-test between hope and fear components of each motivation is presented. Grey squares indicate significant differences

EEG features, ACC ERN, MFN and P300, from money feedback outperformed those features from satisfaction feedback. Also, prefrontal alpha asymmetry can distinguish only two different types of motive situation for both money and satisfaction feedbacks (shown in Fig. 13). In conclusion, the EEG features suggested by the literature all appear to have possibilities for predicting risk-taking attitude, except for ACC TBR (see Fig. 14).

By analysing social attitude using EEG features collected during interaction with each NPC, we summarise the performance of using NPCs to reveal social attitudes in Fig. 15. Data collected during action against and satisfaction feedback from Candy performed worse than data collected during those interactions with the other three characters. Data collected during action against and satisfaction feedback from Vince had better performance than those data collected from other characters.

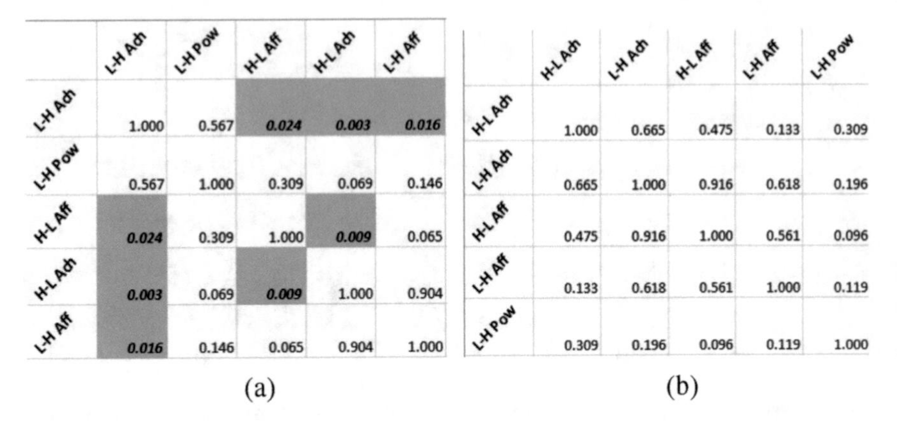

(a)

	L-H Ach	L-H Pow	H-L Aff	H-L Ach	L-H Aff
L-H Ach	1.000	0.567	*0.024*	*0.003*	*0.016*
L-H Pow	0.567	1.000	0.309	0.069	0.146
H-L Aff	*0.024*	0.309	1.000	*0.009*	0.065
H-L Ach	*0.003*	0.069	*0.009*	1.000	0.904
L-H Aff	*0.016*	0.146	0.065	0.904	1.000

(b)

	H-L Ach	L-H Ach	H-L Aff	L-H Aff	L-H Pow
H-L Ach	1.000	0.665	0.475	0.133	0.309
L-H Ach	0.665	1.000	0.916	0.618	0.196
H-L Aff	0.475	0.916	1.000	0.561	0.096
L-H Aff	0.133	0.618	0.561	1.000	0.119
L-H Pow	0.309	0.196	0.096	0.119	1.000

Fig. 11 Results of using ACC MFN in (**a**) money and (**b**) satisfaction feedbacks to reveal risk-taking attitude. The p-value of independent t-test between hope and fear components of each motivation is presented. Grey squares indicate significant differences

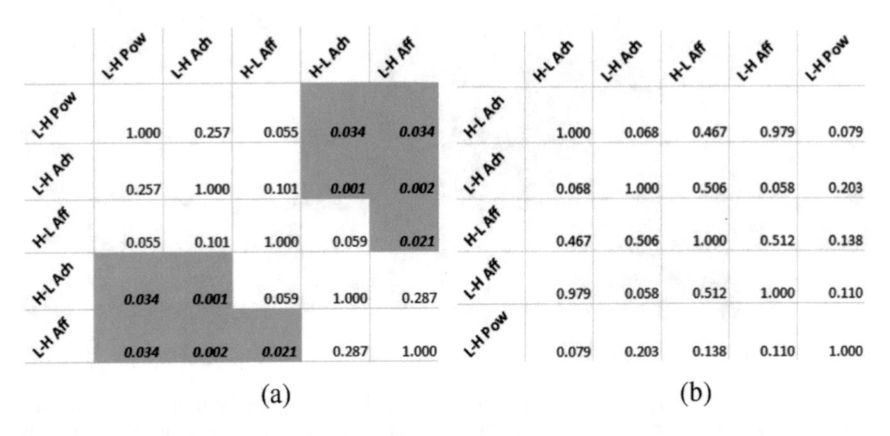

(a)

	L-H Pow	L-H Ach	H-L Aff	H-L Ach	L-H Aff
L-H Pow	1.000	0.257	0.055	*0.034*	*0.034*
L-H Ach	0.257	1.000	0.101	*0.001*	*0.002*
H-L Aff	0.055	0.101	1.000	0.059	*0.021*
H-L Ach	*0.034*	*0.001*	0.059	1.000	0.287
L-H Aff	*0.034*	*0.002*	*0.021*	0.287	1.000

(b)

	H-L Ach	L-H Ach	H-L Aff	L-H Aff	L-H Pow
H-L Ach	1.000	0.068	0.467	0.979	0.079
L-H Ach	0.068	1.000	0.506	0.058	0.203
H-L Aff	0.467	0.506	1.000	0.512	0.138
L-H Aff	0.979	0.058	0.512	1.000	0.110
L-H Pow	0.079	0.203	0.138	0.110	1.000

Fig. 12 Results of using ACC P300 in (**a**) money and (**b**) satisfaction feedbacks to reveal risk-taking attitude. The p-value of independent t-test between hope and fear components of each motivation is presented. Grey squares indicate significant differences

Data collected during satisfaction feedback from Ruby had better performance and data collected during money feedback from Dan had slightly better performance. Overall, the results indicate that Ruby and Vince are good choices for evoking social attitudes, while Candy is not appropriate for achieving this goal.

As shown in Fig. 16, EEG signals collected during action, money and satisfaction feedbacks from Candy do not allow us to significantly differentiate the player motive profile. Data that was collected when participants were taking action against Dan and Vince gave the best performance. In contrast, data that was collected during money feedback from Dan gave the best performance, followed by data collected during satisfaction feedback from Ruby gave the best performance. Overall, Candy is also the least useful character for revealing risk-taking attitudes, while Dan, Ruby

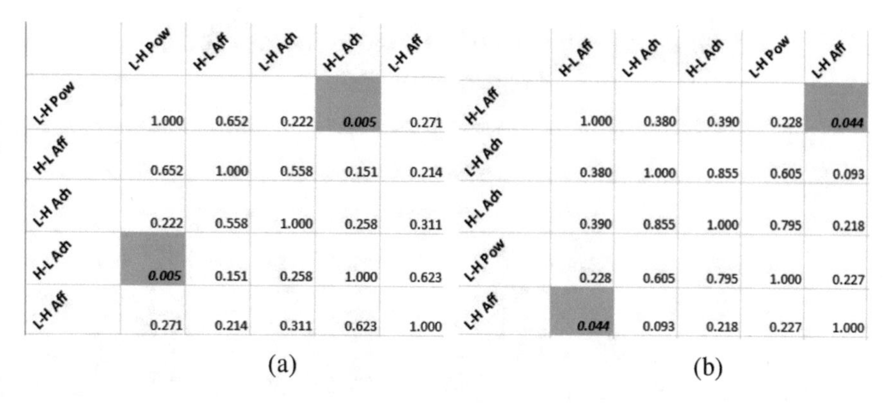

	L-H Pow	H-L Aff	L-H Ach	H-L Ach	L-H Aff
L-H Pow	1.000	0.652	0.222	*0.005*	0.271
H-L Aff	0.652	1.000	0.558	0.151	0.214
L-H Ach	0.222	0.558	1.000	0.258	0.311
H-L Ach	*0.005*	0.151	0.258	1.000	0.623
L-H Aff	0.271	0.214	0.311	0.623	1.000

(a)

	H-L Aff	L-H Ach	H-L Ach	L-H Pow	L-H Aff
H-L Aff	1.000	0.380	0.390	0.228	*0.044*
L-H Ach	0.380	1.000	0.855	0.605	0.093
H-L Ach	0.390	0.855	1.000	0.795	0.218
L-H Pow	0.228	0.605	0.795	1.000	0.227
L-H Aff	*0.044*	0.093	0.218	0.227	1.000

(b)

Fig. 13 Results of using prefrontal alpha asymmetry in (**a**) money and (**b**) satisfaction feedbacks to reveal risk-taking attitude. The *p*-value of independent *t*-test between hope and fear components of each motivation is presented. Grey squares indicate significant differences

	H-L Ach	L-H Ach	L-H Pow	H-L Aff	L-H Aff
H-L Ach	1.000	0.247	0.229	0.665	0.376
L-H Ach	0.247	1.000	0.778	0.108	0.731
L-H Pow	0.229	0.778	1.000	0.143	0.674
H-L Aff	0.665	0.108	0.143	1.000	0.210
L-H Aff	0.376	0.731	0.674	0.210	1.000

(a)

	H-L Ach	L-H Ach	L-H Pow	H-L Aff	L-H Aff
H-L Ach	1.000	0.194	0.145	0.361	0.719
L-H Ach	0.194	1.000	0.221	0.136	0.225
L-H Pow	0.145	0.221	1.000	0.455	0.159
H-L Aff	0.361	0.136	0.455	1.000	0.364
L-H Aff	0.719	0.225	0.159	0.364	1.000

(b)

Fig. 14 Results of using ACC TBR in (**a**) money and (**b**) satisfaction feedbacks to reveal risk-taking attitude. The *p*-value of independent *t*-test between hope and fear components of each motivation is presented. Grey squares indicate significant differences

and Vince are appropriate choices for evoking risk-taking attitudes when players make actions and receive money and satisfaction feedbacks.

EEG analysis of the IPP demonstrates that EEG signals collected during satisfaction feedback indicate social attitude, while EEG signals collected during money feedback appear to reveal risk-taking attitude. The results demonstrate the effectiveness of the NPCs, two dimensions of money and satisfaction at reflecting players' risk-taking and social attitudes.

In addition, we hypothesised that depending on their design, different NPCs would reveal the risk-taking and social attitudes of different motivated players. After analysing the effectiveness of using each NPC in the IPP for revealing risk-taking and social attitudes, we found using EEG signals that Candy is the least useful character for learning human risk-taking and social attitudes. Ruby and Vince are

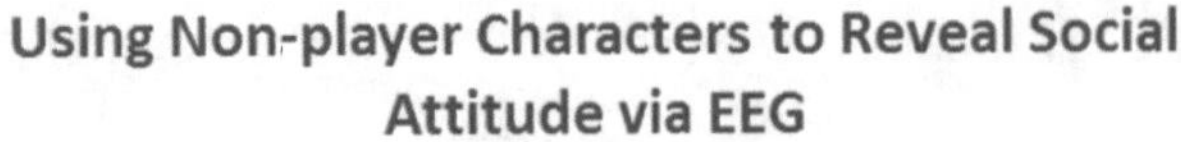

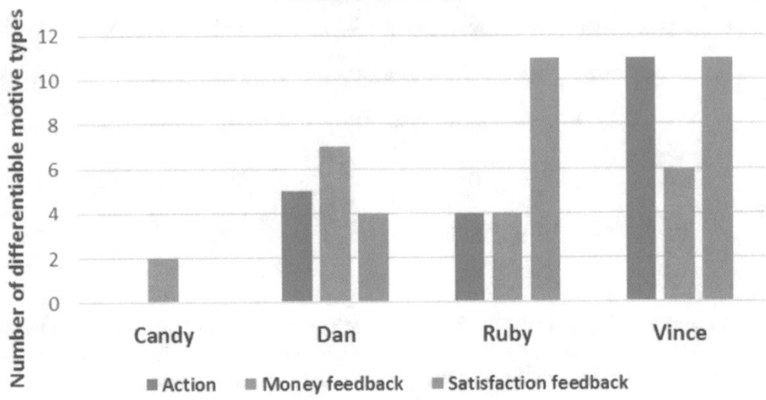

Fig. 15 Using NPCs to reveal social attitude via EEG analysis

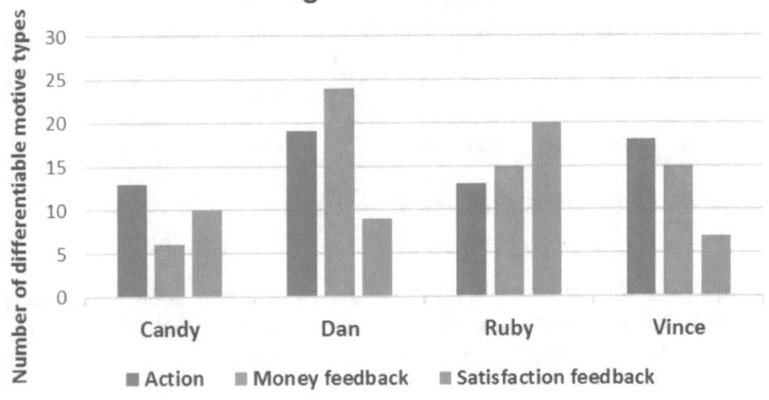

Fig. 16 Using NPCs to reveal risk-taking attitude via EEG analysis

found to be good choices for evoking social attitude, while Dan, Ruby and Vince are found to be useful for evoking risk-taking attitudes.

Candy emerged as the weakest NPC for assessing risk attitude. This could be because players cannot be defected against by Candy, regardless of their actions. Candy always cooperates, which gives an impression of no risk during the interaction. Moreover, Candy does not appear to evoke social attitude, probably because Candy cannot be satisfied even if players cooperate with her (see Fig. 4). This means that players are reluctant to pursue 'friendships' with Candy. In future alternative play strategies should be investigated to reveal more about players.

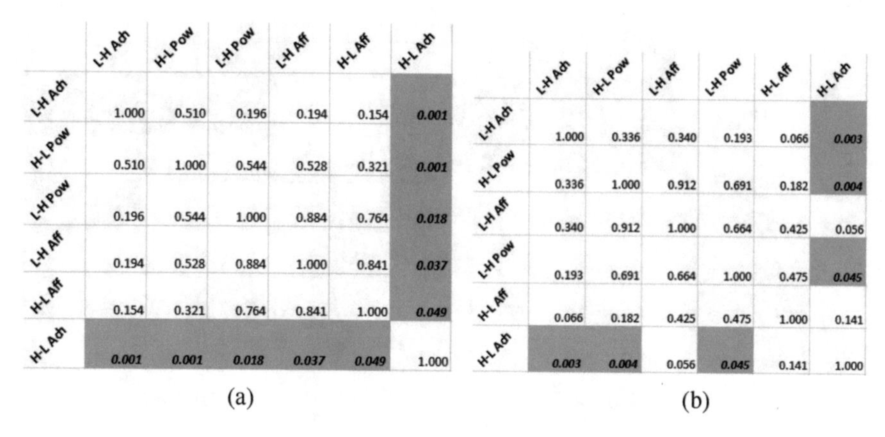

Fig. 17 Results of using ACC mu band in the whole SNP (**a**) and during action (**b**) to reveal social attitude. The p-value of independent t-test between hope and fear components for each motivation is presented. Grey squares indicate significant differences

3.3 Analysing Individual Play and Social Network Phases for Assessing Motivation

The aim of this experiment is to compare the performance of the IPP and SNP in assessing player motivation. Firstly, we analyse the use of the SNP for identifying mental indicators of player motive profiles. Then, we compare the performance of the IPP and SNP by summarising some aforementioned findings.

We hypothesise that both the IPP and the SNP can be used for identifying risk-taking and social attitudes of different player motive profiles.

The corresponding EEG features were extracted for risk-taking and social attitudes (as shown in Table 3). To identify feature differences between motive profiles, an independent t-test was applied and the threshold for statistical significance is that the p-value is less than 0.05 ($P < 0.05$). However, as there is no latency time for money feedback and satisfaction feedback in the SNP (too many stimuli in the interface), we focused our EEG signal analysis during participant actions (time from 0 s to 500 ms) and each trial. This is a preliminary study to explore the possibility of using social network play to evoke risk-taking and social attitudes.

In the SNP, we first assessed the performance of using EEG features to distinguish differences in social attitudes between the H-L and L-H situations of each motivation. Figures 17 and 18 indicate that EEG signals from the SNP have the potential to reveal social attitude. In particular, we focus on the time interval when players make their actions as shown in Figs. 17 and 18. This illustrates that players with different motive profiles expressed different social attitudes when they played against their social network.

We also examined the EEG features from the SNP for assessing risk-taking attitudes. Figures 19, 20, 21 and 22 show that the design of SNP has the potential to evoke risk-taking attitudes via EEG signals. However, the ACC TBR feature (see

(a)

	L-H Aff	H-L Ach	L-H Ach	H-L Pow	H-L Aff	L-H Pow
L-H Aff	1.000	0.367	*0.000*	*0.000*	*0.000*	*0.000*
H-L Ach	0.367	1.000	*0.034*	*0.006*	*0.002*	*0.026*
L-H Ach	*0.000*	*0.034*	1.000	0.698	0.316	0.429
H-L Pow	*0.000*	*0.006*	0.698	1.000	0.529	0.600
H-L Aff	*0.000*	*0.002*	0.316	0.529	1.000	0.928
L-H Pow	*0.000*	*0.026*	0.429	0.600	0.928	1.000

(b)

	L-H Aff	H-L Ach	L-H Pow	L-H Ach	H-L Aff	H-L Pow
L-H Aff	1.000	0.362	*0.010*	*0.011*	*0.000*	*0.000*
H-L Ach	0.362	1.000	0.672	0.452	0.096	0.063
L-H Pow	*0.010*	0.672	1.000	0.624	*0.017*	*0.047*
L-H Ach	*0.011*	0.452	0.624	1.000	0.150	0.105
H-L Aff	*0.000*	0.096	*0.017*	0.150	1.000	0.691
H-L Pow	*0.000*	0.063	*0.047*	0.105	0.691	1.000

Fig. 18 Results of using frontal alpha asymmetry in the whole SNP (**a**) and during action (**b**) to reveal social attitude. The p-value of independent t-test between hope and fear components for each motivation is presented. Grey squares indicate significant differences

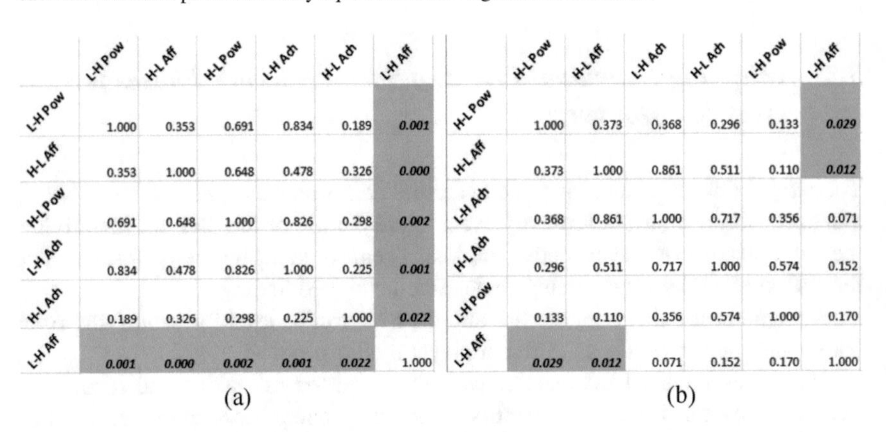

(a)

	L-H Pow	H-L Aff	H-L Pow	L-H Ach	H-L Ach	L-H Aff
L-H Pow	1.000	0.353	0.691	0.834	0.189	*0.001*
H-L Aff	0.353	1.000	0.648	0.478	0.326	*0.000*
H-L Pow	0.691	0.648	1.000	0.826	0.298	*0.002*
L-H Ach	0.834	0.478	0.826	1.000	0.225	*0.001*
H-L Ach	0.189	0.326	0.298	0.225	1.000	*0.022*
L-H Aff	*0.001*	*0.000*	*0.002*	*0.001*	*0.022*	1.000

(b)

	H-L Pow	H-L Aff	L-H Ach	H-L Ach	L-H Pow	L-H Aff
H-L Pow	1.000	0.373	0.368	0.296	0.133	*0.029*
H-L Aff	0.373	1.000	0.861	0.511	0.110	*0.012*
L-H Ach	0.368	0.861	1.000	0.717	0.356	0.071
H-L Ach	0.296	0.511	0.717	1.000	0.574	0.152
L-H Pow	0.133	0.110	0.356	0.574	1.000	0.170
L-H Aff	*0.029*	*0.012*	0.071	0.152	0.170	1.000

Fig. 19 Results of using prefrontal alpha asymmetry in the whole SNP (**a**) and during action (**b**) to reveal risk-taking attitude. The p-value of independent t-test between hope and fear components for each motivation is presented. Grey squares indicate significant differences

in Fig. 14) was unable to differentiate between the H-L and L-H situations of any motive. By examining players' actions in the SNP, we can see from Figs. 19, 20, 21, 22 and 23 that players with different motive types had different levels of risk-taking attitudes when playing against their social network.

The results indicate that EEG signals collected from the SNP can be used as an indicator for identifying player motivation, which supports our hypothesis. Moreover, we found in Sect. 3.2 that players' attitudes towards risk-taking can be revealed with the use of the money feature, while players' social attitude can be learned from the satisfaction feature. Because of the absence of latency of money and satisfaction feedbacks in the SNP, there was not enough information to distinguish aspects of achievement, affiliation and power motivation. In future

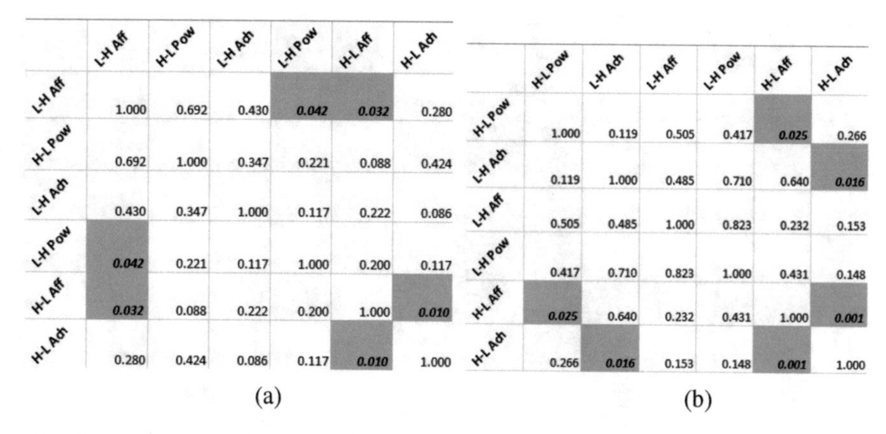

(a)

	L-H Aff	H-L Pow	L-H Ach	L-H Pow	H-L Aff	H-L Ach
L-H Aff	1.000	0.692	0.430	***0.042***	***0.032***	0.280
H-L Pow	0.692	1.000	0.347	0.221	0.088	0.424
L-H Ach	0.430	0.347	1.000	0.117	0.222	0.086
L-H Pow	***0.042***	0.221	0.117	1.000	0.200	0.117
H-L Aff	***0.032***	0.088	0.222	0.200	1.000	***0.010***
H-L Ach	0.280	0.424	0.086	0.117	***0.010***	1.000

(b)

	H-L Pow	L-H Ach	L-H Aff	L-H Pow	H-L Aff	H-L Ach
H-L Pow	1.000	0.119	0.505	0.417	***0.025***	0.266
L-H Ach	0.119	1.000	0.485	0.710	0.640	***0.016***
L-H Aff	0.505	0.485	1.000	0.823	0.232	0.153
L-H Pow	0.417	0.710	0.823	1.000	0.431	0.148
H-L Aff	***0.025***	0.640	0.232	0.431	1.000	***0.001***
H-L Ach	0.266	***0.016***	0.153	0.148	***0.001***	1.000

Fig. 20 Results of using ACC ERN in the whole SNP (**a**) and during action (**b**) to reveal risk-taking attitude. The p-value of independent t-test between hope and fear components for each motivation is presented. Grey squares indicate significant differences

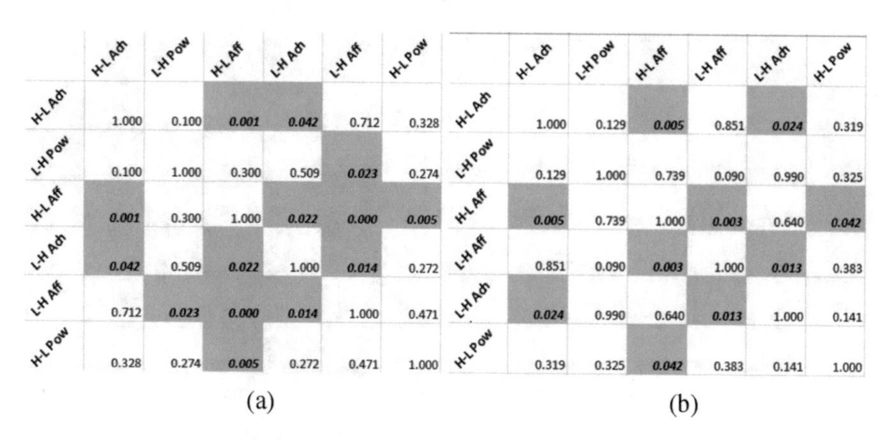

(a)

	H-L Ach	L-H Pow	H-L Aff	L-H Ach	L-H Aff	H-L Pow
H-L Ach	1.000	0.100	***0.001***	***0.042***	0.712	0.328
L-H Pow	0.100	1.000	0.300	0.509	***0.023***	0.274
H-L Aff	***0.001***	0.300	1.000	***0.022***	***0.000***	***0.005***
L-H Ach	***0.042***	0.509	***0.022***	1.000	***0.014***	0.272
L-H Aff	0.712	***0.023***	***0.000***	***0.014***	1.000	0.471
H-L Pow	0.328	0.274	***0.005***	0.272	0.471	1.000

(b)

	H-L Ach	L-H Pow	H-L Aff	L-H Aff	L-H Ach	H-L Pow
H-L Ach	1.000	0.129	***0.005***	0.851	***0.024***	0.319
L-H Pow	0.129	1.000	0.739	0.090	0.990	0.325
H-L Aff	***0.005***	0.739	1.000	***0.003***	0.640	***0.042***
L-H Aff	0.851	0.090	***0.003***	1.000	***0.013***	0.383
L-H Ach	***0.024***	0.990	0.640	***0.013***	1.000	0.141
H-L Pow	0.319	0.325	***0.042***	0.383	0.141	1.000

Fig. 21 Results of using ACC MFN in the whole SNP (**a**) and during action (**b**) to reveal risk-taking attitude. The p-value of independent t-test between hope and fear components for each motivation is presented. Grey squares indicate significant differences

studies, the design of money and satisfaction feedbacks in the SNP needs to be considered (Fig. 23).

4 Conclusion and Future Work

This chapter validates the design of our mini-game using players' behaviour and EEG signals. Results show that motivation profiles can be modelled by the player behaviour with each NPC in two phases of the mini-game. More importantly, social attitude can be revealed by EEG signals collected during satisfaction feedback,

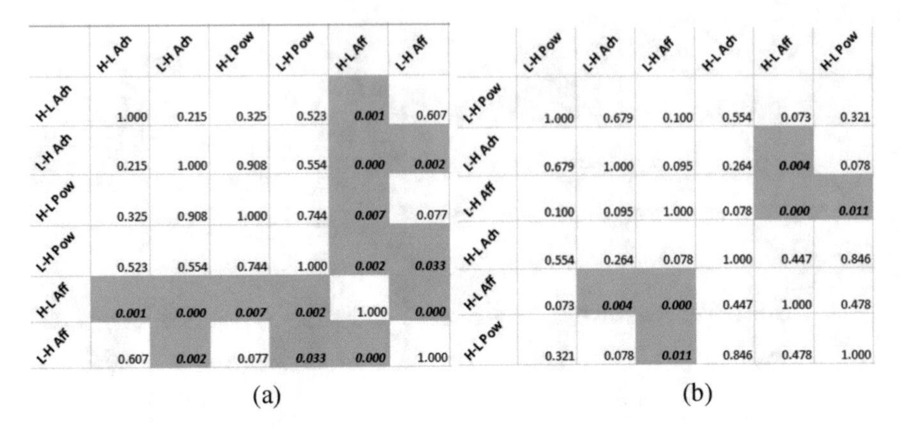

(a)

	H-L Ach	L-H Ach	H-L Pow	L-H Pow	H-L Aff	L-H Aff
H-L Ach	1.000	0.215	0.325	0.523	*0.001*	0.607
L-H Ach	0.215	1.000	0.908	0.554	*0.000*	*0.002*
H-L Pow	0.325	0.908	1.000	0.744	*0.007*	0.077
L-H Pow	0.523	0.554	0.744	1.000	*0.002*	*0.033*
H-L Aff	*0.001*	*0.000*	*0.007*	*0.002*	1.000	*0.000*
L-H Aff	0.607	*0.002*	0.077	*0.033*	*0.000*	1.000

(b)

	L-H Pow	L-H Ach	L-H Aff	H-L Ach	H-L Aff	H-L Pow
L-H Pow	1.000	0.679	0.100	0.554	0.073	0.321
L-H Ach	0.679	1.000	0.095	0.264	*0.004*	0.078
L-H Aff	0.100	0.095	1.000	0.078	*0.000*	*0.011*
H-L Ach	0.554	0.264	0.078	1.000	0.447	0.846
H-L Aff	0.073	*0.004*	*0.000*	0.447	1.000	0.478
H-L Pow	0.321	0.078	*0.011*	0.846	0.478	1.000

Fig. 22 Results of using ACC P300 in the whole SNP (**a**) and during action (**b**) to reveal risk-taking attitude. The p-value of independent t-test between hope and fear components for each motivation is presented. Grey squares indicate significant differences

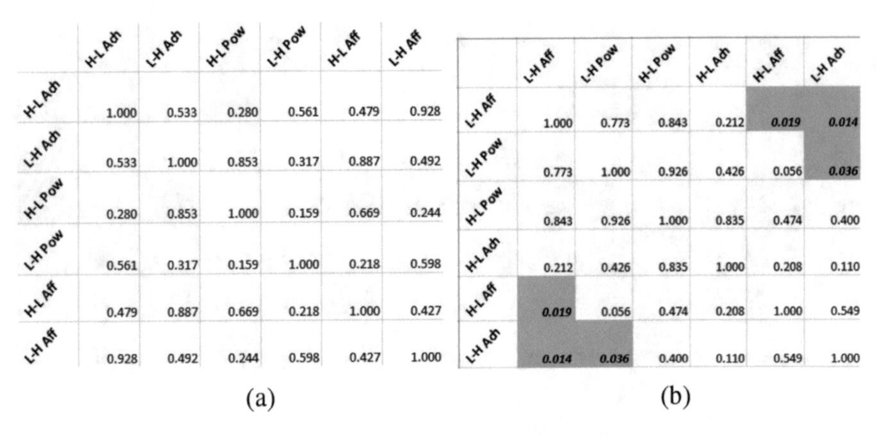

(a)

	H-L Ach	L-H Ach	H-L Pow	L-H Pow	H-L Aff	L-H Aff
H-L Ach	1.000	0.533	0.280	0.561	0.479	0.928
L-H Ach	0.533	1.000	0.853	0.317	0.887	0.492
H-L Pow	0.280	0.853	1.000	0.159	0.669	0.244
L-H Pow	0.561	0.317	0.159	1.000	0.218	0.598
H-L Aff	0.479	0.887	0.669	0.218	1.000	0.427
L-H Aff	0.928	0.492	0.244	0.598	0.427	1.000

(b)

	L-H Aff	L-H Pow	H-L Pow	H-L Ach	H-L Aff	L-H Ach
L-H Aff	1.000	0.773	0.843	0.212	*0.019*	*0.014*
L-H Pow	0.773	1.000	0.926	0.426	0.056	*0.036*
H-L Pow	0.843	0.926	1.000	0.835	0.474	0.400
H-L Ach	0.212	0.426	0.835	1.000	0.208	0.110
H-L Aff	*0.019*	0.056	0.474	0.208	1.000	0.549
L-H Ach	*0.014*	*0.036*	0.400	0.110	0.549	1.000

Fig. 23 Results of using ACC TBR in the whole SNP (**a**) and during action (**b**) to reveal risk-taking attitude. The p-value of independent t-test between hope and fear components for each motivation is presented. Grey squares indicate significant differences

and risk-taking attitude can be revealed in EEG signals collected during money feedback. To be more specific, Ruby and Vince are good choices for evoking social attitude, and Dan, Ruby and Vince are good choices for evoking risk-taking attitudes. However, Candy is the least useful character for reflecting social and risk-taking attitudes. It also appears that player behaviour and EEG data collected from the IPP allow us to assess motivation better than those from the SNP.

There are several reasons why the performance of the SNP is not ideal. First, we can only examine the action and trials in the SNP because there is no latency for money and satisfaction feedbacks. Feedback-related potentials are essential for assessing risk-taking and social attitudes, which are two characteristics of player motivation [3, 8, 12]. Future work needs to consider how to design the money and satisfaction feedbacks stimuli in the SNP to avoid too many stimuli in the interface. In addition, the design of the SNP for evoking social attitude is effective; however, the use of the SNP for risk attitude needs to be enhanced in the future study. Using Ruby, the random character, as the uncertain and risky element in the SNP emerged as inadequate. Finally, the gameplay between players and NPCs is in a pairwise fashion. Further works should consider alternative game mechanics of the social network play, for instance, the n-player iterated prisoner dilemma or the n-player trust game.

In future, we anticipate that mini-game-based approaches such as the one described in this chapter might be incorporated throughout games to detect different aspects of a player's profile. Other computational intelligence techniques can then be used to adapt parts of the game to best suit the player's profile and maximise their enjoyment of the game.

References

1. J.W. Atkinson, G.H. Litwin, Achievement motive and test anxiety conceived as motive to approach success and motive to avoid failure. J. Abnorm. Soc. Psychol. **173**, 166 (1960)
2. R.C. Davis, D.C. McClelland, *The Achieving Society* (JSTOR, Princeton, 1962)
3. J. Hewig, R. Trippe, H. Hecht, M.G. Coles, C.B. Holroyd, W.H. Miltner, Decision-making in blackjack: an electrophysiological analysis. Cereb. Cortex **17**(4), 865–77 (2007)
4. M. Jurišić, D. Kermek, M. Konecki, A review of iterated prisoner's dilemma strategies, in *MIPRO 2012 - 35th International Convention on Information and Communication Technology, Electronics and Microelectronics - Proceedings*, October 2012, pp. 1093–1097
5. D.M. Kuhlman, A.F. Marshello, Individual differences in game motivation as moderators of preprogrammed strategy effects in prisoner's dilemma. J. Pers. Soc. Psychol. **32**(5), 922–931 (1975)
6. X. Liu, K.E. Merrick, H. Abbass, Designing artificial agents to detect the motive profile of users in virtual worlds and games, in *IEEE Symposium Series on Computational Intelligence, Athens, Greece, Presented at IEEE Symposium Series on Computational Intelligence* (2016)
7. X. Liu, K. Merrick, H. Abbass, Towards electroencephalographic profiling of player motivation: a survey. IEEE Trans. Cogn. Dev. Syst. **10**(3), 499–513 (2018)
8. H. Masaki, S. Takeuchi, W.J. Gehring, N. Takasawa, K. Yamazaki, Affective-motivational influences on feedback-related ERPs in a gambling task. Brain Res. **1105**(1), 110–121 (2006)
9. K.E. Merrick, *Computational Models of Motivation for Game-Playing Agents* (Springer, Berlin, 2016)
10. S.K. Mesrobian, M. Bader, L. Goette, A.E.P. Villa, A. Lintas, *Imperfect Decision Making and Risk Taking Are Affected by Personality* (Springer International Publishing, Cham, 2015), pp. 145–184
11. W. Poundstone, *Prisoner's Dilemma* (Doubleday, New York, 1992)
12. B. Schuermann, T. Endrass, N. Kathmann, Neural correlates of feedback processing in decision-making under risk. Front. Hum. Neurosci. **6**, 204 (2012)
13. D. Sirota, D. Klein, *The Enthusiastic Employee: How Companies Profit by Giving Workers What They Want* (FT Press, Upper Saddle River, 2013)
14. K.W. Terhune, Motives, situation, and interpersonal conflict within Prisoner's Dilemma. J. Pers. Soc. Psychol. **8**(3p2), 1 (1968)

Kathryn Kasmarik is an Associate Professor in information technology at the University of New South Wales, Australian Defence Force Academy (UNSW Canberra). She is currently the Deputy Head of School (Teaching) for the School of Engineering and IT. Kathryn completed a Bachelor of Computer Science and Technology at the University of Sydney, including a study exchange at the University of California, Los Angeles (UCLA). She graduated with First Class Honours and the University Medal in 2002. She completed a PhD in computer science through the National ICT Australia and the University of Sydney in 2007. She moved to UNSW Canberra in 2008. She has published over 100 articles in peer reviewed conference and journals. Her research has been funded by the Australian Research Council and Defence Science and Technology Group, among other sources. Before joining UNSW Canberra, Kathryn worked as a computer programmer in the financial services industry, as an intern in digital image processing and as a research assistant at the Key Centre for Design Computing and Cognition at the University of Sydney. Kathryn was interested in maths from a young age and became interested in computer science after attending a summer school at the University of Sydney. She became a researcher because she enjoys the process of finding new problems to solve and pushing science beyond its limits to come up with solutions.

Xuejie Liu is currently a lecturer at School of Physics and Telecommunication Engineering, South China Normal University. Xuejie Liu completed Bachelors of Telecommunication Engineering and International Economy and Trade at LuDong University in 2011. She completed a Master in Acoustics at South China University of Technology in 2014 and a PhD in computer science at University of Engineering and Information Technology, University of New South Wales Canberra Campus in 2019. After that, she moved to South China Normal University. she chose to study in a

STEM field when she was a high school student and became interested in research when she was found herself enjoy learning new things, exploring interesting things and doing meaningful works for the society.

Hussein Abbass is a Professor at the University of New South Wales Canberra, Australia. Prof. Abbass is the Founding Editor-in-Chief of the IEEE Transactions on Artificial Intelligence. He is an Associate Editor of the IEEE Transactions on Neural Networks and Learning Systems, IEEE Transactions on Evolutionary Computation, IEEE Transactions on Cognitive and Developmental Systems, IEEE Transactions on Cybernetics, IEEE Transactions on Emerging Topics in Computational Intelligence, ACM Computing Surveys and four other journals. He was the Vice President for Technical Activities (2016–2019) for the IEEE Computational Intelligence Society. Prof. Abbass is a fellow of the IEEE (FIEEE), a fellow of the Australian Computer Society (FACS), a fellow of the Operational Research Society (FORS), and a fellow of the Institute of Managers and Leaders Australia and New Zealand (FIML). He has published over 300 papers in journals and conferences and four authored books. His current research contributes to trusted human–swarm teaming with an aim to design next generation trusted and distributed artificial intelligence systems that seamlessly integrate humans and machines.

When AI Meets Digital Pathology

Pau-Choo (Julia) Chung and Chao-Ting Li

Acronyms

WSI	whole slide image
CAD	computer-aided detection and/or diagnosis
AI	artificial intelligence
CT	computed tomography
DCNN	deep convolutional neural network
CE	Communate Europpene
FAD	US food and drug administration
CNN	convolutional neural network
H&E	haematoxylin and eosin
IHC	immunohistochemistry
TLI	tumor lymphocytic infiltration
SPM	spatial pyramid matching framework
SVM	support vector machine
DCAN	deep contour-aware network
ASPP	atrous spatial pyramid pooling
MSCN	multiscale convolutional network
FA-MSCN	Feature-Aligned Multiscale Convolutional Network
SSCN	single scale convolutional network
IoU	intersection over Union
CAM	class activation mapping
AIH	autoimmune hepatitis
kNN	k-nearest neighbors
SMOTE	synthetic minority oversampling technique
ConvNet	convolutional network
AL	active learning
HoG	Histogram of Oriented Gradients
SIFT	scale-invariant feature transform

P.-C. (Julia) Chung (✉) · C.-T. Li
National Cheng Kung University, Tainan, Taiwan
e-mail: pcchung@ee.ncku.edu.tw

A. E. Smith (ed.), *Women in Computational Intelligence*, Women in Engineering and Science, https://doi.org/10.1007/978-3-030-79092-9_6

FCN	fully convolution network
MC	Monte Carlo
R-CNN	Region-based CNN
IEAL	imbalance effective active learning
DL	dice loss
DICOM	Digital Imaging and Communications in Medicine
IHE	Integrating the Healthcare Enterprise
ICC	International Color Consortium

1 Evolution of Digital Pathology

Histopathology involves minute microscopic examination of biopsy or bisection organ tissues for the purpose of diagnosing and/or examining diseases of the tissue. The microscope view reveals only a small part of the tissue. However, whole slide images (WSIs) contain a huge number of pixels (in the gigabyte range), and hence, inspecting the tissue manually through a microscope is an extremely long and laborious process. Depending on the particular diagnosis requirements, a pathologist may be required to carefully check all the suspected regions of the WSI, to examine the tissue characteristics in particular regions of the WSI, to compare the characteristics of certain tissue regions with those of their neighbors, and so on. The examination process frequently involves the discrimination of extremely subtle details and is thus not only extremely time-consuming but also heavily reliant on the particular skill and experience of the pathologist. It is therefore not uncommon for different pathologists to reach different diagnoses when observing the same WSI.

To ease the load on pathologists, and improve the objectivity of the diagnosis results, many researchers have examined the feasibility of computer-aided detection and/or diagnosis (CAD) tools. However, realizing such tools involves overcoming two major obstacles. First, pathology specimens are traditionally mounted on glass slides for manual examination under a microscope. Thus, to implement CAD methods, it is first necessary to digitize the glass specimens in some way. Second, biopsy images have an extremely large size; typically of the order of ten or hundred times that of radiology images. Consequently, storing and processing the digitized image contents pose a significant technical challenge.

In practice, to digitize the glass specimens relies on the capabilities of the scanning devices used to convert the original biopsy specimen into a digital image. With modern technology, the time required to scan a whole slide has reduced from several hours to a matter of minutes. Moreover, the use of cartridge systems now makes it possible to accept hundreds of slides for scanning at a time. The scanning resolution and image quality have also improved dramatically in recent years, and with the invention of 3D depth z-stack scanning, it is now possible to build detailed 3D images of the target tissue, thereby further facilitating the diagnosis process.

Another improvement in digital pathology are recent advances in computer technology and machine learning techniques (popularly known as artificial intelligence (AI)). Modern AI techniques are rooted from neural networks, which have been broadly applied to the computer-assisted diagnosis of medical images. Early adoption of the models usually was combined with specially designed features to perform the diagnosis process. For example, the probabilistic-based method proposed in [4] distinguished between hepatoma and hemageoma cells in liver computed tomography (CT) images using fractal features. Meanwhile, the algorithms in [16, 17] performed micro-calcification classification using features computed by Shape-Cognitron, a convolution-like neural network. Finally, the method in [18] performed localization of human body organs in a sequence of CT images using a spatial-constrained neural network based on fuzzy descriptors. In all of these methods, the features used for classification purposes were manually designed. However, some subtle characteristics are hard to be precisely described. Thus, many deep convolutional neural network (DCNN) architectures have been proposed in recent years. Generally speaking, these architectures can automatically learn features which are hard to describe using traditional image-processing or pattern-description methods. As a result, they are capable of recognizing many patterns and features which cannot be detected by traditional algorithms. AI-based CAD systems based on such models have attracted growing attention in the medical field recently, and the number of AI-assistive devices receiving Communate Europpene (CE) or US Food and Drug Administration (FDA) clearance has increased exponentially.

However, in employing AI to analyze pathology images, there are still many key issues remaining to be addressed. One of the most serious challenges is that of data annotation. Histopathology images can be as large as 30–50 GB. Thus, just to delineate the tumor region of a WSI can take several hours, not to mention the time needed to delineate the boundaries of the individual cells. Furthermore, some cells (e.g., plasma cells) are critical to the detection and diagnosis of certain diseases, but appear only very sparsely in WSIs. The resulting data imbalance problem poses significant challenges to AI technology in training reliable models. In addition, the extremely large size of WSIs often precludes their direct processing by convolutional neural network (CNN) models. As a result, some form of pre-processing step, such as patch generation or blank patch exclusion, is generally required before CNN processing can be carried out. Finally, pathology images often exhibit quite significant color variations due to the use of different staining protocols or scanning devices in different hospitals. As a result, there is no guarantee that NN models well-trained on data collected from one hospital or scanned using a particular device will function well when applied to data obtained from another hospital or device.

This chapter commences by describing two of the main issues faced by AI models in performing pathology analysis today, namely, staining-induced differences in the color of scanned WSIs and the huge size of the WSIs required to be processed by the AI model. The chapter then reviews the attempts made in current AI models to mimic the multi-scale approach taken by pathologists in switching between low- and high-magnification settings when examining traditional biopsy slides in order to achieve a tradeoff between the diagnosis accuracy and the diagnosis time. Thereafter, the chapter describes some of the major challenges faced in performing two of the most common tasks in automated pathology analysis, namely, cell detection and cell segmentation. The challenges are described in detail, and the current state-of-the-art proposals for dealing with these challenges are introduced and discussed. The chapter concludes by examining some of the practical issues to be addressed in integrating AI-based solutions for digital pathology into practical applications and discusses future prospective for AI technology in the clinical context.

2 Data Issues and Preprocessing

As described above, digital pathology images are obtained by scanning the related biopsy images using some form of scanning device. However, depending on the particular diagnosis focus, different stains may be applied. Among the various stains in common use, haematoxylin and eosin (H&E) is one of the most frequently used when the aim is to highlight the nucleus and membrane regions of the tissue. Alternatively, Masson stain may be used if the interest of the pathologist lies mainly in identifying possible regions of fibrosis in the image. In immunohistochemistry (IHC) applications, the aim of the pathology process is to observe the antibody binding properties of selective antigens rather than to examine the tissue structure per se. Thus, depending on the particular cellular event to be observed, different antibodies may be applied. Figure 1a–c present typical liver pathology images with H&E, Masson, and IHC staining, respectively. It is apparent that the images have quite different colors and characteristics, and this represents a major challenge in developing AI models capable of application across different tissues with different staining protocols and different diagnosis requirements.

In practical contexts, the staining process is governed by strict protocols which seek to ensure a consistent quality of the stained image, irrespective of the particular institution in which the staining procedure was carried out. Nonetheless, the dicing material component may vary slightly between different hospitals; resulting in inevitable differences in the staining quality of different WSIs. Furthermore, different scanning devices may also reveal different color properties. For example,

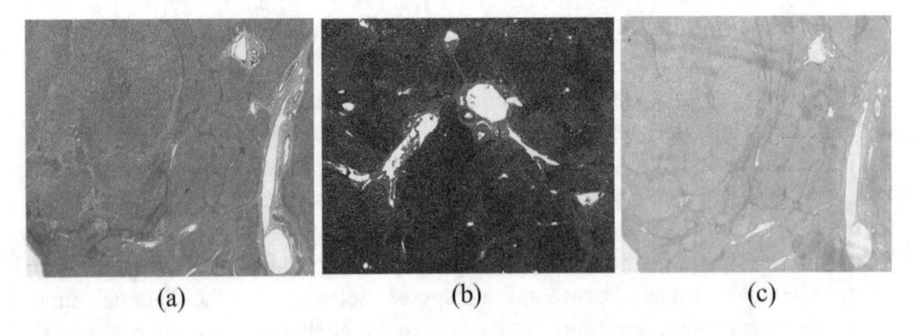

Fig. 1 Pathology images treated using different stains: (**a**) H&E, (**b**) Masson, and (**c**) IHC

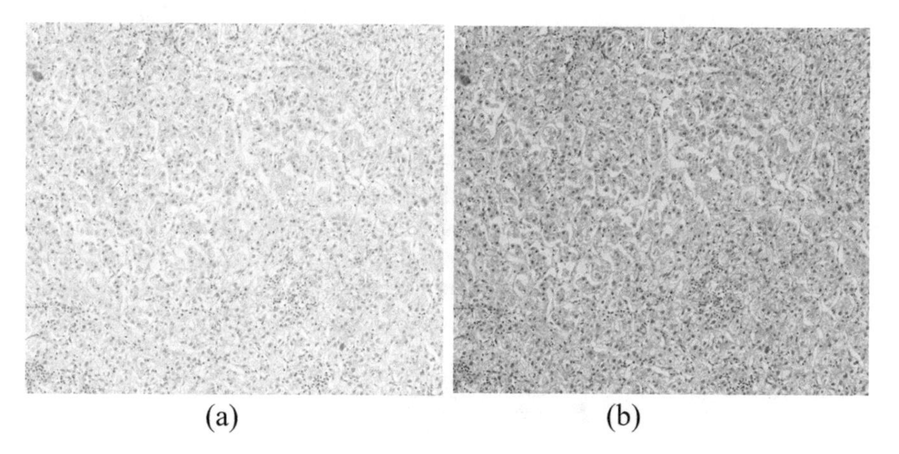

Fig. 2 Images of same H&E stained slice obtained from two different scanning devices: (**a**) Aperio (Leica Biosystems) and (**b**) SR360 (Hamamatsu)

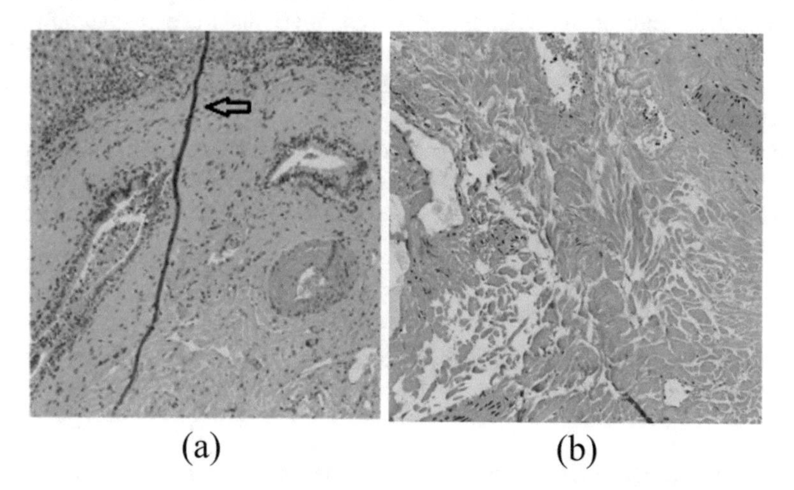

Fig. 3 (**a**) Folded region and (**b**) tattered hepatocyte region in sectioned images

Fig. 2 shows two images of the same H&E stained slice obtained using different scanning devices. The difference in hue of the two images is immediately apparent. This further complicates the development of robust AI models capable of processing WSIs stained and scanned by different institutions. Finally, the sectioning process itself may introduce fold regions (Fig. 3a) or tattered hepatocytes (Fig. 3b), which serve as major sources of interference in the AI analysis process and hence pose a significant challenge in arriving at a correct diagnosis outcome.

Pathology images typically have the form of biopsy images, histopathology images, or blood smear images. Depending on the tissue size, a biopsy image scanned at a magnification of 40× may have a size of around 5 G (uncompressed), while a histopathology image may have a size of up to 40 G. Such large images simply cannot fit into current CNN models. Consequently, some form of data preprocessing is first required. For the case of histopathology images, the preprocessing task commonly commences by trimming the blank regions of the WSI adjacent to the tissue of interest. A patch generation process is then applied to cut the remaining WSI into small overlapping patches with a size typically of 256×256, 512×512 or 1024×1024 pixels using a sliding-window approach.

Once the WSI is divided into patches, a further blank removal operation may be performed to remove any patches containing no (or very little) histopathologic information. The decision as to whether or not to remove a particular patch is made in accordance with a threshold equation, such as that shown in Eq. 1:

$$E(I) = \frac{1}{N^2} \sum_{i=1}^{N} \sum_{j=1}^{N} I_{ij} \geq T \tag{1}$$

where I is the input patch, I_{ij} is the pixel intensity, and N is the patch size. The value of T depends on the staining protocol employed and the particular aim of the pathologist in examining the image. For a typical histopathology liver image stained with H&E, T is assigned a value of around 205, with which no patches containing more than 0.1 tissue area will be removed.

For images in which the background region is smooth and homogeneous (e.g., mammograms and pathology images), the blank regions can alternatively be removed by performing segmentation based on a region-growing operation applied on the background regions [16].

3 Multiscale Convolutional Neural Networks for Tumor Detection

Many cancer prognostic assessments, such as the cancer grade score, cancer stage, and tumor lymphocytic infiltration (TLI), are limited to the tumor region of the tissue. As a result, tumor detection is one of the most commonly performed procedures in pathology image analysis. Many methods have been proposed for automatic tumor detection. For example, Cruz-Roa et al. [7] used a six-layer convolutional neural network (CNN) to detect invasive breast cancer. Sirinukunwattana et al. [27] reviewed the performance of several well-known CNNs, including FCN, AlexNet, U-Net, and LeNet. Zhou et al. [35] developed a spatial pyramid matching framework (SPM) for classifying patches into tumor, normal, and stromal patches using a linear support vector machine (SVM) classifier and a convolutional sparse coding feature extractor. Chen et al. [6] and Xu et al. [32] used a deep contour-aware network (DCAN) and deep multichannel side supervision method, respectively, to perform gland segmentation. Xu et al. [31] used AlexNet as a feature vector extractor and linear SVM as a classifier to identify brain tumor patches as either positive or negative necrosis.

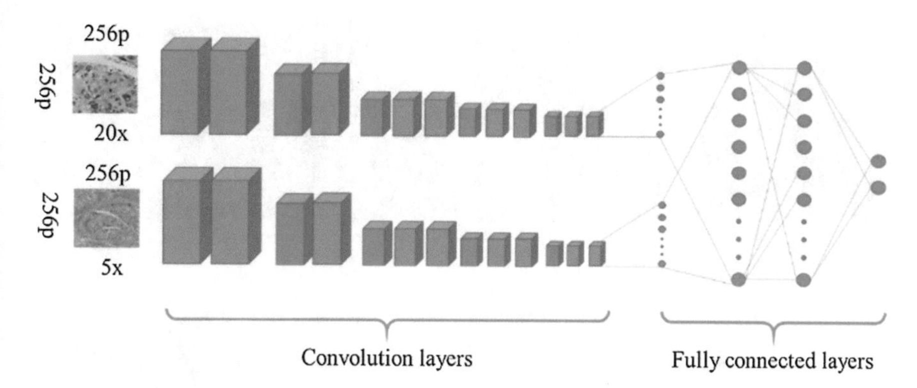

Fig. 4 Multiscale convolutional neural network consisting of two CNNs for feature extraction and fully connected layers

In clinical settings, whole slide images (WSIs) may be obtained at many different magnification scales, e.g., 5×, 10×, 20×, and 40×. Thus, one of the major dilemmas associated with the CNN-based analysis of histopathology images is that of choosing an appropriate magnification scale. Generally speaking, high-magnification patches provide detailed cell-level features, but lack the macroscopic information support required to classify ambiguous patches. Conversely, low-magnification images provide more contextual information, but lack low-level details. For this reason, pathologists tend to switch between low- and high-magnification settings when observing a slide. For example, apoptosis, necrosis, steatosis, and inflammatory cell infiltration phenomena could be very similar in tumor regions and benign regions, respectively. Consequently, having detected the possible existence of such phenomena in a high-level view, a wider macro-view is commonly required to confirm whether or not the suspect region is truly tumorous.

Xiao, Chung et al. in [5, 30] attempted to mimic this multiresolution procedure by means of an atrous spatial pyramid pooling (ASPP) approach, in which convolutional masks of various spatial resolutions were applied to a single-scale image. By contrast, Chen, Papandrous et al. in [28] employed a multiscale deep neural network model consisting of multiple convolutional modules to extract features from image patches with multiple scales. Figure 4 shows a typical network architecture in which two CNNs configured in parallel are used to extract the features from low- and high-magnification input patches, respectively, and fully connected layers are then used to process the two sets of features in order to obtain the final tumor detection result.

Although multiscale CNNs, such as those described above, consider both low- and high-scale features, the features of different scales are supplied to the fully connected layers independently. That is, the spatial relationships among them are ignored in the computations of the fully connected layers. The multiscale convolutional network (MSCN) proposed in [13], termed the Feature-Aligned Multi-Scale Convolutional Network (FA-MSCN), attempted to resolve this problem by using ASPP blocks to integrate the features extracted from feature maps of different magnification scales aligned at the same position of the target image (see Fig. 5).

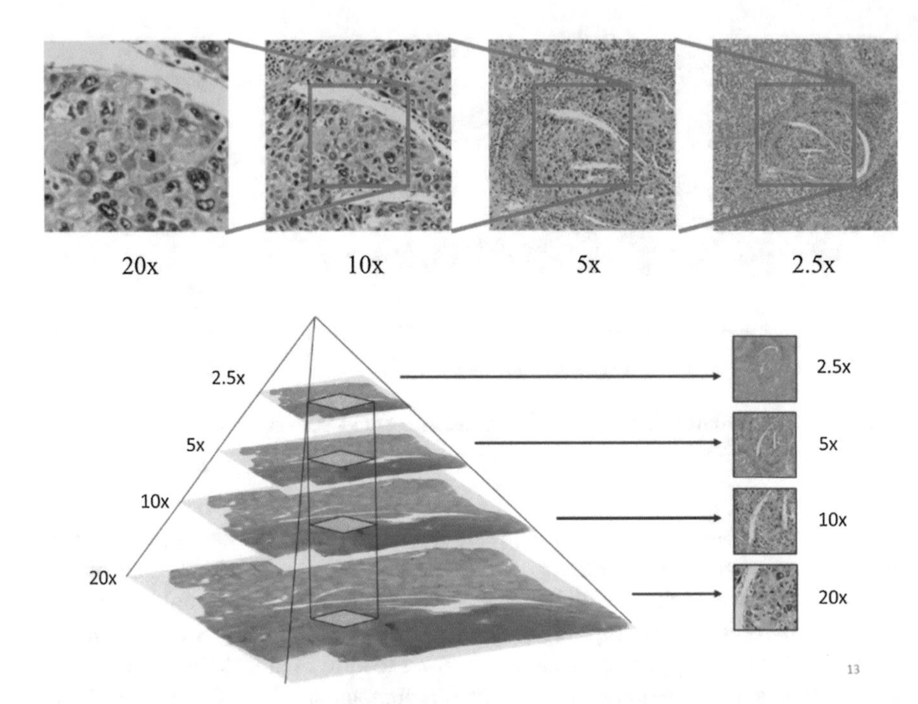

Fig. 5 Feature extraction from spatially aligned feature maps of different scales

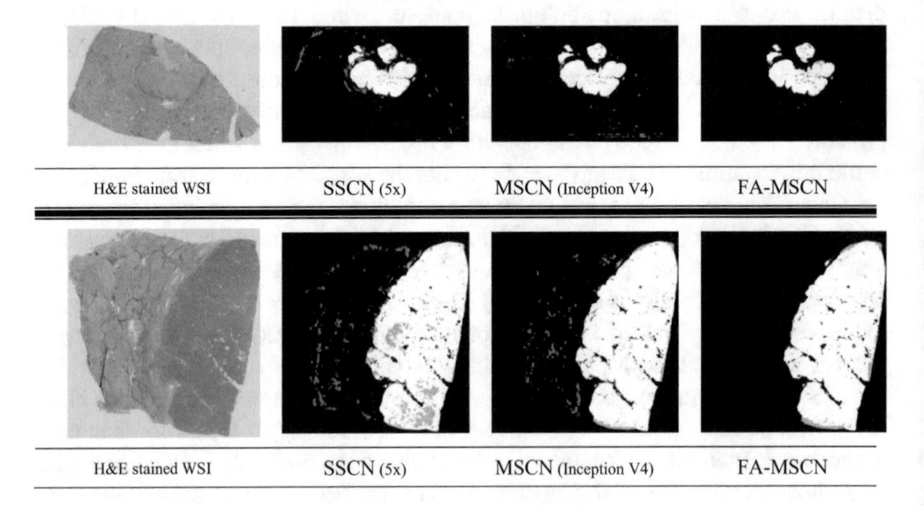

Fig. 6 Typical tumor detection results obtained using different models

Figure 6 presents some typical examples of the tumor detection results obtained using traditional single scale convolutional network (SSCN), MSCN and FA-MSCN, respectively. Table 1 from our preliminary experiments shows the measured

Table 1 Sensitivities of seven different models for seven cases (WSIs)

Model	Case							
	1	2	3	4	5	6	7	mean
VGG16 (20×)	0.992	0.945	0.984	0.929	0.964	0.992	0.959	0.966
VGG16 (5×)	0.993	0.878	0.986	0.914	0.985	0.954	0.964	0.953
Incep V4 (20×)	0.990	0.906	0.985	0.800	0.939	0.964	0.933	0.931
Inception V4 (5×)	0.982	0.996	0.999	0.980	0.983	0.986	0.899	0.975
Multiscale VGG16	0.998	0.986	0.976	0.982	0.981	0.986	0.662	0.939
Multiscale Incep V4	0.992	0.991	0.997	0.986	0.962	0.995	0.958	0.983
Multiscale (FA- aligned)	0.993	0.924	0.968	0.967	0.987	0.984	0.962	0.969

sensitivities of the different models when applied on seven different cases (WSIs). Overall, the results show that single-scale and multiscale networks have a similar sensitivity. However, according to [13], multiscale networks achieve a higher Intersection over Union (IoU). Furthermore, the IoU performance is improved when a spatial-alignment approach is used to extract the features.

4 Cell Detection and Segmentation

Besides tumor detection, cell-level analysis is also a critical step in histopathology for understanding the properties and progression of diseases. Compared to normal cells, tumor cells usually have larger nuclei, irregular patterns, and larger nucleus-to-membrane ratios. Consequently, cell nucleus detection plays a critical role in pathology analysis. Lymphocytes are also of great importance in histopathology. For example, lymphocyte aggregation in the liver portal area is a key indicator of hepatitis, one of the most common disorders of the liver [14]. In clinical practice, hepatitis severity is quantified using the Ishak grading system, which assesses the degree of portal inflammation based on the density and distribution of the aggregated lymphocytes [23]. For example, Fig. 7 shows four histopathology images with lymphocyte densities of normal, mild, moderate, and marked, respectively. Therefore, lymphocyte detection is a critical step in histopathology for grading portal inflammation and assessing the severity of hepatitis accordingly.

Many CNN methods for detecting and segmenting cells in histopathology images have been proposed [9, 38]. These methods implement the CNN as a two-class (i.e., cell and background) classifier and inspect each pixel in the histopathology image in a patch-wise manner using a sliding window. Yang et al. [33] performed lymphocyte detection using a semantic segmentation approach, in which the uncertainty and similarity information provided by a fully convolutional network is used to select the areas for annotation. Among all of the CNN models available for biomedical image segmentation, U-Net is commonly used since it is well-established and consistently provides a state-of-the-art performance [25, 33]. The U-Net model was also used for

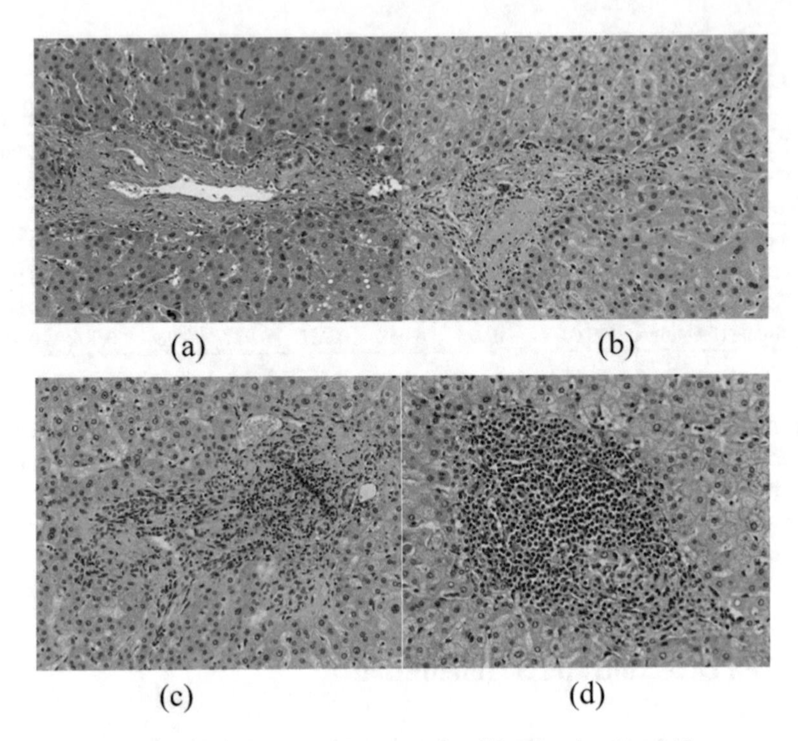

Fig. 7 Grading of portal inflammation: (**a**) normal, (**b**) mild, (**c**) moderate, and (**d**) severe

the segmentation of liver cells of nucleus, lymphocytes, and plasma [19], achieving sensitivities of higher than 92% in lymphocytes and nuclei.

4.1 Dilemmas Associated with Annotation Loading and Data Imbalance

In order to properly train a model for object detection and segmentation, it is first necessary to label a set of training images with the ground truth. However, a typical histopathology whole slide image may contain hundreds of thousands of cells. Thus, for cell level segmentation tasks, delineating the boundaries of all the cells is extremely time-consuming and laborious. The literature therefore contains various methods for reducing the labeling load. For example, the class activation mapping (CAM) method proposed in [36] uses a small amount of data to train a primitive classifier, which is then used to perform classification on an unlabeled image. The classification results are then traced back to the corresponding regions in the original image to perform annotation. In other words, the classified regions are considered as labeled areas in the target. CAM performs reasonably well in detecting relatively large targets such as tumorous areas or other objects occupying a large region of the image frame. However, the cells in pathology images are not only very small but also distributed very widely over the image frame. As a result, CAM is only of very limited use as a cell-level annotation tool.

Class imbalance is another critical problem frequently encountered when performing the detection of nuclei, lymphocytes, and plasma. In liver histopathology images, for example, plasma cells are present only in extremely small quantities compared to nuclei and lymphocytes. Moreover, plasma cell– rich infiltration, which is one of the key indicators of autoimmune hepatitis (AIH), exists only in a very few cases. The resulting class imbalance causes the training process to be strongly biased toward the majority class and therefore degrades the detection accuracy of the minority instances. Existing proposals for dealing with this problem can be classified as either resampling methods or cost-sensitive learning methods, as described in the following.

Resampling involves reconstructing the data samples from the original dataset in such a way as to achieve a more balanced population for training. The resampling process may be performed using either undersampling or oversampling techniques. In the former case, the excess number of majority samples is reduced by randomly selecting a certain number of samples from the majority class equal to the number of samples in the minority class. However, this may lead to the discarding of data potentially useful for classification purposes. Accordingly, the sampling process is commonly performed with the assistance of a k-nearest neighbors (kNN) algorithm to select the more informative samples. Oversampling methods, on the other hand, attempt to balance the skewed class ratio by adding samples to the minority class. This is generally performed by replicating the minority samples. However, such an approach may lead to overfitting and a poor generalization performance in the training process since the test data simply replicate already-existing instances of the minority class. To combat this problem, the synthetic minority over-sampling technique (SMOTE) proposed in [3, 15] generates new data by sampling from the distribution of minority samples and then interpolating new samples based on these samples and their nearest neighbors.

Many loss functions have been proposed for tackling the class imbalance problem by increasing the contribution of the minority class to the loss score. One of the most commonly used loss functions is the weighted cross-entropy function, in which a weighted penalty factor is updated based on the proportion of the minority class. However, the function is inelastic in the sense that it focuses only on the minority samples in the training process, i.e., it ignores the loss contribution of the majority samples. Accordingly, several more elastic loss functions have been proposed in recent years. For instance, the focal loss function proposed in [11] is a dynamically scaled cross-entropy loss function, in which a scaling factor is adjusted based on a predicted probability measure such that the contributions of the easy examples are automatically down-weighted and the training process is thus driven to rapidly focus on the more difficult samples. Ozdemir et al. [24] used a dice loss function for dealing with the class imbalance problem by considering the similarity between the predicted and labeled images.

4.2 Lessening the Burden of Manual Annotation

Manually annotating images with a pixel-level accuracy for multiclass semantic segmentation requires a huge amount of time and effort. Consequently, the literature contains many proposals for lessening the annotation burden.

Among these methods, transfer learning, in which a model is converted from a pretrained knowledge to another domain, has a proven ability to significantly reduce the annotation load. In typical transfer learning methods, a convolutional network (ConvNet) is pretrained on a very large dataset (e.g., ImageNet, comprising 1.2 million images with 1000 categories) and is then used as either an initializer or a fixed feature extractor for the task of interest. This approach is motivated by the observation that the earlier features extracted by a ConvNet are generic and applicable to a wide range of tasks, whereas the later features are progressively more specific to the particular classes contained in the original dataset. He et al. [12] reported that such a pretraining approach is effective for small datasets (e.g., <10 k images). However, for larger datasets, models trained from scratch can achieve a similar accuracy if allowed a greater number of iterations. Nonetheless, Cao et al. [1] found that models pretrained on ImageNet provide a good feature extraction performance for breast histology images.

Active learning (AL) is another promising approach for reducing the labeling workload through the use of an intelligent sample query strategy. In particular, AL selects only the more informative samples for further training (or fine-tuning) the current model, where informative samples are regarded as those samples having a greater effectiveness in improving the accuracy of the model. As shown in Fig. 8, the AL process begins by selecting worthy samples from a large collection of unlabeled images. The selected samples are manually annotated by human experts and are then added to the training dataset for further training (or fine-tuning). The selection and labeling process is repeated iteratively in this way until the specified termination criterion is reached (typically, the accuracy of the trained model reaches the required standard, or the specified number of iterations are completed).

Many AL frameworks have been proposed in recent years for applications such as video image analysis, speech recognition, natural language processing, and so on. Generally speaking, the classifiers in such frameworks are trained with hand-crafted features (e.g. Histogram of Oriented Gradients (HoG) and scale-invariant feature transform (SIFT)) specifically designed for the particular task.

Recently, several CNNs which integrate automatic feature extraction and discriminative classification into a single model have been developed for video recognition and other applications [2, 34]. CNNs have also been incorporated with AL frameworks to learn features and perform classification using a minimum number of labeled samples. Most of these methods focus on image-level classification tasks. For example, Zhou et al. [37] used a pre-trained CNN to select samples with high entropy and high diversity as worthy samples and then used an AL approach to continuously fine-tune the CNN by incorporating newly annotated samples. Wang et al. [29] proposed a heuristic-based algorithm for building a competitive classifier

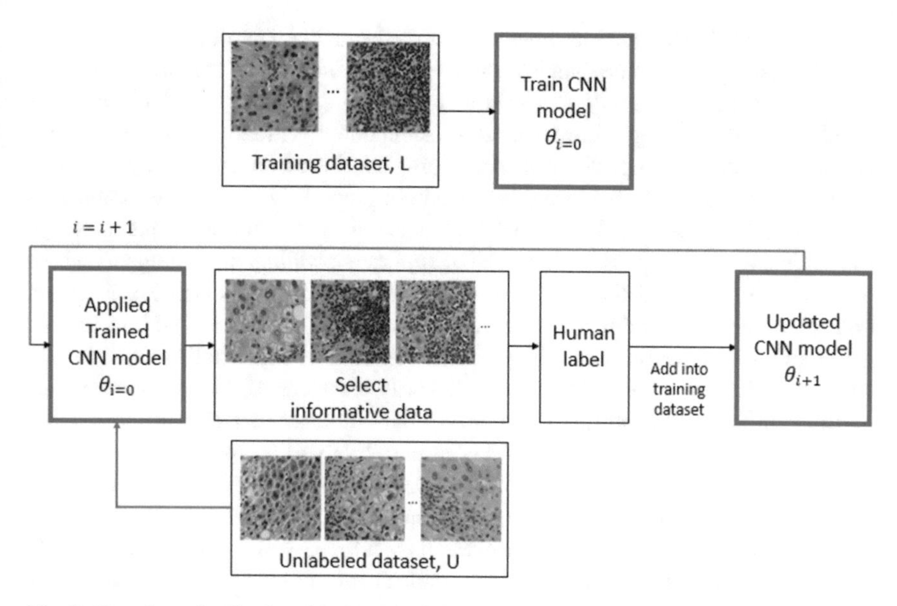

Fig. 8 Flowchart of active learning framework

via a limited amount of labeled data, in which samples with the highest confidence were added to the training set and low confidence samples were subsequently added as pseudo-labels. AL methods have also been applied for semantic segmentation [21] and medical image segmentation [11, 24, 33]. Among the latter methods, Yang et al. [33] proposed an AL framework utilizing the uncertainty and similarity information provided by a fully convolution network (FCN) and formulated a generalized version of the maximum set coverage problem to determine the most representative and uncertain areas for annotation. The potential of the proposed method was demonstrated by segmenting glands in histology images and lymph nodes in ultrasound images. Gorriz et al. [11] and Ozdemir et al. [24] used a Monte Carlo (MC) dropout method to sample uncertain samples and then selected representative samples based on their content information using a specially designed penalty scheme to maximize the information at the network layer.

However, when AL is used to select informative samples for labeling, the following question arises, "Can the strategy be designed in such a way as to favor the minority data and reduce the imbalance?" In addressing this question, several approaches for favoring minority samples during the AL process have been proposed [8, 10]. In general, the results have shown that AL is innately a good choice for overcoming the class imbalance problem. Sadafi [26] measured the uncertainty in AL through the dropout, which means randomly ignore neurons during forward or backward process, in a fast Region-based CNN (R-CNN) and solved the imbalanced data problem by lowering the classification threshold for the minority instances (red blood cells) so as to increase their classification probability.

Li et al. [19] developed an imbalance effective active learning (IEAL) method for addressing the extreme data imbalance problem inherent in the classification of sparse plasma cells. For a typical WSI with a size of around 30 G and an enormous number of cells, finding informative patches containing “virtually zero” plasma cells is extremely challenging. Standard AL approaches based on uncertainty sampling simply disregard these classes and hence cannot properly solve the class imbalance problem. By contrast, IEAL performs both uncertainty measurement and class type estimation on the data pool. In particular, it applies uncertainty measurement on both the overall dataset and the dataset of each individual type of class. Furthermore, both randomness sampling and minority oversampling are embedded into the AL procedure, where the over-sampling minority operation intentionally selects further samples which are predicted to contain minority instances. IEAL additionally flips and hue and color transforms the minority data in order to augment the number of minority samples. As a result, IEAL is more likely to select samples that truly contain minority instances, thereby mitigating the extreme data imbalance problem.

In Li et al. [19], U-Net with dropout is applied for the segmentation of nuclei, lymphocytes, and plasma cells. However, to train the U-Net model, an appropriate loss function must first be defined. For pathology images, it is frequently necessary to perform small object segmentation. This can be regarded as one form of data imbalance problem, and consequently, the loss function must be chosen accordingly. Several studies have reported that the dice loss (DL) function, which focuses on tuning the IoU of the positive samples for binary segmentation, may represent a suitable choice [20, 22]. Mathematically, the DL function is computed as

$$\mathrm{DL} = 1 - \mathrm{DSC}\left(p, g\right), \tag{2}$$

where

$$\mathrm{DSC}\left(p, g\right) = \frac{2\sum_{i}^{N} p_i g_i}{\sum_{i}^{N} p_i^2 + \sum_{i}^{N} g_i^2}. \tag{3}$$

However, while the DL function performs well for images with an imbalanced distribution of binary positive (foreground) and negative (background) samples, it performs less well for multiclass segmentation problems in which the positive samples (e.g., the nuclei, lymphocytes, and plasma cells) themselves include class imbalance. For such problems, the DL function must be calculated for each class separately and then averaged to obtain a final DLC score (Eq. (4)). In this way, each class provides an equal contribution toward the final loss and the imbalance effect between classes is mitigated:

$$\mathrm{DLC} = \frac{1}{C}\sum_{c=0}^{C}\left(1 - \mathrm{DSC}\left(p_c, g_c\right)\right). \tag{4}$$

Figure 9 shows a distribution of the nucleus, lymphocyte, and plasma cells in the patches of a plasma-rich infiltrated AIH WSI from our experiment. It is seen

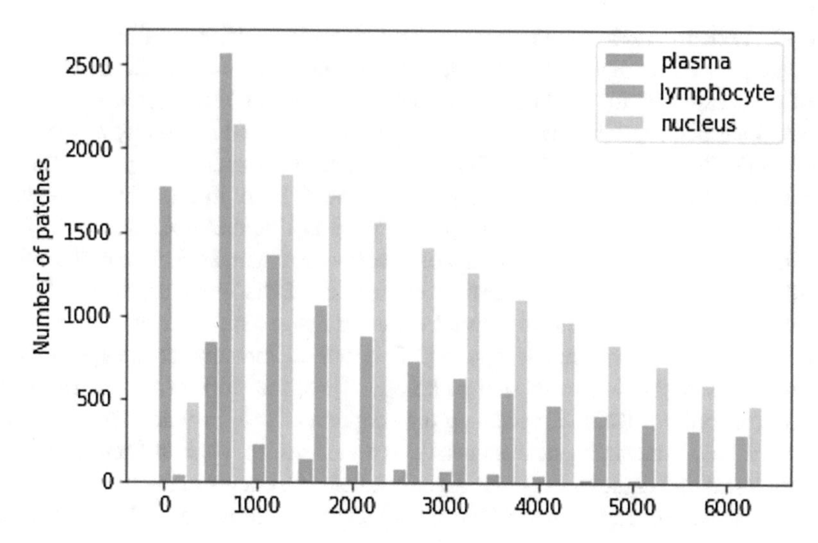

Fig. 9 Illustrative pixel distribution histogram for nucleus, lymphocyte, and plasma cells within a patch [19]

that most of the patches do not contain plasma cells. By contrast, the nuclei are abundant in almost all of the patches. Overall, the pixel ratio for the plasma cells, lymphocytes, and nuclei is equal to around 1:8:10.

The IEAL method presented in [19] was evaluated using 43 H&E stained liver histopathology slides collected from National Cheng Kung University Hospital in Taiwan. The slides were scanned using two different devices, namely, an OLYMPUS BX51 scanner at 20× optical magnification (equivalent to 0.32 μm/pixel) and a Leica scanner at 40× optical magnification (equivalent to 0.25 μm/pixel). The data scanned by the Leica device were downsampled to 0.32 μm/pixel to obtain the same resolution as the OLYMPUS data. Both sets of data were then segmented using the IEAL model. The experimental results showed that the system achieved the same accuracy with 50% samples for plasma and lymphocytes and 55% samples for the nuclei.

5 System Integration and Interoperability

Artificial intelligence has immense potential for the development of digital pathology. However, the success of digital pathology ultimately depends on achieving a seamless integration of the system into the daily clinical workflow. For example, in current pathology scanning systems, the WSIs are stored in a proprietary format, which hinders the integration of systems from different vendors. While open-source software is available for reading such commercial data formats, this software provides the capability only for accessing the image pixel values. In other words, the metadata relating to the clinical context remain inaccessible. Accordingly, the need to establish a pathology-specific Digital Imaging and Communications in Medicine (DICOM) standard for facilitating the interoperability of WSIs between different systems has been specifically recognized by both the Digital Pathology Association and the DICOM working group. The resulting DICOM standardization is expected to facilitate the work of the Integrating the Healthcare Enterprise (IHE) in achieving not only within-hospital integration but also cross-institution integration.

For AI to achieve truly platform-agnostic software, a standardization of the image quality is also required. As described earlier, the WSIs obtained from different commercial scanners frequently reveal different color tones even when the same stain and staining protocol is applied. This poses immense challenges to any software-based solution for digital pathology. Thus, the work of the International Color Consortium (ICC) in promoting the adoption of open, vendor-neutral, cross-platform color management systems is expected to be vital in facilitating the development and deployment of AI digital pathology systems in the future.

Current AI software aims to help pathologists by prescreening WSIs so as to reduce the time required for diagnosis, increase the diagnosis accuracy, and decrease interobserver variability. However, in the future, AI is expected to be extended to the actual prognosis estimation task itself. Clearly, such an assistive operation must be tightly coupled with the daily diagnosis and treatment evaluation process. As a result, extensive and ongoing cooperation between the AI software development community, international standardization organizations, medical institutions, and hardware/software suppliers is essential in unlocking the undoubted potential of AI assistive tools for digital pathology.

Acknowledgments The authors would like to thank Drs. Hung-Wen Tsai and Nan-Haw Chow of National Cheng Kung University Hospital and Tsung-Lung Yang of Kaohsiung Veteran General Hospital for their data support and provision of domain knowledge. The authors would also like to thank Wei-Che Huang and Qi-En Xiao of National Cheng Kung University for making their experimental data available. The authors would like to thank National Center for High Performance Computing, Taiwan, for providing the computing power. Finally, the authors wish to acknowledge the financial support provided to this work by the Ministry of Science and Technology (MOST), Taiwan, under Grant No. MOST 108-2634-F-006 -004.

References

1. H. Cao, S. Bernard, L. Heutte, R. Sabourin, Improve the performance of transfer learning without fine-tuning using dissimilarity-based multi-view learning for breast cancer histology images, in *International Conference Image Analysis and Recognition*, (2018), pp. 779–787
2. C.H. Chan, T.T. Huang, C.Y. Chen, C.C. Lee, M.Y. Chan, P.C. Chung, Texture-map based branch-collaborative network for oral cancer detection. IEEE Trans. Biomed. Circ. Syst. **13**(4), 766–780 (2019)
3. N.V. Chawla, K.W. Bowyer, L.O. Hall, W.P. Kegelmeyer, SMOTE: synthetic minority over-sampling technique. J. Artif. Intell. Res. **16**, 321–357 (2002)
4. E.L. Chen, P.C. Chung, C.L. Chen, H.M. Tsai, C.I. Chang, An automatic diagnostic system for CT liver image classification. I.E.E.E. Trans. Biomed. Eng. **45**(6), 783–794 (1998)

5. L.C. Chen, G. Papandreou, I. Kokkinos, K. Murphy, A.L. Yuille, DeepLab: semantic image segmentation with deep convolutional nets, Atrous convolution, and fully connected CRF. IEEE Trans. Pattern Anal. Mach. Intell. **4**(4), 834–848 (2016)
6. H. Chen, X. Qi, L. Yu, Q. Dou, J. Qin, P.A. Heng, DCAN: deep contour-aware networks for object instance segmentation from histology images. Med. Image Anal., 496–504 (2017)
7. A. Cruz-Roa, H. Gilmore, A. Basavanhally, M. Feldman, S. Ganesan, N.N. Shih, A. Madabhushi, Accurate and reproducible invasive breast cancer detection in whole-slide images: a deep learning approach for quantifying tumor extent. Sci. Rep. **7**(46450) (2017). https://doi.org/10.1038/srep46450
8. S. Ertekin, J. Huang, L. Bottou, L. Giles, Learning on the border: active learning in imbalanced data classification, in *Proceedings of the Sixteenth ACM Conference on Conference on Information and Knowledge Management*, (2007), pp. 127–136
9. W.H. Fridman, F. Pages, C. Sautes-Fridman, J. Galon, The immune contexture in human tumours: impact on clinical outcome. Nat. Rev. Cancer **12**(4), 298–306 (2012)
10. C. Fu, W. Qu, Y. Yang, Actively learning from mistakes in class imbalance problems. IFAC Proc. Vol. **46**(13), 341–346 (2013)
11. M. Gorriz, X. Giro-i-Nieto, A. Carlier, E. Faure, Cost-effective active learning for melanoma segmentation, in *ML4H: Machine Learning for Health Workshop at NIPS*, (2017)
12. K. He, R. Girshick, P. Dollár, Rethinking Imagenet Pre-training. arXiv preprint arXiv:1811.08883
13. W.C. Huang, P.C. Chung, H.W. Tsai, N.H. Chow, Y.Z. Juang, C.H. Wang, Automatic HCC detection using convolutional network with multi-magnification input images, in *2019 IEEE International Conference on Artificial Intelligence Circuits and Systems (AICAS)*, (2019), pp. 194–198
14. K. Ishak, A. Baptista, L. Bianchi, F. Callea, J. De Groote, F. Gudat, H. Denk, V. Desmet, G. Korb, R.N. MacSween, et al., Histological grading and staging of chronic hepatitis. J. Hepatol. **22**(6), 696–699 (1995)
15. M. Kubat, S. Matwin, Addressing the curse of imbalanced training sets: one-sided selection, in *Proceedings of the Fourteenth International Conference on Machine Learning (ICML)*, (1997), pp. 179–186
16. S.K. Lee, C.S. Lo, C.M. Wang, P.C. Chung, C.I. Chang, C.W. Yang, P.C. Hsu, A computer-aided design mammography screening system for detection and classification of microcalcifications. Int. J. Med. Inform. **60**, 29–57 (2000)
17. S.K. Lee, P.C. Chung, C.I. Chang, C.-S. Lo, T. Lee, G.C. Hsu, C.W. Yang, Classification of clustered microcalcifications using a shape cognitron neural network. Neural Netw. **16**, 121–132 (2003)
18. C.C. Lee, P.C. Chung, H.M. Tsai, Identifying multiple abdominal organs from CT image series using a multimodule contextual neural network and spatial fuzzy rules. IEEE Trans. Inf. Technol. Biomed. **7**(8), 208–217 (2003)
19. C.T. Li, H.W. Tsai, T.L. Yang, K.S. Cheng, N.H. Chow, P.C. Chung, Imbalance-effective active learning in nucleus, lymphocyte and plasma cell detection, in *Interpretable and Annotation-Efficient Learning for Medical Image Computing, MICCAI-LABEL 2020, Lecture Notes in Computer Science*, vol. 12446, (2020), pp. 223–232
20. T.Y. Lin, P. Goyal, R. Girshick, K. He, P. Dollár, Focal loss for dense object detection, in *Proceedings of the IEEE International Conference on Computer Vision*, (2017), pp. 2980–2988
21. R. Mackowiak, P. Lenz, O. Ghori, F. Diego, O. Lange, C. Rother, Cereals-cost-effective Region-based Active Learning for Semantic Segmentation, arXiv preprint arXiv:1810.09726 (2018)
22. F. Milletari, N. Navab, S.A. Ahmadi, V-net: fully convolutional neural networks for volumetric medical image segmentation, in *2016 Fourth International Conference on 3D Vision (3DV)*, (2016), pp. 565–571
23. https://tpis.upmc.com/tpislibrary/schema/mHAI.html on Modified HAI Scoring System

24. F. Ozdemir, Z. Peng, C. Tanner, P. Fuernstahl, O. Goksel, Active learning for segmentation by optimizing content information for maximal entropy, in *Deep Learning in Medical Image Analysis and Multimodal Learning for Clinical Decision Support*, (2018), pp. 183–191
25. O. Ronneberger, P. Fischer, T. Brox, U-net: convolutional networks for biomedical image segmentation, in *International Conference on Medical Image Computing and Computer-Assisted Intervention*, (2015), pp. 234–241
26. A. Sadafi, N. Koehler, A. Makhro, A. Bogdanova, N. Navab, C. Marr, T. Peng, Multiclass deep active learning for detecting red blood cell subtypes in brightfield microscopy. Med. Image Comput. Comput. Assist. Intervent. (MICCAI), 685–693 (2019)
27. K. Sirinukunwattana, J.P. Pluim, H. Chen, X. Qi, P.A. Heng, Y.B. Guo, A. Böhm, Gland segmentation in colon histology images: the glas challenge contest. Med. Image Anal. **35**, 489–502 (2017)
28. H. Tokunaga, Y. Teramoto, A. Yoshizawa, R. Bise, Adaptive weighting multi-field-of-view CNN for semantic segmentation in pathology, in *Proceedings of the IEEE Conference on Computer Vision and Pattern Recognition*, (2019), pp. 12597–12606
29. K. Wang, D. Zhang, Y. Li, R. Zhang, L. Lin, Cost-effective active learning for deep image classification. IEEE Trans. Circ. Syst. Video Technol. **27**(12), 2591–2600 (2016)
30. Q.E. Xiao, P.C. Chung, H.W. Tsai, K.S. Cheng, N.H. Chow, Y.Z. Juang, H.H. Tsai, C.H. Wang, T.A. Hsieh, Hematoxylin and Eosin (H&E) stained liver portal area segmentation using multi-scale receptive field convolutional neural network. IEEE J. Emerg. Select. Top. Circ. Syst. **9**(4), 623–634 (2019)
31. Y. Xu, Z. Jia, Y. Ai, F. Zhang, M. Lai, E.I.C. Chang, Deep convolutional activation features for large scale brain tumor histopathology image classification and segmentation, in *ICASSP, IEEE International Conference on Acoustics, Speech and Signal Processing – Proceedings*, (2015), pp. 947–951
32. Y. Xu, Y. Li, M. Liu, Y. Wang, M. Lai, E.I.C. Chang, Gland instance segmentation by deep multichannel side supervision, in *Lecture Notes in Computer Science (Including Subseries Lecture Notes in Artificial Intelligence and Lecture Notes in Bioinformatics)*, (2016), pp. 496–504
33. L. Yang, Y. Zhang, J. Chen, S. Zhang, D.Z. Chen, Suggestive annotation: a deep active learning framework for biomedical image segmentation, in *International Conference on Medical Image Computing and Computer-Assisted Intervention*, (2017), pp. 399–407
34. W.J. Yang, Y.T. Cheng, P.C. Chung, Improved lane detection with multilevel features in branch convolutional neural networks. IEEE Access **7**, 173148–173156 (2019)
35. Y. Zhou, H. Chang, K. Barner, P. Spellman, B. Parvin, Classification of histology sections via multispectral convolutional sparse coding, in *Proceedings of the IEEE Computer Society Conference on Computer Vision and Pattern Recognition*, (2014), pp. 3081–3088
36. B. Zhou, A. Khosla, A. Lapedriza, A. Oliva, A. Torralba, Learning deep features for discriminative localization, in *2016 IEEE Conference on Computer Vision and Pattern Recognition (CVPR)*, (2016), pp. 2921–2929
37. Z. Zhou, J. Shin, L. Zhang, S. Gurudu, M. Gotway, J. Liang, Fine-tuning convolutional neural networks for biomedical image analysis: actively and incrementally, in *Proceedings of the IEEE Conference on Computer Vision and Pattern Recognition*, (2017), pp. 7340–7351
38. R.X. Zhu, W.K. Seto, C.L. Lai, M.F. Yuen, Epidemiology of hepatocellular carcinoma in the Asia-Pacific region. Gut Liver **10**(3), 332–339 (2016)

Pau-Choo (Julia) Chung (S'89-M'91-SM'02-F'08) received her PhD degree in electrical engineering from Texas Tech University, USA, in 1991. She then joined the Department of Electrical Engineering, National Cheng Kung University (NCKU), Taiwan, in 1991 and became a full professor in 1996. She served as Head of Department of Electrical Engineering (2011–2014), Director of Institute of Computer and Communication Engineering (2008–2011), Vice Dean of College of Electrical Engineering and Computer Science (2011), Director of the Center for Research of E-life Digital Technology (2005–2008), and Director of Electrical Laboratory (2005–2008), NCKU. She was elected Distinguished Professor of NCKU in 2005 and received the Distinguished Professor Award of Chinese Institute of Electrical Engineering in 2012. She also served as Program Director of Intelligent Computing Division, Ministry of Science and Technology (2012–2014), Taiwan. She was Director General of the Department of Information and Technology Education, Ministry of Education (2016–2018), Taiwan. Currently she is Director of Computing Center of NCKU.

Dr. Chung has been with IEEE for 32 years. Particularly, she has been involved with IEEE Computational Intelligence Society (CIS) and IEEE Circuits and Systems Society (CASS). She served as Vice President for Members Activities, IEEE CIS (2015–2018) and currently is Chair of High School Outreach Subcommittee. Dr. Chung was Chair of IEEE Computational Intelligence Society (CIS) (2004–2005) in Tainan Chapter, Chair of the IEEE Life Science Systems, and Applications Technical Committee (2008–2009). She was a member in BoG of CAS Society (2007–2009, 2010–2012). She served as an IEEE CAS Society Distinguished Lecturer (2005–2007) and Chair of CIS Distinguished Lecturer Program (2012–2013). She served on two terms of ADCOM member of IEEE CIS (2009–2011, 2012–2014) and as Chair of IEEE CIS Women in CI (2014). She is Member of Phi Tau Phi honor society and is an IEEE fellow since 2008.

Dr. Chung's research interests include computational intelligence, medical image analysis, video analysis, and pattern recognition. Her research journey to computational intelligence started from the analysis of Hopfield associative memories robustness in 1987, when backpropogation learning was developed. After her graduation in 1991, she expanded her researches on neural network model analysis to practical applications. She envisioned the use of the learning to train a neural network model for handling challenging diagnosis, like an experienced doctor. According to her interactions with doctors, she developed neural network models for the processing and analysis of medical images. She also extended her researches to video understanding applied on human behavior analysis and video surveillance. Currently, she is working on pathology image analysis using computational intelligence techniques, especially the deep learning models. She also targets on researches to tackle the essential and challenging issues faced in pathology image analysis.

Prof. Chung's involvement with IEEE started during her PhD. She joined IEEE as a student member in order to obtain the benefit of discount rate for subscribing to IEEE journals. She continued her association with IEEE after gradation and became an IEEE member. She founded IEEE CIS Tainan Chapter, invited CIS senior speakers, and organized activities. She was then invited to serve on subcommittee chairs, run for the ADCOM member, participate with strategy planning, etc. She frequently attends CIS-sponsored conferences, especially IJCNN and WCCI. She also served as Publicity Co-Chair of WCCI 2014, SSCI 2013, SSCI 2011, and WCCI 2010. She also served as Associate Editor of IEEE Transactions on Neural Network and Learning Systems (2013–2015) and Associate Editor of IEEE Transactions on Biomedical Circuits and Systems. Through her participation with society activities, she makes more friends to work with in the IEEE societies. Based on her experiences, participating with academic societies such as IEEE CIS not only broadens her view on researches but also makes her attending conferences more enjoyable. It gives the excitement of meeting friends through the conference occasion.

Chao-Ting Li received her masters' degree in Computer and Communication Engineering from National Cheng Kung University, Taiwan, 2019. During her MS study, she majored in image processing and pattern recognition and published a conference paper in IEEE International Computer Symposium (ICS 2018). She hold a bachelor's degree in Communication Engineering from National Central University, Taiwan, 2017. During that time, she engaged in a project of wireless network and published a conference paper in Wireless, Ad Hoc and Sensor Network (WASN 2016).

During her study in NCKU under the supervision of Prof. Pau-Choo Chung, she focused on medical image analysis and pattern recognition. During her bachelor's and master's studies, she took many courses related to communication and computer science, such as engineering mathematics, digital signal processing, machine learning, and graphic theorem. Thanks to these courses, she have a more solid background and knowledge for the later research on the area of computational intelligence. Her research associated with computational intelligence started from the analysis of histopathological images. Her master's thesis focused on the challenge of pathology image with deep learning, for example, class imbalance and insufficient medical training data. Thus, she proposed a method to resolve the problem. The challenge of pathology analysis is an interesting issue, and she also gained valuable experiences from it.

Linguistic Intelligence as a Base for Computing Reasoning

Daniela López De Luise

1 Introduction

Although Natural Language Processing is usually focused in textual data, some traditional methods include statistical and heuristic processing of utterances and discourse analyses. This provides a certain quality of linguistic reasoning to an automated system.

From a language perspective, every linguistic level depends to some degree on morphology. Morphosyntactics describe the set of rules that govern linguistic units whose properties are definable by both morphological and syntactic paradigms. It may be thought of as a basis for spoken and written language that guides the process of externally encoding ideas produced in the mind. But semantic and ontological elements of speech become apparent though linguistic pragmatics. In fact, words distribution has been proven to be related to semantics and ontology. Words are a powerful lens of human thought and other manifestations of collective human dynamics.

Perhaps one of the first big contributions in this perspective is Zipf's Empirical Law [10]. It was formulated using mathematical statistics. It refers to the fact that many types of data studied in the physical and social sciences can be approximated with a Zipfian distribution, one of a family of related discrete power law probability distributions, as in Fig. 1.

Zipf's law applies to most languages. Given some corpus of natural language utterances, the frequency of any word is inversely proportional to its rank in the frequency table. Eq. 1 describes the behavior.

D. López De Luise (✉)
CI2S Labs, CAETI, IDTI Labs, Sociedad Científica Argentina, Buenos Aires, Argentina
e-mail: daniela_ldl@ieee.org

A. E. Smith (ed.), *Women in Computational Intelligence*, Women in Engineering and Science, https://doi.org/10.1007/978-3-030-79092-9_7

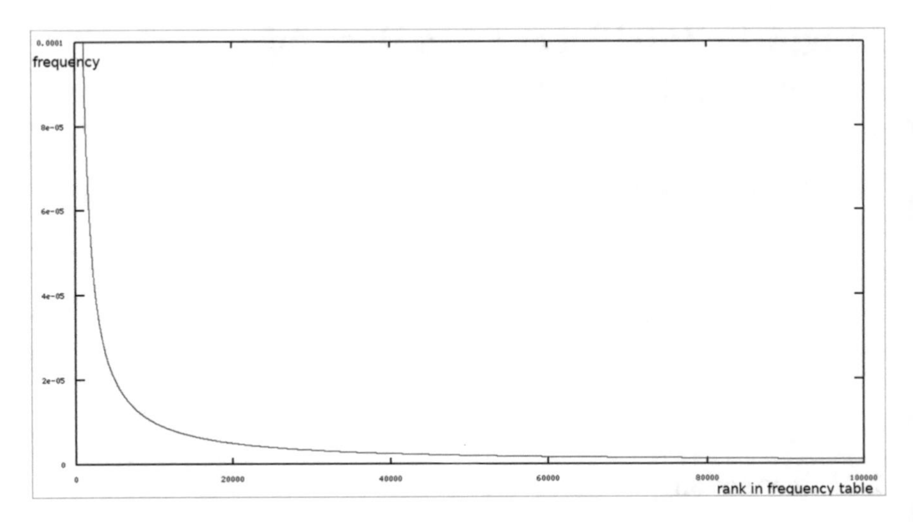

Fig. 1 Zipf's law for texts in English

$$P_n \sim 1/n^a \tag{1}$$

Here $\boldsymbol{P_n}$ stands for the word positioned as the $\boldsymbol{n}$-th most frequent, and $\boldsymbol{a}$ is some real number. Experts explain this linguistic phenomenon as a natural conservation of effort in which speakers and hearers minimize the work needed to reach understanding, resulting in an approximately equal distribution of effort consistent with the observed Zipf's distribution.

Following this concept, speech can be processed taking the utterances and their energy distribution. This chapter explains a Computational Intelligence approach for modeling language production, using heuristics and data mining tools: The **M**orphosyntactic **L**inguistic **W**avelets (***MLW***). This provides a possibility to approximate an abstraction process with reasonable precision during learning. It defines a process as a progressive sequence of filtering and selection steps. The approach is named ***MLW*** due to its management of granularity, analogous to traditional wavelets.

In the rest of this chapter, language is presented as a tool for communication (Sect. 2) which includes an introduction to ***MLW*** and an explanation of how utterances in a dialogue can be modeled by fractals, and some applications are presented. In Sect. 3 are the conclusions and current and future work.

2 Language as a Tool for Communication

Language is one of the main encodings for human interactions. There are several channels to interchange language information, but it is usually reduced to visual or auditory representations. Many times, visual expression consists of a set of numbers or a specific set of sign, icons, or ideograms. Among them only words and numbers are linked to specific sounds. That is the reason of their success and relevance as a communication tool.

This section next introduces ***MLW***, an automated approach to model language usage in current natural language pragmatics, and the relationship between fractals and oral dialogue.

2.1 MLW

Since 2003, **C**omputational **I**ntelligence and **I**nformation **S**ystems **L**abs (CI2S Labs) has published a set of preliminary analysis [7] explaining how humans express themselves using an encoding that is based on mathematical equations. The set of equations, and steps to interrelate them, were collected into a heuristic called **M**orphosyntactic **L**inguistic **W**avelets (***MLW***).

The brain can be compared with a unique device that can perform highly complex processing and at the same time is able to change itself with plasticity. This ability must be also modeled in order to generate an accurate representation.

Data-driven learning is a type of Machine Learning that linguists use to understand the way people build sentences, that is, their natural language bias. But it has many problems to focus deep inside linguistics. That may happen due to the statistical bias, or because the heuristic bias is too much.

However, ***MLW*** is an approach to heuristically model semantics without using man-made ensembles of language productions like dictionaries or special ontological metadata. Nowadays, many data heuristics in deep learning claim to be suitable for linguistic processing and are quite similar to some part of the already existing ***MLW***.

This approach constitutes a mimic of classical wavelets used for signal processing (typically sound or image data) as they are very powerful since they can compress and successfully represent diverse and complex analogical data.

The first and main challenge for ***MLW*** is the systematic and automatic derivation of mathematical representations from something that is not numeric. Other challenges are to automatically represent subjectivity and context and derive commonsense parameters. To overcome these tasks means to be able to model successfully not only the physical and ontological bias[1] of words but also sentences and context.

Using a wavelet-like approach, it is possible to process texts and expressions and gain understanding of more subtle issues like early bilingualism and deaf effects on learning [6].

[1]Bias: in statistics. A systematic as opposed to a random distortion of a statistic as a result of sampling procedure. I have added it to the references.

Table 1 *MLW* rules for symbolic representation in Spanish

ID	RULE OF APPLICATION	SYMBOLIC REPRESENTATION
1	A de B [y | o C[y |o ...]]	A ... C B
4	A y|e B [XXX]	A B y XXX
7	A o B [XXX]	A B o XXX
31	A se B	A→B

One interesting derivation of this technology is symbolic language, an alternate representation of spoken words that entails the language flow of intended expressions in a universal way. Table 1 shows a few symbols and the rules of its usage for Spanish.

The translation from textual to symbolic representation can be performed automatically. When it was tested with native volunteers, they were able to understand and rebuild original texts almost with the same wording as the original text. Figure 2 shows the representation of a short text in Spanish. It is the explanation of the topaz (topacio in Spanish), taken from Wikipedia in 2007, with 245 words.

Most of ***MLW*** consists of a progressive parameterization of current textual expressions following simple heuristics. This way, it is possible to generate a set of parameters named "descriptors" that can be used for semantic clustering using structures of the sentence. It is expected that future work can use this approach also to generate sentences as part of speech in natural language.

The underlying hypothesis is that the fewer handmade structures applied, the better language dynamics bias is taken into consideration. Therefore the inherent subtle devices of language production are part of the model.

One interesting finding is that word dynamics and concept handling can be exhibited as graphical 3D representations. For example, in Fig. 3 the red line in curve on the right side represents the specific way a volunteer uses verbs, nouns, adverbs, etc. In contrast, the red curve on the left is a distribution obtained for a random set of words (i.e., a text consisting in random words). Every person has a different curve.

It is interesting to note that MLW confirms the Zipfian law, but in dialogues. Another interesting derivation is the ability to approximate the sound derivation of oral dialogues. This will be described in the following subsection.

2.2 Sounds and Utterances Behavior

MLW is a unique approach suitable for both spoken and written language. This section describes the relevance in **N**atural **L**anguage **P**rocessing (NLP) of sound

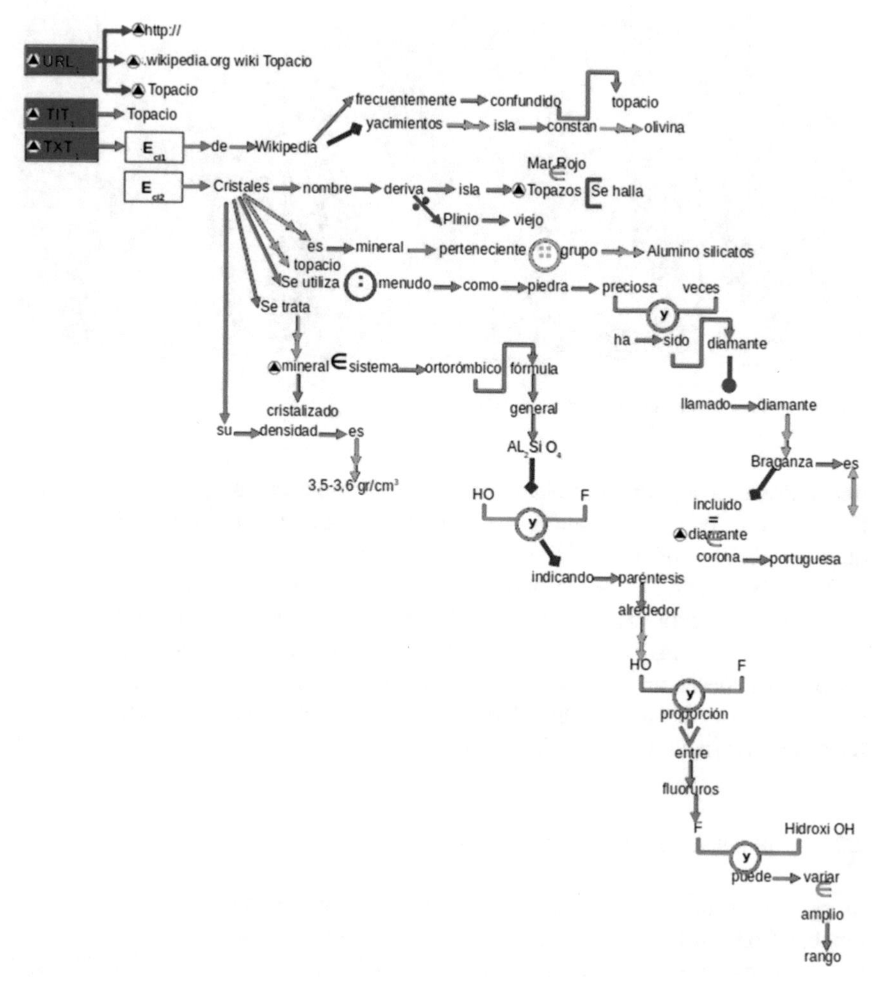

Fig. 2 Symbolic representation of topacio's text

processing and phonetics and its role in human knowledge from a Verbal Behavior perspective as published in several papers. For more details see [8].

Verbal **B**ehavior (***VB***) was proposed by B. F. Skinner [12]. He stated that language may be thought as a set of functional units, with certain types of components named "operants" acting during language acquisition and production. An *operant* is a functional unit of language with a specific function. He described a group of verbal *operants*: Echoics, Mands, Tacts, and Intraverbals. These components of language are necessary for effective verbal communication and should be the focus of intervention intended to teach language.

It can be said that ***VB*** is shaped and sustained by a verbal environment. In this context, a dialogue is an interaction between a speaker and a listener, with the

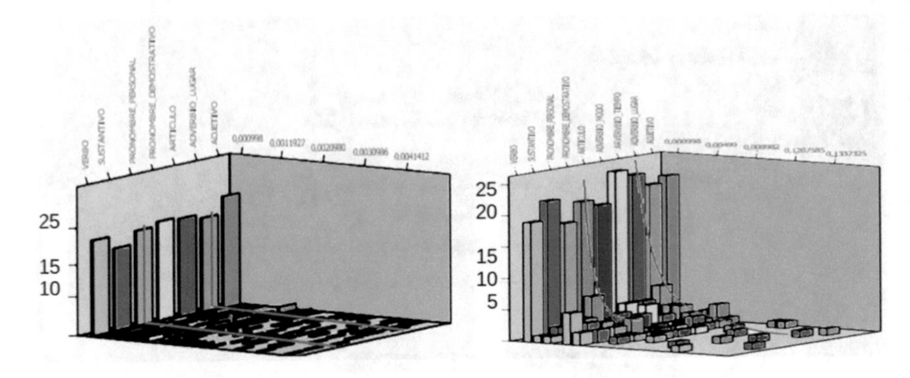

Fig. 3 Symbolic representation of topacio's text

production of a certain set of phenomena. Some examples are a gesture of surprise, the change in body temperature, the feeling of tension when faced with a certain question, the tone of voice of the interlocutor, etc.

There is growing evidence pertaining to the onto-genetic sources of language and its development [5]. There is also increased interest in the theory that verbal capability in humans is a result of evolution and onto-genetic development. It can be said that speech and sounds have an important relation to ***VB*** [3] and may serve as a way to identify certain disorders. An example is the case of patients suffering from ASD (autistic spectrum disorder).

2.2.1 Spoken Language and L-Systems

In general, audible sounds are analog signals (the opposite to digital signals), ranging from 20 Hz to 20 kHz. But the range may vary for every individual and age (known as presbyaudia). This range covers 10 octaves ($2^{10} = 1024$). Beyond the audible spectrum, there is ultrasound (acoustic waves with frequency over 20 KHz).

Approaches for sound processing range from the classical Fourier transforms and spectrograms (time-frequency analysis) to modern heuristics like dynamical spectrum, wavelets, Mel coefficients, Teager operators, and signal intensity. Many of them are explained in [1].

The relationships between sound, utterances, and speech [3] can be modeled and become useful tools for tracking patients with several disorders. For instance, in children with language acquisition anomalies, there is language degradation and perhaps even complete absence of speech.

Sound can be represented by curves and approximated by recursive methods such as Lindenmayer systems, also known as L-Systems. An L-system is a parallel rewriting system (i.e., a non-sequential system of production) and a type of formal grammar. It consists of an alphabet of symbols that can be used to make strings and a collection of production rules. These grammar rules indicate the way certain well-defined symbols may expand into a larger string of symbols. There is also an initial *axiom* string from which to begin construction and a mechanism for translating the generated strings into geometric structures.

L-systems have also been used to model the morphology of a variety of organisms and can be used to generate self-similar fractals [2] such as iterated function systems.

For the spoken language study case, the symbols to be considered are $S = \{X, Y, F, -, +, \varepsilon\}$, and the following are grammar rules:

```
R1: X → X+YF+
R2: Y → -FX-Y
R3: F → ε
```

And the axiom: ***FX***.

Let symbol ε stand for the empty string, which means that we will delete any ***F*** from the result when the process is expanding the string using rules ***R1*** to ***R3***.

When symbols are evaluated with:

```
F: move forward one step
-: turn right 90°
+: turn left 90°
```

The resulting graph is a Dragon curve.

2.2.2 Dragon Curves

The Dragon curve is a family of self-similar fractal curves. It is an **I**terated **F**unction **S**ystem (IFS) made up of the union of several copies of itself, each copy being transformed by function:

$$f_1(z) = \frac{(1+i) * z}{2} \tag{2}$$

$$f_2(z) = 1 - \frac{(1-i) * z}{2} \tag{3}$$

With starting points {0, 1}, it can be graphically represented with its iterations 1, 2, 3, 4, 5, 6, 7, and 8, as in Fig. 4.

2.2.3 Dialogue Content and Its Relation with Energy Distribution

Mining the energy distribution of sound tracks leads to information that might be related to statistical interpretations and express the manner in which humans use spoken language. Time-related features of the samples are closely related to simple

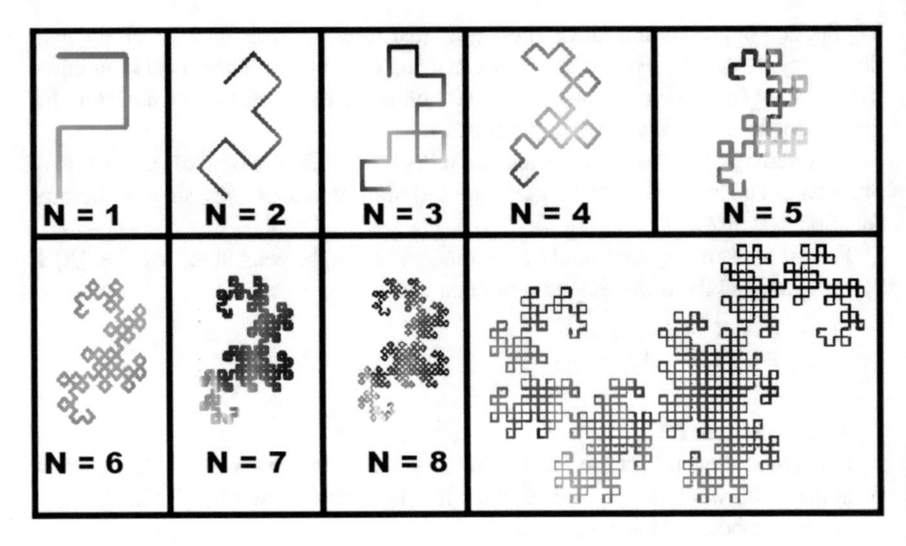

Fig. 4 Dragon plot

changes in the signal energy. As a consequence, these changes may be used to distinguish recordings. This type of analysis is a very simple way to characterize the way every person speaks, but they lack strong precision and must be combined with frequency information.

If we split the sound slice 'frame$_i$' into $x_i(n)$; $n = 1 \ldots N$, each of length N. Energy for each frame$_i$ is given by Eq. 4:

$$E(i) = \frac{1}{N} \sum_{n=1}^{N} |x_i(n)|^2 \tag{4}$$

This is used to automatically detect silences and to discriminate different types of sounds. Figure 5 presents distributions for both music and speech, making the large differences evident between them.

In general, voice productions have many silences between periods of high energy. As a consequence, the convolution curve[2] has many changes, and its standard deviation σ^2 may be used to detect speech. Traditionally, the equation (σ^2/μ) is used to get energy of this curve.

[2]In general, convolution is a mathematical operation on two functions (*f* and *g*) that produces a third function (*f* * *g*) that expresses how the shape of one is modified by the other.

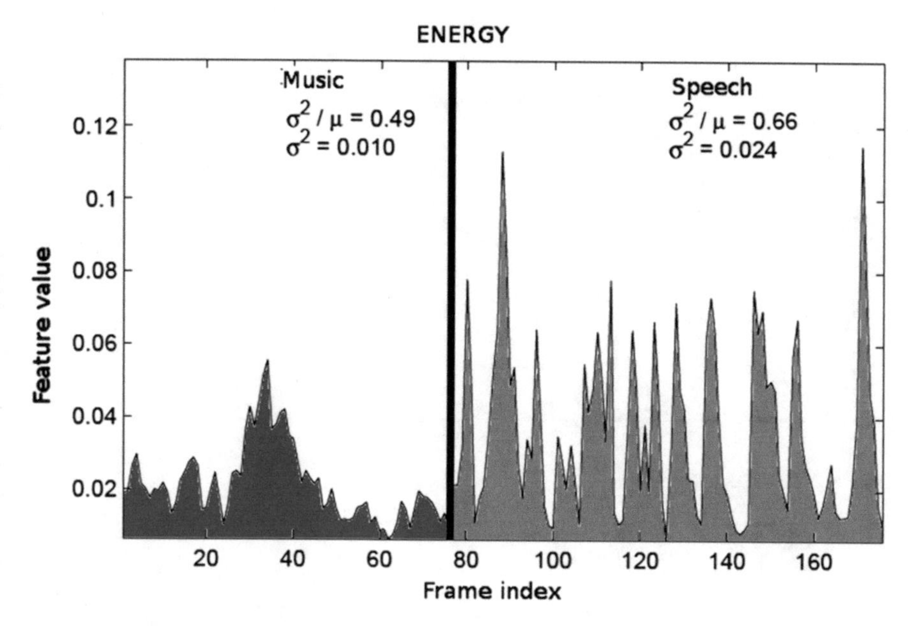

Fig. 5 Comparison of energy distribution for speech and music

2.2.4 Spoken Language Analyses

Spoken language (consisting mainly of riddles, puzzles, locutions, and dialogues) can be modeled using certain rules to produce utterances in natural language. Relevant information is contained in the **S**hort **T**ime **E**nergy, the energy distribution framed values, in the following denoted as ***STE***.

Coefficients from the fractal Dragon and those derived from ***STE*** for sound records exhibit a close relation. Figure 6 is the plot for both distributions (***STE*** and DRAGON) in platform Infostat (c) [4]. We see a combination of linear correlation with some extra dots (it may be a sampling error).

Linear regression confirms that the correlation is statistically relevant. When each STE is projected on DRAGON, the figure suggests a small difference between them (see Fig. 7). Something similar happens when Zipfian fractal is used to test regression, but with plain written texts instead of spoken dialogues.

This shows that variation in data due to residual values is not significant. After some technical verification (see [8] for more details), it is found that this similarity is due to the fact that Dragon is a better model to approximate STE (the energy distribution of the speech) in the case of spoken language in the context of dialogues. Zipf fractal remains as a better model for texts in natural language.

This explains the relevance of sound processing for evaluating linguistics performance from a fractal point of view. Since spoken language relates to written texts, it can also be used to analyze text productions. Fractal behavior has an energy

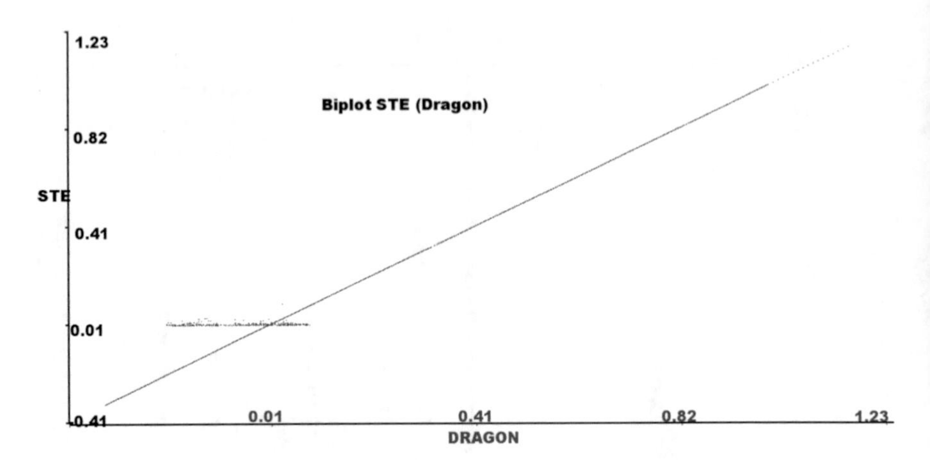

Fig. 6 STE vs. Dragon plot

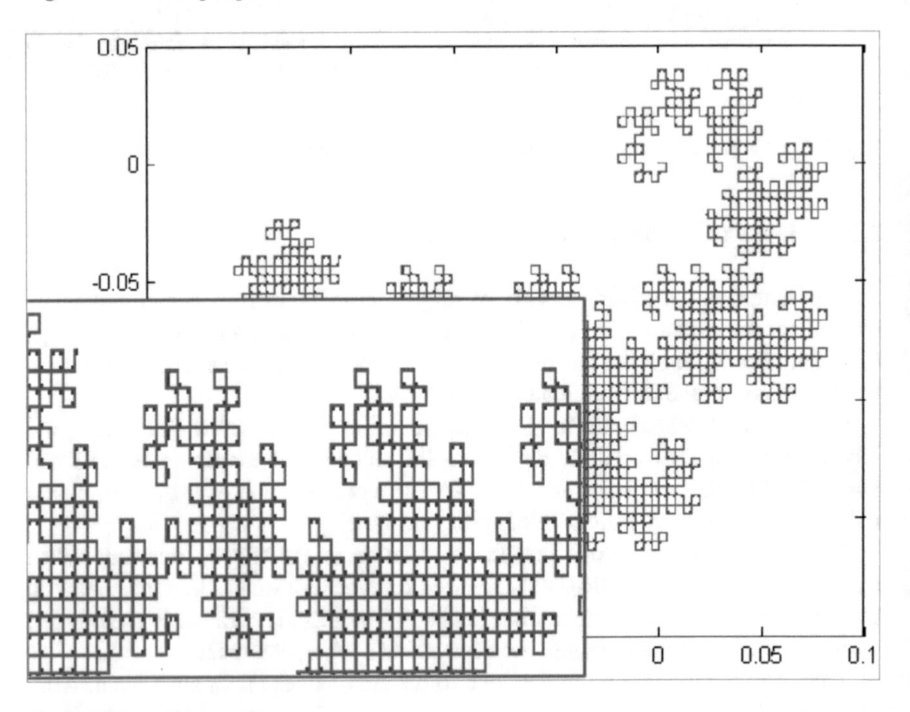

Fig. 7 STE vs. Dragon plot

distribution according to a Dragon curve. The author is currently working to adjust and explain the small differences and the pattern behind the differences between them.

3 Conclusions and Future Work

Language is a key part of daily performance. It is not only a set of vocalizations or texts. Its main goal is to transport information from a sender to a receiver, functioning as an encoding. But there are other ways to encode information, other "languages" that may be very diverse. Those languages sometimes can be easily related to objects and events, but sometimes it is hard to find the link between them. For instance, R. Berwick described verbal behavior as body manifestations as an alternate encoding of language. In fact music itself is an encoding that can be described by a special type of mathematics named harmony [11].

Some work performed in CI2S Labs has been focused on how to transform visual information to acoustic information. As part of the project named ***HOLOTECH*** [9], a prototype in Android was developed to help blind people outdoors with a sound language. The goal is to prevent the individual against cars, obstacles, or any other object in the surrounding. This type of translation between a visual to an auditory encoding is very hard since sight receives information in a very efficient, synthetic, and compact way, but the ear processes the information sequentially.

To summarize, information is everywhere and encoded in very diverse ways: movements, music, speech, noise, drawings, shapes, colors, dog barking, whale songs, etc. Sometimes, its expressions are more subtle than others. Also, information flows constantly, but just a few encodings are considered by formal linguistics.

Furthermore, there are many very well-known approaches to quantify and track spoken and written information, but there is still a long way to fully understand the relationship between energy (closely related in this context to entropy), context, and language dynamics. For instance, Natural Language is still hard to produce automatically. Other pending works are how information can flow, new transmission channels, how reasoning manages and creates sentences, and how entropy relates to reasoning with information. This is an exciting world that is not the property of humans, but of nature.

References

1. J. Alm, J. Walker, Time-frequency analysis of musical instruments. SIAM **44**(3), 457–476 (2002)
2. J.R. Azor Montoya, J.R. Perez, *Análisis del comportamiento autosimilar con distribución Pareto del tráfico Ethernet*. (Anales del XVI Congreso Argentino de Ciencias de la Computación, 2010), pp 118–124.
3. P. Bustamante, A. Lafalla, D. López De Luise, C. Párraga, R. Azor, J. Moya, J.L. Cuesta, *Evaluación sistemática del comportamiento somático y oral como respuesta a estímulos sonoros en pacientes con TA*. CLAIB (Congreso Latino Americano de Ingeniería Biomédica, Latin America Congress of Bioengineering, 2014).
4. J.A. Di Rienzo, F. Casanoves, M.G. Balzarini, L. Gonzalez, M. Tablada, C.W. Robledo, *InfoStat*, (Córdoba University, Argentina, 2013). URL http://www.infostat.com.ar

5. G.D. Douglas, The ontogenetic selection of verbal capabilities. Int. J. Psychol. Psychol. Ther. **8**(3), 363–386 (2008)
6. D. López De Luise, MLW and bilingualism. Advanced research and trends in new technologies, software, human-computer interaction, and communicability. IGI Global, 555–586 (2013)
7. D. López De Luise, *El uso de Soft Computing para el modelado del razonamiento*. (Argentina Scientific Society (SCA) Transactions of Sociedad Científica Argentina, 2019), pp. 12–18.
8. D. López De Luise, R. Azor, Sound model for dialogue profiling. Int. J. Adv. Intell. Paradig. Archive **9**(5–6), 623–640 (2017)
9. N. Park, D. López De Luise, D. Rivera, L. Bustamante, J. Hemanth, Multi-neural networks object identification. SOFA (International workshop on soft computing applications, Arad, Romania). Adv. Intell. Syst. Comput. (2018)
10. S.T. Piantadosi, Zipf's word frequency law in natural language: A critical review and future directions. Psychon. Bull. Rev. **21**, 1112–1130 (2015)
11. B.F. Skinner, *Chapter 1: A Functional Analysis of Verbal Behavior. Verbal Behavior* (Copley Publishing Group. ISBN 978-1-58390-021-5, Acton, 1957)
12. M.L. Sundberg, A brief overview of a behavioral approach to language assessment and intervention for children with ASD. Assoc. Behav. Anal. Newsl. **30**(3) (2007)

Daniela López De Luise was born in 1967 Argentina, in a small city near Buenos Aires named Lomas de Zamora. His father was a physician and researcher specialized in gastroenterology, and he led her first steps in informatics.

One day, in the mid '80s, he suggested that she study the Basic programming language, as new devices called 'personal computers' were appearing that promised to be useful in many different applications. So, being 14 years old, she began her journey through the world of computers, and became so interested that she decided to pursue a university degree in that field. She decided to finish high school quickly in order to be able to accomplish her parent's restriction and move to Buenos Aires with her brother, and approved the last two high school years in one. As a fact of interest, the last student who had done the same at that school had been her father, 39 years earlier.

In 1985, the first year implementing the CBC, UBA was flooded with about 90.000 freshmen. Thus, with 16 years old, she had to get used to the largest city in Argentina and overcome the initial collapse produced in the university due to its sudden increase of the number of students. While studying at the university, she collaborated with her father's research in bioengineering. Along with him and an Electronics Engineer she built the first electrogastrograph in the world, a device to collect and manage electrical signals from several parts of the digestive system. She patented it many years later, in 1992, as MEGASOFT MD/92. Soon she changed her expectations towards a mysterious new field: intelligent systems.New approaches related to operations research, neural networks and other physics-like topics became more interesting to her than the traditional information systems.

She completed 2 years in one again and ended her career in 1988, receiving her diploma in Sytems in August 1989, after a 6-month bureaucratic delay.

In those years she realized the need to have her own space to accomplish her objectives, so she started several small research laboratories and interest groups to work on Artificial Intelligence.

Simultaneously, from 1989 to 2000, she had to spent several years working for small and large companies, trying to fill positions with tasks related to those of her dreams.

Being a member of the IEEE since his student days, she saw that the Institute could be the source of up-to-date and first-line knowledge in Computational Intelligence (CI), facilitating her access to its greatest experts and the most relevant publications. So she decides to be also a volunteer. One of her first actions was to be one of the founders of the Argentine Chapter of the then Neural Networks Society (established in 2002) and now Computational Intelligence Society (CIS). She was its first Chapter Chair (2003-2004), then in 2011, 2012 and 2019, and held other positions on the Executive Board of the Chapter along the years.

In 2004 she creates the oldest local IC event, a biennial contest of original papers produced by students and young professionals known initially as TAR, then becoming TRIC. Its ninth edition is being organized in December 2020. At the Argentine Section, she was Secretary in 2011-2012 and President in 2013-2014.

Over the years she became interested in intelligent approaches to reasoning and linguistics and their applications. She started out working alone. Her boss at the University of Palermo (UP, Buenos Aires, Argentina), engineer Esteban Di Tada, encouraged her to continue and deepen these activities. The activities grew under her direction and over 10 years, resulting in 9 research projects and some 50 people involved (graduates, graduates, doctors, teachers and practitioners). The activity in the UP ended in 2013 with the retirement of the engineer Di Tada, which motivated Daniela to create CI2S Labs as a home for her activities in 2013.

During those years, he resumed contact with his former Professor; engineer Adrian Quijano, who encouraged her to resume her way to the doctorate. He insisted that it was a key step towards her objectives, and he took her to visit the National University of La Plata (UNLP), where he participated years ago in the creation of the Department of Informatics. Her authorities, A. De Giusti, M. Naiouf and G. Rosso, agreed to help her complete her doctorate there.

She obtained her PhD in 2008, together with the first person from UNLP to obtain a doctorate and being the first to obtain a doctorate without belonging to UNLP. In that year she joined the Society for Industrial and Applied Mathematics (SIAM).

During those years she was able to meet many world-renowned professionals, which inspired her. Some of them are B. Widrow (Stanford University) that visited Argentina in 2005 and lectured about associative memories, R. Berwick (MIT, 2009) working with Chomsky's theory features on generative grammar, U. Rammamurthy (Memphis Univ., 2010) in consciousness theory and W. Pedricz (Univ. of Alberta, 2015) in granular computing,

In those years, she started a short collaboration between her team and S. Franklin, exploring the application of consciousness to autonomous mobile robots for disaster zones. Years later she extended that seminal work with her current bacteria reasoning, developed an hybrid morphosyntactic theory using sets and machine learning, called Morphosyntactic Linguistics Wavelets (MLW), and two other theories: Harmonics Systems (HS) and self expansion of statements as part of linguistic reasoning during a Natural Language dialog. Her work in novel approaches to develop human resources in science was recognized in 2012 with the Argentine Sadosky Award. Part of that work derived in her current STEAM (Science, Technology, Engineering, Art and Mathematics) project.

In 2013 the IEEE established the ad-hoc Humanitarian Activities Committee in Argentina, which among I feel passionate by nature since I was very young. Though many obstacles in my world made my path to my dreams difficult, I did not give up. Once a professor told me: keep hamming the same nail, until you get it. I followed his advice and now I find myself driven to explore the world we live in the way I dreamed: as a scientist.

Women have been making significant contributions to science since the earliest times. But sometimes what really makes the difference is to know how simple persons can do a lot. This book includes the history of many females working in Computational Intelligence, as well as mine. We hope it could serve to encourage the next generation of female scientists into the field.

Its functions provided support to 'humanitarian projects' carried out through the Special Interest Groups in Humanitarian Technologies (SIGHT).

Being tools for her work, she became involved with videogames (VG) and found that it was an immature area in the region. So, aimed to solve the problem, she started the local IEEE CIS GTC, Games Technical Committee.

At the 2016 Regional Meeting, IEEE Region 9 Latin America awarded her the Eminent Engineer Award, "for her outstanding contributions to computational intelligence in the field of applied linguistic reasoning."

In 2017 the Argentine Scientific Society (SAC) invited her to collaborate in an initiative to stimulate 'STEAM' (Science, Technology, Engineering, Art and Mathematics). The National Academy of Sciences in Buenos Aires asked her to found and lead the CETI (Center for Intelligent Technologies).

Today, she continues to work on frontier themes of the mathematical foundations of reasoning, the abstraction process in learning and the language incidence in several human activities. She directs three research laboratories (CI2S Labs, IDTI Lab and L-IN-CIE-VIS), has directed more than 10 degree theses and 8 research projects, and is an international consultant in her field for industry and academia. She published several books and chapters and more than 150 publications of various kinds, was also a member of the editorial committee of 3 international journals and transactions and invited speaker at numerous IEEE activities for about a decade.

Message:

I feel passionate by nature since I was very young. Though many obstacles in my world made my path to my dreams difficult, I did not give up. Once a professor told me: keep hammering the same nail, until you get it. I followed his advice and now I find myself driven to explore the world we live in the way I dreamed: as a scientist.

Women have been making significant contributions to science since the earliest times. But sometimes what really makes the difference is to know how simple persons can do a lot. This book includes the history of many females working in Computational Intelligence, as well as mine. We hope it could serve to encourage the next generation of female scientists into the field.

Part II
Learning

Intrusion Detection: Deep Neural Networks Versus Super Learning

Simone A. Ludwig

1 Introduction

Computer security is an important issue especially when thinking about the ever growing use of computer technology. Anyone using a computer is at risk of an intrusion even if the computer is not connected to the Internet or any other network. For example, if the computer is left unattended then anyone can try to access and/or misuse the system. However, the thread of an intrusion is far greater when the computer is connected to the internet or a network. Potentially any user from around the world can reach the computer remotely (to some extent) and might attempt to access private and confidential information or launch some type of attack to bring the system to a halt or make it cease to function.

An intrusion to a computer system does not need to be done manually. The system might be attacked automatically using "engineered" software. A well-known example is the Slammer worm (aka Sapphire), which performed a global Denial of Service (DoS) attack in 2003. Slammer exploited a vulnerability in Microsoft's SQL server that allowed it to disable the database servers and to overload the networks [26]. According to the report, Slammer was referred to as "the fastest computer worm in history" that infected 75,000 computer systems around the world within 10 minutes. Furthermore, the worm caused network outages and events such as canceled airline flights as well as ATM failures, etc.

Private computer users but also professional companies and governmental organizations are at risk. For example, in 2009 the US power grid had been infiltrated by an intruder leaving malware behind that had the potential to shut down the entire US grid [6]. In 2010, a major spy network (GhostNet) was discovered [7]. GhostNet

S. A. Ludwig (✉)
Department of Computer Science, North Dakota State University, Fargo, ND, USA
e-mail: simone.ludwig@ndsu.edu

A. E. Smith (ed.), *Women in Computational Intelligence*, Women in Engineering and Science, https://doi.org/10.1007/978-3-030-79092-9_8

is claimed to have compromised more than 1,000 computers around the world with victims such as foreign ministers and embassies. Another incident that involved the government was in 2008 when the Russian military was accused of launching DoS (Denial of Service) attacks against Georgia during the war over South Ossetia [12].

Given the need to make computer systems securer, the concept of an intrusion detection system (IDS) was introduced. An IDS has the charge to detect when the behavior of a user conflicts with the intended use of the computer or computer network. Before the 1990s, intrusion detection was done by system administrators manually analyzing logs of user behavior and system messages with very limited success of being able to detect intrusions in progress [18]. This has been addressed with early works of Anderson [4] and Denning [13] developing software to automatically analyze data for system administrators. The first IDS to perform intrusion detection in real-time was in the early 1990s [18]. However, intrusion detection still remains an issue due to the increase in the number of computers used as well as the large volume of data in computer networks.

Many different Artificial Intelligence (AI) techniques have been applied in IDS systems. Initially, rule-based systems were employed successfully and are still used mainly as part of larger IDSs. Rule-based systems automatically filter network traffic to analyze user data in order to identify patterns of known intrusions. Intrusions that might be compromising can be reported to an administrator explaining the rules that lead to the intrusion alert. A significant drawback of a rule-based system is that it cannot detect new intrusions or variations of known intrusions [21, 27].

Other AI techniques that have been applied are machine learning techniques that perform classification or clustering. These techniques are often applied to learn to perform intrusion detection automatically from a training set with examples of user behavior or network traffic. The benefit is that these machine learning techniques are able to generalize well from known attacks to variations of those and to even detecting entirely new types of intrusions.

IDSs that scan for known attacks fall into the category of misuse detection. Anomaly detection is another category. Basically, the machine learning techniques learn what normal behavior looks like and then variations from the normal behavior are considered a potential intrusion. Thus, these types of IDSs are able to detect entirely new types of attacks [14, 20].

2 Related Work

For monitoring the data flow in networks in order to prevent temporary and permanent damages caused by unauthorized access, many different systems have been built. However, all these systems cannot detect all types of intrusions since attack permutations are inevitably occurring over time. Different machine learning algorithms have been applied in the past to classify normal and anomalous behavior in a network.

Related work which made use of K-means and K-nearest neighbor algorithms is outlined in [9, 37]. A centroid function was used to choose the average and closest grouping of new instances in order to group similar training examples together. This effectively produces a classifier to classify the attack and non-attack classes.

A Support Vector Machines (SVM) approach for an IDS was used for classification in [1, 16]. Another SVM approach was used to perform feature selection and thus improve the classification rate [19]. A CSV-ISVM algorithm was proposed in order to apply incremental SVM to select candidate support vectors demonstrating the advantages in real-time network intrusion detection [11].

Signature matching anomaly detection, also known as threshold-based anomaly detection, has been widely applied to model network traffic in the past. To provide an example, [15] argues that traditional network-based profile models are not sufficient enough to satisfy user profiles in the environment. Therefore, a genetic algorithm approach was applied to find signatures of pattern detection rules via permutations of parent signatures. In [28], the authors proposed a core-plus-module framework (STAT), which is based on the state transition analysis technique. The state transition analysis technique is used to tailor the design of an IDS to specific traffic types and environments. Another research study investigated a genetic algorithm approach and compared it with other approaches such as Naive Bayes and K-nearest neighbor [1].

Different deep learning methods have been applied to intrusion detection. For example, in [2] a deep belief networks (DBN) approach is proposed where a two-layer restricted Boltzmann machine (RBM) is used to train the network in an unsupervised fashion. This is followed by a feedforward layer whereby backpropagation is used to train this layer in a supervised fashion. The classification accuracy is reported on comparing their approach with SVM, and a hybrid version of DBN with SVM. Another research where the DBN algorithm was implemented is given in [3]. The authors applied the DBN algorithm on the NSL-KDD data set and evaluated their approach in terms of classification accuracy, TP (true positives), FP (false positives), TN (true negatives), and FN (false negatives).

A hybrid approach based on autoencoder and DBN is implemented in [22]. The autoencoder is used to reduce the dimensionality of the data. Afterwards, DBN learning is applied to detect anomalies. The DBN learning is done in two steps. First, the RBMs are trained using an unsupervised approach. Second, supervised training is then performed with the feedforward layer. The evaluation measures reported are TPR (true positive rate), FPR (false positive rate), accuracy, and CPU time.

Since fine-tuning of DBN is a very time-consuming task and the method also suffers from the possibility of only reaching a local optimum, an improved version of DBN is proposed in [23]. In particular, ELM (Extreme Learning Machines) was applied during the training process in order to improve accuracy and efficiency. The improved approach was compared with the vanilla DBN and achieved an improvement of 0.6% (detection rate). It also resulted in the reduction of the execution time by half.

A parallel version of deep neural networks (DNN) was proposed as an accelerated DNN in [30]. Since the training phase is a very time-consuming task, the parallel implementation is used to accelerate the training. The training phase consists of several forward and backward passes. The experimental evaluation consisted of reporting on the accuracy.

In [33], a deep learning approach for flow-based anomaly detection is introduced. In particular, a Software Defined Networking (SDN) environment is used whereby the DNN model is built using just six basic features. The evaluation conducted reported on accuracy while varying different learning rates.

Research on intrusion detection of multiple attack classes using a deep neural net ensemble was presented in [24, 25]. The investigation used the NSL-KDD data set for the experiments.

3 Approaches

This section describes the different approaches that were used. In particular, DNN, Gradient boosting, random forest, XGBoost, and the stacked ensemble algorithms are described.

3.1 Deep Neural Networks

Deep learning is a category in machine learning that involves artificial neural networks. Architectures that belong to deep learning are DNN, DBN, recurrent neural networks, and convolutional neural networks. These have been widely applied to many different research areas such as speech recognition, natural language processing, audio recognition, computer vision, bioinformatics, gaming, and many more. A DNN [32] consists of an input, several hidden layers, and an output layer. The network is trained using backpropagation in order to minimize the error between the actual output and the desired output.

3.2 Super Learner

Ensemble methods are heavily used in the machine learning community due to their success primarily in classification tasks. Ensemble methods can be described as a technique that trains multiple learning algorithms, which achieve significantly higher accuracy than a single learner [31]. The common methods that are used are boosting, bagging, stacking, and a combination of base learners.

Boosting uses a model that was trained on data and then incrementally constructs new models that focus on the errors in the classification made by the previous model. An example of boosting is XGBoost [10].

Bagging involves the training of models on random subsamples. Then each model votes with equal weight on the classification. For example, random forest uses a bagging approach to allow the selection of a random set of features at each internal node to be used [36].

Stacking takes that output of a set of models and feeds them into another algorithm that combines them in order to make a final prediction. For this, any set of base learners and combiner algorithm can be used. The combiner algorithm takes the predictions of the models and combines them with a simple or weighted average approach.

Super learning is a combination method that finds the optimal weights during the calculation of the final prediction [34]. In particular, super learning optimizes the weights of the base learners by minimizing a loss function given the cross-validated output of the learners. Since super learning finds the optimal set of weights for the learners, this guarantees that the performance will be at least as good as the best base learner [34].

Having diversity among the component learners is essential for the performance when constructing an ensemble, however, not only for the performance but also to provide strong generalization ability [38]. The super learner should be able to adapt to various problems given a set of diverse base learners since the weights of the components are optimized for the problem. In addition, there is flexibility in the choice of base learners to be used depending on the requirements or constraints of the problem as well as the computational resources at hand. Figure 1 outlines the architecture of the super learner.

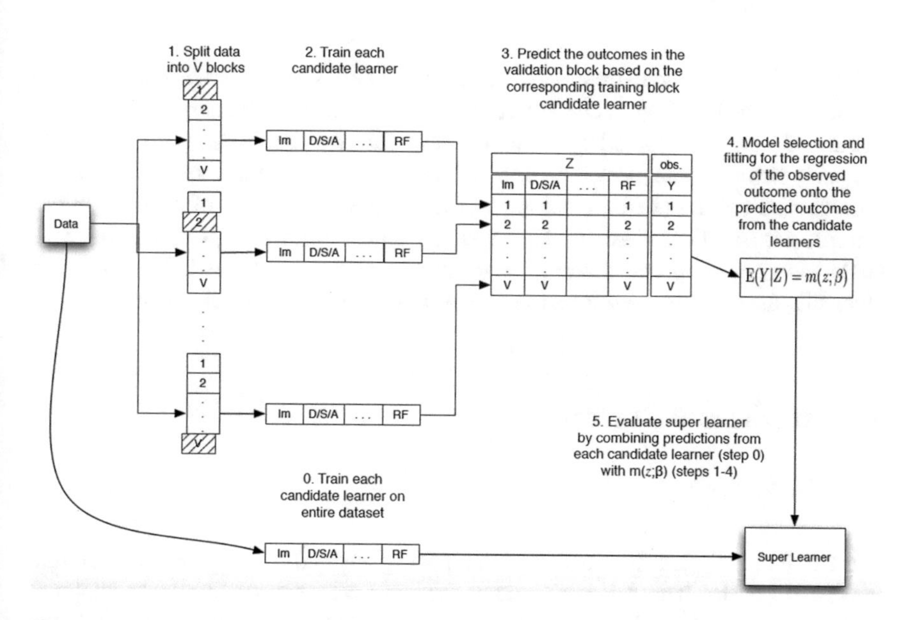

Fig. 1 Super learner diagram [29]

3.3 Gradient Boosting Estimator

The gradient boosting estimator [35] uses an ensemble of weak prediction models to produce a prediction model. This model is built in a stage-wise manner using as a generalization function an arbitrary differentiable loss function. The gradient boosting estimator is one of the most powerful classification and regression algorithm available today.

3.4 Random Forest Estimator

The random forest estimator [35] is a powerful classification as well as regression algorithm. The algorithm generates a forest of classification (or regression) trees, instead of only a single classification (or regression) tree. Each of these trees resembles a weak learner, thus, more trees will reduce the variance. Both the classification task and the regression task use the average prediction over all trees in order to make the final classification/prediction.

3.5 XGBoost

XGBoost [10] refers to extreme gradient boosting and is an efficient implementation of the gradient boosting decision tree algorithm [10]. In particular, XGBoost was optimized as a distributed gradient boosting library to be designed as highly efficient, flexible, and portable. Both XGBoost and gradient boosting machines (GBMs) are ensemble tree methods that apply the mechanism of boosting weak learners via gradient descent. However, XGBoost further improves upon base GBM via system optimization and algorithmic enhancements [10]. The XGBoost algorithm used for this investigation is part of the AutoML [5] framework.

4 Data Set and Environment

4.1 Data Set

The CICIDS2017 data set was used which was generated by the Canadian Institute for Cybersecurity. The data set contains the most recent types of attacks captured during five days in July 2017, and consists of eight CSV files, which are the network traffic analysis results that were collected using the CICFlowMeter tool [8]. Each CSV file contains one type of attack along with normal traffic (Benign). Moreover, each record consists of 78 features describing the behavior of 25 users.

Table 1 Properties of data sets

	Number of features	Label and instance count	Number of instances
Bot net	4	Normal=188,955 BotNet=1956	190,911
Brute force	3	Normal=431,813 BruteForce=13,832	445,645
Port scan	5	Normal=127,292 PortScan=158,804	286,086
Web attacks	4	Normal=168,051 WebAttacks=2180	170,231

For our experiments, we considered the attacks Bot Net, Brute Force, Port Scan, and Web Attack. Table 1 shows the data sets including the number of instances, the percentage of the majority class, and the percentage in terms of total number of instances in the data set. From the table, we can see that three out of the four data sets are highly imbalanced.

Instances that had a missing value were removed. All multi-label data sets were transformed into binary data sets. For example, for the Brute Force data set, "SSH-Patator" and "FTP-Patator" labels were mapped to one label named "Brute Force." For the Web Attack data set, "Web Attack-Sql," "Web Attack-XSS," and "Web Attack-Brute Force" labels were mapped to "Web Attack."

4.2 Evaluation Measures

The performance measures used to evaluate the ensemble classifier are the following:

- Confusion matrix: contains the number of actual and predicted classifications achieved by the classifier.
- False positives (FP): defines the number of detected attacks that are actually normal behavior.
- False negatives (FN): are wrong predictions where instances that are attacks are classified as normal.
- True positives (TP): instances that are correctly classified as normal.
- True negatives (TN): attack instances that are correctly classified.
- Accuracy or True positive rate (TPR): percentage of correct predictions compared to all predictions calculated as:

$$\frac{TP}{TP + TN} \tag{1}$$

- Area Under Curve (AUC): describes the curve between TPR and FPR and the area under the curve; FPR is calculated as :

$$\frac{FP}{TN + FN} \tag{2}$$

- Detection Rate: The ratio of detecting the attacks is calculated as:

$$\frac{TP}{TP + FN} \tag{3}$$

- False Alarm Rate: The ratio of the wrong prediction of normal behavior is calculated as:

$$\frac{FP}{TN + FP} \tag{4}$$

- MAE: mean absolute error.
- MSE: mean squared error.
- RMSE: root mean squared error.
- Logloss: Logarithmic loss (related to cross-entropy) measures the performance of a classification model with the prediction input being a probability value between 0 and 1.

4.3 H2O Environment

H2O [17] is an open source tool/environment that has been used for this investigation. H2O is a scalable machine learning platform that allows one to easily run the super learner as well as other ML algorithms (supervised and unsupervised algorithms such as Deep Learning, Tree Ensembles, and GLRM). The developer has made H2O scalable using in-memory and distributed computation. The parallelization is done using the Map/Reduce framework that uses DFS (Distributed File Storage).

5 Experiments and Results

5.1 Parameter Setup

The parameters used for the experiments are listed below. In addition, 10-fold cross-validation was used.
Gradient Boosting Estimator:

- learn_rate = 0.01
- sample_rate = 0.6
- col_sample_rate_per_tree = 0.7

- ntrees = 50
- max_depth = 6
- balance_classes = True

Random Forest Estimator:

- sample_rate = 0.6
- col_sample_rate_per_tree = 0.7
- ntrees = 50
- max_depth = 6
- balance_classes = True

Deep Neural Network:

- activation = rectifier
- epochs = 10
- l_1 = 1e-3
- l_2 = 1e-3

H2O Stacked Ensemble Estimator: The default values were used.

5.2 *Experiments with DNN*

Tables 2, 3, 4, 5 show the MSE, RMSE, logloss, AUC, accuracy, detection rate, and false alarm rate for the four data sets based on the test set. Different hidden network structures were experimented with. These were: one layer (1, 20, 30, 40 nodes), two layers (10-5, 20-10, 20-5), three layers (20-10-5), and four layers (20-15-10-5).

Table 2 shows the results for the Bot Net data set. The best values achieved for MSE, RMSE, logloss, AUC, and accuracy were for the network structure with three hidden layers of size 20, 10, and 5. The best accuracy is however also achieved by other networks structures. The best values for the detection rate and the false alarm rate were achieved by the four hidden layer structure of size 20, 15, 10, and 5. The values achieved were 99.95% and 6.06%, respectively.

Table 3 shows the results for the Brute Force data set. The best MSE, RMSE, and logloss were achieved by the two hidden network structure of size 10 and 5; however, in terms of AUC this network is outperformed by all the other network structures. The best accuracy, detection rate, and false alarm rate was achieved by the one hidden node structure of size 10.

Table 4 shows the results for the Port Scan data set. The best values for MSE and RMSE were achieved by the 10-node hidden network structure. The best logloss was achieved by the 10-5 network structure whereas the best AUC was obtained by the 30-node network structure. In terms of accuracy, detection rate, and false alarm rate two network structures, the one with 30 hidden nodes and the other one with 40 hidden nodes, achieved the best values (accuracy=44.81%; detection rate=99.94%; false alarm rate=0.05%).

Table 2 DNN results for different network structures for the Bot net data set

	10	20	30	40	10-5	20-10	20-5	20-10-5	20-15-10-5
MSE	0.010210	0.006514	0.006564	0.006599	0.006209	0.006799	0.006729	**0.005921**	0.006555
RMSE	0.101046	0.080710	0.081018	0.081236	0.078797	0.082460	0.082036	**0.076948**	0.080963
LogLoss	0.053107	0.027141	0.027079	0.026926	0.026029	0.027475	0.027577	**0.025448**	0.027107
AUC	81.55%	97.86%	97.85%	97.74%	97.85%	97.83%	97.87%	**0.979146**	0.978486
Accuracy	99.03%	**99.27%**	**99.27%**	**99.27%**	**99.27%**	**99.27%**	**99.27%**	**99.27%**	99.24%
Detection rate	99.70%	99.94%	99.94%	99.94%	99.94%	99.93%	99.93%	99.94%	**99.95%**
False alarm rate	95.63%	8.07%	8.07%	7.80%	8.07%	8.86%	8.86%	8.07%	**6.06%**

Table 3 DNN results for different network structures for the Brute force data set

	10	20	30	40	10-5	20-10	20-5	20-10-5	20-15-10-5
MSE	0.029957	0.029927	0.029942	0.029927	**0.029983**	0.029955	0.029932	0.029922	0.029928
RMSE	0.173080	0.172995	0.173037	0.172993	**0.173157**	0.173076	0.173008	0.172979	0.172996
LogLoss	0.132727	0.132312	0.132639	0.132331	**0.133158**	0.132774	0.132453	0.132315	0.132374
AUC	**0.995666**	**0.995666**	**0.995666**	**0.995666**	0.995638	**0.995666**	**0.995666**	**0.995666**	**0.995666**
Accuracy	**96.86%**	96.13%	96.13%	96.13%	96.13%	96.13%	96.13%	96.13%	96.13%
Detection rate	**100.00%**	99.22%	99.22%	99.22%	99.22%	99.22%	99.22%	99.22%	99.22%
False alarm rate	**19.55%**	**19.55%**	**19.55%**	**19.55%**	19.60%	**19.55%**	**19.55%**	**19.55%**	**19.55%**

Table 4 DNN results for different network structures for the Port scan data set

	10	20	30	40	10-5	20-10	20-5	20-10-5	20-15-10-5
MSE	**0.009283**	0.008524	0.007807	0.008300	0.008840	0.008824	0.008813	0.009155	0.008651
RMSE	**0.096351**	0.092325	0.088360	0.091103	0.094023	0.093938	0.093878	0.095684	0.093011
LogLoss	0.040629	0.037845	0.038211	0.039319	**0.041084**	0.038686	0.038687	0.039534	0.039097
AUC	0.998541	0.998484	**0.998818**	0.998807	0.998275	0.998463	0.998460	0.998455	0.998455
Accuracy	44.56%	44.78%	**44.81%**	**44.81%**	44.76%	44.78%	44.78%	44.78%	44.78%
Detection rate	99.00%	99.88%	**99.94%**	**99.94%**	99.84%	99.88%	99.88%	99.88%	99.88%
False alarm rate	0.08%	0.10%	**0.05%**	**0.05%**	0.13%	0.10%	0.10%	0.10%	0.10%

Table 5 DNN results for different network structures for the web attacks data set

	10	20	30	40	10-5	20-10	20-5	20-10-5	20-15-10-5
MSE	**0.007279**	0.006643	0.006618	0.006854	0.006767	0.006607	0.006688	0.006655	0.006581
RMSE	**0.085314**	0.081503	0.081351	0.082787	0.082260	0.081285	0.081780	0.081578	0.081121
LogLoss	**0.028757**	0.027390	0.027500	0.027944	0.027756	0.027424	0.027477	0.027390	0.027306
AUC	0.968191	0.967891	**0.971063**	0.967568	0.970303	0.970655	0.967648	0.967760	0.968082
Accuracy	**98.94%**	98.41%	98.41%	98.40%	98.30%	98.40%	98.27%	98.28%	98.36%
Detection rate	**99.80%**	99.46%	99.45%	99.45%	99.35%	99.44%	99.31%	99.33%	99.41%
False alarm rate	34.31%	**33.63%**	33.83%	34.12%	37.95%	34.16%	39.13%	38.63%	35.63%

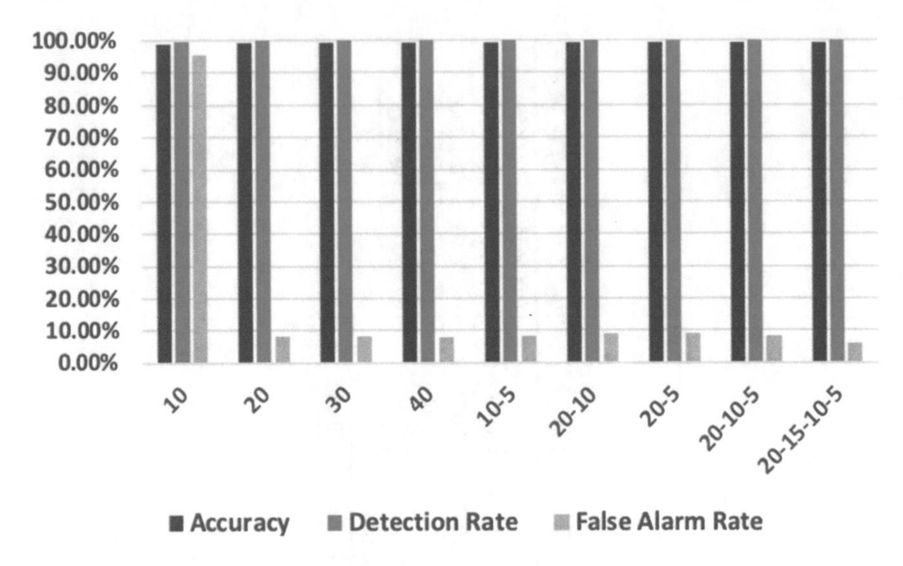

Fig. 2 Accuracy, detection rate, and false alarm rate for different DNN network structures for the Bot net data set

Table 5 shows the results for the Web Attacks data set. Best values for MSE, RMSE, logloss, accuracy, and detection rate were achieved by the "smallest" network of size 10. The best AUC was achieved by the 30-node network structure. The best false alarm rate was obtained using the 20-node network structure.

Figure 2 shows the accuracy, detection rate, and false alarm rate in pictorial format for the Bot Net data set. Figure 3 shows the confusion matrices of all four data sets. As can be seen from the matrices is that for both the Port Scan and the Brute Force data set shows close to ideal values. The Web Attacks and Bot Net data sets however have a high false negative rate of 16% and 36%, respectively.

Given that the detection rate and false alarm rate are the two most important measures for IDSs, the following hidden layer structures of the DNN were chosen for the following experiments:

- Bot Net: 20-15-10-5
- Brute Force: 10
- Port Scan: 30
- Web Attacks: 10

5.3 *Experiments with Superlearner*

Tables 6, 7, 8, 9 shows the results of the super learner. Other comparison algorithms are GBM, DRF, DNN, and AutoML's version of XGBoost. Overall, the XGBoost

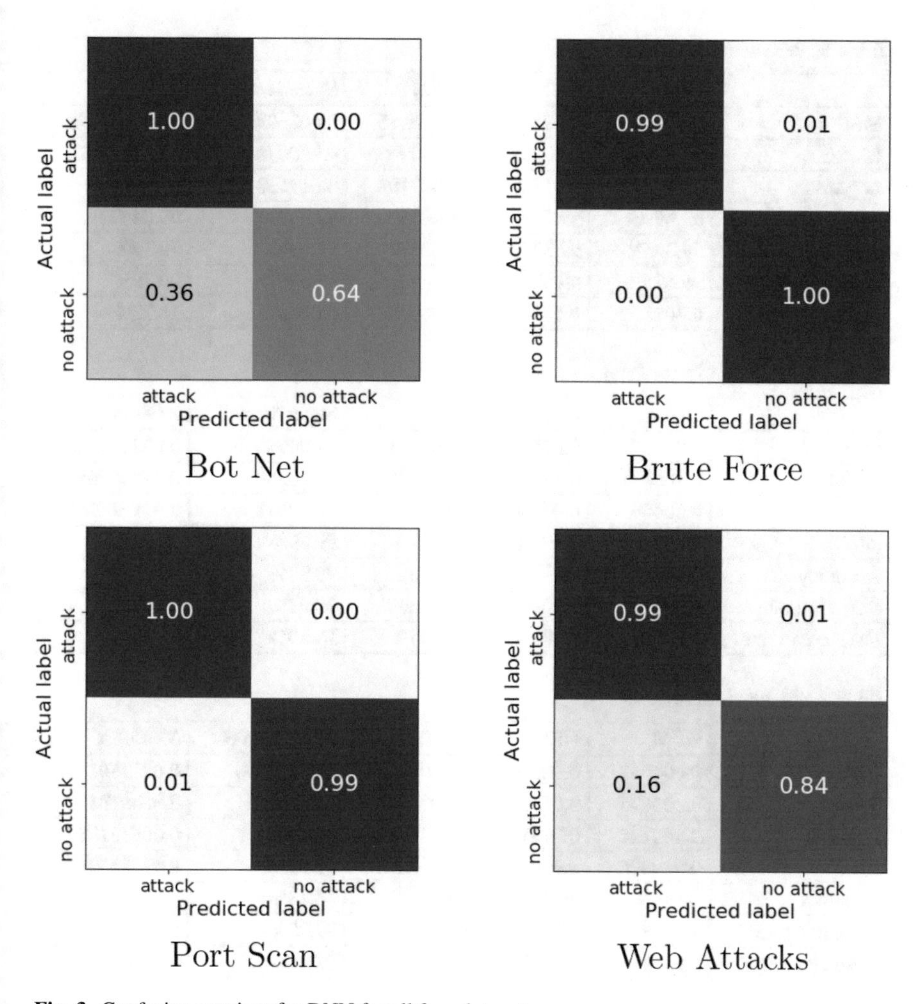

Fig. 3 Confusion matrices for DNN for all four data sets

outperforms all other algorithms on all data sets in terms of MSE, RMSE, logloss, and AUC. As for the other measures, we have to look at the individual tables.

From the results in Table 6 we can see that an accuracy of 99.93% as well as a perfect false alarm rate of 0.00% is achieved by three algorithms, namely GBM, DRF, and SuperLearner. The best detection rate was obtained by XGBoost.

From the results in Table 7 we can see that GBM achieves the best value of 96.97%, and DNN achieves a perfect 100.00% as the detection rate. The lowest false alarm rate of 12.89% is obtained by both DRF and the SuperLearner.

From the results in Table 8 we can see that DNN achieved the highest accuracy. The highest detection rate was achieved by XGBoost whereas the best results in terms of the false alarm rate were obtained by GBM and DRF with 0.04%.

Table 6 Results for Bot net

	GBM	DRF	DNN	Super learner	XGBoost
MSE	0.009994	0.005261	0.006555	0.003609	**0.002627098**
RMSE	0.099972	0.072531	0.080963	0.060078	**0.051255228**
LogLoss	0.045380	0.017800	0.027108	0.017138	**0.008929**
AUC	99.69%	99.69%	97.85%	99.66%	**99.84%**
Accuracy	**99.33%**	**99.33%**	99.28%	**99.33%**	99.17%
Detection rate	99.62%	99.62%	99.63%	99.62%	**99.79%**
False alarm rate	**0.00%**	**0.00%**	6.06%	**0.00%**	18.22%

Table 7 Results for Brute force

	GBM	DRF	DNN	Super learner	XGBoost
MSE	0.027879	0.025914	0.029957	0.004955	**0.003420689**
RMSE	0.166971	0.160979	0.173080	0.070389	**0.058486654**
LogLoss	0.104543	0.085444	0.132727	0.019518	**0.011028277**
AUC	99.87%	99.91%	99.57%	99.89%	**99.92%**
Accuracy	**96.97%**	96.95%	96.86%	96.95%	96.90%
Detection rate	99.90%	99.92%	**100.00%**	99.92%	99.97%
False alarm rate	12.96%	**12.89%**	19.55%	**12.89%**	12.97%

Table 8 Results for Port scan

	GBM	DRF	DNN	Super learner	XGBoost
MSE	0.094062	0.002907	0.007807	0.003968	**0.001150102**
RMSE	0.306695	0.053915	0.088360	0.062992	**0.033913152**
LogLoss	0.363122	0.013152	0.038211	0.017609	**0.005331357**
AUC	0.999826	0.999807	0.998818	0.999781	**0.999845313**
Accuracy	44.47%	44.49%	**44.57%**	44.18%	44.32%
Detection rate	99.52%	99.41%	99.00%	99.90%	**99.95%**
False alarm rate	**0.04%**	**0.04%**	0.05%	0.73%	0.25%

From the results in Table 9 we can see that DNN achieves the highest accuracy of 98.94%. The highest detection rate was achieved by GBM, and the lowest false alarm rate with 0.00% was obtained by the SuperLearner.

Figure 4 shows the confusion matrices for all data sets for the SuperLearner. The values for the Bot Net data set are identical whereas for the other three data sets (Web Attacks, Port Scan, and Brute Force) the values are slightly different but not significantly compared to Fig. 3.

Table 9 Results for web attacks

	GBM	DRF	DNN	Super learner	XGBoost
MSE	0.011858	0.003141	0.007279	0.001739	**0.000428**
RMSE	0.108894	0.056042	0.085314	0.041706	**0.020684**
LogLoss	0.051296	0.012183	0.028757	0.011616	**0.001925**
AUC	99.94%	99.93%	0.968191	99.81%	**99.99%**
Accuracy	98.78%	98.80%	**98.94%**	98.90%	98.79%
Detection rate	**99.97%**	99.95%	99.80%	99.85%	99.96%
False alarm rate	4.07%	1.35%	34.31%	**0.00%**	0.77%

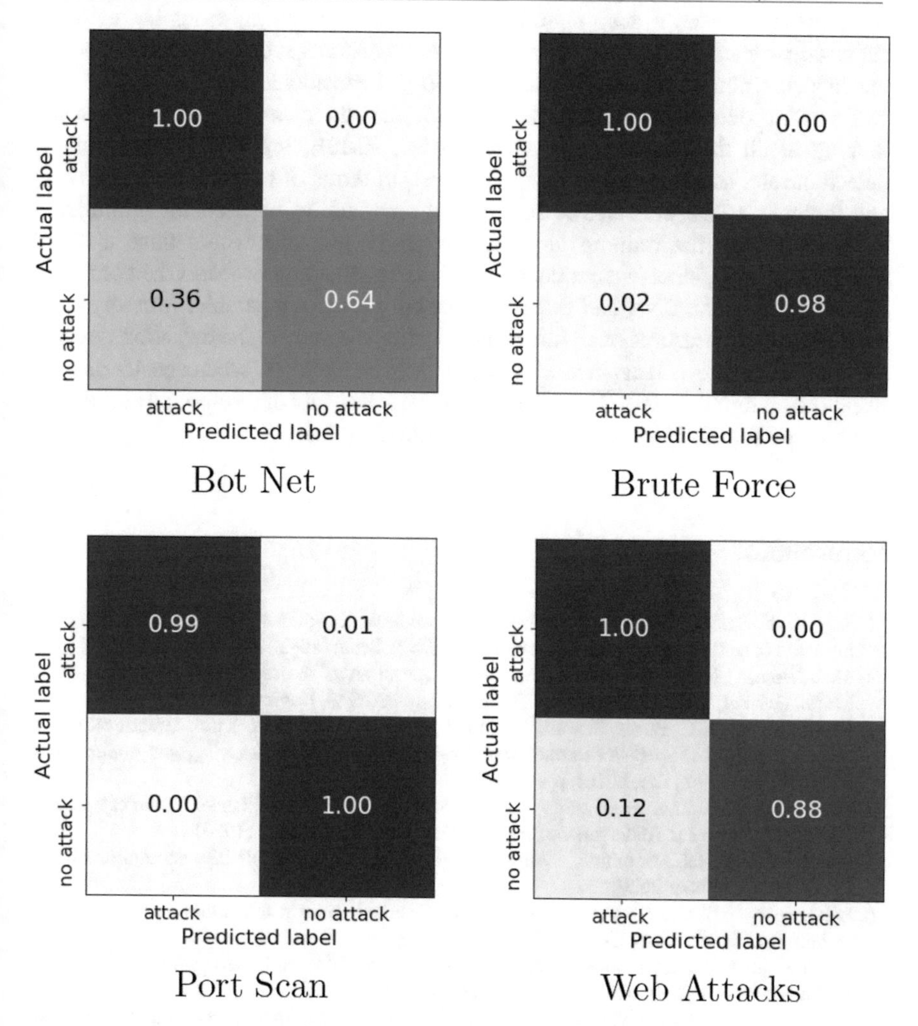

Fig. 4 Confusion Matrix for SuperLearner for all four data sets

6 Conclusion

Intrusion detection is an important issue that needs to be addressed in order to make computer systems secure and prevent access and/or misuse from unauthorized users. In this chapter, the CICIDS2017 data set was investigated, which contains the most up-to-date common attacks resembling true real-world data. The data was pre-processed and transformed into binary data sets, thus we used four data sets namely Bot Net, Brute Force, Port Scan, and Web Attacks. The data sets were investigated applying different machine learning approaches. In particular, a comparison between a deep neural network approach and an ensemble methods called super learner was done as well as other algorithms such as gradient boosting machine, distributed random forest, and XGBoost considered.

From the overall results, we can conclude that there is no clear winner when looking at all the evaluation measures (MSE, RMSE, logloss, AUC, accuracy, detection rate, and false alarm rate). However, in terms of reducing MSE, RMSE, and logloss, XGBoost seems to be the best approach to be used for minimizing the error during the training process. For regression, this would have a direct impact. However, since we are dealing with a classification problem the confusion matrices show that DNN and SuperLearner are pretty comparable. Moreover, the most important measures of an intrusion detection system are the detection rate and the false alarm rate, and again no clear winner can be identified when considering the results of all four data sets. Thus, the conclusion is that any of the applied algorithms can be used for this intrusion detection classification task.

References

1. A. Ali, A. Saleh, T. Ramdan, Multilayer perceptrons networks for an intelligent adaptive intrusion detection system. Int. J. Comput. Sci. Netw. Secur. **10**(2), 275–279, (2010)
2. M.Z. Alom, V. Bontupalli, T.M. Taha, Intrusion detection using deep belief networks, in *2015 National Aerospace and Electronics Conference (NAECON)*, Dayton, OH (2015)
3. K. Alrawashdeh, C. Purdy, Toward an online anomaly intrusion detection system based on deep learning, in *15th IEEE International Conference on Machine Learning and Applications (ICMLA)* (Anaheim, CA, 2016), pp. 195–200
4. J.P. Anderson, Computer Security Threat Monitoring and Surveillance. Technical Report 79F296400. James P. Anderson Co., Fort Washington, Pennsylvania (1980)
5. AutoML, Available online: https://docs.h2o.ai/h2o/latest-stable/h2o-docs/automl.html, [Accessed December 2020]
6. BBC, Spies 'infiltrate US power grid' (2009c). Available online: https://news.bbc.co.uk/1/hi/technology/7990997.stm, [Accessed December 2020]
7. BBC, Major cyber spy network uncovered (2009b). Available online: https://news.bbc.co.uk/1/hi/world/americas/7970471.stm, [Accessed December 2020]
8. Canadian Institute for Cybersecurity, Available online: https://www.unb.ca/cic/datasets/ids-2017.html, [Accessed December 2020]
9. I. Chairunnisa, Lukas, H.D. Widiputra, Clustering base intrusion detection for network profiling using k-means, ecm and k-nearest neighbor algorithms, in *Konferensi Nasional Sistem dan Informatika*, pp. 247–251 (2009)

10. T. Chen, C. Guestrin, XGBoost: A scalable tree boosting system, in *Proceedings of the 22nd ACM SIGKDD International Conference on Knowledge Discovery and Data Mining*, August 13–17, 2016, ed. by B. Krishnapuram, M. Shah, A.J. Smola, C.C. Aggarwal, D. Shen, R. Rastogi (ACM, San Francisco, CA, USA, 2016), pp. 785–794
11. R. Chitrakar, C. Huang, Selection of candidate support vectors in incremental SVM for network intrusion detection. Comput. Secur. **45**, 231–241 (2014)
12. CNet News, Georgia accuses Russia of coordinated cyberattack (2008). Available online: https://news.cnet.com/8301-1009_3-10014150-83.html, [Accessed December 2020]
13. D.E. Denning, An intrusion-detection model. IEEE Trans. Softw. Eng. **13**, 222–232 (1987). ISSN 0098-5589
14. P. Dokas, L. Ertoz, V. Kumar, A. Lazarevic, J. Srivastava, P. Tan, Data mining for network intrusion detection, in *Proceedings of the NSF Workshop on Next Generation Data Mining*, Baltimore, MD (2002)
15. F. Giroire, J. Chandrashekar, G. Iannaccone, K. Papagiannaki, E.M. Schooler, N. Taft, The cubicle vs. the coffee shop: Behavioral modes in enterprise end-users, in *Proceedings of the 2008 Passive and Active Measurement Conference* (Springer, 2008), pp. 202–211
16. N. Gornitz, M. Kloft, K. Rieck, U. Brefeld, Active learning for network intrusion detection, in *2nd ACM Workshop on Security and Artificial Intelligence*, pp. 47–54 (2009)
17. H2O, Available online: https://docs.h2o.ai/h2o/latest-stable/h2o-docs/welcome.html, [Accessed December 2020]
18. R.A. Kemmerer, G. Vigna, Intrusion detection: A brief history and overview. Computer **35**, 27–30 (2002)
19. M. Kloft, U. Brefeld, P. Dussel, C. Gehl, P. Laskov, Automatic feature selection for anomaly detection, in *Proceedings of the 1st ACM Workshop on Workshop on AISec*, pp. 71–76 (2008)
20. C. Kruegel, F. Valeur, G. Vigna, *Intrusion Detection and Correlation: Challenges and Solutions* (Springer, Telos, 2004)
21. L.M. Lewis, A case-based reasoning approach to the resolution of faults in communication networks, in *Proceedings of the IFIP TC6/WG6.6 Third International Symposium on Integrated Network Management with participation of the IEEE Communications Society CNOM and with Support from the Institute for Educational Services* (North-Holland, 1993), pp. 671–682
22. Y. Li, R. Ma, R. Jiao, A hybrid malicious code detection method based on deep learning. Int. J. Secur. Appl. **9**(5), 205–216 (2015)
23. Y. Liu, X. Zhang, Intrusion detection based on IDBM, in *IEEE 14th Intl Conf on Dependable, Autonomic and Secure Computing*, Auckland, pp. 173–177 (2016)
24. S.A. Ludwig, Intrusion detection of multiple attack classes using a deep neural net ensemble, in *IEEE Symposium Series on Computational Intelligence (SSCI)*, Honolulu, HI, USA, pp. 1–7 (2017)
25. S.A. Ludwig, Applying a neural network ensemble to intrusion detection. J. Artif. Intell. Soft Comput. Res. **9**(3), 177–188 (2019)
26. D. Moore, V. Paxson, S. Savage, C. Shannon, S. Staniford, N. Weaver, Inside the Slammer-Worm. IEEE Secur. Priv. **1**, 33–39 (2003)
27. S.F. Owens, R.R. Levary, An adaptive expert system approach for intrusion detection. Int. J. Secur. Netw. **1**, 206–217 (2006). ISSN 1747-8405
28. M. Pillai, J. Eloff, H. Venter, An approach to implement a network intrusion detection system using genetic algorithms, in *Proceedings of South African Institute of Computer Scientists and Information Technologists* (Western Cape, South Africa, 2004), pp. 221–228
29. E.C. Polley, Super Learner. Ph.D. Thesis, Fall 2010 (2010). Available online: https://digitalassets.lib.berkeley.edu/etd/ucb/text/Polley_berkeley_0028E_10767.pdf, [Accessed December 2020]
30. S. Potluri, C. Diedrich, Accelerated deep neural networks for enhanced intrusion detection system, in *IEEE 21st International Conference on Emerging Technologies and Factory Automation (ETFA)*, Berlin, pp. 1–8 (2016)
31. G. Seni, J.F. Elder, Ensemble methods in data mining: improving accuracy through combining predictions, in *Synthesis Lectures on Data Mining and Knowledge Discovery*, ed. by R. Gross-

man (Morgan & Claypool, 2010)
32. P. Stetsenko, Machine Learning with Python and H2O. H2O. ai Inc. (2016). https://docs.h2o.ai/h2o/latest-stable/h2o-docs/booklets/PythonBooklet.pdf, [Accessed December 2020]
33. T.A. Tang, L. Mhamdi, D. McLernon, S.A. Raza Zaidi, M. Ghogho, Deep learning approach for network intrusion detection in software defined networking, in *International Conference on Wireless Networks and Mobile Communications (WINCOM)*, Fez, Morocco (2016)
34. M.J. Van der Laan, E.C. Polley, A.E. Hubbard, Super learner. Stat. Appl. Genet. Mol. Biol. **6**(1), 1–22 (2007)
35. J. Vanerio, P. Casas, Ensemble-learning approaches for network security and anomaly detection, in *Proceedings of the Workshop on Big Data Analytics and Machine Learning for Data Communication Networks* (ACM, 2017), pp. 1–6
36. J. Xie, V. Rojkova, S. Pal, S. Coggeshall, A combination of boosting and bagging for KDD Cup 2009 - Fast scoring on a large database. J. Mach. Learn. Res. (JMLR) **7**, 35–43 (2009)
37. S. Zanero, S.M. Savaresi, Unsupervised learning techniques for an intrusion detection system, in *SAC '04: Proceedings of the 2004 ACM symposium on Applied computing*, New York, NY, USA (2004), pp. 412–419
38. Z.H. Zhou, *Ensemble Methods: Foundations and Algorithms*. Machine Learning & Pattern Recognition Series (Chapman & Hall/CRC. Boca Raton, FL, 2012). https://tjzhifei.github.io/links/EMFA.pdf, [Accessed December 20]

Simone A. Ludwig is a Professor and Interim Chair in the Department of Computer Science at North Dakota State University, USA. Prior to joining NDSU she worked at the University of Saskatchewan (Canada), Concordia University (Canada), Cardiff University (UK), and Brunel University (UK). Dr. Ludwig received her Ph.D. degree and M.Sc degree with distinction from Brunel University in 2004 and 2000, respectively. Before starting her academic career, she worked several years in the software industry. Dr. Ludwig's research interests lie in the area of computational intelligence including swarm intelligence, evolutionary computation, (deep) neural networks, and fuzzy reasoning. Example application areas are data mining/machine learning (including big data), image processing, intrusion detection, cryptography, and cloud computing. She started doing research in the computational intelligence area after completing her PhD degree and has ever since been fascinated by the vastness of this area. Dr. Ludwig has served as symposium chair, program chair, track chair, and committee member for numerous conferences. She is an Associate Editor of the Journal of Swarm and Evolutionary Computation (Elsevier) since 2016. In addition, Dr. Ludwig has been serving in various roles and on different committees for the Computational Intelligence Society (IEEE CIS).

Lifelong Learning Machines: Towards Developing Optimisation Systems That Continually Learn

Emma Hart

1 Introduction

Optimisation is vital in industries to improve processes in order to maximise value for money. For example, in the UK, the manufacturing industry is estimated to benefit from productivity gains of 3% per year through use of optimisation [8], while in the logistics sector, it has been estimated that a 15% reduction in costs would contribute approximately £7.5bn or 0.45% in equivalent growth for the UK economy [7].

To achieve these projected gains, the research community has endeavoured to develop optimisation algorithms for many years, resulting in methods stemming from multiple fields (e.g., mathematics, operations research, meta-heuristics) and which take a wide variety of approaches. Although the available methods range in complexity, speed, and expected performance, they typically have much in common from a methodological perspective: following a design process, algorithm(s) is tuned to work well on a set of example instances that one hopes will adequately represent the instances to be faced in practice. Following deployment, the optimisation software usually remains in place until it requires upgrading due to decreased performance, or becomes obsolete as a result of changed user requirements. Unfortunately, this approach has significant weaknesses. First, systems do not improve as they are exposed to more instances. This is clearly inefficient and, from a practical perspective, reduces potential economic gains if systems are performing sub-optimally. Second, they are unable to adapt to a changing operating environment—for example, reacting to new optimisation criteria, or to shifts in instance characteristics that range from one-off changes (e.g., schedule an

E. Hart (✉)
Edinburgh Napier University, Edinburgh, UK
e-mail: e.hart@napier.ac.uk

A. E. Smith (ed.), *Women in Computational Intelligence*, Women in Engineering and Science, https://doi.org/10.1007/978-3-030-79092-9_9

“untypical” order) to longer-term trends resulting from gradually changing customer requirements. At best, this means that systems need to be periodically redesigned, while at worst, systems can become completely obsolete after some period of time.

These issues within optimisation design are symptomatic of a deep-rooted problem in the wider field of AI, where the need for research to shift towards designing systems that are capable of recognising and reacting to situations beyond those for which they were specifically designed is now widely recognised [25]. In machine-learning and robotics tasks, this is commonly referred to as *lifelong learning*. Here, we extend this concept to the field of optimisation, describing a lifelong learning optimiser that when faced with a continual stream of problems to optimise: (a) refines an existing set of algorithms so that they improve over time as they are exposed to more examples and (b) automatically generates new algorithms as required. The paper summarises a line of work that is described in detail in a number of previous publications [9, 27–30] and to which the reader is referred for a comprehensive overview.

2 Background and Motivation

To motivate the approach, consider a practical example of a company that provides a service cutting parts from metal sheets. The company wishes to optimise the cutting process in order to minimise material usage. Meta-heuristic algorithms provide a pragmatic method for providing fast, high-quality solutions, but as any algorithm can be expected to have a unique “footprint” (i.e., a subset of instances on which it performs best [31]), the company designs two algorithms to cope with its two largest customers.

In this case, the shapes to be cut can be characterised by two features: the average area of the shape and the number of vertices it has. Assume that customer A’s orders are characterised by pieces with few vertices and small area, while customer B’s orders by pieces with many vertices and large area. This is shown in Fig. 1. A specialised algorithm is designed, implemented, and tuned for each customer. However, over time, the requirements of customer A gradually change such that they require pieces in which both the area and number of vertices have slightly increased compared to previous examples: the algorithm designed for customer A no longer produces optimal results on the new instance and must be re-tuned. Furthermore, a new customer C is acquired whose instances are unlike either of the previous customers in terms of their characteristics. No algorithm is available for customer C. At best, a new algorithm must be selected from an existing pool that is likely to have compromised performance, and at worst, a new algorithm must be manually designed and then tuned.

Similar situations occur in many scenarios, e.g., factory scheduling, or vehicle routing, in that optimisation algorithms are being run within a dynamically changing environment with dynamically changing user requirements. As a result, any optimisation algorithm is unlikely to be fit for purpose after some period of time, hence

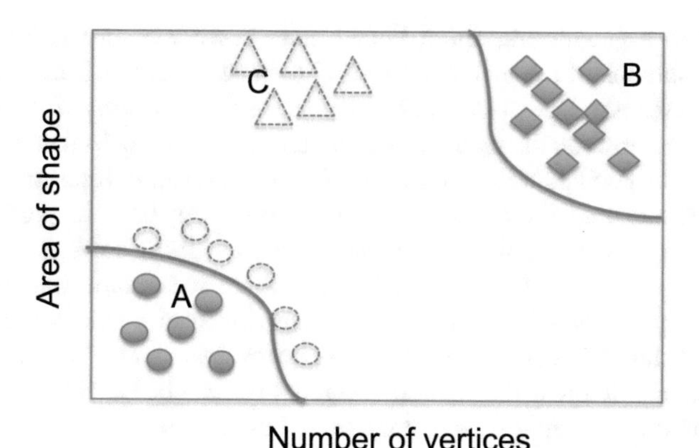

Fig. 1 Characteristics of shapes to be cut in an example stock-cutting problem. Circles represent typical orders from one customer optimised by Algorithm A. Dashed circles show characteristics of new instances as customer A's requirements change over time. Diamonds represent instances of another customer B solved by Algorithm B, which require a different algorithm for optimal performance. Dashed triangles represent a future customer orders for which there is no suitable existing algorithm

motivating the desire for optimisation systems that autonomously improve and adapt over time to maximise their economic benefits.

2.1 Related Work

The concept of lifelong learning machines (LML) was introduced by Silver et al. [25] who noted *it is now appropriate for the AI community to move beyond learning algorithms to more seriously consider systems that are capable of learning over a lifetime*. They go on to describe several examples that cover supervised, unsupervised, and reinforcement learning methods. Ruvolo and Eaton [23] propose an efficient lifelong learning algorithm dubbed ELLA that focuses on multi-task learning in machine-learning applications relating to prediction and recognition tasks; ELLA is able to transfer previously learned knowledge to a new task and integrate new knowledge through the use of shared basis vectors in a manner that is proved to be computationally efficient. A lifelong learning multi-agent system approach is described in [32] in the context of a robotic toy that learns over time to interact with a child through repeated interaction. The concept is also gaining traction in the domain of natural-language processing, e.g., applied to sentiment analysis [5, 6].

Despite this, there appears to be little literature in the optimisation field. One exception is work by Louis et al. [17] on Case-Injected Genetic Algorithms—CIGAR algorithms. Conjecturing that a system that combined a robust search algorithm with an associative memory could learn by experience to solve problems, they combine a genetic algorithm with a case-based memory of past problem solutions. Solutions in the memory are used to seed the population of the genetic algorithm; cases are chosen on the basis that similar problems will have similar solutions, and therefore, the memory will contain building blocks that will speed up the evolution of good solutions. Using a set of design and optimisation problems, [17] demonstrate that their system learns to take less time to provide quality solutions to a new problem as it gains experience from solving other similar problems. The system differs from other case base reasoning systems in that it does not use prior knowledge based on a comparison to previously encountered problems but simply injects solutions that are deemed similar to those already in the GA population. Although the CIGAR methodology can be considered as an LML technique as it builds up a case history over time, a weakness is that it does not enable "forgetting" but simply maintains a memory that increases in size as time passes. A more recent example is provided by Ortiz-Bayliss et al. [21] who describe a hyper-heuristic model for constraint satisfaction that is inspired in the idea of a lifelong learning process in which a hyper-heuristic to continually improve the quality of its decisions by incorporating information from every instance it solves. As previously noted, the obvious lack of focus regarding use of lifelong learning in optimisation has motivated our own interest in this field, with a line of work described in a series of publications [8, 25–28] and which is summarised within this chapter.

3 An Immune-Inspired Approach to Lifelong Learning

Silver et al. [25] identify three essential components of a lifelong learning system:

- It should be able to retain and/or consolidate knowledge, i.e., incorporate a long-term memory: the system should be computationally efficient when storing learned knowledge in its long-term memory, and ideally, retention should occur online.
- It should selectively transfer prior knowledge when learning new tasks.
- It should adopt a system approach that ensures the effective and efficient interaction of the elements of the system.

They proposed a framework for a generic LML design that encompasses supervised, unsupervised, and reinforcement learning techniques, with a view to developing test applications in the robotics and agents domains. In contrast, here we turn to biology for inspiration in building an LML for optimisation purposes, and in particular to the *natural immune system*, noting that it has properties that fulfil the three requirements for an LML system listed above. The field of Artificial Immune Systems (which takes inspiration from the immune system to develop computational applications) was first developed in the early 1990s and has since inspired applications as diverse as anomaly detection, fault tolerance in robotics, classification, and optimisation (see [10]); however, this is the first time it has been used in this context. A brief introduction on the immune system is given below to illustrate the salient concepts.

3.1 A Brief Primer on the Immune System

Perhaps surprisingly, there is no clear consensus among the immunological community as to precisely how the immune system operates. Of most relevance to this work is the *idiotypic network model* of immunity, proposed by Jerne [14] as a possible mechanism for explaining long-term memory. B-cells are the basic unit of an immune response: a B-cell produces antibodies that specifically recognise and bind to proteins on an invading bacteria or virus. Each B-cell produces a single type of antibody; hence, to ensure that a wide variety of pathogens can be recognised, the bone marrow continually produces a stream of B-cells through random gene recombination, therefore creating diversity. From this continuous stream, a selection process proliferates *useful* B-cells at any given time based on their ability to produce an antibody capable of recognising and binding to pathogens present in the body. Challenging existing thinking at the time that suggested that antibodies that are not stimulated by reacting with pathogens are removed from circulation, Jerne [14] proposed that antibodies interact with each other (even in the *absence* of pathogens) to form a self-sustaining network of collaborating cells. This network can be thought of as collectively embodying a memory of previous responses. The network integrates new elements overtime and also enables elements to be removed, thereby providing adaptation to a time-varying environment. The network encapsulates a *memory* of prior responses, therefore enabling it to respond rapidly when faced with pathogens it has previously been exposed to. It can *selectively adapt prior knowledge* via clonal-selection mechanisms that can rapidly replicate existing antibodies with small mutations, which produce new variants of previous pathogens. Finally, it embodies a systemic approach by maintaining a repertoire of antibodies that collectively cover the space of potential pathogens, providing an efficient response.

3.2 NELLI: Mapping to a Lifelong Learning Optimiser

It is straightforward to map this to an optimisation context. Instead of a stream of pathogens entering a body, we consider a continual stream of instances that need to be optimised entering an optimisation system. In place of bone marrow that generates a continual supply of novel antibodies in the natural system, we use a form of evolutionary algorithm that generates a stream of novel *algorithms* for solving instances. Each new algorithm is evaluated in terms of its ability to optimise one or more instances. The algorithms are integrated into a network in which algorithms compete against each other to "win" instances, that is, provide a better solution than

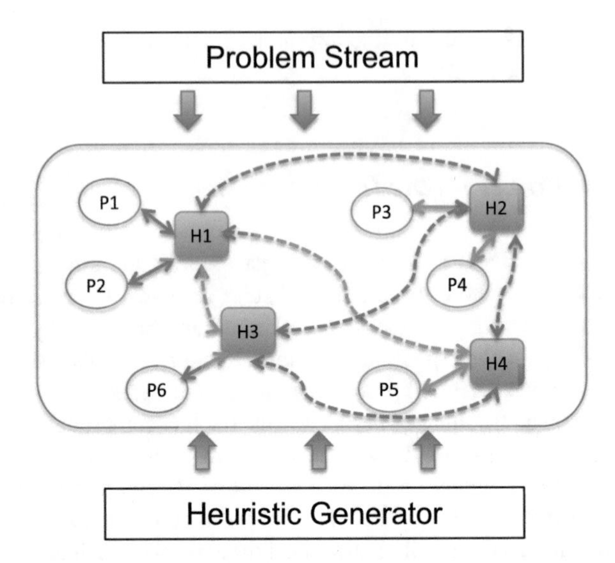

Fig. 2 The generator continually generates new heuristics that enter the system (H1, H2, H3, H4). Problem instances are also continually injected. A heuristic "wins" a problem instance (P1–P6) if it delivers the best result on that instance. Dotted lines between heuristics indicate that the heuristics interact in a competition for instances

any other algorithm in the network. Successful algorithms are integrated into the network; those that do not win any instances are rejected. Algorithms disappear from the network over time even if once successful if they do not continue to prove useful, for example, if the characteristics of the problem stream change over time and instances of a type seen in the past are no longer relevant. This enables a form of forgetting within the system, just as in the real immune system, protection afforded by vaccines against some diseases wears off over time. A conceptual model of the system is illustrated in Fig. 2.

We dub the system NELLI: Network for Lifelong Learning. NELLI continuously generates new knowledge in the form of novel deterministic heuristics that produce solutions to problems; these are integrated into a network of interacting heuristics and problems. The problems incorporated in the network provide a minimal representative map of the problem space, while the heuristics generalise over the problem space, each occupying its own niche. Memory is encapsulated in the network and is exploited to rapidly find solutions to new problems. The network is plastic both in its contents and its topology: this means that both the number and type of algorithm change over time, and that the region of the instance space covered by a single algorithm expands or contracts as new algorithms are introduced.

The immune network sustains a minimal repertoire of heuristics need to cover the instance space: only heuristics that provide a unique contribution in that they produce a better result on at least one problem than any other heuristic are retained. It also sustains a minimal repertoire of problems that provide a representative map

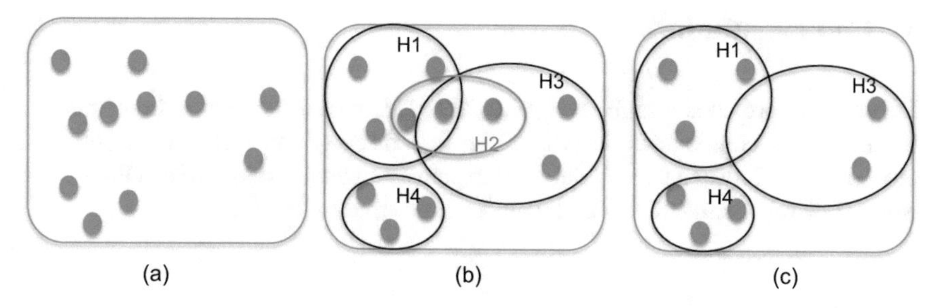

Fig. 3 Diagram (**a**) shows the problems that the system is currently exposed to $\mathcal{E}$. The middle diagram (**b**) shows a set of generated heuristics that cover the problems in $\mathcal{E}$. The problems $P1$ and $P2$ shown are equally solved by one or more heuristics and therefore not required to map the problem space. Heuristic H2 shown in red is redundant as it does not have a niche. The right-hand diagram (**c**) shows the resulting *network* $\mathcal{N}$ that sustains the minimal set of problems and heuristics required to describe the space

of the problem space to which the system has been exposed over its lifetime. It is not necessary for the system to contain representatives of *all* problems from the problem stream shown in Fig. 2—in contrast, a representative subset is maintained that contains examples indicative of other instances in a particular region.

This is represented conceptually in Fig. 3. In this diagram, Figure (a) shows a set of problems $\mathcal{E}$ that the system is currently exposed to. Figure (b) shows a set of heuristics $\mathcal{H}$ that collectively cover the problems in $\mathcal{E}$. The problems $P1$ and $P2$ are solved equally by two or more heuristics. H2 is subsumed in that it cannot solve *any* problem better than another heuristic. In Figure (c), H2 is removed as it does not have a niche in solving problems; problems P1 and P2 are removed as they do not have a niche in describing the problem space.[1] A competitive exclusion effect is observed between heuristics (and also between problems), which results in efficient coverage of the problem space. A key aspect of the compression is that it significantly decreases the computation time of the method. Hence to summarise, the system has the following features:

- It rapidly produces solutions to new problems instances that are similar in structure to previous problem instances that the system has been exposed.
- It responds by generating new heuristics to provide solutions to new problems that differ from those previously seen.

[1] Although these problems have been removed from the network, they can still be solved by the system as heuristics H1 and H3 remain in the network.

4 Implementation

This section provides a high-level overview of how the approach is implemented. Each of three components shown in Fig. 2 is discussed in turn. The description provided is mainly at a conceptual level to convey a general sense of how the system works. For full technical details the reader is referred to [9, 30].

4.1 Application: Bin-Packing

We consider an application in the domain of bin-packing, a field that has many practical applications, for example, relating to packing of objects in warehouses or delivery vehicles.

A simplified version of packing often studied in the academic literature is the one-dimensional bin-packing problem (BPP). The goal is to find a packing that minimises the number of containers b of fixed capacity c required to accommodate a set of n items with weights $\omega_j : j \in \{1 \ldots n\}$ falling in the range $1 \leq \omega_j \leq c, \omega_j \in \mathbb{Z}$ while enforcing the constraint that the sum of weights in any bin does not exceed the bin capacity c. The goal of a bin-packing heuristic is to determine an appropriate sequence by which to pack items combined with a method of assigning an item to the most appropriate bin. In the scenario considered, instances of bin-packing problems arrive in stream, which can be either one at a time, or in batches.

4.2 Heuristic Generation

A key component of the approach shown in Fig. 2 is the ability to generate a stream of novel heuristics. The field of Automated Algorithm Design (ADA) is rapidly [3] maturing and provides a number of methods by which algorithms can be automatically generated. One such approach is to use a template-based design [1, 16] in which an algorithm searches over possible assignments of variables in the template. However, the most popular approaches have used either Grammatical Evolution [20] or Genetic Programming (GP) [15]. The latter has been used to build new components of an existing algorithm [11] as well as to design new heuristics/algorithms, particularly in the domain of hyper-heuristics [4, 18, 22]. We have investigated the effectiveness of various forms of GP to generate heuristics, including Single Node Genetic Programming (SNGP) [12, 13] that has a benefit that heuristics are of fixed maximum size (in terms of the number of nodes). These results are described in [9, 26]. However, further investigation revealed that a representation describing a linear sequence of heuristic components was most effective [27]. The idea is illustrated in Fig. 4. Each node causes one or more items to be packed into a bin according to a heuristic specified by the node. A pointer

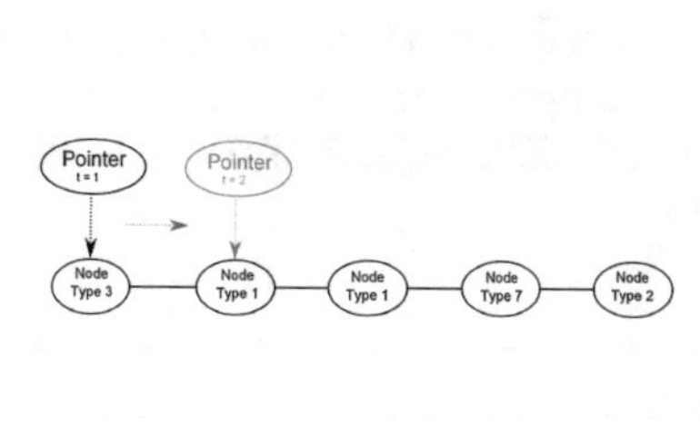

Node Type	Description
1	Packs the single largest item into the current bin
2	Packs the largest combination of exactly 2 items into the current bin
3	Works as for 1 but packs exactly 3 items
4	Works as for 2 but packs exactly 4 items
5	Works as for 2 but packs exactly 5 items
6	Packs the largest combination of up to 2 items into the current bin giving preference to sets of lower cardinality.
7	As for 5 but considers sets of up to 3 items
8	As for 5 but considers sets of up to 4 items
9	As for 5 but considers sets of up to 5 items

Fig. 4 Heuristics are represented as linear sequences of nodes. A pointer keeps track of which node to apply next. The sequence restarts from the beginning after the last node is processed

advances along the sequence one step at a time, applying the components in the order specified. If all items have not been packed when the pointer reaches the end, the pointer returns to the start. Hence, the goal of the system is to generate novel sequences that represent effective heuristics. The heuristic components used to generate sequences are described in Fig. 4.

4.3 Network

The network $\mathcal{N}$ attempts to sustain a set of interacting heuristics and problems in which heuristics are *directly* stimulated by problems and vice versa. Heuristics and problems need remain stimulated in order to survive; hence, the stimulation level provides competition between heuristics, which forces a heuristic to find an individual niche in which to operate.

A heuristic h can be stimulated by one or more problems. The total stimulation of a heuristic is the sum of its *affinity* with each problem in the set $\mathcal{P}$ currently in the network $\mathcal{N}$. A heuristic h has a non-zero affinity with a problem $p \in \mathcal{P}$ *if and only if* it provides a solution that uses fewer bins than any other heuristic currently in $\mathcal{H}$. If this is the case, then the value of the affinity $p \leftrightarrow h$ is equal to the improvement in the number of bins used by h compared to the next-best heuristic. If a heuristic provides the best solution for a problem p but one or more other heuristics give an equal result, then the affinity between problem p and the heuristic h is zero. If a heuristic h uses more bins than another heuristic on the problem, then the affinity between problem p and the heuristic h is also zero. In short, in order to survive, a heuristic must be able to solve at least one problem in the system better than any other heuristic currently in the system. Although no quantitative value is calculated for heuristic$\leftrightarrow$heuristic interactions, there is essentially an indirect interaction between them as they are competing against each other to win problems.

As the affinity between a problem and a heuristic is symmetrical, then the stimulation of a problem is simply the affinity between the problem and the heuristic that best solves it. A problem for which the best solution is provided by more than one heuristic receives zero stimulation. The motivation for this choice lies in the fact that a problem that is equivalently best solved by multiple heuristics is on one hand "easy" to solve, while, on the other, does not reflect niche regions of the instance space that require specialised heuristics.

Algorithm 1: NELLI pseudo code

Require: $\mathcal{H} = \emptyset$:The set of heuristics
Require: $\mathcal{P} = \emptyset$:The set of current problems
Require: $\mathcal{E} = \mathcal{E}_{t=0}$:The set of problems to be solved at time t
1: **repeat**
2: *optionally* replace $\mathcal{E} : \mathcal{E}^* \leftarrow \mathcal{E}^* \cup \mathcal{E}$
3: Add n_h randomly generated heuristics to $\mathcal{H}$ with concentration c_{init}
4: Add n_p randomly selected problem instances from $\mathcal{E}$ to $\mathcal{P}$ with concentration c_{init}
5: calculate $h_{stim} \forall h \in \mathcal{H}$
6: calculate $p_{stim} \forall p \in \mathcal{P}$
7: increment all concentrations (both $\mathcal{H}$ and $\mathcal{P}$) that have concentration $< c_{max}$ and stimulation > 0 by Δ_c
8: decrement all concentrations (both $\mathcal{H}$ and $\mathcal{P}$) with stimulation ≤ 0 by Δ_c
9: Remove heuristics and problems with $concentration \leq 0$
10: **until** $stopping\ criteria\ met$

The pseudocode for the algorithm driving the system is given in Algorithm 1. The reader is referred again to [30] for a detailed description. At each step, the environment can change through one or more instances being injected into the system or by the current set of instances being replaced with an entirely new set. A fixed number of new heuristics are generated and added to the system and assigned an initial concentration value. A sample of instances from the environment is added to the system. Each heuristic computes its affinity to each problem instance and vice versa, i.e., its stimulation level. Heuristics and problem instances with a positive stimulation increase their concentration in proportion to the level of stimulation. Those with zero stimulation have their concentration reduced. Finally, any heuristics or problems with no concentration are removed from the system.

5 Demonstration

In order to demonstrate that NELLI functions effectively as a continuous learning system, it must be tested in a dynamically changing problem environment. This is necessary to demonstrate that it is responsive to new problems and exhibits the plasticity required for network to adapt over time as the nature of those problems changes.

An experiment is conducted in which the stream of instances input to the system toggles between two different datasets every 200 iterations. The first dataset known as $ds1$ [24] contains 720 problems, in which the optimal solutions have on average 3 items per bin. The second dataset $ds2$ with 480 instances (also introduced by Scholl et al. [24]) has widely variable item weights, and optimal solutions have between 3 and 9 items per bin. It is well known that due to the different characteristics of each set, heuristics that perform well on one dataset are not expected to perform well on the other. Typically, $ds1$ problems are also easier to solve.

Each dataset is presented in sequence, i.e., $ds2, ds1, ds2, ds1$ following a "start-up" epoch in which the system is initialised from scratch with $ds1$ and run for 200 iterations. For each dataset, we record the total number of bins more than optimal summed over all instances at the start of an epoch (b_s), and again at the end of each epoch (b_e). Given this is a minimisation problem, we expect b_e to be lower than $b - s$.

We then test the following hypotheses to determine whether the system is capable of learning; whether it has memory; and whether it can continue to learn over time:

- *Hypothesis 1* If the system is capable of learning, then for any given epoch, there *should* be a significant difference between b_s and b_e (marked as points A,B in Fig. 5).
- *Hypothesis 2* If the system has retained some memory, then when a dataset is re-introduced, we expect b_s at epoch t to be similar to b_e at epoch $t - 2$), i.e., the start point of an epoch is similar to the end point at the last epoch the dataset was present (points B,C in Fig. 5).
- *Hypothesis 3* If the system continues to learn over its lifetime, there *should* be a significant difference between b_e at the first epoch the dataset appears, and b_e at the last epoch it occurs (points A,D in Fig. 5).

The results are shown in Table 1 and graphically in Fig. 5, averaged over 20 runs in each case, which also compares results to those obtained by a set of four well-known heuristics from the literature (FFD, DJD, ADJD, and DJT, [30]) in which the best heuristic for each problem is selected using a greedy approach.

With respect to *hypothesis 1*, t-tests between the values of b_s and b_e at the two epochs where $ds2$ appears both give p-values <0.0001, confirming that the observed difference in performance is significant. The same result is found for $ds1$ at epochs 2 and 4. Thus, we confirm that learning occurs over an epoch. With respect to *hypothesis 2*, t-tests conducted on the values obtained at the end of epoch 1 and the start of epoch 3 for $ds2$, and epochs 2 and 4 for $ds1$ show no significant difference between results ($P = 0.581, 0.581$), i.e., there is no evidence to reject the null hypothesis that the two distributions are the same, suggesting that memory has been maintained.[2] Finally, we compare the value of b_e at epoch 1 with b_e at epoch 3 ($P < 0.0001$) and similarly with epochs 2 and 4 ($P = 0.0138$) proving the ability of NELLI to learn over time.

[2]That is, that statistics show performance has not significantly deteriorated or improved.

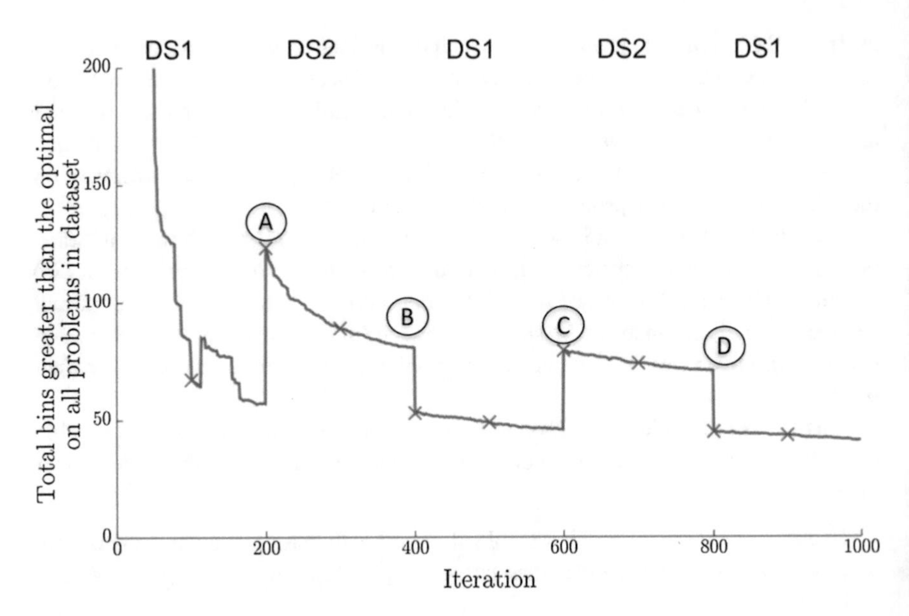

Fig. 5 The average *total* number of bins greater than the optimal alternating every 200 iterations between ds1 (summed over 720 problems) and ds2 (summed over 480 problems)

Table 1 Bins greater than the known optimal at epochs. Averaged over 20 runs

	Epoch 1 DS2		Epoch 2 DS1		Epoch 3 DS2		Epoch 4 DS1	
Set	Start	End	Start	End	Start	End	Start	End
NELLI* $\mathcal{E}$	123.28	80.61	52.94	45.78	79.39	70.72	44.83	41.28
Greedy $\mathcal{E}$	129	129	75	75	129	129	75	75
NELLI* $\mathcal{U}$	312.89	259.17	257.67	247.22	246.89	233.94	234.44	229.17
Greedy $\mathcal{U}$	364	364	364	364	364	364	364	364

To further confirm these observations, an additional experiment is conducted in which the environment is switched to a set of 685 instances randomly selected from a large set of 1370 problems in every 200 instances. The results are shown in Fig. 6. The figure clearly shows that learning continues over the 1000 iterations the system runs for. The difference between b_e and b_s is less defined at each epoch when compared to the previous graph, as some of the problems in the subset presented at each epoch are likely to overlap with those present in the randomly selected set drawn for other epochs. However, it is worth observing that the magnitude of $|b_e - b_s|$ decreases over time (where b_e is measured at the end of epoch t and b_s at the start of epoch $t+1$): this is a direct result of the system augmenting its memory over time that enables it quickly adapt to previously seen instances.

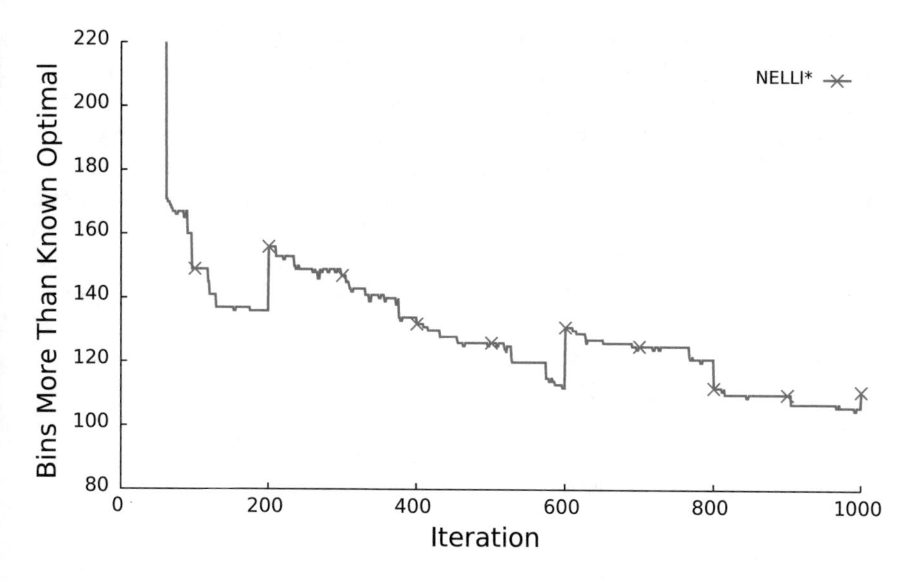

Fig. 6 Every 200 iterations, a random set of 685 problems drawn from a set of 1370 instances is injected into the system. The graph shows the average *total* number of bins greater than the optimal summed over the 685 problems

6 Conclusion

Many industries and organisations can benefit from the application of optimisation algorithms to improve processes, usually leading to economic benefits. However, commissioning software to provide this function usually requires the use of experts, which can be a costly process. Given that most organisations operate in a changing environment, it is clear that software that is not designed to cope with such change can quickly become obsolete.

In this chapter, we have described a method to redress this situation in the field of optimisation, describing a system dubbed NELLI that is capable of continual learning. At a high level, the processes captured in NELLI are inspired by those observed in the natural immune system, evolved over thousands of years to protect us due to its ability to continually learn.

This method is illustrated using an example in the domain of bin-packing, in which a system has to solve a continual stream of instances, which may change with respect to their characteristics over time. In other works, we have shown that the method also transfers to the job-shop scheduling domain [27]. In the version described, the optimisation algorithms generated by NELLI are composed of simple heuristic methods. However, there remains much scope for replacing this element of the system with a more complex algorithm generation method, drawing on the latest research in the field of Automated Algorithm Design [3] that would facilitate the generation of more complex algorithms. On the other hand, it is becoming clear that in industry, organisations do not have time to generate globally optimum solutions and therefore place higher importance on finding robust, quality solutions that can be generated quickly due to changes in the environment [2, 19]; hence, systems like NELLI that *rapidly* produce high-quality solutions clearly have a role to play.

Despite the rapid advances in AI technologies seen in recent years, many so-called intelligent systems make no attempt to retain learned knowledge and re-use it in future learning. In comparison, humans are capable of learning effectively from few examples by drawing on previously accumulated knowledge retained in long-term memory. It is clear that AI will need to address this issue in the future if it is to deliver the potential benefits that are so often cited—not just in the domain of optimisation, but in any kind of intelligent system designed to operate and remain fit for purpose over long periods of time. While NELLI makes some advances in this respect in the field of optimisation in its ability to deal with a continual stream of instances as well as auto-generate new algorithms in direct response to the stream, there remains much scope for future work.

Acknowledgments A significant part of this work was done in conjunction with Dr Kevin Sim who was instrumental in helping to refine my original ideas in the topic of lifelong learning, played a considerable part in the design of the system, and named the resulting algorithm. He also produced all of the code and experimental data. I also gratefully acknowledge funding from EPSRC EP/J021628/1 that enabled this work to be conducted.

References

1. L.C. Bezerra, M. López-Ibánez, T. Stützle, Automatic component-wise design of multiobjective evolutionary algorithms. IEEE Trans. Evol. Comput. **20**(3), 403–417 (2015)
2. M.R. Bonyadi, Z. Michalewicz, Evolutionary computation for real-world problems, in *Challenges in Computational Statistics and Data Mining* (Springer, Berlin, 2016), pp. 1–24
3. J. Branke, S. Nguyen, C.W. Pickardt, M. Zhang, Automated design of production scheduling heuristics: a review. IEEE Trans. Evol. Comput. **20**(1), 110–124 (2015)
4. E.K. Burke, M.R. Hyde, G. Kendall, G. Ochoa, E. Ozcan, J.R. Woodward, Exploring hyper-heuristic methodologies with genetic programming, in *Computational Intelligence* (Springer, Berlin, 2009), pp. 177–201
5. Z. Chen, B. Liu, Lifelong machine learning for natural language processing, in *Proceedings of the 2016 Conference on Empirical Methods in Natural Language Processing: Tutorial Abstracts* (2016)
6. Z. Chen, N. Ma, B. Liu, Lifelong learning for sentiment classification (2018). Preprint, arXiv:1801.02808
7. I.T. Commission, Improving the efficiency of freight movements: the contribution to UK economic growth (2014). http://www.theitc.org.uk/wp-content/uploads/2011/03/ITC-Freight-interim-report-July-14.pdf
8. Deloitte, Measuring the economic benefits of mathematical science research in the UK (2012). https://www.lms.ac.uk/sites/lms.ac.uk/files/Report%20EconomicBenefits.pdf
9. E. Hart, K. Sim, On the life-long learning capabilities of a NELLI*: a hyper-heuristic optimisation system, in *International Conference on Parallel Problem Solving from Nature* (Springer, Berlin, 2014), pp. 282–291

10. E. Hart, J. Timmis, Application areas of AIS: the past, the present and the future. Appl. Soft Comput. **8**(1), 191–201 (2008)
11. L. Hong, J.H. Drake, J.R. Woodward, E. Özcan, A hyper-heuristic approach to automated generation of mutation operators for evolutionary programming. Appl. Soft Comput. **62**, 162–175 (2018)
12. D. Jackson, A new, node-focused model for genetic programming, in *Genetic Programming*, ed. by A. Moraglio, S. Silva, K. Krawiec, P. Machado, C. Cotta. Lecture Notes in Computer Science, vol. 7244 (Springer, Berlin, 2012), pp. 49–60
13. D. Jackson, Single node genetic programming on problems with side effects, in *Parallel Problem Solving from Nature - PPSN XII*, ed. by C. Coello, V. Cutello, K. Deb, S. Forrest, G. Nicosia, M. Pavone. Lecture Notes in Computer Science, vol. 7491 (Springer, Berlin, 2012), pp. 327–336. https://doi.org/10.1007/978-3-642-32937-1_33
14. N.K. Jerne, Towards a network theory of the immune system. Ann. Immunol. (Paris) **125C**(1-2), 373–89 (1974)
15. J.R. Koza, *Genetic Programming: On the Programming of Computers by Means of Natural Selection* (MIT Press, Cambridge, 1992)
16. M. López-Ibáñez, F. Mascia, M.É. Marmion, T. Stützle, A template for designing single-solution hybrid metaheuristics, in *Proceedings of the Companion Publication of the 2014 Annual Conference on Genetic and Evolutionary Computation* (2014), pp. 1423–1426
17. S. Louis, J. McDonnell, Learning with case-injected genetic algorithms. IEEE Trans. Evol. Comput. **8**(4), 316–328 (2004)
18. J. MacLachlan, Y. Mei, J. Branke, M. Zhang, Genetic programming hyper-heuristics with vehicle collaboration for uncertain capacitated arc routing problems. Evol. Comput. **48**(4), 563–593 (2020)
19. Z. Michalewicz, Ubiquity symposium: evolutionary computation and the processes of life: the emperor is naked: evolutionary algorithms for real-world applications. Ubiquity **2012**, 1–13 (2012)
20. M. O'Neill, C. Ryan, Grammatical evolution. IEEE Trans. Evol. Comput. **5**(4), 349–358 (2001)
21. J.C. Ortiz-Bayliss, H. Terashima-Marín, S.E. Conant-Pablos, Lifelong learning selection hyper-heuristics for constraint satisfaction problems, in *Mexican International Conference on Artificial Intelligence* (Springer, Berlin, 2015), pp. 190–201
22. N. Pillay, W. Banzhaf, A genetic programming approach to the generation of hyper-heuristics for the uncapacitated examination timetabling problem, in *Portuguese Conference on Artificial Intelligence* (Springer, Berlin, 2007), pp. 223–234
23. P. Ruvolo, E. Eaton, ELLA: an efficient lifelong learning algorithm. J. Mach. Learn. Res. **28**(1), 507–515 (2013)
24. A. Scholl, R. Klein, C. Jürgens, BISON: a fast hybrid procedure for exactly solving the one-dimensional bin packing problem. Comput. Oper. Res. **24**(7), 627–645 (1997)
25. D. Silver, Q. Yang, L. Li, Lifelong machine learning systems: beyond learning algorithms, in *AAAI Spring Symposium Series* (2013)
26. K. Sim, E. Hart, Generating single and multiple cooperative heuristics for the one dimensional bin packing problem using a single node genetic programming island model, in *Proceedings of GECCO 2013* (ACM, New York, 2013)
27. K. Sim, E. Hart, An improved immune inspired hyper-heuristic for combinatorial optimisation problems, in *Proceedings of the 2014 Annual Conference on Genetic and Evolutionary Computation* (2014), pp. 121–128
28. K. Sim, E. Hart, A novel heuristic generator for JSSP using a tree-based representation of dispatching rules, in *Proceedings of the Companion Publication of the 2015 Annual Conference on Genetic and Evolutionary Computation* (2015), pp. 1485–1486
29. K. Sim, E. Hart, B. Paechter, Learning to solve bin packing problems with an immune inspired hyper-heuristic, in *Proceedings of ECAL 2013, 12th European Conference on ALife* (MIT Press, Cambridge, 2013)
30. K. Sim, E. Hart, B. Paechter, A lifelong learning hyper-heuristic method for bin packing. Evol. Comput. **23**(1), 37–67 (2015)

31. K. Smith-Miles, B. Wreford, L. Lopes, N. Insani, Predicting metaheuristic performance on graph coloring problems using data mining, in *Hybrid Metaheuristics*, ed. by E.G. Talbi. Studies in Computational Intelligence, vol. 434 (Springer, Berlin, 2013), pp. 417–432
32. N. Verstaevel, J. Boes, J. Nigon, D. d'Amico, M. Gleizes, Lifelong machine learning with adaptive multi-agent systems, in *Proceedings of the 9th International Conference on Agents and Artificial Intelligence - Volume 1: ICAART*. INSTICC (SciTePress, Setubal, 2017), pp. 275–286

Emma Hart gained a 1st Class Honours Degree in Chemistry from the University of Oxford, followed by an MSc in Artificial Intelligence from the University of Edinburgh. Her PhD, also from the University of Edinburgh, explored the use of immunology as an inspiration for computing, examining a range of techniques applied to optimisation and data classification problems.

She moved to Edinburgh Napier University in 2000 as a lecturer, and was promoted to a Chair in 2008 where she leads a group in Nature-Inspired Intelligent Systems, specialising in optimisation and learning algorithms applied in domains that range from combinatorial optimisation to swarm robotics. Her own research focuses on developing systems that are capable of adapting autonomously over time to cope with changes in their operating environment, for example re-writing their own software in response to observed or predicted change. She has published widely in the fields of Evolutionary Computing, Hyper-Heuristics and Operations Research. Currently, she is applying ideas of autonomous lifelong learning to developing robotic ecosystems in which robots continually adapt both their form and function, as well as to challenging optimisation problems in the scheduling and logistics industries, incorporating methods from the field of machine-learning to augment those from evolutionary computation.

She was appointed as Editor-in-Chief of Evolutionary Computation (MIT Press) in 2017. She has been invited to give keynotes on her work at major international conferences including CLAIO 2020, IEEE CEC 2019, EURO 2016 and UKCI 2015 and was General Chair of PPSN 2016, and as a Track Chair at GECCO for several years. She is an elected member of the Executive Board of the ACM SIG on Evolutionary Computation.

More broadly, she is an invited member of the UK Operations Research Society Research Panel, and in Scotland, co-led the Artificial Intelligence theme within Scottish Informatics and Computer Science Alliance, a pooling of all CS departments in all Scottish institutions. She has been appointed as a panel member for the upcoming REF2021 (UoA11 Computer Science). In 2020, she was appointed to the Steering Committee developing Scotland's new AI Strategy, led by the Scottish Government. She has a sustained track record of obtaining funding from the EU, EPSRC and of engaging with industry via KTP projects and consultancy, and participates enthusiastically in public-engagement activity.

Message:
It was a pleasure to contribute a chapter to this book. I hope it inspires you to find out a bit more about the field.

I first got interested in computing as a teenager in the mid-80s when the original home PCs such as the Microsoft ZX81 were first introduced to the market. I learned to write very simple programs, sharing knowledge with my teenage male peers—I was definitely a lone female at the time! I was always fascinated by science and went to university to study Chemistry. Computing was not part of the degree at the time, but in my final year, I managed to find a dissertation project with a Professor who was interested in exploring computing to model gas solubility. I wrote programs that were sent off to a mainframe to compile, usually coming back with errors a few hours later! Despite this, I knew I was hooked. On the suggestion of a friend, I applied to do an MSc in Artificial Intelligence at the University of Edinburgh, more out of interest than anything else. It turned out to be a fantastic course, introducing a broad range of topics in AI. After the first lectures in Evolutionary Computing, I knew I had found my "thing": I completed my Master's and started a PhD, where by chance I came across the new subject of Artificial Immune Systems, which became the focus of my PhD. Much of my inspiration for this came from the work of Prof. Stephanie Forrest in the USA, a pioneering female in figure in Computational Intelligence, who I still keep in touch with today. However, I have also been lucky to have great male mentors from the field along the way, who have supported and encouraged me throughout, as well as the unwavering support of my husband and children. I am delighted to have been able to push some boundaries during my career—becoming the first female Editor-in-Chief of Evolutionary Computation and being one of the early members of the Women@GECCO workshops at the ACM GECCO conference. The field is male-dominated for sure—but that does not mean that it is hostile or that that you cannot thrive. I have made lifelong friends of all genders and many nationalities, and I guess I am lucky that Evolutionary Computing has many strong female role-models. I strongly encourage you to take a risk, dive in and find out for yourself!

Reinforcement Learning Control by Direct Heuristic Dynamic Programming

Jennie Si and He (Helen) Huang

1 Introduction

In this chapter, our reinforcement learning control entails integrated approaches of reinforcement learning and optimal control. Specifically, we conceptualize the optimal control problem within a dynamic programming solution setting, and we use reinforcement learning as a computational method of learning to approximately and iteratively solve an optimal control problem that achieves Bellman optimal performance objective.

Optimal control is a well-established branch in control theory. It is a dynamic optimization method that aims at finding an optimal control trajectory of a dynamic system so that an optimal performance measure is achieved. It has found many important applications in science and engineering since its introduction in the 1950s [2] which traditionally include minimizing fuel consumption of a spacecraft or minimizing unemployment rate by developing good monetary policy. Recent applications include modeling of biological movement systems [19] or optimal drug scheduling for cancer chemotherapy [17]. Classical approaches to solving the optimal control problem are either by the minimum principle of Pontryagin or the dynamic programming method that iteratively solves the Bellman optimality equation. Approximations are required in either solution approach [2]. Reinforcement learning control is a means of providing optimal control problems with feasible and

J. Si (✉)
School of Electrical, Computer and Energy Engineering, Arizona State University, Tempe, AZ, USA
e-mail: si@asu.edu

H. (Helen) Huang
The NC State/UNC Department of Biomedical Engineering, North Carolina State University, Raleigh, NC, USA
e-mail: hhuang11@ncsu.edu

A. E. Smith (ed.), *Women in Computational Intelligence*, Women in Engineering and Science, https://doi.org/10.1007/978-3-030-79092-9_10

practical solutions especially when the problem is complex and of large scale and/or if it is difficult or impossible to model the dynamics of the system by a differential or difference equation. The problem solutions are based on approximately solving the Bellman optimality equation [1, 14, 22].

Reinforcement learning is closely related to the theory of optimal control and dynamic programming as it has been demonstrated with strong mathematical foundations and several important applications [5, 6, 8, 13, 16, 21, 23, 25]. Within this context, reinforcement learning control is also referred to as approximate/adaptive dynamic programming. Among the many different reinforcement learning design methods, the actor-critic architecture utilizes the temporal-difference method to approximate the "critic" or the value function of the reinforcement learning controller, and it utilizes a trial-and-error method to update the parameters in the approximating functions of the critic and the actor. Many of the reinforcement learning algorithms, including the actor-critic method, can be data-driven, without an explicitly or independently identified system model, which is required in classical control system designs. It is worth noting however that prior knowledge such as high fidelity mathematical models of the system should help increase the selection of design tools and can be expected to improve system performance when well-integrated into the reinforcement learning controller.

The direct heuristic dynamic programming (dHDP) was proposed as an online reinforcement learning method that learns from interacting with the environment through making associations between state measurements and control policy to approximately solve the Bellman optimal value function in the meantime [20]. It is an actor-critic structure that both the value function and the control policy are approximated by neural networks. As such, it has the promise of generalizing to complex and large-scale problems. In this chapter, we will first introduce the basic dHDP structure and its learning algorithm for controller design, followed by a summary of analytical properties that the dHDP possesses. We will then describe how to use dHDP for solving learning control problems especially those without system models or when the system models are difficult or impossible to obtain. We will illustrate the design approach using a robotic knee control design problem as an example. Results of dHDP controlled robotic knee with human in the loop will be presented. Finally, we will discuss some ideas to further improve the robustness of the dHDP control design.

2 The direction Heuristic Dynamic Programming (dHDP) as an Actor-Critic Type Reinforcement Learning Controller

An actor-critic RL controller entails an identification via estimation of the performance measure of the current control policy to perform policy evaluation. That information is then used to update the control policy for policy improvement. Figure 1 is an illustration of an actor-critic RL controller structure that the dHDP utilizes.

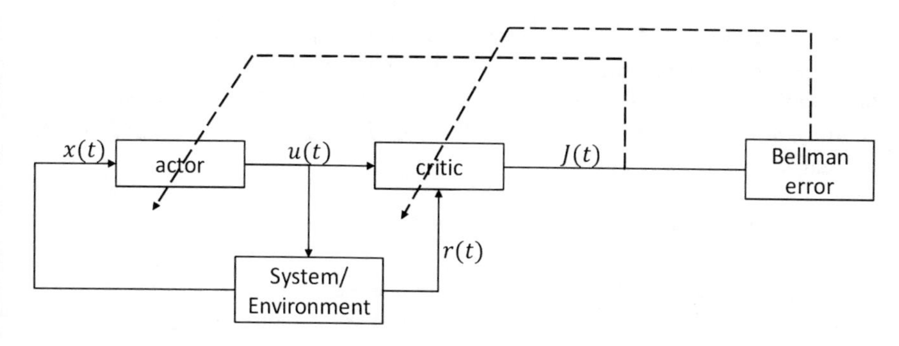

Fig. 1 Schematic of actor-critic reinforcement learning controller that the dHDP utilizes

In dHDP, the actor and the critic are modeled as deterministic functions for real-time control applications. The actor is a deterministic policy that maps a system state to a control action. The critic estimates a performance index for a given policy, and such performance index is usually constrained by the Bellman equation which is updated to achieve Bellman optimal solutions of the performance and control policy.

Note that the dHDP is an online learning algorithm and, as such, the resulted policy in dHDP may be nonstationary. That is to say that the actor in dHDP can represent a class of nonstationary deterministic policy. A similar actor-critic controller, neural fitted Q with continuous action (NFQCA), may be viewed as a batch version of the dHDP according to the authors [9].

2.1 *Building Blocks of the dHDP*

The dHDP was formulated as an online, data-driven, learning-based optimal control with a goal to achieve optimality in terms of reducing the error between a desired state and the controlled variable aiming at approaching the desired state, energy expenditure, and time and data sample efficient. The approximations implemented in the actor and the critic are important to derive forward in time solutions of optimal control. It is necessary that the learning controller seeks optimal solutions while maintaining overall system stability during learning. In other words, since the actor and critic functions are implemented by neural networks, the weights of the actor and the critic are necessary to converge to optimal solutions while maintaining system stability.

Refer to Fig. 1. The reinforcement signal $r(t)$ is provided from the system or the external environment as an instantaneous reflection of system performance which is also referred to as the primary reinforcer in computer science literature or stage cost in controls literature. We use the latter as we are concerned about control applications. Here the stage cost can be simple, discrete scaler values, for example, a "0" for success or a "-1" for failure. Or it can be a sum of squares in states $x(t)$ (or state errors) and controls $u(t)$, i.e., $r(t) = x(t)^T Qx(t) + u(t)^T Ru(t)$ which is a nonnegative measure reflecting state error and control energy consumption.

We consider an infinite horizon future cost $R(t)$, or cost-to-go, based on the instantaneous cost $r(t)$:

$$R(t) = r\,(t+1) + \alpha r\,(t+2)\, r(t+2)^2 + \alpha^2 r\,(t+3) + \ldots \tag{1}$$

where in (1), α $(0 < \alpha < 1)$ is a discount factor for the infinite-horizon problem.

The output of the critic network, $J(t)$, in Fig. 1 is to approximate the discounted total future cost $R(t)$.

Given the infinite horizon future cost as the performance index and letting $J(t)$, the critic output, be a neural network approximation of the future cost, $R(t)$, we have the following deterministic Bellman equation:

$$J\,(t) = r\,(t+1) + \alpha J\,(t+1)\,. \tag{2}$$

The critic network output $J(t)$ as that of a neural network is updated as the critic network weights adjusted. The critic weights are adjusted to minimize the Bellman error:

$$e_c(t) = \alpha J(t) - \{J\,(t-1) - r(t)\}\,. \tag{3}$$

Note that Eq. (3) is Eq. (2) with a time delayed by 1 unit. By doing so, we will need to save the respective data from the previous time step. The Bellman error is a measure of consistency of the Bellman equation. The Bellman equation has to be satisfied at each time step or the Bellman error has to be reduced at each time step. The solution to this equation is the value that can be viewed as the expected future cost under actuator noise. If the actuator is subject to zero mean white noise, then the value can be viewed as the critic network output.

The critic network weights are adjusted so that the Bellman error square below can be minimized:

$$E_c(t) = \frac{1}{2}\,\{\, e_c(t)\}^2. \tag{4}$$

The actor in Fig. 1 is to perform policy improvement. The principle in adapting the weights in the actor network is to (indirectly) minimize the error between the desired ultimate objective and the approximate future cost. Ideally this error is zero for future costs formulated for control applications. The weight updating in the actor can be formulated as follows. Let

$$E_a(t) = \frac{1}{2}\,\{\, e_a(t)\}^2. \tag{5}$$

In (5), $e_a(t)$ stands for the error between the estimated future cost from the critic and the desired future cost which ideally is zero for control applications.

Training of the weights of the actor and critic networks can be carried out using backpropagation of the error gradients [20].

2.2 Properties of dHDP Learning Controller

The dHDP is an online learning mechanism that can be applied to real-time control problems. As there are two essential approximation structures in the dHDP, it is important to know if the parameters/weights in the approximators converge during online learning. Additionally, it is important to ensure that the states and the controls are bounded so to guarantee the controlled system stability. The first direct view on the training of the actor and critic networks can treat the problem as stochastic approximation. As such, we can first view the weights in the actor and the critic, respectively, approaching convergence in a statistical sense if they are, respectively, treated as a statistical approximation algorithm.

Convergence of Weights in Statistical Sense Robbins and Monro developed and analyzed a recursive procedure for finding the root of a real-valued function $g(w)$ of a real variable w [18]. The function is not known, but noise-corrupted observations could be taken at values of w selected by the experimenter. Mathematically $g(w) = Ex\{f(w)\}$, where $Ex\{.\}$ is the expectation operator. In this formulation, $f(w)$ is not known but its samples are available. Robbins-Monro algorithm can be used to iteratively seek the root $w*$ of the function $g(w)$:

$$w\,(t+1) = w(t) - l(t)f\,[w(t)]\,. \tag{6}$$

Specifically, under some very mild conditions, which can be easily met in our actor-critic approximation setting, such as $g(w)$ is monotone with a single root, the variance between $g(w)$ and the sample function is finite, and $g(w)$ satisfies a sector-like condition, and if the learning $l(t)$ in (6) satisfies the following conditions: (1) $\lim_{t\to\infty} l(t) = 0$, (2) $\sum_{t=0}^{\infty} l(t) = \infty$, and (3) $\sum_{t=0}^{\infty} l(t)^2 = \infty$, all of which are also easily met during neural network training of the actor and the critic, then $w(t)$ will converge toward $w*$ in the mean square error sense and also with probability one, i.e.,

$$\lim_{t\to\infty} Ex\left\{\left|w(t) - w^*\right|^2\right\} = 0, \tag{7}$$

$$Prob\left\{\lim_{t\to\infty} w(t) = w^*\right\} = 1. \tag{8}$$

Applying the Robbins-Monro algorithm to training the weights in the actor and critic neural networks, we let $g(w) = \partial E/\partial w$, where E is the respective actor or critic training objective function. If E has a local optimum at w^*, $g(w)$ will satisfy the first Robbins-Monro condition locally at w^*. If E has a quadratic form, $g(w)$ will satisfy the same condition globally. Based on the Robbins-Monro algorithm, we can use neural network training samples to represent $f(w)$ and obtain a convergence guarantee in statistical sense as stated in Robbins-Monro algorithm. But note that, as an online learning algorithm, the dHDP error reduction during training of the actor and critic networks is to reduce sample-based error objectives but not the expected error objectives as in the Robbins-Monro [20]. We therefore pursued further results to analyze the behavior of the dHDP during learning.

Uniformly Ultimately Boundedness (UUB) of Weights Consider the dHDP learning controller with its actor and critic in Fig. 1 implemented by neural networks, which are universal approximators typically realized by two layers of weights and hyperbolic type thresholding functions in the hidden layer and linear units in the output layer. Under mild conditions we have shown that the weights in the actor-critic dHDP learning controller are uniformly ultimately bounded (UUB) [15].

We first assume that the true optimal weights exist for the actor and the critic networks, respectively, and they are finite. This can be considered reasonable because of how we constructed the neural networks and the weights are usually trained by gradient descent type algorithms with small learning rates. For the dHDP learning controller, if the actor network weights and the critic network weights are trained based on the formulations of Eqs. (4)–(6) with specific weight update equations given in [20], but we constrain the learning rates for the actor and the critic as follows, (1) $l_c < \frac{1}{\alpha^2 \| \varnothing_c \|^2}$, (2) $l_a < \frac{1}{\| \varnothing_a \|^2}$, where $\frac{1}{\sqrt{2}} < \alpha < 1$, $\varnothing_a$ and $\varnothing_c$ are actor and critic hidden unit output vectors, respectively, then we can show that the weights in the actor and the critic networks are uniformly ultimately bounded [15]. Note that in this result, we have assumed that the weights in the first layer are randomly generated initially and remain there during learning. Also note that neural networks in this case still are universal approximators.

Additionally, we can show similar UUB results as above under similar condition as above but with the following updated conditions on learning rates: (1) $l_c < \frac{1}{\alpha^2 N_{hc}}$, (2) $l_a < \frac{1}{N_{ha}}$, where $\frac{1}{\sqrt{2}} < \alpha < 1$, N_{ha} and N_{hc} are number of hidden units in the actor and critic networks, respectively. The proofs of this result and the above can be found in [15].

The dHDP as a Stabilizing Controller As previously discussed results focused on the convergence of the weights, for dHDP as an online learning controller, it is necessary to show that the controlled system is stable in the UUB sense. We address this issue in [27]. Since this result also relies on weight convergence of the actor and the critic networks, we will be using similar conditions on the actor and critic neural networks as stated before. Additionally, as we now consider the controlled system stability, we will need some very mild conditions on the system.

We assume that the system is controllable, 0 is an equilibrium of the system, and the system stays at 0 without additional control. This assumption can be satisfied by many controlled systems, including the human-robot system that we will discuss next. Besides the previously specified conditions on the actor and critic weights, we will need another mild condition that the nonlinear thresholding function of the actor and the critic is Lipschitz continuous. Typical universal approximators meet this requirement without a problem.

Our proved result on dHDP as a stabilizing controller is as follows [27]. Given the above-described conditions, which can be met without a problem under the dHDP framework involving an actor and a critic neural networks, if the learning rates in the gradient descent-based learning algorithm are constrained as (1) $l_c < \frac{1}{\|\varnothing_c\|^2}$, and (2) $l_a < \frac{1}{\|\varnothing_a\|^2}$, then the control $u(t)$, which is the output of the actor network, is a stabilizing control to guarantee the UUB of the nonlinear system under the training algorithm based on the formulation described in Eqs. (4)–(6). We recently obtained the same stabilizing control result [26] but under a more relaxed condition that the weights of the first layer of the actor and the critic networks can be initialized randomly as above but can also be updated by backpropagating the error gradients.

3 Applications of the dHDP for Automatic Tuning of Prosthesis Control Parameters

As shown above, the dHDP is an online learning controller possessing some qualitative properties that are necessary for solving real engineering problems. We are now in a position to show that the dHDP design can be applied to solving real-world, challenging problems that may not have readily available solutions from classical control or robotic engineering.

3.1 The Robotic Knee Prosthesis Control Problem and Current Approaches

The robotic knee prosthesis control is not a "typical" control problem as this robot is worn by a human and thus, there is a human in the control loop. Clearly the goal for robotic prosthesis control is for individuals with lower limb amputations to regain normative walking. The prosthesis control of such a wearable robotics relies on a large number of configurable parameters. These control parameters need to be customized for the prosthesis user due to differences in individual body construct, physical ability, and personal preference. Impedance control framework is the most accepted structure for lower limb prosthesis control [10]. Under this framework, a total of 12 control parameters are to be tuned for each gait cycle. The state-of-the-art practice in a clinic now is to manually tune each parameter at a time. As a result, the coordination among the parameters is not considered, and only a subset of the parameters are tuned. Re-tuning is often required. This procedure costs time, money, and inconvenience.

Automatic control of the robotic knee prosthesis has been attempted but with limitations. For example, the idea of using the intact leg as a reference for the robotic knee seems intuitive and natural, but it is yet to be validated for feasibility [3]. The response surface optimization [7] and cyber expert system [11] methods configure wearable robot control parameters basically in an open-loop fashion, and they are not adaptable to new situations or new users. The design there is not efficient. The authors of [12] proposed to use joint patterns as virtual constraints as was done in bipedal robots. It is unclear however if such a control model is used by amputee subjects. A reinforcement learning controller is naturally appealing as it learns from measurements of the human-robot system. As shown previously, dHDP design relies on the system state and control variables which are measured as real-time signals for learning. The future cost is to be estimated using a critic network. This makes dHDP and similar algorithms unique and advantageous as they are data-driven and do not rely on a mathematical description of the complex dynamics to be controlled. In applications such as robotic prosthesis control, such a mathematical description prior to controller design is not feasible because of a human in the loop. On the other hand, classical control designs do rely on an accurate system dynamics model which prevents these classical methods from applying to wearable robot application problems.

3.2 The dHDP for Automatic Tuning of Robotic Prosthesis

Reinforcement learning or adaptive dynamic programming is perhaps the most promising approach to solving large-scale, complex, nonlinear control problems under uncertainty [21]. Such problems may be characterized by a lack of mathematical model of the underlying dynamic process to be controlled, the dynamics may change over time, and the control objectives may also change over time, and so on. In our previous work, we have demonstrated those important properties of RL controllers, or specifically the dHDP, in several challenging problems [4–6, 8, 16]. All these problems have continuous values of states and controls, which create additional challenges to typical reinforcement learning algorithms that are based on table look-up methods such as the original Q-learning. Function approximations will be needed for those methods in order to deal with such problems, which eventually resort to actor-critic type controllers.

We approach the online control of the robotic knee prosthesis problem as follows.

Refer to Fig. 2. First notice that the robotic knee control is realized by a well-established FS-IC framework [10], within which the three impedance control parameters are to be tuned either manually or by a control algorithm. We are the first to apply reinforcement learning control toward automatic tuning of the impedance

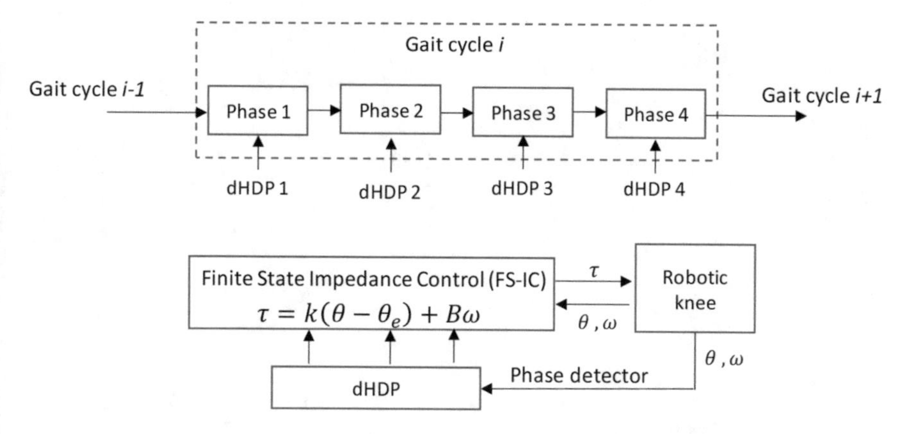

Fig. 2 Schematic of the dHDP within a finite state impedance control (FS-IC) framework for automatic tuning of the impedance control parameters

parameters. Specifically, we used a dHDP control to tune the three impedance parameters. The dHDP is integrated into the FS-IC as follows. The FS-IC views a gait cycle as multiple phases, which is four in our realization. The initial state of each phase is the terminal state of the previous phase (Fig. 2, top panel). Within each of the four phases of the FS-IC control framework, the three impedance parameters (k, θ_e, ω) are adjusted to make the robotic knee fit comfortably to an amputee user to help restore normative walking. The control torque τ is generated according to Eq. (9):

$$\tau = k\,(\theta - \theta_e) + B\omega \tag{9}$$

which is to enable the joint angle movement of the robotic knee. The bottom panel of Fig. 2 illustrates how the dHDP is integrated into each phase of the FS-IC controller, i.e., each phase of the FS-IC employs an identical dHDP controller, however, with different control policies or action networks.

Refer to Fig. 3; the four FS-IC phases are differentiated by the unique landmarks of a knee gait profile – the first and the second peaks correspond to the first and the third phase, respectively, and the first and the second troughs correspond to the second and the fourth phase, respectively. For each phase, the two state variables in the dHDP controller are (1) the difference between the real and the target knee angles (also referred to as the peak error) and (2) the time difference between the real and time target phase transition landmarks (also referred to as the duration error) [24].

The dHDP controller for each phase aims at optimizing the future cost of the form in Eq. (10):

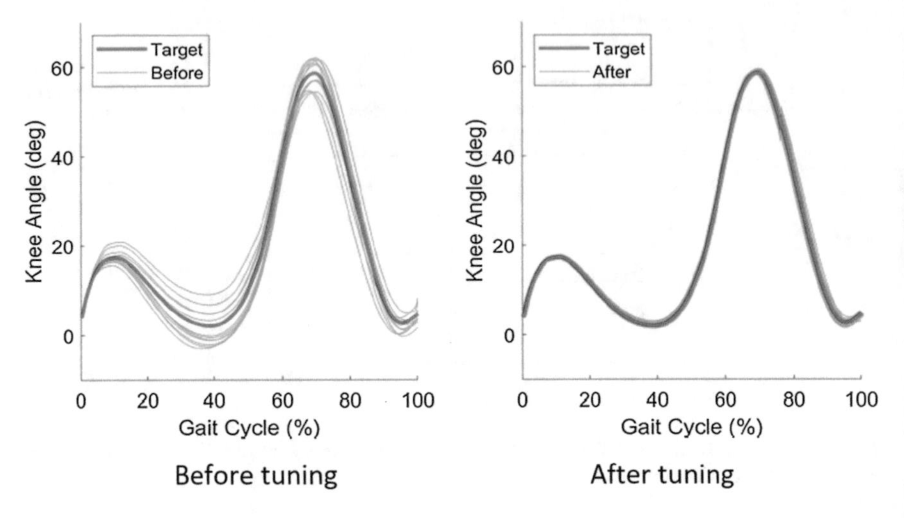

Fig. 3 Automatic tuning result of the robotic knee angle profile

$$\sum_{t=0}^{\infty} \alpha^t r_t, \quad 0 < \alpha < 1 \tag{10}$$

where r_t is the stage cost. Given that the problem involves human subject, we therefore impose safety constraints in terms of the peak error and the duration error. The stage cost is constructed based on those errors: failure corresponds to state going beyond the safety boundary ($r_t = -1$), and success corresponds to state staying within a small error tolerance. We also utilized an intermediate value between success and failure as feedback when training the dHDP controller [24].

The dHDP online learning control takes effort as follows. The FS-IC impedance parameters are first initialized to a setting that will not cause the states to go beyond safety limits. The actor and the critic networks are initialized to random weights. At each gait cycle, the state variables are measured and used in training the dHDP controller. Once the state variables reach the designated target range, learning pauses. The resulting policy, i.e., the output of the actor, can then be used to enable continuous walking of the subject.

Figure 3 (left) is a set of knee profiles from the initial impedance parameters. They approach the target knee profile after learning; refer to Fig. 3 (right). Additional performance evaluations and statistical summaries of dHDP learning outcomes can be found in [24]. After testing dHDP control of robotic knee using simulations, we tested the design on human subject successfully [25].

4 Conclusion

Since the introduction of the dHDP algorithm, we have developed new findings and new applications for dHDP to demonstrate its potential for real-world applications. Reinforcement learning-based designs offer a new approach to the important class of problems involving the design of effective and efficient control algorithms. While classical control theory has demonstrated its prowess in many applications especially those that enable our everyday life, the information age has introduced new challenges that may not be readily addressed by classical control theory. Reinforcement learning therefore may be the most promising candidate to bridge the gap. In this chapter, we have used dHDP as a specific class of reinforcement learning controller to illustrate the great potential of reinforcement learning-based control.

References

1. D.P. Bertsekas, *Reinforcement Learning and Optimal Control* (Athena Scientific, Belmont, MA, 2019)
2. A.E. Bryson, *Applied Optimal Control: Optimization, Estimation and Control* (CRC Press, 1975)
3. M.F. Eilenberg, H. Geyer, H. Herr, Control of a powered ankle-foot prosthesis based on a neuromuscular model. IEEE Trans. Neural Syst. Rehabil. Eng. (2010)
4. R. Enns, J. Si, Apache helicopter stabilization using neural dynamic programming. J. Guid. Control. Dyn. **25**(1), 19–25 (2002)
5. R. Enns, J. Si, Helicopter trimming and tracking control using direct neural dynamic programming. IEEE Trans. Neural Netw. (2003)
6. R. Enns, J. Si, "Helicopter flight-control reconfiguration for main rotor actuator failures," J. Guid. Control. Dyn., 2003
7. W. Felt, J.C. Selinger, J.M. Donelan, C.D. Remy, 'Body-in-the-loop': Optimizing device parameters using measures of instantaneous energetic cost. PLoS One **10**(8), 1–21 (2015)
8. W. Guo, F. Liu, J. Si, D. He, R. Harley, S. Mei, Online supplementary ADP learning controller design and application to power system frequency control with large-scale wind energy integration. IEEE Trans. Neural Networks Learn. Syst. (2016)
9. R. Hafner, M. Riedmiller, Reinforcement learning in feedback control. Mach. Learn. (2011)
10. N. Hogan, Impedance control: An approach to manipulation: Part III—Applications. J. Dyn. Syst. Meas. Control. **107**(1), 17 (1985)
11. H. Huang, D.L. Crouch, M. Liu, G.S. Sawicki, D. Wang, A cyber expert system for auto-tuning powered prosthesis impedance control parameters. Ann. Biomed. Eng. **44**(5), 1613–1624 (2016)
12. S. Kumar, A. Mohammadi, N. Gans, R.D. Gregg, Automatic tuning of virtual constraint-based control algorithms for powered knee-ankle prostheses. IEEE Conf. Control Technol. Appl. (2017)
13. F.L. Lewis, D. Vrabie, K.G. Vamvoudakis, Reinforcement learning and feedback control: Using natural decision methods to design optimal adaptive controllers. IEEE Control. Syst. (2012)
14. F. L. Lewis, D. Liu (eds.), *Reinforcement Learning and Approximate Dynamic Programming for Feedback Control* (John Wiley & Sons, 2013)

15. F. Liu, J. Sun, J. Si, W. Guo, S. Mei, A boundedness result for the direct heuristic dynamic programming. Neural Netw. (2012)
16. C. Lu, J. Si, X. Xie, Direct heuristic dynamic programming for damping oscillations in a large power system. IEEE Trans. Syst. Man, Cybern. Part B Cybern. (2008)
17. R.B. Martin, *Optimal Control Drug Scheduling of Cancer Chemotherapy* (Automatica, 1992)
18. H. Robbins, S. Monro, A stochastic approximation method. Ann. Math. Stat. (1951)
19. S. Schaal, P. Mohajerian, A. Ijspeert, Dynamics systems vs. optimal control – a unifying view, in *Progress in Brain Research*, ed. by A. P. Cisek, T. Drew, J. F. Kalaska, vol. 165, (Elsevier B.V, 2007), pp. 425–445
20. J. Si, Y.T. Wang, On-line learning control by association and reinforcement. IEEE Trans. Neural Netw. **12**(2), 264–276 (2001)
21. J. Si, A. Barto, W. Powell, D. Wunsch, Handbook of Learning and Approximate Dynamic Programming. 2004
22. R.S. Sutton, A.G. Barto, *Reinforcement Learning, An Introduction*, 2nd edn. (MIT Press, Cambridge, MA, 2018)
23. Y. Wen, J. Si, X. Gao, S. Huang, H.H. Huang, A new powered lower limb prosthesis control framework based on adaptive dynamic programming. IEEE Trans. Neural Networks Learn. Syst. (2017)
24. Y. Wen, J. Si, X. Gao, S. Huang, H. Huang, S. Member, A new powered lower limb prosthesis control. IEEE Trans. Neural Networks Learn. Syst. **28**(9), 2215–2220 (2017)
25. Y. Wen, J. Si, A. Brandt, X. Gao, H.H. Huang, Online reinforcement learning control for the personalization of a robotic knee prosthesis. IEEE Trans. Cybern. (2020)
26. Z. Yao, J. Si, R. Wu, J. Yao, "Toward Reliable Designs of Data-Driven Reinforcement Learning Tracking Control for Euler-Lagrange Systems," https://arxiv.org/abs/2101.00068
27. Q. Zhao, J. Si, J. Sun, Online reinforcement learning control by direct heuristic dynamic programming: From time-driven to event-driven. IEEE Trans. Neural Networks Learn. Syst

Jennie Si received the B.S. and M.S. degrees from Tsinghua University, Beijing, China, and the Ph.D. degree from the University of Notre Dame, Notre Dame, IN, USA. She has been a faculty member in the School of Electrical, Computer and Energy Engineering at Arizona State University, Tempe, AZ since 1991. Her research focuses on reinforcement learning control utilizing tools from optimal control theory, machine learning, and neural networks. She is also interested in fundamental neuroscience of the frontal cortex and its role in decision and control processes. Dr. Si was a recipient of the NSF/White House Presidential Faculty fellow Award in 1995 and the Motorola Engineering Excellence Award in 1995. She is a Distinguished Lecturer of the IEEE Computational Intelligence Society and an IEEE fellow. She consulted for Intel, Arizona Public Service, and Medtronic. She has served on several professional organizations' executive boards and international conference committees. She was an Advisor to the NSF Social Behavioral and Economical Directory. She served on several proposal review panels. She is a past Associate Editor of the IEEE Transactions on Semiconductor Manufacturing, the IEEE Transactions on Automatic Control, and Action Editor of Neural Networks. She is a current Associate Editor of the IEEE Transactions on Neural Networks.

He (Helen) Huang received her B.S. from the School of Electronics and Information Engineering at Xi'an Jiao-tong University (China) and M.S. and Ph.D. in Biomedical Engineering from Arizona State University. Her post-doctoral training is in Neural Engineering and Rehabilitation Engineering at the Rehabilitation Institute of Chicago/Northwestern University. Currently Dr. Huang is the Jackson Family Distinguished Professor in the NCSU/UNC Joint Department of Biomedical Engineering at North Carolina State University and the University of North Carolina at Chapel Hill. She is also the Director of the Closed-Loop Engineering for Advanced Rehabilitation (CLEAR) Core.

Dr. Huang's research interest lies in neural-machine interfaces for robotic prosthetic limbs and assistive exoskeletons, wearer-robot interaction, intelligent control of wearable robots, human movement control, and rehabilitation engineering. Her main interest in computation intelligence is its application in biomedical and rehabilitation engineering. Her previous related work includes development of advanced machine learning approaches to decode neuromuscular signals for neural control of robotic prosthesis arms and legs and using computer vision to enable environment-aware wearable robot control. Currently she collaborates with Prof. Jennie Si on reinforcement learning control to personalize assistive wearable robots for optimal movement performance in individuals with disabilities.

Dr. Huang's research has been sponsored by federal agencies (such as NSF, NIH, DOD, DARPA, and NIDILRR) and private companies (such as Össur and BionX). She was a recipient of the Delsys Prize for Innovation in Electromyography, the Mary E. Switzer fellowship with the National Institute on Disability, Independent Living, and Rehabilitation Research (NIDILRR), and a NSF CAREER Award. Dr. Huang serves(d) as an Associate Editor for IEEE Transactions on Neural System and Rehabilitation, IEEE Transaction on Biomedical Engineering, Journal of Neuroengineering and Rehabilitation, and IEEE International Conference on Robotics and Automation. She is also a guest associate editor for IEEE Transactions on Robotics. She is a member for the Delsys Prize Review Board, VA/Rehab Research and Development Service Scientific Merit Review Board, and Programmatic Panel member for DOD Orthotics and Prosthetics Outcomes program. She is a senior member of IEEE and a member of the Society for Neuroscience, Biomedical Engineering Society, AAAS, and American Society of Biomechanics.

Distributed Machine Learning in Energy Management and Control in Smart Grid

Kumar Utkarsh and Dipti Srinivasan

1 Introduction

From the beginning of the industrialization age to the recent past, electric power generation, transmission, and distribution have followed a conventionally simple architecture. This architecture involved a unidirectional flow of electricity—from large generating stations generating power at high voltages, to the transmission system at stepped-up extra-high to ultra-high voltages, and finally to various distribution systems at reduced voltages. In this architecture, the generating stations and some power flow control devices, along with substation switchgear present in generation, transmission, and distribution systems, were the only controllable entities [26]. However, this power architecture is rapidly changing as smart grid evolves, and several distributed energy resources (DERs) are being integrated into the power system, mainly at medium-to-low voltage distribution networks. DER is a broad terminology for generation and storage resources typically located behind the energy meter of a consumer or a prosumer. These resources include solar photovoltaics (PV), wind turbines, combined heat and power systems, and energy storage systems (ESSs). Further, the DER terminology is also mostly broadened to include demand response of controllable loads, electric vehicles (EVs), as well as energy efficiency. Therefore, the basic notion of a DER is a controllable device mostly on the distribution side of the power system and that provides an extra degree of freedom to the system. One special type of DERs are microgrids, which are themselves a set of several other DER types including several renewable distributed generators (RDGs), battery energy storage systems (BESSs), and controllable loads. They are usually categorized as the part of the power network that is controlled and

K. Utkarsh · D. Srinivasan (✉)
National University of Singapore, Singapore, Singapore
e-mail: dipti@nus.edu.sg

A. E. Smith (ed.), *Women in Computational Intelligence*, Women in Engineering and Science, https://doi.org/10.1007/978-3-030-79092-9_11

operated by some entity, and they are often regarded as building blocks of the future smart grid.

The increasing integration of DERs (especially renewable energy sources) is driven mainly due to concerns of global warming, which has resulted in governmental policies encouraging their adoption. The real benefits of increased integration of DERs may not be fully appreciated unless proper control strategies are developed to appropriately utilize the newly formed extra degrees of freedom in the future smart power grid.

Another aspect of control of DERs is their desired level of intelligence or controllability. In the future smart grid, the number of controllable DER entities, such as RDGs, controllable loads, BESSs, and EVs, will significantly outnumber those that are present today in the conventional power system [32]. Thus, new control strategies need to be developed, which move away from the contemporaneous centralized control strategies and toward increasingly distributed strategies, such that the RDGs and other DER entities can be controlled autonomously with high levels of redundancy and low communication requirements. This automation must be implemented by utilizing state-of-the-art information and communication technologies [32].

First, there is a need for improved and smarter distributed control algorithms and approaches over the present literature that would enable efficient integration of a large number of DER entities in the power network [10]. Additionally, it is expected that distributed energy management and control strategies provide computationally efficient improvements over centralized ones due to reduction in memory usage as well as computation runtime of distributed strategies.

Distributed algorithms are a well-studied subject in the algorithmic literature. However, their extensive applications to power systems have significantly emerged only in the last decade in the context of smart grid [33]. Moreover, there is a need to develop distributed algorithms that would significantly reduce the requirement for linear or convex approximations. Further, the distributed algorithms in the present literature operate on only one probable solution [30] and iteratively evolve that solution toward either local or global optimality. However, if multiple solutions are concurrently evaluated and evolved iteratively, a globally optimal (or near-optimal) solution can be obtained in a relatively shorter amount of time [19].

This chapter presents a distributed algorithm that would require significantly reduced approximations, with good convergence, privacy-preserving and optimality-approaching properties, and that would also evaluate multiple solutions using a population-based computational intelligence (CI) technique [19], for example, particle swarm optimization (PSO), Genetic Algorithm (GA), or Differential Evolution (DE).

The primary objective of this chapter is to present effective and smart coordinated control strategies for the operation of DERs in medium-to-low-voltage distribution networks. The primary objectives are summarized as follows:

1. The first goal is to develop a computationally efficient distributed control strategy for coordination of a large number of DERs in a power network. The distributed

strategy is to be designed such that the requirements of linearization or convexification of power flow equations and other objectives are almost eliminated. Further, the strategy must also have reduced communication requirements while obtaining global network states and must preserve the privacy of individual DER entities.
2. The second goal is to apply the distributed control strategy for integration of DERs in the power system considering different network topologies. The inclusion of network topologies entails consideration of power line impedances in the problem formulation and the effect they have on communication infrastructure required for the individual DER controllers executing the distributed control strategy. Therefore, the result is a limit on the operation of DERs such that the network's voltage constraints are not violated.
3. The third goal is to perform a comprehensive analysis of cooperation and competition scenarios among multiple interconnected microgrids with several DERs, corresponding to two cases, where (1) the microgrids may be owned by a single entity or (2) there are multiple competing entities owning individual microgrids. These two scenarios have significant ramifications on the utilization of BESSs, the quantum of active/reactive power to be exchanged and trading prices, and finally on the effective utilization of renewable sources present in the microgrids so as to reduce utilization of fossil-fuel generated power from the external grid.

This chapter is organized as follows. Section 2 presents emerging trends in distributed energy management and control in smart grids. Sections 4 and 3 give an overview on energy management and distributed control algorithms in grid-connected microgrids. Section 5 outlines the proposed distributed algorithm for optimal control of smart grids. Section 6 presents results and discussion, and conclusion follows in the end.

2 Emerging Trends in Distributed Energy Management and Control in Smart Grids

The increasing adoption of highly variable renewable energy sources along with controllable loads, such as electric vehicles, is a positive step toward a sustainable future, but they will also strain the power networks in terms of increased difficulty in network voltage regulation, increased power losses, and coordination issues with existing conventional dispatchable generators [35, 53]. Therefore, in order to make the future power system more secure and reliable, efficient control strategies for proper coordination of such DER entities need to be designed.

DER control can be performed at different levels depending on the control objective and required response time. The control strategies used in the power system optimization literature can be typically divided into the following three categories [37], as also depicted in Fig. 1:

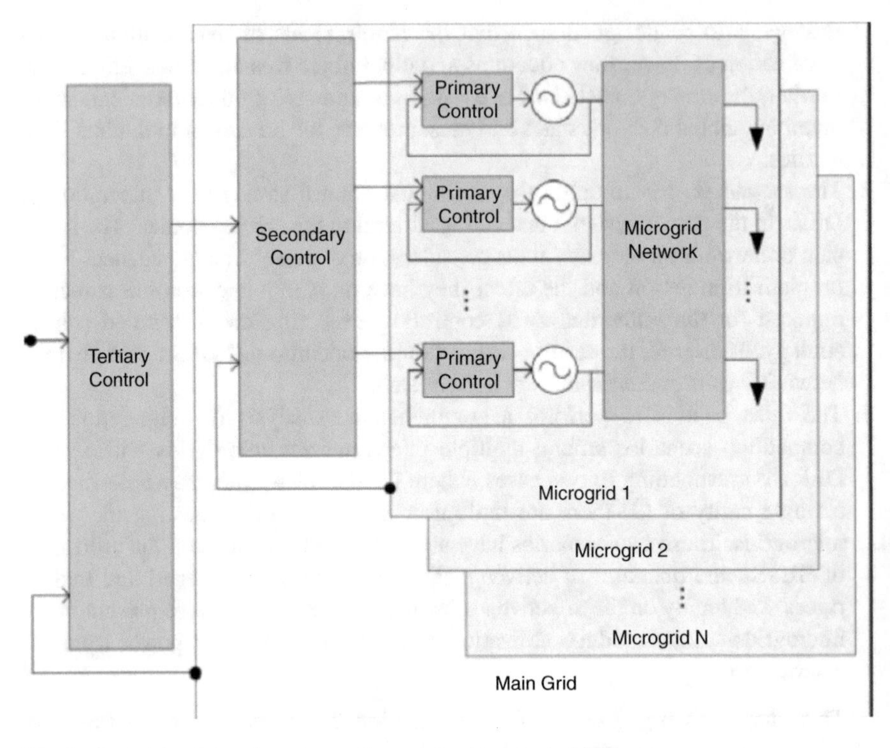

Fig. 1 Control levels—primary, secondary, and tertiary control [37]

1. **Primary control:** Primary control is the first level of control closest to a DER and with the fastest response. It relies only on local system state measurements with no communication to other DERs. Some examples of the functions of primary control are grid-side converter control, power-sharing control, and detection of islanding [25].
2. **Secondary control:** Secondary control is the next upper level of control with a slightly delayed response time than primary control. It relies on communication with the power network's nodal devices about network states, cost functions, operational limits, etc. Its main objectives are economically secure and reliable operation of the power network. This control level is also referred to as the energy management system (EMS) of the network. Apart from the network's economic operation, the security and reliability parameters include frequency stability, voltage stability, power balance, line flow constraints, etc. [16].
3. **Tertiary control:** The highest level of control in a power network is tertiary control with the slowest response time in the order of several minutes. It is essentially responsible for coordinating the operation of the power network it is managing via communication with tertiary controllers of other networks. Such coordination requirements may include voltage support, frequency regulation, and other ancillary services [52].

It is worth noting that these three control categories also interact with each other, thereby making a complete control architecture for the power network. The tertiary-level controllers provide signals to secondary-level controllers and other subsystems that form the full power network. On the other hand, secondary controllers send signals to and coordinate with primary controllers of the DERs within the microgrids. Finally, primary controllers operate independently with their inner voltage and current control loops and react to any local events. The primary control acts almost instantaneously with the fastest response time, while the secondary control typically operates in the range of a few seconds to a couple of minutes. However, this operation rate becomes particularly challenging in case of microgrids with a high share of renewables such that the dispatch command of the secondary control needs to be at a high-enough rate so as to follow the net load fluctuations appropriately, and thus needs to be operated in the order of a few seconds [37, 54].

The work presented in this chapter deals with the secondary control level (i.e., the EMS) of the power system, and that "control" of DERs present in a microgrid refers to obtaining optimal set points of active/reactive power through the use of the proposed algorithms (i.e., the EMS). These set points are then sent to the power electronic converters (i.e., the actuators) of the DERs as references, which are then followed by the voltage/current loops of the primary control level present in the actuators. The secondary control strategies are usually divided into the following four significant domains:

1. **Centralized control strategy:** A centralized control strategy involves a central controller that gathers all the relevant information about various DERs and non-controllable loads present in the power system. This information may include the cost functions, the system's network parameters, as well as the forecasts of loads and RDGs. This information is then processed by the central controller, and the optimal set points are dispatched to the controllable DER entities in the system for implementation. Although centralized control strategies usually provide the best optimal solutions for a problem, they will be unable to efficiently handle a large number of control variables introduced by an increasing number of DERs, resulting in computation/communication bottlenecks [10, 64].
2. **Decentralized control strategy:** A decentralized control strategy performs in an antithetical manner to a centralized control strategy, in that there is no central controller involved to coordinate the operation of DERs. Each DER has its own local controller, which operates autonomously by utilizing only local measurements to determine its optimal set point based on an objective function. Decentralized control strategies provide the highest level of autonomy to the controllable entities while also preserving their privacy and providing a high level of redundancy. However, since the local controllers do not communicate with each other, the obtained optimal set points usually do not correspond to those obtained by a centralized controller having access to the global information of the power network [3].

3. **Hierarchical control strategy:** The next domain is that of hierarchical control strategy, which lies midway between decentralized and centralized strategies. In a hierarchical strategy, there are two or more layers of controllers, wherein controllers in a particular layer have some level of autonomy but have to coordinate their operation with controllers in the next layer. This strategy provides some level of relief from computation/communication bottlenecks that would normally be encountered in centralized control strategies [8].
4. **Distributed control strategy:** The final domain is that of distributed control strategies, which are similar to decentralized strategies in that each controllable entity has its own local controller with a high level of autonomy, privacy, and redundancy characteristics. However, unlike decentralized strategies, the local controllers communicate with each other in this strategy based on some communication topology. Therefore, the local controllers are able to determine the global states of the system and obtain their optimal set points accordingly. Thus, they also eliminate the disadvantages of decentralized control strategies and reach an optimal solution close to those obtained by centralized control strategies [1, 4, 5, 7]. Due to the aforementioned reasons, distributed control strategies are considered to be an important and integral part of the future control architecture in power systems and are, therefore, adopted throughout this chapter.

It is further noted that the guarantee of stability is not trivial in the case of distributed/decentralized algorithms and that to guarantee stability, different methodologies can be adopted. One such methodology that has been adopted in this chapter is first to obtain the state transition matrix of the system and then to analyze its eigenvalues to determine convergence [36]. Another popular methodology is called as Lyapunov method for stability, and it involves first forming the Lyapunov function and then determining if its gradient is less than or equal to zero for all values of the control variables [32].

Remark 1 The availability of system measurements, communication, and high-speed computational facilities are some of the challenges for the microgrids aiming to operate in either grid-connected or islanded mode. These considerations favor either hierarchical or distributed, decentralized approaches. However, for geographically smaller microgrids, hierarchical control architecture may be quite appealing than distributed/decentralized architecture given the smaller number of devices involved, where fast dynamics are handled by the primary control level in the output controls and slower dispatch commands are handled by the central EMS in the secondary control level. However, the usage of DERs with different response rates, such as batteries, flywheel energy storage, or even compressed air storage, along with electric vehicles and photovoltaics, will make the integration of their primary-level controls with the secondary-level control more difficult for geographically larger microgrids. As such, a distributed approach will be more suited in coordinating the DERs while also preserving privacy of individual DER owners, as compared to a hierarchical approach [37].

Remark 2 In the context of distributed control strategy adopted in this chapter, the term "local controller" of a DER is used to describe a controller that has the following properties: (1) ability to take sensory inputs from the DER as well as from the power network; (2) ability to react to changes in such inputs and to drive the DER set point toward optimality; and (3) ability to communicate and exchange information with other controllers. These three attributes of the local controllers also correspond to that of an intelligent agent, as proposed by Wooldridge [56]. Therefore, the terms "local controller" and "agent" have been used interchangeably throughout this chapter.

3 Distributed Control Algorithms in Smart Grids

Since most distributed optimization techniques rely on consensus-based approaches [20, 29, 36], first, this section briefly reviews foundations of the consensus technique. Later, this section provides an overview of some of the well-established distributed algorithms in the existing power systems literature [33]. The algorithms discussed are dual decomposition, Alternating Direction Method of Multipliers, and Optimality Condition Decomposition (OCD) techniques. These algorithms are well suited for operation optimization of power systems and are essentially based on Lagrangian function, augmented Lagrangian function, or distributed estimation of Karush–Kuhn–Tucker (KKT) conditions of the original problem. Finally, a brief review of the population-based distributed optimization technique is provided.

3.1 Consensus-Based Approach

Consensus-based optimization forms a very important part of the distributed computing literature. In consensus problems, the primary objective is for a group of individuals with varying opinions to reach a common understanding. Consensus is especially important in situations where individuals value privacy and are reluctant to share their local information with a centralized entity. In networked systems, consensus plays a key role in enabling individual entities to get an independent estimate about the global value of a desired variable. As such, averaging consensus algorithms have been extensively used in the literature for estimation of such desired variables [36].

Consensus-based distributed optimization algorithms find use in a wide variety of areas, such as in intelligent transportation systems [40], energy management in smart grids [60, 65], synchronization of coupled oscillators [39], and control of multi-robot systems [57], among others. An important application of consensus-based distributed algorithms is in the optimization of large-scale systems with a high number of control variables. Hence, consensus-based algorithms provide a platform in which the need for a centralized optimizer is obviated and computational effort is evenly spread across multiple entities. Using average-consensus algorithms [45], an agent in a network can communicate with its neighboring agents and iteratively arrive at a common value, given by the average of the agents' initial estimates. For continuous-time systems, the consensus algorithm to be executed by an agent i can be written as follows:

$$\dot{x}_i(t) = \sum_{j \in N_i} a_{ij} \left(x_j(t) - x_i(t) \right), \tag{1}$$

where $x_i(t)$ is the estimated parameter of agent i at time t, and $\dot{x}_i(t)$ is its gradient. N_i represents the set of agents connected to agent i for which the communication weights are $a_{ij} \neq 0$, otherwise $a_{ij} = 0$ [36]. Equivalently, for discrete-time systems, a typical consensus algorithm can be written for each discrete iteration time k as follows:

$$x_i(k+1) = x_i(k) + \sum_{j \in N_i} a_{ij} \left(x_j(k) - x_i(k) \right). \tag{2}$$

Several alternate forms of the consensus algorithm have also been developed, in which the communication weights a_{ij} are replaced with a function depending on the number of connected agents N_i, for example:

$$\dot{x}_i(t) = \frac{1}{N_i} \sum_{j \in N_i} \left(x_j(t) - x_i(t) \right), \tag{3}$$

for continuous-time systems, and

$$x_i(k+1) = x_i(k) + \sum_{j \in N_i} a_{ij} \left(x_j(k) - x_i(k) \right), \tag{4}$$

for discrete-time systems [36]. An important consideration for consensus algorithms is their convergence rate, which can be easily manipulated using different values of the communication weights [59].

3.2 *Dual Decomposition*

Dual decomposition techniques take advantage of the Lagrangian functions of problems with a separable structure [11, 12, 15]. Let an optimization problem be of the form:

$$\min_{\forall x_i} \sum_{i=1}^{N} f_i(x_i) \\ \text{subject to} \sum_{i=1}^{N} A_i x_i = b, \tag{5}$$

where $f_i(\cdot)$ is a cost function, x_i is a decision variable vector, A_i is a matrix, and b is a vector (with respect to power systems, the physical meaning of b can be equated to, for example, net power purchased from the upstream grid, net voltage deviation at the nodes, etc.). The Lagrangian function with dual variable y can now be constructed as follows:

$$L(x, y) = \sum_{i=1}^{N} L_i(x_i, y), \tag{6}$$

$$L_i(x_i, y) = f_i(x_i) + y^T A_i x_i - (1/N) y^T b. \tag{7}$$

Since this form is decomposable, dual decomposition methods perform "dual ascent," which is defined iteratively as follows:

$$x_i^{k+1} = \arg\min_{x_i} L_i\left(x_i, y^k\right), \tag{8}$$

$$y^{k+1} = y^k + \alpha^k \left(\sum_{i=1}^{N} A_i x_i^{k+1} - b\right), \tag{9}$$

where k is the iteration number and α^k is the step size. It is noted that Eq. (8) can be performed in a fully distributed manner. However, Eq. (9) requires some sort of hierarchical coordination between the local controllers. This can be easily handled by utilizing consensus algorithms to update the dual variables y.

3.3 Optimality Condition Decomposition

In OCD techniques, an optimization problem is divided into sub-problems with coupled variables, and an agent then solves its own sub-problem considering its non-coupled variables only. The coupled variables are further added as penalties in each sub-problem in the Lagrangian function. Each agent then iteratively solves the KKT conditions of the Lagrangian function of its own sub-problem while also sharing primal and dual values with other agents in its communication vicinity [7, 18].

Another variation of the OCD technique is the Consensus + Innovation approach [20, 24] where, unlike conventional OCD, all the decision variables evolve iteratively to determine parameters of the optimal KKT conditions of the sub-problems.

3.4 Population-Based Distributed Algorithms

In comparison to aforementioned consensus-based distributed optimization algorithms, population-based distributed algorithms work with multiple solutions (called as a population of solutions) at once and have the potential to provide a near-optimal solution for high-dimensional problems within fewer iterations of the algorithm [17]. This is a critical requirement for operation of smart grids with a high penetration of DERs. Therefore, as a part of distributed population-based techniques to deal with such problems of high dimensions, distributed evolutionary algorithms (EAs) have been developed in the literature.

Distributed EAs have been further classified as population-distributed or dimension-distributed EAs [17]. Population-distributed EAs assume each agent to consist of a probable solution vector, say a particle in PSO or a chromosome in GA, i.e., $x \in \mathbb{R}^D$, where D is the number of control variables involved in optimization. Therefore, in population-distributed EAs, each agent needs information about all the D control variables. However, in a real scenario, the D control variables may belong to separate private entities, and therefore, it is much more reasonable to associate an agent with each of the D control variables instead. This is the rationale of dimension-distributed EAs, in which agents correspond to the dimensions (i.e., control variables) of the solution vector. Thus, dimension-distributed EAs are truly distributed implementations with respect to a "practical problem," in which each dimension is controlled by a separate local controller or agent.

Existing works on dimension-distributed EAs can be further classified as utilizing either coevolution models [38] or game-theoretic multi-agent models [41]. The game-theoretic multi-agent model requires global objectives to be expressed as a sum of local objectives of agents, which may not always be possible [22, 41, 51]. In the coevolution model, on the other hand, an agent needs to know the variables of other agents as well, which may lead to a violation of privacy [44]. Also, an agent performs several generations of evolution of its primary variable (corresponding to its own dimension) while considering secondary variables (corresponding to other agents' dimensions) to be clamped. This results in the evolution being performed with respect to outdated secondary variables. Then, in the communication phase, each agent updates its secondary variables, which need to be broadcast by other agents, resulting in extra communication time overhead [46, 62].

The distributed techniques mentioned in Sect. 3 are well studied in the power systems literature to develop distributed secondary control architectures for coordinated management of DERs with the aim to optimize the power network(s) based on several economic and technical objectives as well as constraints. These objectives and constraints are critical for efficient, reliable, and secure operation of the power network(s) with high penetration of DERs. Therefore, Sect. 4 now provides a brief review of the existing literature for DER management considering different power network configurations.

4 Energy Management and Control in Grid-Connected Microgrids

The EMS of a microgrid takes data of the power network as inputs, and outputs DERs' optimal set points obtained after solving an optimal power flow (OPF) problem [37]. An OPF problem essentially optimizes the total cost of the power system with several controllable DER entities while respecting physical as well as technical/engineering constraints of the power network as well as of the DERs. The response time of this optimization depends on the type of OPF problem being solved. For example, for real-time systems, the OPF response time should be less than 4–5 s [54], while for EMS and tertiary control systems the response time can vary from a few seconds to a few minutes [37].

The input data required by the EMS may include [47]:

- Forecasted power output of non-dispatchable or renewable generators
- Forecasted local load
- State of charge of the BESSs
- Operational limits of dispatchable generators (such as conventional distributed generators) and BESSs
- Security and reliability constraints of the microgrid
- Utility grid ancillary service requirements
- Forecasted grid energy prices

The optimal DER outputs obtained after OPF optimization are then sent to the primary control systems of the DERs as active/reactive reference power values to be tracked by inner voltage and current control loops of their fast-responding and dynamically efficient power electronic converters [31].

Further, frequency-regulation-related concerns are not present in grid-connected microgrids as the network frequency is handled by the upstream high-inertia grid [10, 37]. Therefore, some of the important objective metrics utilized by the EMS of a grid-connected microgrid for optimal control of DERs can be listed as follows.

4.1 *Operation Cost*

The microgrid is considered as a set of various DERs, such as solar PVs, wind turbines, conventional distributed generators, BESSs, and controllable loads. In a scenario where a microgrid entity only operates the power network, it needs to compensate the independent power generators and earns revenues from the power consumers [27]. On the other hand, if a microgrid entity operates the power network as well as owns the DERs and the loads, its objective could simply be to optimize the utilization of the DERs and the loads [21]. Therefore, for both the scenarios, the objective of operation cost optimization is usually separable across the decision variables (the DERs) and can be written as follows:

$$\min_{\forall x_i} \sum_{i=1}^{N} f_i(x_i), \tag{10}$$

where $f_i(x_i)$ is the cost/revenue related to active/reactive power x_i of a DER i among N DERs.

4.2 Power Losses

Power losses in the microgrid power network subsequently lead to buying of extra power from the external grid or from the DERs (for the scenario of a microgrid entity only operating the power network and not owning any DERs) and, therefore, need to be minimized [28, 64]. These losses essentially result from the phasorial voltage difference across the power lines and are, therefore, related to the voltage regulation of the network.

$$\min_{\forall V_i, i \neq 1} \sum_{i=1}^{N} \sum_{k=1}^{N} V_i V_k G_{ik} \cos \delta_{ik}, \tag{11}$$

where V_i is the voltage magnitude at node i, and G_{ik} and δ_{ik} are conductance and angle difference, respectively, between nodes i and k.

4.3 Network Voltage Regulation

The microgrid's network voltages must be kept inside their band constraints while being optimized for power loss and other objectives [48].

$$V_{\min} < V_i < V_{\max} \quad \forall i. \tag{12}$$

4.4 Line Limits

Apart from voltage limits being a measure of stability in grid-connected microgrids, another stability metric is power line thermal limits [42], which may be exceeded in case of current overloading of the lines.

$$I_{\min} < I_i < I_{\max} \quad \forall i. \tag{13}$$

4.5 Power Flow Modeling Constraints

Power flow constraints arise out of the physics of current flow in a power network. They are basically virtual constraints, in the sense that they will always be physically fulfilled in any given power network; however, they need to be handled explicitly when designing control strategies so that the obtained solutions closely conform to the actual physical operation of the power network [33]. Power flow modeling constraints of the network can be handled in several ways depending upon the desired level of complexity.

4.6 Reactive Power Limit Constraints

The reactive power availability of distributed generation source (DG) or controllable loads (CL) converters is constrained.

$$abs\left(Q_i[z]\right) \leq abs\left(Q_{i,\max}\right). \tag{14}$$

4.7 Reactive Power Ramp-Rate Constraints

Reactive power change between two consecutive iterations needs to be constrained.

$$Q_i[z-1] - QR_i \leq Q_i[z] \leq Q_i[z-1] + QR_i. \tag{15}$$

5 Proposed Distributed Algorithm for Optimal Control of Smart Grids

In this section, the proposed consensus-based distributed algorithm for optimal control of smart grids is presented in the context of a generic large-scale global optimization problem. First, the consensus algorithm is discussed, which is used by the agents to estimate the states of the system. Second, the rationale for the choice of PSO is described on which the proposed distributed CI algorithm is based, and finally, the proposed distributed algorithm is presented.

5.1 Consensus Algorithm

Consider a connected graph G with N nodes and E edges, in which each node is assumed to be controlled by an agent. Any two agents are considered "neighbors" *iff* they have a common edge $e \in E$, and each agent can communicate with its neighboring agents only. Now, let $\xi_k[z] \in \mathbb{R}$ denote the estimate of a parameter maintained by agent k at an iteration z. The value ξ_k may represent any quantity that the agents are trying to reach a consensus on. A network of N agents is considered to have reached consensus *iff* $|\xi_k - \xi_l| \leq \mu, \forall k, l = (1 \ldots N), k \neq l$, where μ is a very small positive number. According to the consensus algorithm, each agent k updates its estimates using the following discrete-time model:

$$\xi_k[z+1] = \sum_{l=1}^{N} d_{k,l}\xi_l[z], \tag{16}$$

where $d_{k,l}$ represents the communication coefficient between agents k and l. It is noted that $d_{k,l}$ is non-zero only if the two agents are neighbors. Thus, the combined consensus algorithm can be written as

$$\bar{\boldsymbol{\xi}}[z+1] = \mathfrak{D} * \bar{\boldsymbol{\xi}}[z], \tag{17}$$

where $\mathfrak{D}$ is a matrix composed of elements $d_{k,l}, \forall k, l = (1, \ldots, N)$, and $\bar{\boldsymbol{\xi}}$ is the column vector containing all ξ_k. According to the proof in [36] if the matrix $\mathfrak{D}$ is doubly stochastic, i.e., all its rows and columns individually sum up to one, and if the eigenvalues of $\mathfrak{D}$ are less than or equal to 1, then all the $\xi_k's$ will converge to a common estimate as given below:

$$\xi_k[\infty] = \frac{1}{N}\sum_{k=1}^{N} \xi_k[0], \forall k = (1, \ldots, N). \tag{18}$$

The matrix $\mathfrak{D}$ can be made to fulfill the aforementioned conditions by using the *metropolis* algorithm [36, 58].

5.2 Choice of PSO as Basis for the Proposed Distributed CI Algorithm

CI techniques [55] are adaptive mechanisms to enable or facilitate intelligent behavior in complex, uncertain, and changing environments. These adaptive mechanisms include those bio-inspired and artificial intelligence paradigms that exhibit an ability to learn or adapt to new situations. Thus, CI techniques are essentially nature-inspired methodologies for addressing complex problems where traditional rigorous methods are ineffective or infeasible. CI techniques primarily consist of Artificial Neural Networks, fuzzy logic systems, evolutionary computation, swarm intelligence, and immune systems. They have also been widely applied to many challenging real-world problems in signal processing, power systems control, communication, robotics, etc.

Of the various CI techniques, EAs [17, 46] have been shown to perform very well in dynamic and complex environments [41, 50]. Some well-known examples of EAs are PSO [14], GA [43], and DE [13]. However, among these, PSO and its variants have been shown to efficiently solve large-scale non-linear problems while not suffering from the curse of dimensionality or slow convergence [2, 9, 23, 49]. Further, PSO has the following favorable properties [2]:

1. PSO is easier to implement and has fewer parameters to be tuned.
2. In PSO, each control variable has its own self-knowledge component and group-knowledge component, thereby having a better memory capability than other algorithms, such as GA or DE.
3. PSO allows faster convergence thanks to the momentum effects on particle movement. Also, it can offer more variety and diversity of the population of solutions.

These points prompted the development of a distributed CI-based algorithm with PSO chosen as the preferred alternative.

PSO is a direct search and stochastic optimization technique that is based on the behavior of a swarm of social group of animals. In PSO, the decision of an ith particle $\Theta_i \in \mathbb{R}^N$, where N is the dimension of the solution hyperspace, is influenced by three factors: the particle's own velocity, the best-known solution found by the particle, and the best-known solution found among all the particles. The basic equations that the standard PSO uses for a particle are as follows:

$$\begin{aligned} v_i[z] = w \cdot v_i[z-1] + \varphi_1.\text{rand1}.\left(\text{pbest}_i - \Theta_i[z-1]\right) \\ + \varphi_2.\text{rand2}.\left(\text{gbest} - \Theta_i[z-1]\right), \end{aligned} \tag{19}$$

$$\Theta_i[z] = \Theta_i[z-1] + v_i[z], \tag{20}$$

where z is the present iteration, v_i and Θ_i are velocity and position, respectively, of particle i, w is the inertia factor, φ_1 and φ_2 are two positive numbers, rand1 and rand2 are two random numbers uniformly distributed in the range [0,1], pbest_i is the best solution found so far by particle i, and g best is the best solution found among all the particles.

5.3 Proposed Consensus-Based and Dimension-Distributed PSO-Based Algorithm

In the proposed algorithm, N agents are considered corresponding to N control variables in a system, with each agent i corresponding to a control variable $x_i \in \mathbb{R}$ (i.e., reactive power output Q_i of a DG or CL converter in the formulated optimal smart grid control problem). Each agent also maintains P probable values of its control variable in the form of a column vector. Hereafter, each such value of the vector is referred to as an element. Now, for each of its P elements, an agent has to estimate its contribution toward the state vector $\boldsymbol{y}$ of the system. Specifically, an agent i, for each element p, will estimate its contribution as $\boldsymbol{T}(:, i) * x_i^p$ and will share this estimate with neighboring agents using consensus algorithm to get an updated estimate of state vector $\boldsymbol{y}$. In the context of the formulated optimal smart grid control problem, each agent i, for each element p, will estimate its contribution toward the voltage vector $\boldsymbol{V}$ in the system using $\bar{\boldsymbol{Y}}(:, i) * Q_i^p$ and will share this estimate with neighboring agents using consensus algorithm to get an updated estimate of the voltage vector $\boldsymbol{V}$. This framework is also depicted in Fig. 2 where the element vectors and the estimated state vectors of two agents i and j are shown.

It is noted that the optimal control problem in smart grids is a dynamic optimization problem, in the sense that the system parameters (i.e., active power generation/consumption of DGs and CLs) vary with time. To this end, let a time duration (of seconds or minutes) be considered for which the system can be assumed to be static, i.e., the system parameters or the active power generation/consumption of DGs and CLs remain constant. This time duration is denoted as a *control cycle*

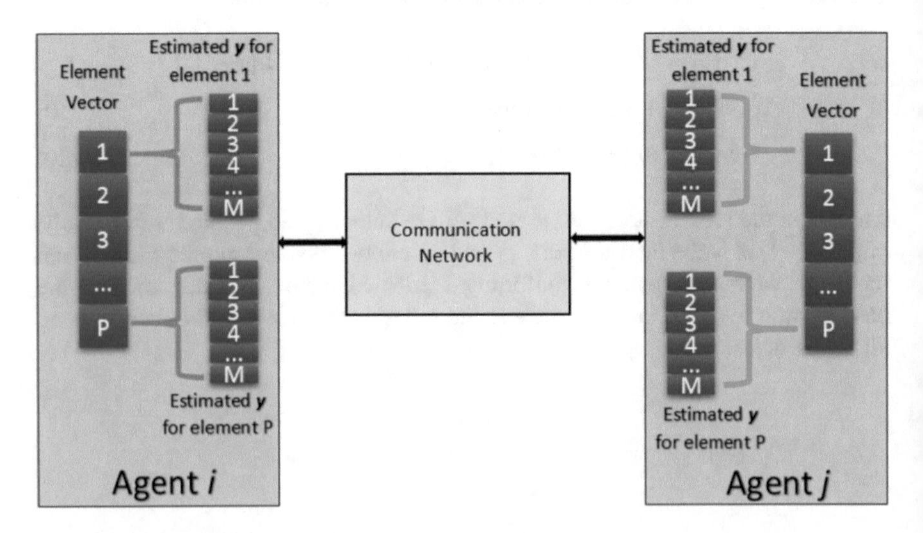

Fig. 2 Framework of agents for the proposed distributed algorithm

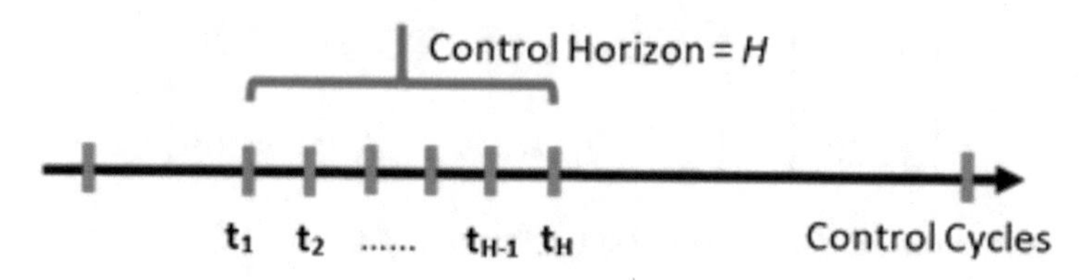

Fig. 3 Control cycles and Control horizon

(t) in this chapter. A collection of H number of such consecutive control cycles is denoted as a *control horizon* ($\mathcal{H}$), and it is assumed that the system parameters vary when the current control cycle ends and the next control cycle begins. This concept is also shown pictorially in Fig. 3.

Further, it is considered that in each control cycle t (out of the H cycles) the agents perform several iterations of the proposed algorithm until convergence to optimize the system, before moving on to the next control cycle $t + 1$. It is noted that the time duration of each control cycle is upon the user to decide based on the system's requirements and the convergence time of the algorithm.

Considering the dynamics in the system, the execution steps of the algorithm, to be performed by each agent i in the control horizon $\mathcal{H}$, are presented next.

Step 0 (Commencement of Control Horizon $\mathcal{H}$) Consider this step as the starting point of the control horizon $\mathcal{H}$, i.e., this step is the beginning of control cycle $t = 1$.

Step 1 (Initialization) The agent i initializes a column vector of its probable values, according to the following:

$$\tilde{x}_i[0] = -x_{i,\max} + 2 * x_{i,\max} * \text{rand}(P, 1), \tag{21}$$

where $\tilde{x}_i = \left[x_i^1, \ldots, x_i^P\right]^T$ is a P-element column vector, $x_{i,\max}$ is the maximum allowed value of the control variable x_i, and rand($\cdot$) generates a column vector of random numbers between 0 and 1. It is noted that x_i is analogous to the reactive power Q_i of DG and CL converters. For each element p in the column vector obtained in Step 1, the agent then initializes its estimates of the state vector (or equivalently the voltage vector $\boldsymbol{V}$) as follows:

$$\begin{bmatrix} y_1^{i,p}[0] \\ \cdots \\ y_M^{i,p}[0] \end{bmatrix} = N * \boldsymbol{T}(:, i) * x_i^p[0]. \tag{22}$$

Step 2 (Synchronization) The agent then communicates with its neighboring agents and updates its state estimates as follows:

$$\begin{bmatrix} y_1^{ip}[0] \\ \cdots \\ y_M^{i,y}[0] \end{bmatrix} = \begin{bmatrix} \sum_{j=1}^{N} d_{i,j} * N * \boldsymbol{T}(1, j) * x_j^p[0] \\ \cdots \\ \sum_{j=1}^{N} d_{i,j} * N * \boldsymbol{T}(M, j) * x_j^p[0] \end{bmatrix}. \tag{23}$$

The agent now calculates the objective function f as in

$$\min_{\boldsymbol{x} \in \mathbb{R}^N} f(\boldsymbol{x}) \equiv \min_{\boldsymbol{y} \in \mathbb{R}^M} f(\boldsymbol{y}), \tag{24}$$

where N is the dimension of the control variable vector, $\boldsymbol{x} \in \mathbb{R}^N$. Although the global objective function $f(x)$ is generalized enough to capture all possible functions, it is more convenient to write the function f in terms of a state vector say vector $\boldsymbol{y} \in \mathbb{R}^M$, of dimension M.

The objective function f is equivalent to power loss in the formulated smart grid problem Eqs. (11)–(15) for each element p. It then finds out the *element-best* value for each element p (i.e., the value of the control variable x_i^p corresponding to the best-known objective function found so far by the element p) and the *global-best* value (i.e., the value of the control variable x_i^p corresponding to the best-known objective function found so far among all the P elements). It is noted that the l-dimensional *element-best* and *global-best* values of an agent correspond to the N-dimensional *particle-best* and *global-particle-best* values in the standard PSO. The velocities of the elements are also randomly initialized in this step. The main part of the algorithm follows from Step 3.

Step 3 (Evolution of Elements) Consider this step as the start of iteration z. To eliminate the phenomenon of *stagnation* in the proposed fully distributed algorithm, two important modifications are made in this heuristic module:

1. Velocity and position (i.e., element value) update equations of the *global-best* element are integrated with the Guaranteed Convergence PSO (GCPSO) [6]. In GCPSO, the modified velocity update equation of the *global-best* element is as follows:

$$v_{i,g}[z] = w.v_{i,g}[z-1] - x_{i,g}[z-1] + \text{pbest}_{i,g} + \rho[z]r, \tag{25}$$

where $v_{i,g}$ and $x_{i,g}$ are velocity and position (i.e., element value), respectively, of the *global-best* element, and $\text{pbest}_{i,g}$ is its *element-best* value. w is a weighting coefficient, r is a random number sampled from the interval (−1,1), and $\rho(z)$ is a scaling factor determined as follows.

$$\rho[0] = 1$$
$$\rho[z+1] = \begin{cases} 2\rho[z], & \text{if success count } > s_c \\ 0.5\rho[z], & \text{if failure count } > f_c \\ \rho[z], & \text{otherwise } e, \end{cases} \tag{26}$$

where s_c and f_c are thresholds. The thresholds are such defined that whenever the *global-best* element improves its *element-best* value pbest $_{i,g}$, the success count is incremented, and the failure count is set to 0 and vice versa. However, the success and failure counts are both set to 0 whenever the *global-best* element itself changes. These modifications ensure that the *global-best* element keeps performing a random search around its position, resulting in its escape from a local minimum.

2. The velocities of all elements, except the *global-best* element, are checked if they have fallen below a certain threshold. If yes, then they are reset to random values in the allowable range. This allows the elements to keep looking for better solutions, while at the same time retaining knowledge learned by the *global-best* element.

Step 4 (Adaptation and Synchronization) For each element p, the agent now estimates the new state vector, due to change in the element's value at the present iteration, and communicates the state vector to its neighboring agents as follows:

$$\begin{aligned} \begin{bmatrix} y_1^{i,p}[z] \\ \cdots \\ y_M^{i,p}[z] \end{bmatrix} &= \begin{bmatrix} y_1^{i,p}[z-1] \\ \cdots \\ y_M^{i,p}[z-1] \end{bmatrix} + N * \boldsymbol{T}(:,i) * \left(x_i^p[z] - x_i^p[z-1]\right) \\ \begin{bmatrix} y_1^{i,p}[z] \\ \cdots \\ y_M^{i,p}[z] \end{bmatrix} &= \begin{bmatrix} \sum_{j=1}^{N} d_{i,j} y_1^{j,p}[z] \\ \cdots \\ \sum_{j=1}^{N} d_{i,j} y_M^{j,p}[z]. \end{bmatrix}. \end{aligned} \tag{27}$$

The agent now calculates the objective function f as in Eq. (24) or Eqs. (11)–(15), for each element p, and updates the *element-best* and *global-best* values.

Step 5 (Implementation of the Obtained Solution) In the proposed algorithm, each agent immediately applies its *global-best* element value found at each iteration to its respective node. This strategy ensures that the agents respond as fast as possible to changing system conditions and do not wait until convergence is achieved to implement the obtained solutions.

Step 6 (Check for Convergence) The algorithm is assumed to have converged if the agents ascertain that the change in their *global-best* element values is below a threshold for a certain number of consecutive iterations, or if a maximum number of iterations are reached. If the algorithm has not converged, the control is transferred to Step 3 again, and the iteration counter z is also incremented to $z+1$. Further, if the algorithm has converged and the present control cycle t has ended, it is incremented to $t+1$ and the control is transferred to Step 7.

Step 7 (Population Update for Dynamic Optimization) It has previously been noted in this section that the optimal control problem in smart grids is a dynamic optimization problem, in which the system parameters vary with time.

To deal with such dynamics, an obvious approach could be to restart the entire algorithm for each control cycle. However, this restart-based algorithm would be very inefficient, because all the knowledge gained from the previous control cycle would be essentially lost, which could have otherwise been utilized to reach the new optimal solution faster [61]. Based on this logic and to make the algorithm more flexible, evolutionary dynamic optimization (EDO) technique is integrated in the proposed algorithm. The integrated EDO-based approach is able to effectively manage system dynamics, in that a fast convergence to the optimal solution is achieved after a change in system conditions [61].

In this integrated EDO-based approach, at the end of each control cycle, each agent i modifies the final converged vector of its control variable (i.e., $\tilde{x}_i$) in the following manner. First, each agent randomly selects 10% of its elements and replaces them with random immigrants, which basically are randomly chosen values in the allowable domain of its control variable. Second, each agent randomly selects 40% of the elements and replaces them with elite immigrants, which are random mutations within ±20% of the previously obtained global-best element value. After these modifications, the control is again transferred to Step 4.

6 Simulation Results and Discussion

The proposed algorithm is implemented using MATLAB on two distribution test systems: (1) the standard IEEE 30-node system, which is a meshed network as shown in Fig. 4, and which is modified to simulate a distribution system with a total load of 10 MW and a nominal voltage of 11 kV [66]; and (2) an 11 kV radial distribution system as shown in Fig. 5 with a total load of 14 MW and with 119 nodes [63], which effectively simulates a large-scale system to test scalability of the proposed algorithm.

These networks have been extensively studied in the existing power systems literature, see for example, [10, 18, 32, 53], and are therefore well suited to perform solution benchmarks for the proposed distributed algorithm. The voltage variations at all the nodes are limited to ±3% of the nominal value. For both the systems, DGs and CLs are assumed to have been placed after proper siting and sizing studies, with their active power generation/consumption ratings being 5 MW each and the rating of the converters being 5 MVA each. Each such DG and CL is controlled by an agent, and the network of such agents forms the communication topology, wherein an agent communicates with other agents using power line communication (PLC) technology. Other technologies as well as communication topologies can very well be incorporated in the proposed algorithm, with the effect being that the convergence time will decrease for a more densely connected topology and vice versa. It is also assumed that the PLC technology is capable of a 10 kHz communication rate, i.e., for each iteration of the proposed algorithm, any two neighboring agents take approximately 0.1 ms to exchange their state information. Further, the time duration of each control cycle is chosen as 5 min.

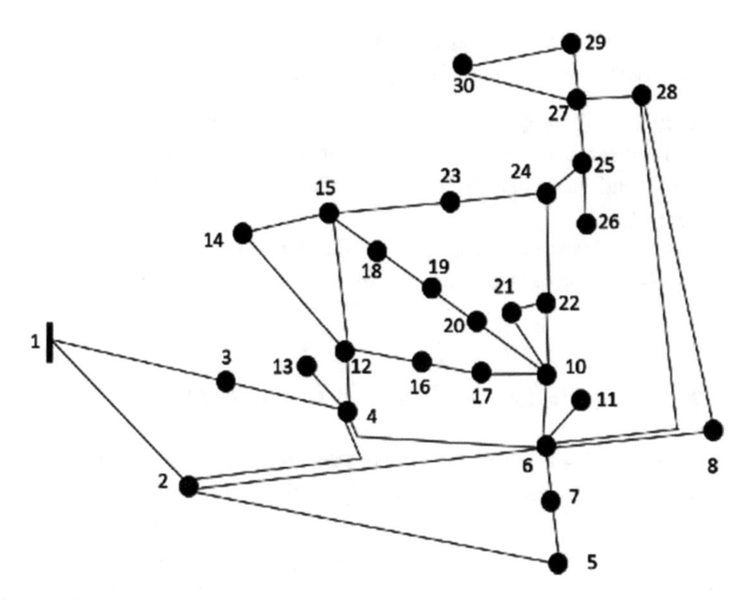

Fig. 4 The 30-node test system

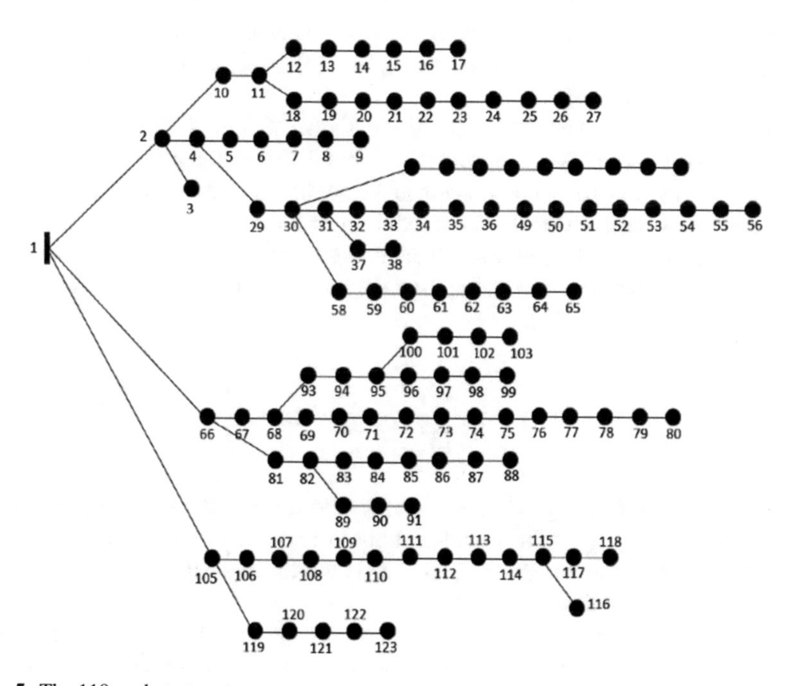

Fig. 5 The 119-node test system

In this chapter, two case studies each were conducted for both 30-node and 119-node systems to demonstrate the efficacy of the proposed distributed algorithm:

- Case Study 1—Convergence and adaptability of the proposed algorithm, under dynamic renewable outputs and loads for a control horizon $\mathcal{H} = 10$, are investigated and benchmarked with a reference consensus-based distributed algorithm [32].
- Case Study 2—Optimality of the proposed algorithm is investigated with respect to a centralized-controller-based algorithm for a control horizon $\mathcal{H} = 10$.

6.1 Simulation Studies on 30-Node Test System

6.1.1 Case Study 1—Convergence, Adaptability, and Performance Benchmark for the 30-Node Test System

In this case study, convergence and adaptability of the proposed algorithm are investigated for changing conditions, i.e., variability of renewables and loads, for the given control horizon $\mathcal{H} = 10$. Further, comparisons are drawn among (1) the EDO-based proposed approach, (2) the restart-based proposed approach (in which after every control cycle, each agent reinitializes its elements' values and velocities for optimization of the next control step), and (3) the reference consensus-based distributed algorithm in [32].

The results of the simulation study are shown in Figs. 6, 7, and 8. Figure 6 displays power loss progression for the aforementioned three algorithms for the control horizon $\mathcal{H} = 10$, and Fig. 7 provides a zoomed-in view of the second control cycle for visual clarity. It can be observed from both the figures that the EDO-based algorithm is able to find better solutions and converges faster than the other two algorithms. This property can be appropriately quantified in terms of mean best-of-generation value F_{BOG} [34].

$$F_{BOG} = \frac{1}{G}\sum_{i=1}^{G}\left(\frac{1}{R}\sum_{j=1}^{R} F_{BOG,ij}\right), \tag{28}$$

where G is the number of generations (i.e., iterations), R is the number of runs of the algorithm, and $F_{BOG,ij}$ is the global-best power loss value obtained at ith iteration and jth run. Thus, the lower the F_{BOG}, the better the algorithm is in finding good solutions in a shorter time.

The mean power losses, over 10 control cycles and for 20 runs of the algorithm, expressed as a percentage of the total load are presented in Table 1. Referring to the table, it is seen that the EDO-based proposed algorithm has the lowest F_{BOG}, thus outperforming the other two algorithms. It is also observed that the percentage power losses for the 119-node system are lower than those for the 30-node system. This can be attributed to the fact that, even though the number of DGs and CLs

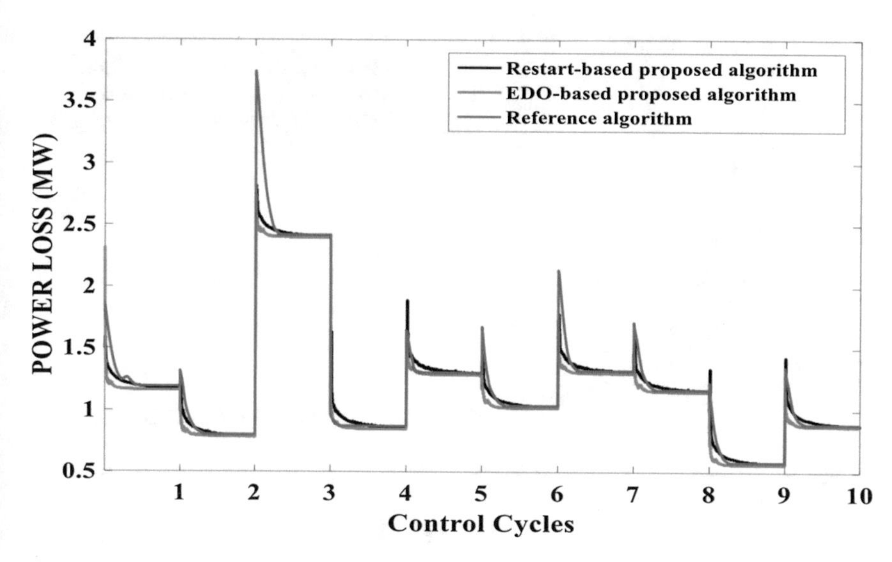

Fig. 6 Power loss plots for control horizon $\mathcal{H} = 10$ for the 30-node system

in the 119-node system is in the same ratio as in the 30-node system, the total load itself of the 119-node system is only 40% higher. Therefore, the available DG capacity per unit load is approximately 6 MVA/MW for the 30-node system, while it is 17 MVA/MW for the 119-node system. This peculiarity of the simulated 119-node system means that its DGs have a higher available capacity and can be further utilized to minimize power losses as compared to the 30-node system. Further, Fig. 8 shows that the EDO-based algorithm is able to effectively maintain node voltages inside the prescribed limits of $\pm 3\%$ in a fully distributed manner.

6.1.2 Case Study 2—Performance Benchmark with Centralized Controller for 30-Node Test System

In this case study, the optimality of the solution reached by the EDO-based algorithm is investigated with respect to the centralized-controller-based algorithm. The results are presented in Fig. 9, which shows average converged values for 20 runs of both the algorithms. It can be seen that the EDO-based algorithm is able to find a solution almost similar to the centralized algorithm. In fact, the average error is found to be 0.94% for the given control horizon $\mathcal{H} = 10$, which indicates convergence to near optimality by the EDO-based proposed algorithm.

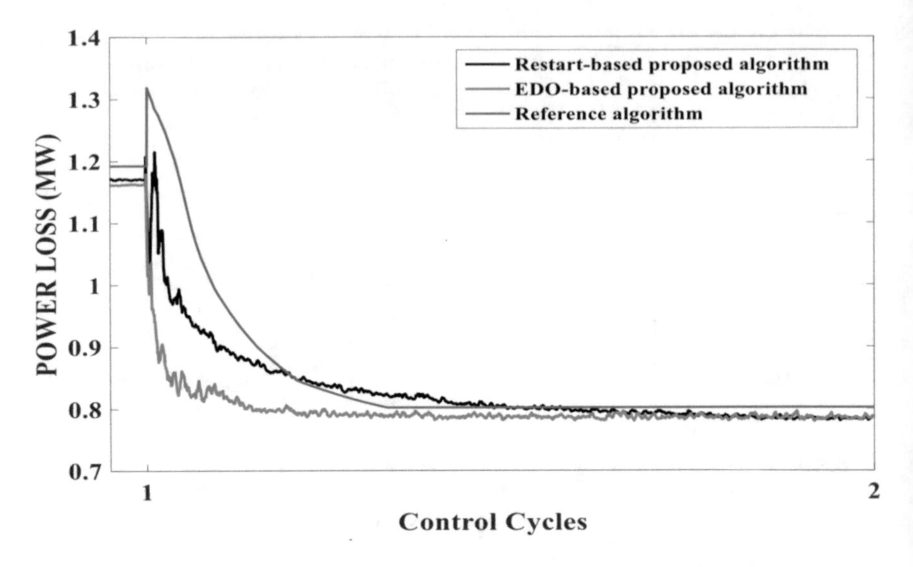

Fig. 7 Enlarged view of the second control-cycle plots for the 30-node system

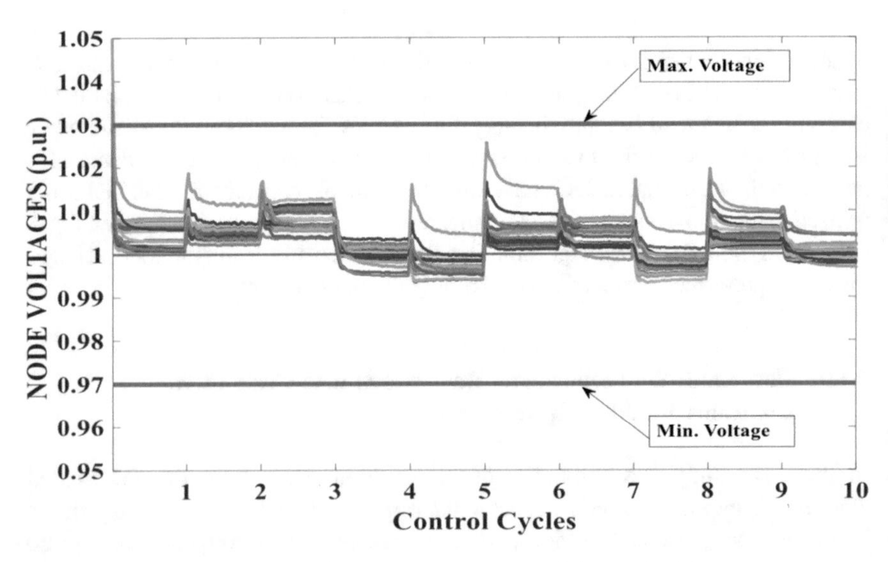

Fig. 8 Node voltage plots for the 30-node system for the EDO-based algorithm

6.2 Simulation Studies on the 119-Node Test System

In this section, the scalability aspect of the proposed algorithm is investigated by conducting simulation studies on a large 119-node distribution test system (Fig. 5). The DG penetration is set to 57.6% in the following studies, with 48 DGs and 32 CLs in the system. Thus, the number of control variables involved is

Table 1 Mean best-of-generation value(F_{BOG}) or power loss, for $R = 20$

System	EDO-based algorithm	Restart-based algorithm	Reference algorithm
30-node	11.62%	12.28%	12.31%
119-node	6.96%	7.23%	8.11%

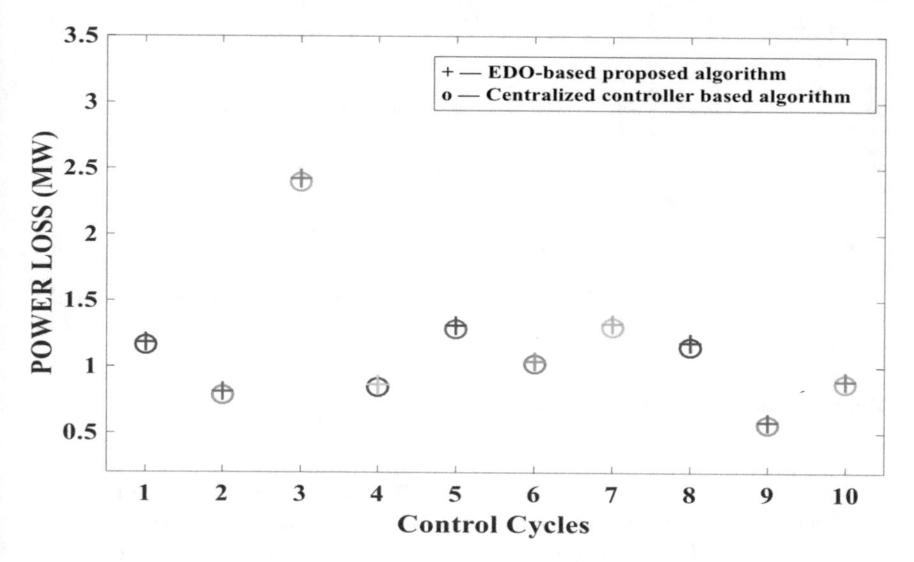

Fig. 9 Converged power loss values for Case Study 2 for the 30-node system

80, which represents a sufficiently large-scale system. It is noted that consensus on a large-scale system may take a bit longer to be achieved, especially if the network graph is sparsely connected. This may result in delays in the convergence of the algorithm. Therefore, as discussed earlier, the entire power system can be divided into two sets A_i and B_i for each node i. For example, Fig. 10 shows the sensitivity values (magnitude of $\bar{Y}$) of nodes 27 and 73. Therefore, the effect of power variation at node 27, for example, is assumed to be limited to the nodes in set-A_{27} only. This implies that even a very large system can be broken down into smaller subsystems for estimating the effect due to a node i, and, therefore, enables distributed optimization of such large systems effectively. However, it must be noted that too much restriction on this set of nodes may lead to a solution that is much farther from the optimal solution, despite having converged faster.

6.2.1 Case Study 1—Convergence, Adaptability, and Performance Benchmark for 119-Node Test System

In this case study, convergence and adaptability of the proposed algorithm are investigated for changing conditions, i.e., variability of renewables and loads, for the given control horizon $\mathcal{H} = 10$. As for the 30-node system comparisons are drawn

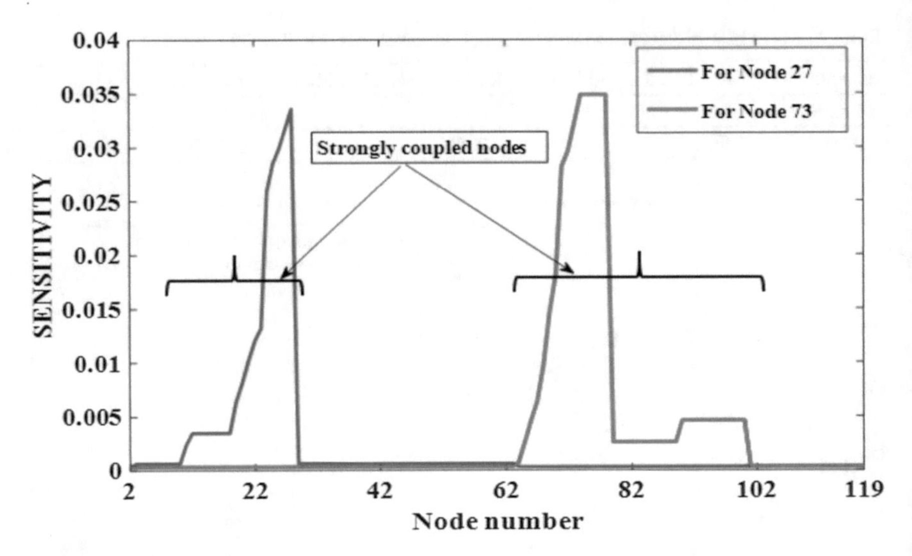

Fig. 10 Sensitivity values of nodes 27 and 73, for the 119-node system

here among (1) the EDO-based proposed approach, (2) the restart-based proposed approach, and (3) the reference consensus-based distributed algorithm in [32].

The results of the simulation study are shown in Figs. 11, 12, and 13. Figure 11 displays power loss progression for the aforementioned three algorithms for the given control horizon $\mathcal{H} = 10$, and Fig. 12 provides a zoomed-in view of the second control cycle for visual clarity.

It can be observed from both the figures that the EDO-based algorithm is able to find better solutions and converges faster than the other two algorithms. Also, referring to Table 1, it is seen that the EDO-based proposed algorithm has the lowest F_{BOG} value, thus outperforming the other two algorithms.

Further, Fig. 13 shows that the EDO-based algorithm is able to effectively maintain node voltages inside the prescribed limits of $\pm 3\%$ in a fully distributed manner, even for a large 119-node system. It is noted from Fig. 11 that the reference algorithm is not performing well, especially for control cycles 5–10. This is due to some of its assumptions, such as not considering phase angles of the nodes, which result in the accumulation of a significant error for the larger 119-node system.

6.2.2 Case Study 2—Performance Benchmark with Centralized Controller for 119-Node Test System

In this case study, the optimality of the solution reached by the EDO-based algorithm is compared with the centralized-controller-based algorithm. The results are presented in Fig. 14, which shows average converged values for 20 runs of both the algorithms. It can be seen that the EDO-based algorithm is able to find a solution

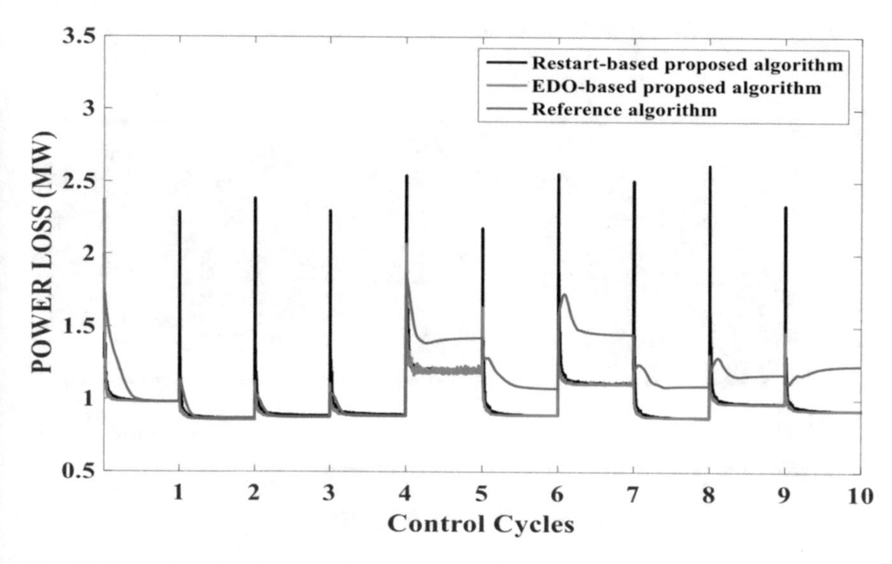

Fig. 11 Power loss plots for control horizon $\mathcal{H} = 10$ for the 119-node system

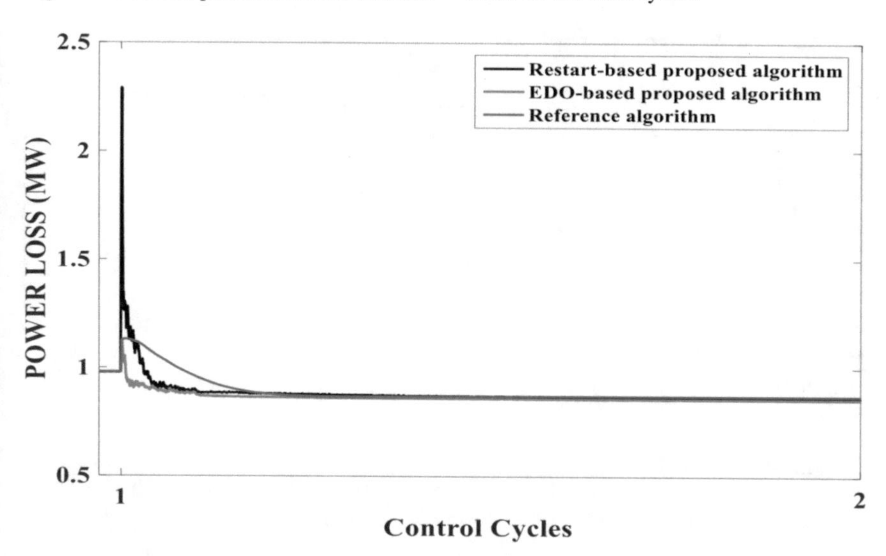

Fig. 12 Enlarged view of the second control-cycle plots for the 119-node system

almost similar to the centralized algorithm, and the average error is found to be 2.31% for the given control horizon $\mathcal{H} = 10$. It is noted that this increase in error percentage, as compared to the 30-node system, can be attributed to the tradeoff of using the ε-based decomposition method.

From the presented results it can be seen that, compared to the benchmark algorithm, the proposed algorithm is well suited for implementation since it

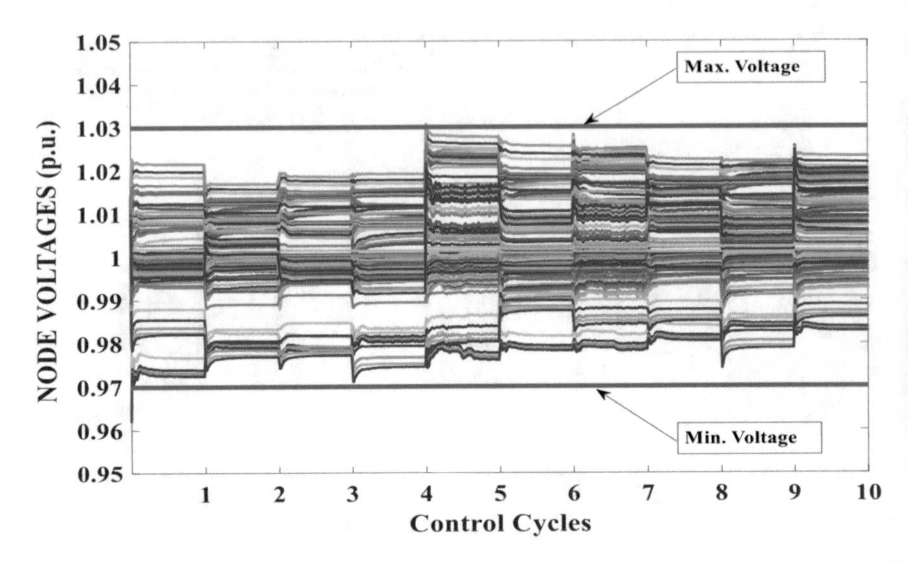

Fig. 13 Node voltage plots for the 119-node system for the EDO-based algorithm

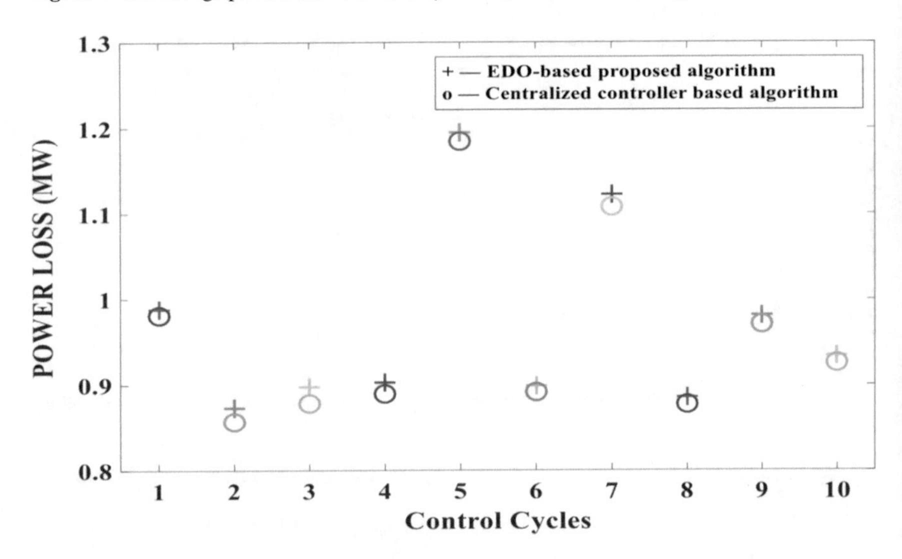

Fig. 14 Converged power loss values for Case Study 2 for the 119-node system

models the non-convex power flow equations using a differential state estimation model, which is demonstrated to provide significantly accurate distributed state estimates, comparable to those provided by well-known solvers such as Newton–Raphson. Further, compared to a centralized-controller-based solution, the proposed algorithm provides a near-global-optimal solution that proves the effectiveness of the approach.

7 Conclusions

This chapter proposes a consensus-based dimension-distributed PSO-based algorithm for optimal control of smart distribution grids. The algorithm is able to handle a large number of controllable DER entities, such as DGs or CLs, which will be a necessary requirement for future smart power distribution networks. The algorithm is also designed with consideration of the privacy-preserving requirements of individual controllable entities. Due to its approach of population-based solution search combined with the robust searching properties of PSO, the proposed algorithm is able to reach (near-)optimal solutions faster as compared to conventional distributed algorithms. Further, the convergence properties of the algorithm are analytically derived and empirically analyzed, thereby providing actual performance guarantees for the approach. To assess the performance of the proposed algorithm, its convergence properties were also analyzed with respect to a state-of-the-art benchmark algorithm. The results show a superior performance of the proposed algorithm vis-à-vis convergence, optimality, and adaptability to dynamic system conditions, for both 30-node and 119-node distribution test systems.

There are several directions in which the work proposed in this chapter can be extended. One possible future research avenue is behavioral modeling, and the ways in which it can be integrated into the existing solutions. For example, in the work presented in this chapter, and in most of the literature, it is assumed that the DER entities are rational beings and that they perform only in the most logical way so as to minimize their costs of operation. However, this is not the case in real-world scenarios. Hence, behavioral modeling is an interesting area of further research. Another possible research direction can be related to neural network integration. In the work proposed in this chapter, coordination of DERs is performed using a technique based on distributed computational intelligence, which is modeled on the swarm behavior of living animals, e.g., birds. This technique results in a significant improvement over conventional distributed techniques. Nevertheless, the proposed strategy still has to iterate toward an optimal solution. This iterative behavior of distributed (and centralized) algorithms could be avoided if Artificial Neural Networks are trained to reach a (near-)optimal solution almost instantly even in a distributed situation, similar to how human brains react to uncertain and incomplete information by building and correcting models of their own environment.

References

1. A. Abessi, V. Vahidinasab, M.S. Ghazizadeh, Centralized support distributed voltage control by using end-users as reactive power support. IEEE Trans. Smart Grid **7**, 178 (2016)
2. M.N. Ab Wahab, S. Nefti-Meziani, A. Atyabi, A comprehensive review of swarm optimization algorithms. PLoS ONE **10** (2015)
3. C. Ahn, H. Peng, Decentralized voltage control to minimize distribution power loss of microgrids. IEEE Trans. Smart Grid **4**, 1297 (2013)

4. K.E. Antoniadou-Plytaria, I.N. Kouveliotis-Lysikatos, P.S. Georgilakis, N.D. Hatziargyriou, Distributed and decentralized voltage control of smart distribution networks: models, methods, and future research. IEEE Trans. Smart Grid **8**, 2999 (2017)
5. O. Ardakanian, S. Keshav, C. Rosenberg, Real-time distributed control for smart electric vehicle chargers: from a static to a dynamic study. IEEE Trans. Smart Grid **5**, 2295 (2014)
6. M. Assadian, M.M. Farsangi, H. Nezamabadi-pour, GCPSO in cooperation with graph theory to distribution network reconfiguration for energy saving. Energy Convers. Manag. **51**, 418 (2010)
7. K. Baker, J. Guo, G. Hug, X. Li, Distributed MPC for efficient coordination of storage and renewable energy sources across control areas. IEEE Trans. Smart Grid **7**, 992 (2016)
8. A. Bidram, A. Davoudi, Hierarchical structure of microgrids control system. IEEE Trans. Smart Grid **3**, 1963 (2012)
9. D.W. Boeringer, D.H. Werner, Particle swarm optimization versus genetic algorithms for phased array synthesis. IEEE Trans. Antennas Propag. **52**, 771 (2004)
10. S. Bolognani, R. Carli, G. Cavraro, S. Zampieri, Distributed reactive power feedback control for voltage regulation and loss minimization. IEEE Trans. Autom. Control **60**, 966 (2015)
11. S. Boyd, N. Parikh, E. Chu, B. Peleato, J. Eckstein, Distributed optimization and statistical learning via the alternating direction method of multipliers. Found. Trends® Mach. Learn. **3**, 1 (2011)
12. A.J. Conejo, E. Castillo, R. Minguez, R. Garcia-Bertrand, *Decomposition Techniques in Mathematical Programming: Engineering and Science Applications* (Springer, Berlin, 2006)
13. S. Das, P.N. Suganthan, Differential evolution: a survey of the state-of-the-art. IEEE Trans. Evol. Comput. **15**, 4 (2011)
14. Y. del Valle, G.K. Venayagamoorthy, S. Mohagheghi, J. Hernandez, R.G. Harley, Particle swarm optimization: basic concepts, variants and applications in power systems. IEEE Trans. Evol. Comput. **12**, 171 (2008)
15. H. Everett III, Generalized lagrange multiplier method for solving problems of optimum allocation of resources. Oper. Res. **11**, 399 (1963)
16. R. Firestone, C. Marnay, Energy manager design for microgrids. Technical report, Lawrence Berkeley National Lab. (LBNL), Berkeley, CA (2005)
17. Y.-J. Gong, W.-N. Chen, Z.-H. Zhan, J. Zhang, Y. Li, Q. Zhang, J.-J. Li, Distributed evolutionary algorithms and their models: a survey of the state-of-the-art. Appl. Soft Comput. **34**, 286 (2015)
18. J. Guo, G. Hug, O.K. Tonguz, Intelligent partitioning in distributed optimization of electric power systems. IEEE Trans. Smart Grid **7**, 1249 (2016)
19. R.G. Harley, J. Liang, Computational intelligence in smart grids, in *IEEE Symposium Series on Computational Intelligence (SSCI)*, vol. 1 (2011), pp. 1–8
20. G. Hug, S. Kar, C. Wu, Consensus + innovations approach for distributed multiagent coordination in a microgrid. IEEE Trans. Smart Grid **6**, 1893 (2015)
21. A. Hussain, V. Bui, H. Kim, A resilient and privacy-preserving energy management strategy for networked microgrids. IEEE Trans. Smart Grid **9**, 2127 (2018)
22. M.K. Jalloul, M.A. Al-Alaoui, A distributed particle swarm optimization algorithm for block motion estimation using the strategies of diffusion adaptation, in *2015 International Symposium on Signals, Circuits and Systems (ISSCS)* (IEEE, Piscataway, 2015), pp. 1–4
23. V. Kachitvichyanukul, Comparison of three evolutionary algorithms: Ga, pso, and de. Ind. Eng. Manag. Syst. **11**, 215 (2012)
24. S. Kar, G. Hug, J. Mohammadi, J.M.F. Moura, Distributed state estimation and energy management in smart grids: a consensus+ innovations approach. IEEE J. Sel. Top. Signal Process. **8**, 1022 (2014)
25. H. Karimi, H. Nikkhajoei, R. Iravani, Control of an electronically-coupled distributed resource unit subsequent to an islanding event. IEEE Trans. Power Delivery **23**, 493 (2008)
26. F. Katiraei, R. Iravani, N. Hatziargyriou, A. Dimeas, Microgrids management. IEEE Power Energy Mag. **6**, 54 (2008)

27. V. Kekatos, G. Wang, A.J. Conejo, G.B. Giannakis, Stochastic reactive power management in microgrids with renewables. IEEE Trans. Power Syst. **30**, 3386 (2015)
28. I. Khan, Y. Xu, H. Sun, V. Bhattacharjee, Distributed optimal reactive power control of power systems. IEEE Access **6**, 7100 (2018)
29. J. Khazaei, Z. Miao, Consensus control for energy storage systems. IEEE Trans. Smart Grid **9**, 3009 (2018)
30. Y. Liu, Z. Qu, H. Xin, D. Gan, Distributed real-time optimal power flow control in smart grid. IEEE Trans. Power Syst. **32**, 3403 (2017)
31. J. Ma, G. Geng, Q. Jiang, Two-time-scale coordinated energy management for medium-voltage dc systems. IEEE Trans. Power Syst. **31**, 3971 (2016)
32. A. Maknouninejad, Z. Qu, Realizing unified microgrid voltage profile and loss minimization: a cooperative distributed optimization and control approach. IEEE Trans. Smart Grid **5**, 1621 (2014)
33. D.K. Molzahn, F. Dörfler, H. Sandberg, S.H. Low, S. Chakrabarti, R. Baldick, J. Lavaei, A survey of distributed optimization and control algorithms for electric power systems. IEEE Trans. Smart Grid **8**, 2941 (2017)
34. T.T. Nguyen, S. Yang, J. Branke, Evolutionary dynamic optimization: a survey of the state of the art. Swarm Evol. Comput. **6**, 1 (2012)
35. T. Niknam, M. Zare, J. Aghaei, Scenario-based multiobjective volt/var control in distribution networks including renewable energy sources. IEEE Trans. Power Delivery **27**, 2004 (2012)
36. R. Olfati-Saber, J.A. Fax, R.M. Murray, Consensus and cooperation in networked multi-agent systems. Proc. IEEE **95**, 215 (2007)
37. D.E. Olivares, A. Mehrizi-Sani, A.H. Etemadi, C.A. Cañizares, R. Iravani, M. Kazerani, A. H. Hajimiragha, O. Gomis-Bellmunt, M. Saeedifard, R. Palma-Behnke, G. A. Jiménez-Estévez, N. D. Hatziargyriou, Trends in microgrid control. IEEE Trans. Smart Grid **5**, 1905 (2014)
38. M.A. Potter, K.A.D. Jong, Cooperative coevolution: An architecture for evolving coadapted subcomponents. Evol. Comput. **8**, 1 (2000)
39. V.M. Preciado. G.C. Verghese, Synchronization in generalized erdö s-rènyi networks of nonlinear oscillators, in *Proceedings of the 44th IEEE Conference on Decision and Control* (2005), pp. 4628–4633
40. M. Saeednia, M. Menendez, A consensus-based algorithm for truck platooning. IEEE Trans. Intell. Transp. Syst. **18**, 404 (2017)
41. F. Seredynski, A.Y. Zomaya, P. Bouvry, Function optimization with coevolutionary algorithms, in *Intelligent Information Processing and Web Mining* (Springer, Berlin, 2003), pp. 13–22
42. A. Sharma, D. Srinivasan, A. Trivedi, A decentralized multiagent system approach for service restoration using DG islanding. IEEE Trans. Smart Grid **6**, 2784 (2015)
43. M. Srinivas, L.M. Patnaik, Genetic algorithms: a survey. Computer **27**, 17 (1994)
44. R. Subbu, A.C. Sanderson, Modeling and convergence analysis of distributed coevolutionary algorithms. IEEE Trans. Syst. Man Cybern. Part B (Cybern.) **34**, 806 (2004)
45. Q. Sun, R. Han, H. Zhang, J. Zhou, J.M. Guerrero, A multiagent-based consensus algorithm for distributed coordinated control of distributed generators in the energy internet. IEEE Trans. Smart Grid **6**, 3006 (2015)
46. K.C. Tan, Y.J. Yang, C.K. Goh, A distributed cooperative coevolutionary algorithm for multiobjective optimization. IEEE Trans. Evol. Comput. **10**, 527 (2006)
47. W. Tushar, B. Chai, C. Yuen, D.B. Smith, K.L. Wood, Z. Yang, H.V. Poor, Three-party energy management with distributed energy resources in smart grid. IEEE Trans. Ind. Electron. **62**, 2487 (2015)
48. A. Vaccaro, G. Velotto, A.F. Zobaa, A decentralized and cooperative architecture for optimal voltage regulation in smart grids. IEEE Trans. Ind. Electron. **58**, 4593 (2011)
49. R.D. Villarroel, D.F. García, M.A. Dávila, E.F. Caicedo, Particle swarm optimization vs genetic algorithm, application and comparison to determine the moisture diffusion coefficients of pressboard transformer insulation. IEEE Trans. Dielectr. Electr. Insul. **22**, 3574 (2015)
50. J.G. Vlachogiannis, K.Y. Lee, Quantum-inspired evolutionary algorithm for real and reactive power dispatch. IEEE Trans. Power Syst. **23**, 1627 (2008)

51. Y. Wakasa, S. Nakaya, Distributed particle swarm optimization using an average consensus algorithm, in *2015 54th IEEE Conference on Decision and Control (CDC)* (IEEE, Piscataway, 2015), pp. 2661–2666
52. H. Wang, J. Huang, Incentivizing energy trading for interconnected microgrids. IEEE Trans. Smart Grid **9**, 2647 (2018)
53. Z. Wang, J. Wang, Review on implementation and assessment of conservation voltage reduction. IEEE Trans. Power Syst. **29**, 1306 (2014)
54. G. Wenzel, M. Negrete-Pincetic, D.E. Olivares, J. MacDonald, D.S. Callaway, Real-time charging strategies for an electric vehicle aggregator to provide ancillary services. IEEE Trans. Smart Grid **9**, 5141 (2018)
55. P.J. Werbos, Computational intelligence for the smart grid-history, challenges, and opportunities. IEEE Comput. Intell. Mag. **6**, 14 (2011)
56. M. Wooldridge, *Intelligent Agents* (MIT Press, Cambridge, 1999), pp. 27–77
57. X. Xi, E.H. Abed, Formation control with virtual leaders and reduced communications, in *Proceedings of the 44th IEEE Conference on Decision and Control* (2005), pp. 1854–1860
58. L. Xiao, S. Boyd, S.-J. Kim, Distributed average consensus with least-mean-square deviation. J. Parallel Distrib. Comput. **67**, 33 (2007)
59. Y. Xu, W. Liu, Novel multiagent based load restoration algorithm for microgrids. IEEE Trans. Smart Grid **2**, 152 (2011)
60. Y. Xu, Z. Yang, W. Gu, M. Li, Z. Deng, Robust real-time distributed optimal control based energy management in a smart grid. IEEE Trans. Smart Grid **8**, 1568 (2017)
61. S. Yang, H. Cheng, F. Wang, Genetic algorithms with immigrants and memory schemes for dynamic shortest path routing problems in mobile ad hoc networks. IEEE Trans. Syst. Man Cybern. Part C (Appl. Rev.) **40**, 52 (2010)
62. J. Zhang, H.S.H. Chung, W.L. Lo, Pseudocoevolutionary genetic algorithms for power electronic circuits optimization. IEEE Trans. Syst. Man Cybern. Part C (Appl. Rev.) **36**, 590 (2006)
63. D. Zhang, Z. Fu, L. Zhang, An improved TS algorithm for loss-minimum reconfiguration in large-scale distribution systems. Electr. Pow. Syst. Res. **77**, 685 (2007)
64. W. Zhang, W. Liu, X. Wang, L. Liu, F. Ferrese, Distributed multiple agent system based online optimal reactive power control for smart grids. IEEE Trans. Smart Grid **5**, 2421 (2014)
65. C. Zhao, J. He, P. Cheng, J. Chen, Consensus-based energy management in smart grid with transmission losses and directed communication. IEEE Trans. Smart Grid **8**, 2049 (2017)
66. R.D. Zimmerman, C.E. Murillo-Sánchez, R.J. Thomas, MATPOWER: steady-state operations, planning, and analysis tools for power systems research and education. IEEE Trans. Power Syst. **26**, 12 (2011)

Kumar Utkarsh is currently pursuing a Doctor of Philosophy degree in the Department of Electrical & Computer Engineering at the National University of Singapore, Singapore. He has worked extensively in designing computational intelligence-based and conventional control strategies for power systems with high levels of integration of renewables and distributed energy resources. His research interests are in micro-grid and smart-grid energy management systems, distributed energy resource integration studies, frequency/voltage stability, generation/demand side management, and power system resilience quantification and enhancement.

Dipti Srinivasan is a Professor in the Department of Electrical & Computer Engineering at the National University of Singapore, where she also heads the Centre for Green Energy Management & Smart Grid (GEMS). Her recent research projects are in the broad areas of optimization and control, wind and solar power prediction, electricity price prediction, deep learning, and development of multi-agent systems for system operation and control. Her current research focuses on the development of novel computational intelligence-based models and methodologies to aid the integration of the new Smart Grid technologies into the existing infrastructure so that power grid can effectively utilize pervasive renewable energy generation and demand-side management programs, while accommodating stochastic load demand. She is the author of four books and over 400 journal and conference papers in the field of computational intelligence, renewable energy and smart grids which have been published in top tier journals.

Dipti is a fellow of IEEE. She was awarded the IEEE PES Outstanding Engineer award in 2010 and IEEE Singapore Outstanding Volunteer Award in 2020. Dipti is an Associate Editor of IEEE Transactions on Smart Grid, IEEE Transactions on Sustainable Energy, IEEE Transactions on Evolutionary Computation, IEEE Transaction on Neural Networks and Learning Systems, and IEEE Transactions on Artificial Intelligence. At the ECE department of National University of Singapore, she teaches courses in the areas of Sustainable Energy systems, Smart Grid, and computational intelligence methods. She is the recipient of NUS Annual Teaching Excellence Award and Engineering Educator Award.

Part III
Modeling

Fuzzy Multilayer Perceptrons for Fuzzy Vector Regression

Sansanee Auephanwiriyakul, Suwannee Phitakwinai, and Nipon Theera-Umpon

1 Introduction

One of the problems in data science is regression. A feature vector used in the problem is usually a vector of real numbers and is analyzed by a numeric algorithm. However, sometimes uncertainties occur in the process of collecting a data set, e.g., imprecision of data collector or imprecision of measurement, etc. Hence, a feature can be described with a linguistic variable. A feature vector can also be described as a noninteractive linguistic vector called a linguistic vector or a fuzzy vector [11].

S. Auephanwiriyakul (✉)
Department of Computer Engineering, Faculty of Engineering, Chiang Mai University, Chiang Mai, Thailand

Biomedical Engineering Institute, Chiang Mai University, Chiang Mai, Thailand
e-mail: sansanee@eng.cmu.ac.th

S. Phitakwinai
Department of Computer Engineering, Faculty of Engineering, Chiang Mai University, Chiang Mai, Thailand

Department of Computer Engineering, Faculty of Engineering, Rajamangala University of Technology Lanna, Chiang Mai, Thailand
e-mail: suwannee_phi@rmutl.ac.th

N. Theera-Umpon
Biomedical Engineering Institute, Chiang Mai University, Chiang Mai, Thailand

Department of Electrical Engineering, Faculty of Engineering, Chiang Mai University, Chiang Mai, Thailand
e-mail: nipon.t@cmu.ac.th

A. E. Smith (ed.), *Women in Computational Intelligence*, Women in Engineering and Science, https://doi.org/10.1007/978-3-030-79092-9_12

There are several research works involving linear or nonlinear regression with fuzzy inputs and fuzzy outputs [1, 4, 5, 14, 23, 33, 34, 38]. The optimization methods used in these algorithms are the extension of the least mean square method, the extension of gradient descent, or a genetic algorithm. Some other research works involve numeric inputs and fuzzy outputs, for example, a fuzzy support vector machine with particle swam optimization [25] and a fuzzy inference system (ANFIS) in fuzzy nonparametric regression function [8].

The fuzzy neural network with fuzzy inputs was introduced and explained in [12]; however, they did not calculate fuzzy derivative. The fuzzy neural networks work with real number weights and fuzzy number weights [6, 15–18, 28, 31] with either a direct extension of the weight update equations or a fuzzy weight update equation from fuzzy derivative equation. Although this methodology seems to be correct, there are some uncertainties in the computation of the equation. The derivative equation with fuzzy variables might not be an exact answer as in the derivative equation with numeric variables. Hence, there are some other works choosing other optimization algorithms, e.g., the genetic algorithm (GA) [2, 21, 22, 29] or particle swarm optimization [30, 37]. There are also recurrent fuzzy neural networks with the evolutionary algorithm (EA) [3]. We developed the fuzzy multilayer perceptrons with cuckoo search (CS-FMLP) for fuzzy inputs and fuzzy outputs in [32]. However, all the mentioned algorithms are used for classification problems, not for regression problems. Hence, in this chapter, we will introduce this CS-FMLP for the regression problem. We implement this algorithm on three real data sets, the yacht hydrodynamics data set [35], the energy efficiency data set [35], and the upper Ping river data set [36].

2 Fuzzy Multilayer Perceptron with Cuckoo Search

Although the multilayer perceptron [13] is extended for fuzzy inputs and fuzzy outputs in [12] as shown in Fig. 1, the weights and biases are still numeric numbers. A noninteractive fuzzy vector [24] used in this work is a vector of fuzzy numbers [19, 20, 26]. The fuzzy operation done in this work is based on Zadeh's extension principle [40]. To prevent irregular and inaccurate results from the calculation, interval arithmetic and decomposition theorem [20, 27] are used. Also, to prevent multiple occurrences of the parameters problem, we also used Dong's method [9, 10].

The following is the brief detail of the fuzzy multilayer perceptrons with cuckoo search (CS-FMLP) [32]. Let $\mathbf{X} = \left\{ \vec{\boldsymbol{X}}_j \middle| 1 \le j \le N \right\}$ be a set of noninteractive fuzzy vectors in p-dimensional space, $\vec{\boldsymbol{X}}_j = \left(\boldsymbol{X}_{j1}, \ldots, \boldsymbol{X}_{jp}\right)^t \in [\Im(R)]^p$. Suppose a weight of node k is defined as $\vec{\mathbf{W}}_k = \left(w_{k1}, \ldots, w_{kp+1}\right)^t$ where w_{kp+1} is a bias. The fuzzy output of node k is

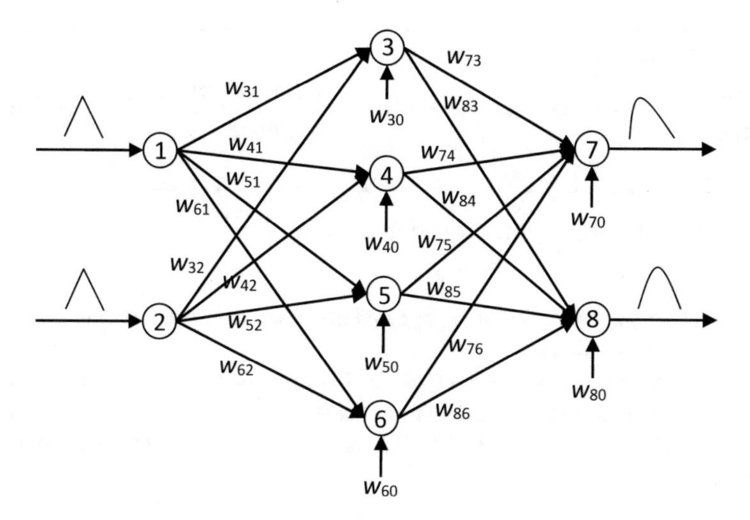

Fig. 1 Fuzzy multilayer perceptrons

$$\boldsymbol{O}_k = f\left(\boldsymbol{Net}_k\right) = \frac{1}{1+\exp\left(-\boldsymbol{Net}_k\right)} = \underset{\alpha\in[0,1]}{\cup}\left(\left[\frac{1}{1+\exp\left(-\boldsymbol{Net}_k\right)}\right]_\alpha\right) \tag{1}$$

where

$$\boldsymbol{Net}_k = \sum_j \overrightarrow{\mathbf{W}}_{kj}^t \overrightarrow{\boldsymbol{X}}_j = \underset{\alpha\in[0,1]}{\cup}\left(\left[\sum_j \overrightarrow{\mathbf{W}}_{kj}^t \overrightarrow{\boldsymbol{X}}_j\right]_\alpha\right) = \underset{\alpha\in[0,1]}{\cup}\left(\left[\boldsymbol{Net}_k\right]_\alpha\right). \tag{2}$$

The inner product is defined as

$$\overrightarrow{\mathbf{W}}_{kj}^t \overrightarrow{\boldsymbol{X}}_j = \underset{\alpha\in[0,1]}{\cup}\left(w_{k1}\left[\boldsymbol{X}_{j1}\right]_\alpha + \cdots + w_{kp}\left[\boldsymbol{X}_{jp}\right]_\alpha + w_{p+1}[\mathbf{1}]_\alpha\right). \tag{3}$$

The lower and upper bounds of each α–cut interval calculation are used to compute the lower and upper bounds of the result of Eq. (1) because of the increasing function nature of the sigmoid function

$$f\left(\boldsymbol{Net}_k\right) = \frac{1}{1+\exp\left(-\mathbf{Net}_k\right)}. \tag{4}$$

Suppose, $[\boldsymbol{Net}_k]_\alpha = \left[\underline{net}_{k\alpha}, \overline{net_{k\alpha}}\right]$ $\forall\alpha \in [0, 1]$, hence

$$[\boldsymbol{O}_k]_\alpha = \left[f\left(\underline{net}_{k\alpha}\right), f\left(\overline{net}_{k\alpha}\right)\right] \ \ \forall\alpha \in [0, 1]. \tag{5}$$

w_{31}, w_{32}, w_{41}, w_{42}, w_{51}, w_{52}, w_{61}, w_{62},	$w_{30}, w_{40}, w_{50}, w_{60}$	$w_{73}, w_{74}, w_{75}, w_{76}$, $w_{83}, w_{84}, w_{85}, w_{86}$,	w_{70}, w_{80}

Fig. 2 Cuckoo solution arrangement of the network in Fig. 1

This feed forward calculation is repeated until it reaches each output node where the fuzzy output is calculated.

In our work, the cuckoo search (CS) [39] is used as an optimization tool to find the optimal weights and biases of the network. This algorithm is based on Lévy flights behavior of some birds and fruit flies. There are three idealized rules of CS as follows:

- Each cuckoo lays one egg at a time and dumps its egg in a randomly chosen nest.
- The best nests with high quality of eggs will carry over to the next generations.
- The number of available host nests is fixed, and the egg laid by a cuckoo is discovered by the host with a probability $p_a \in [0, 1]$.

In this case, a vector of weights and biases of the networks arranged from left to right and top to bottom of the network as shown in Fig. 2 is set as a solution **x**. For example, for the 2–4–2 network, the dimension of **x** is 22 in total (8 (from 2–4) + 4(biases) +8(from 4–2) +2(biases)).

The new solution $\mathbf{x}^{(t+1)}$ for cuckoo i is

$$\mathbf{x}_i^{t+1} = \mathbf{x}_i^t + \alpha \oplus \text{Levy}\,(\beta)\,, \tag{6}$$

where α is the step size scaling factor and normally $\alpha > 0$. The step length s of a random walk of Lévy flights is

$$s = \eta \times 0.01 \frac{u}{|v|^{1/\beta}} \left(\mathbf{x}_j^t - \mathbf{x}_i^t\right), \tag{7}$$

where $\mathbf{x}^t{}_j$ is the best solution and $\eta \sim N(0, 1)$, $u \sim N(0, {\sigma_u}^2)$, and $v \sim N(0, {\sigma_v}^2)$, are random numbers. We also set ${\sigma_v}^2 = 1$ and $\sigma_\mu^2 = \left\{\frac{\Gamma(1+\beta)\sin(\pi\beta/2)}{\Gamma[(1+\beta)/2]\beta 2^{(\beta-1)/2}}\right\}^{1/\beta}$ where $\beta = 1.5$ and Γ is the standard Gamma function.

We set the errors in every epoch to be the fitness value where a small value indicates high fitness or vice versa. The mean absolute error (MAE) is used to evaluate the errors in the output nodes, i.e.,

$$\boldsymbol{MAE} = \frac{1}{N}\sum_{i=1}^{N}\left|\mathrm{y}_i - \hat{\mathbf{y}}_i\right| = \bigcup_{\alpha\in[0,1]} \frac{1}{N}\sum_{i=1}^{N}\left|\mathrm{y}_{i_\alpha} - \hat{\mathbf{y}}_{i_\alpha}\right|, \tag{8}$$

where N is the total number of samples, $\mathbf{y}_i$ is the single fuzzy desired output, and $\hat{\mathbf{y}}_i$ is the program output of sample i.

MAE is a fuzzy number; therefore, Choobineh and Li's algorithm [7] is utilized to rank the ***MAE***. The ranking of fuzzy number $\boldsymbol{A}$ is

$$R(\mathbf{A}) = a + \left[\frac{1}{2}\left(h_A - \frac{D\left(\mu_A, \mu_{U_A}\right) - D\left(\mu_A, \mu_{L_A}\right)}{d - a}\right) \times (d - a)\right], \tag{9}$$

where h_A is the height of $\boldsymbol{A}$, a and d are left and right boundaries,

$$D\left(\mu_A, \mu_{U_A}\right) = \sum_{i=0}^{h_A} (\mu_{i+1} - \mu_i)(d - g_A(\alpha)), \tag{10}$$

and

$$D\left(\mu_A, \mu_{L_A}\right) = \sum_{i=0}^{h_A} (\mu_{i+1} - \mu_i)(f_A(\alpha) - a). \tag{11}$$

The CS-FMLP algorithm is shown below:

```
begin
      Initialize n host nests
      Generate random weight vectors  between Lb and Ub for
      each nest
      Get the best nest ni and its fitness Fi for a current
      best nest
     while (t < MaxGeneration) or (Fi < stop criterion) do
            for all cuckoo nests do
                for all training data do
                  Feed forward the neural network
                end for
              Calculate the mean absolute errors (Fj)
            end for
            Get a cuckoo randomly by Lévy flights and evaluate
            its quality/fitness Fi
            Choose a nest among n
            if (Fj <Fi)
                Replace Fi by Fj and keep nj for new best nest
            end if
            A fraction (pa) of worse nests are abandoned and
            new ones are built;
            Keep the best solutions (or nests with quality
            solutions);
            Rank the solutions and find the current best
         end while
         return the best nest (weight vector)
     end
```

We would like to minimize the calculation time in the testing process and to make it simpler in the result comparison between our fuzzy model and numeric models. Hence, in all of the experiments, we defuzzify a fuzzy number ***MAE*** of each test sample using its centroid value instead of its ranking number as in the training process. This is because, in the testing process, we do not compare the ***MAE*** fuzzy numbers.

3 Experiment Results

We implement the CS-FMLP on three real data sets, the yacht hydrodynamics data set [35], the energy efficiency data set [35], and the upper Ping river data set [36]. In all three data sets, there might be uncertainties occurring in feature measurements. Therefore training a fuzzy model might be more suitable than a numeric model on these data sets.

We implement k-fold cross validation on the training data set. The remaining part of the data set is kept as a blind test set (used for testing process only). Since we use k-fold cross validation scheme, in each fold, there are training set (used to train the model) and validation set (used to test the trained model from the particular fold). One might also be curious of how comparable this algorithm with the regular multilayer perceptrons is; hence, we also show the result of the multilayer perceptrons with cuckoo search (CS-MLP). The results shown in each experiment are the one providing the best validation set result from both CS-FMLP and CS-MLP in that particular structure of multilayer perceptrons. The details of each experiment are as follows.

3.1 Yacht Hydrodynamics Data Set

The yacht hydrodynamics data set [35] was created by Ship Hydromechanics Laboratory, Maritime and Transport Technology Department, Technical University of Delft. This data set consists of 308 samples with 6 parameters of the Delft yacht hull ship-series. The descriptions of six attributes are as follows:

- Longitudinal position of the center of buoyancy
- Prismatic coefficient
- Length-displacement ratio
- Beam-draught ratio
- Length-beam ratio
- Froude number

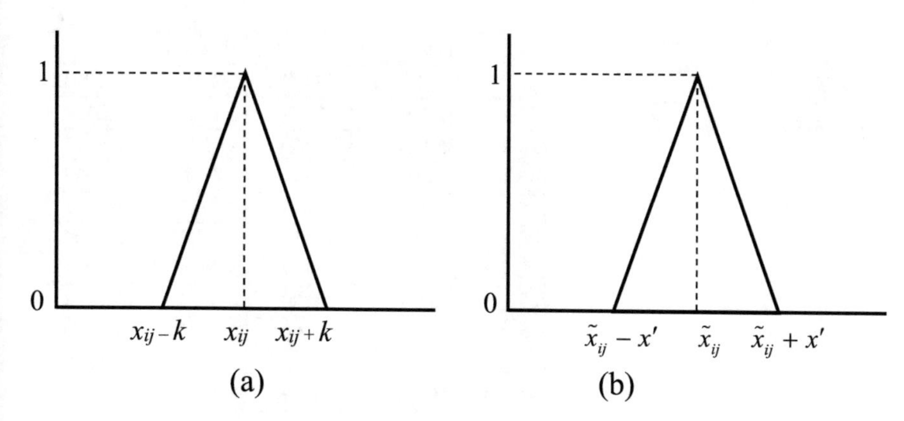

Fig. 3 (**a**) Fuzzy number X_{ij} and (**b**) normalized $\tilde{X}_{ij}$

The target variable is the measured residuary resistance per unit weight of displacement. Sample numbers 1 to 280 are set to be a training data set. Sample numbers 281 to 308 are set to be a blind test data set. We implement fourfold cross validation of the training data set to prevent overtraining.

We fuzzify each dimension of each sample to be a triangular with $k = 0.02$ as shown in Fig. 3a. Then a fuzzy number at the *j*th dimension of the *i*th sample $\left(\tilde{X}_{ij}\right)$ is normalized as shown in Fig. 3b using

$$\tilde{x}_{ij} = \frac{x_{ij} - \overline{x}_j}{\sigma_j} \tag{12}$$

and

$$x'_j = \frac{k}{\sigma_j} \tag{13}$$

where $\overline{x}_j$ and σ_j are an average value and standard deviation of feature in the *j*th dimension and $k = 0.01$, respectively.

In this experiment, we set $p_a = 0.2$, $Ne = 20$, $\alpha = 0.01$, MaxGeneration $= 1.5 \times 10^4$, and $\varepsilon = 0.001$. $[L_A, U_A]$ is $[-20,20]$. We randomly initialize weights to be a single fuzzy vector for the CS-FMLP and CS-MLP. We also vary the structure of the MLP. Table 1 shows the results from the best validation set for each model. We can see that the best validation set result from CS-FMLP is $MAE = 3.002$ with a model of 6-2-2-1, whereas CS-MLP gives *MAE* of 4.389. We run this model on the blind test data set, and the result is in Table 2. The *MAE* from the CS-FMLP is 3.43, whereas that from the CS-MLP is 3.975.

Table 1 *MAE* of training and validation sets from CS-FMLP and CS-MLP for the yacht hydrodynamics data set

Model	Training set		Validation set	
	CS-FMLP	CS-MLP	CS-FMLP	CS-MLP
6-2-1	3.6103	4.3973	3.1810	4.0454
6-3-1	3.9620	5.8730	4.4523	6.4173
6-4-1	7.5005	5.1451	6.4056	6.4663
6-5-1	6.4415	6.9688	4.1717	9.0607
6-6-1	6.9100	8.2614	5.4773	5.5606
6-2-2-1	**2.7646**	**3.3854**	**3.0022**	**4.3893**
6-3-2-1	6.0742	5.9139	4.3027	5.7258
6-4-2-1	3.0418	3.5499	3.4523	3.8100
6-5-2-1	3.2242	2.7515	3.3538	3.5193
6-6-2-1	5.6214	3.6353	4.1584	4.3848

Table 2 *MAE* of the blind test set from 6-2-2-1 CS-FMLP and CS-MLP for yacht hydrodynamics data set

CS-FMLP		CS-MLP
MAE	MAE_ε	*MAE*
3.4303	1.8043	3.9749

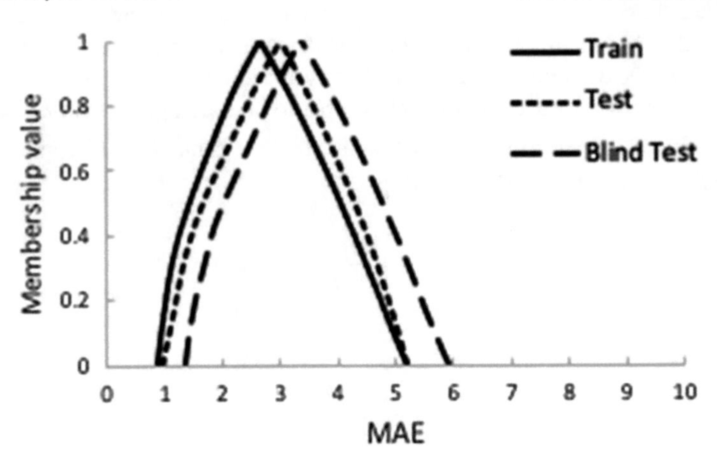

Fig. 4 ***MAE*** of training, validation (test), and blind test sets for yacht hydrodynamics data set

Figure 4 shows the ***MAE*** of the training, validation, and blind data sets. We see that the ***MAE*** of training and validation sets is closer to the ***zero*** fuzzy number than that of the blind test set. We confirm that by showing the predicted values against the target values in Fig. 5. Since the predicted value from CS-FMLP is a fuzzy number, we will show the core and the support of the predicted fuzzy number against the target value. We can see that the support of the predicted fuzzy number covers the target value most of the time. One can view this as an interval of the predicted value. Or one can say that there is a possibility that the predicted value should be between these numbers. Whereas the predicted value from CS-MLP (numeric in the figure) is only one value, if there is an incorrect prediction, there is no chance to give an alternative result.

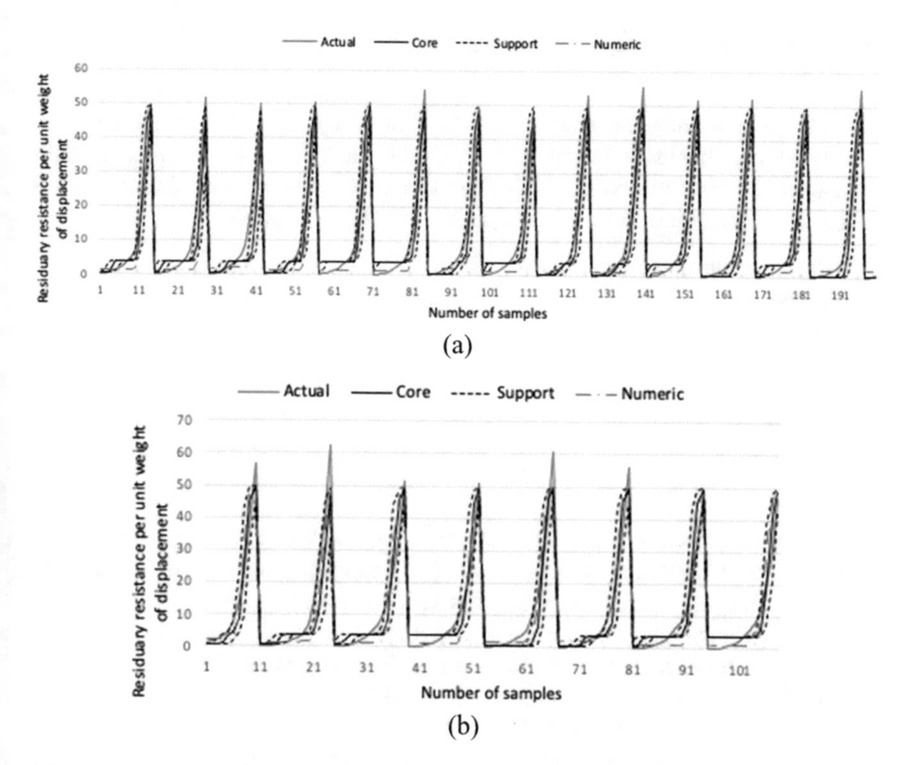

Fig. 5 Predicted value against target value (**a**) training and (**b**) blind test sets for yacht hydrodynamics data set

To prove the advantage of the result from CS-FMLP, we compute the ε-insensitive loss function as

$$\mathbf{diff}_{Y_i} = \begin{cases} 0 & \text{if} \quad \underline{Y}_i \le ActualValue_i \le \overline{Y}_i \\ |ActualValue_i - \boldsymbol{Y}_i| & \text{otherwise} \end{cases} \tag{14}$$

where support of fuzzy number $\boldsymbol{Y}_i$ is $\left(\text{supp}\,(\mathbf{Y}_i) = \left[\underline{Y}_i, \overline{Y}_i\right]\right)$. And, then

$$\boldsymbol{MAE}_{\varepsilon} = \frac{1}{N}\sum_{i=1}^{N} \mathbf{diff}_{\mathbf{Y}_i}. \tag{15}$$

Again, the centroid is used to defuzzify the $\boldsymbol{MAE}_{\varepsilon}$. The defuzzified value of $\boldsymbol{MAE}_{\varepsilon}$ of the blind test data set from the best CS-FMLP is 1.804 as shown in Table 2.

When we look at the results shown in Fig. 5b, we can see that, although the support of the result from CS-FMLP does not cover the target value as in the result from training data set, it is closer to the target value than that from CS-MLP. It is also confirmed by MAE_{ε} from the CS-FMLP shown in Table 2.

3.2 Energy Efficiency Data Set

The energy efficiency data set [35] was created by Angeliki Xifara and was processed by Athanasios Tsanas, Oxford Centre for Industrial and Applied Mathematics, University of Oxford, UK. This data set consists of 768 samples with 8 parameters. The descriptions of the eight attributes are as follows:

- Relative compactness
- Surface area
- Wall area
- Roof area
- Overall height
- Orientation
- Glazing area
- Glazing area distribution

The target attributes in this data set are heating load and cooling load. Sample numbers 1 to 600 are our training data set, and sample numbers 601–768 are our blind test data set. We run fourfold cross validation on the training data set as in the yacht hydrodynamics data set. We also normalize the data set using the same scheme as in the previous data set. However, as the ranges of input attributes are very different, we utilize k for each dimension as follows: $k_1 = 0.1$, $k_2 = 20$, $k_3 = 10$, $k_4 = 10$, $k_5 = 0.5$, $k_6 = 0$, $k_7 = 0.1$, $k_8 = 0$.

Since there are two target values, we use a model for each target result. In the experiment, we set $p_a = 0.25$, $Ne = 20$, $\alpha = 0.01$ (in the heating model), $\alpha = 0.1$ (in cooling model), MaxGeneration $= 1 \times 10^4$, and $\varepsilon = 0.001$. $[L_A, U_A]$ is $[-10,10]$ in both models. The singleton weight fuzzy vector is initialized with the same method as in the previous data set. The results of the heating model are shown in Tables 3 and 4. We can see that the model 8-3-2-1 CS-FMLP gives the best validation result with $MAE = 2.362$, whereas that from CS-MLP provides MAE of 2.65 on the validation

Table 3 *MAE* for heating load of training and validation sets from CS-FMLP and CS-MLP

	Training set		Validation set	
Model	CS-FMLP	CS-MLP	CS-FMLP	CS-MLP
8-2-1	2.5755	2.6786	2.5909	2.6488
8-3-1	2.6517	2.6131	2.6001	2.6601
8-4-1	2.8872	2.7273	2.7002	2.8400
8-5-1	2.7823	2.6790	2.6414	2.6450
8-6-1	2.7529	2.8093	2.5518	2.7665
8-2-2-1	2.8261	2.6923	2.5917	2.7488
8-3-2-1	**2.6326**	**2.4954**	**2.3620**	**2.6500**
8-4-2-1	2.6425	2.6206	2.5228	2.5705
8-5-2-1	2.4704	2.5854	2.6774	2.6924
8-6-2-1	2.5052	2.5891	2.6053	2.6197

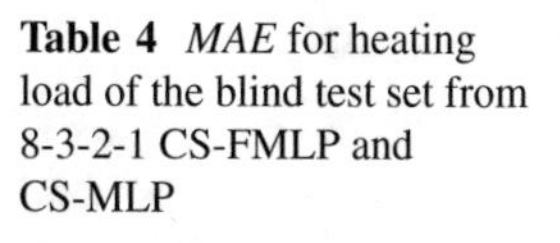

Table 4 *MAE* for heating load of the blind test set from 8-3-2-1 CS-FMLP and CS-MLP

CS-FMLP		CS-MLP
MAE	MAE_{ε}	*MAE*
2.4805	0.7980	2.8982

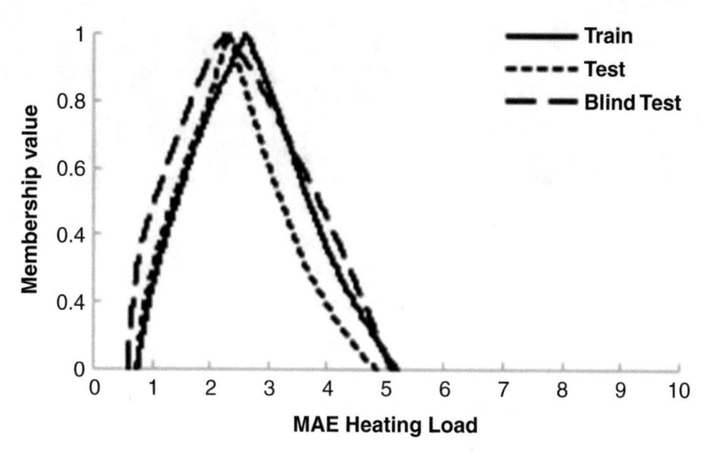

Fig. 6 ***MAE*** for heating load of training, validation (test), and blind test sets

set. This 8-3-2-1 CS-FMLP model provides *MAE* of 2.48 and MAE_{ε} of 0.798 on the blind test data set, whereas that of CS-MLP is 2.898.

Figure 6 shows the ***MAE*** fuzzy number from the training, validation (test), and blind test data sets. Again, to confirm the advantage of the predicted fuzzy number from CS-FMLP over the exact predicted number from CS-MLP, we show the predicted results against the target value from the training and blind data sets in Fig. 7.

Tables 5 and 6 show the results of the cooling model from the CS-FMLP and CS-MLP. In this case, CS-FMLP and CS-MLP provide *MAE* of 1.975 and 2.271, respectively, on the validation test data set. The best model in this case is 8-3-1. The *MAE* and MAE_{ε} for this model on the blind test data set are 1.991 and 1.012, respectively, for the CS-FMLP. For the CS-MLP, the MAE is 2.436. Figure 8 shows the ***MAE*** fuzzy numbers from 8-3-1 CS-FMLP on training, validation, and blind test data sets. Again, Fig. 9 confirms the advantage of the CS-FMLP over CS-MLP as in the aforementioned cases.

3.3 Ping River Data Set

One of the rivers flowing to the Chao Phraya river system, which is the main river of Thailand, is the Ping river. It flows southward from Chiang Mai province and passes through many provinces. The Ping river basin contains major watersheds at a nearby district. The upstream of the river at the Chiang Mai station is a high mountainous

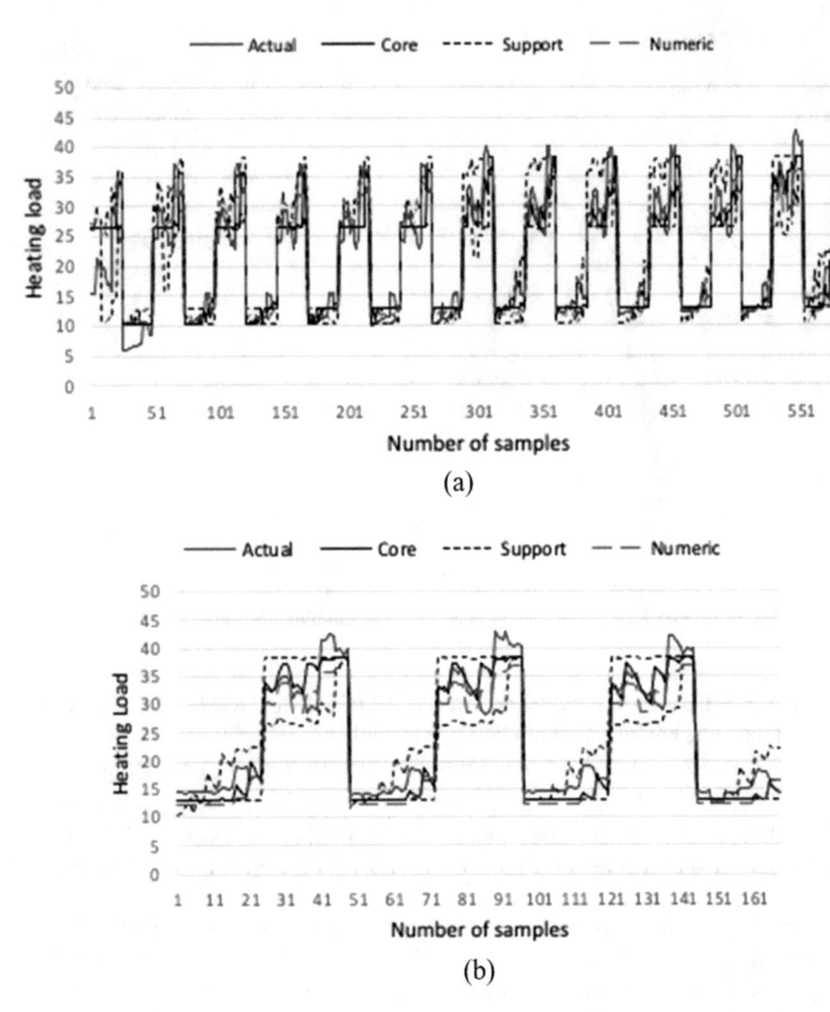

Fig. 7 Predicted value against target value for heating load (**a**) training and (**b**) blind test sets

area with steep slopes. It is the cause of the downstream flooding within a few hours after heavy rainfall. The Ping river causes central Chiang Mai to flood in the rainy season each year, especially from August to September.

The gauging station and other details for the area of interest are shown in Fig. 10. Chiang Mai city downtown will be flooded if the river level at the UP.06 station exceeds 3.70 meters [36]. The duration of the prediction period usually depends on the distance and the flow rate of the water between two stations. For example, the distance between the UP.04 and UP.06 stations is 32 km along the water channel. The time lag (about 6 to 7 h) of the water body between these two stations depends on the river level at the UP.04 station.

Table 5 *MAE* for cooling load of training and validation sets from CS-FMLP and CS-MLP

Model	Training set		Validation set	
	CS-FMLP	CS-MLP	CS-FMLP	CS-MLP
8-2-1	2.4039	2.5814	2.2316	2.4214
8-3-1	**2.1770**	**2.4074**	**1.9752**	**2.2714**
8-4-1	2.0884	2.1847	2.0870	2.1200
8-5-1	2.2984	2.2902	2.2473	2.2955
8-6-1	2.1580	2.3629	1.9896	2.1583
8-2-2-1	2.5799	2.5514	2.4221	2.5183
8-3-2-1	2.1836	2.1573	2.0002	2.2435
8-4-2-1	2.6154	2.3510	2.2438	2.2191
8-5-2-1	2.1806	2.2068	2.2273	2.1448
8-6-2-1	2.4905	2.3548	2.0533	2.4727

Table 6 *MAE* for cooling load of the blind test set from 8-3-2-1 CS-FMLP and CS-MLP

CS-FMLP		CS-MLP
MAE	MAE_ε	*MAE*
1.9906	1.0121	2.4362

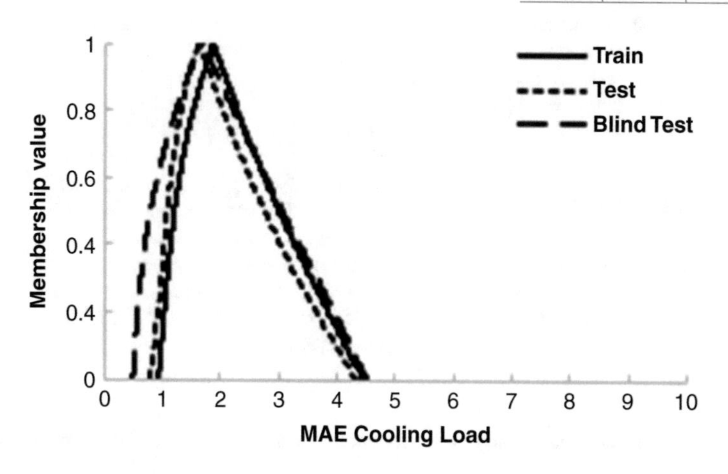

Fig. 8 *MAE* for cooling load of training, validation (test), and blind test data sets

The hourly measured river levels used in this research were collected at two gauging stations, UP.04 and UP.06. The UP.04 station uses a bubble gauge to measure the river level, whereas that of the UP.06 station uses a floating gauge. The data set was collected from nine different crucial flooding events, five from the year 2005, two from the year 2006, and two from the year 2010 as shown in Table 7.

We train our system with the data set numbers 1 to 7, and data set numbers 8 and 9 are used as the blind test data set. The data are divided into seven groups consisting of the training set and blind test set as shown in Table 8. We implement fourfold cross validation on each training data group.

We fuzzify the data set using the same scheme as in the previous data sets. However, according to the information from the Upper Northern Region Irrigation

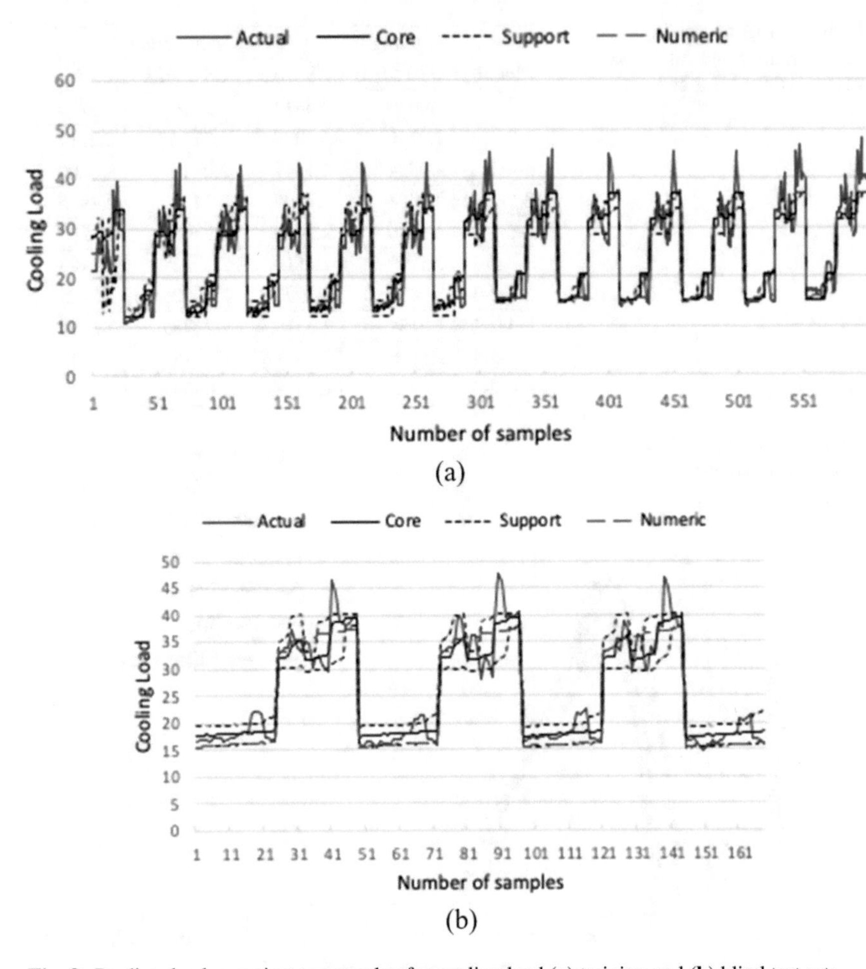

Fig. 9 Predicted value against target value for cooling load (**a**) training and (**b**) blind test sets

Hydrology Center, Chiang Mai Thailand, the imprecision of the data is ± 2 centimeters. Hence k in eq. 13 for this data set is 2. We perform 7-h-ahead flood prediction at the UP.06 station with a 2-h model as shown in Table 9. We utilized the data at time lags of 0 and -1 h at the UP.04 and UP.06 stations. Please note that time lag of 0 is the current time. The target is the data at time lag $+7$ at the UP.06 station.

For this data set, we calculate the mean absolute percentage error (MAPE) because water level data set tends to be in a large range (it can be from 100 centimeters to 800 centimeters). To show the goodness of the prediction across the data set, the MAPE is more suitable than the MAE.

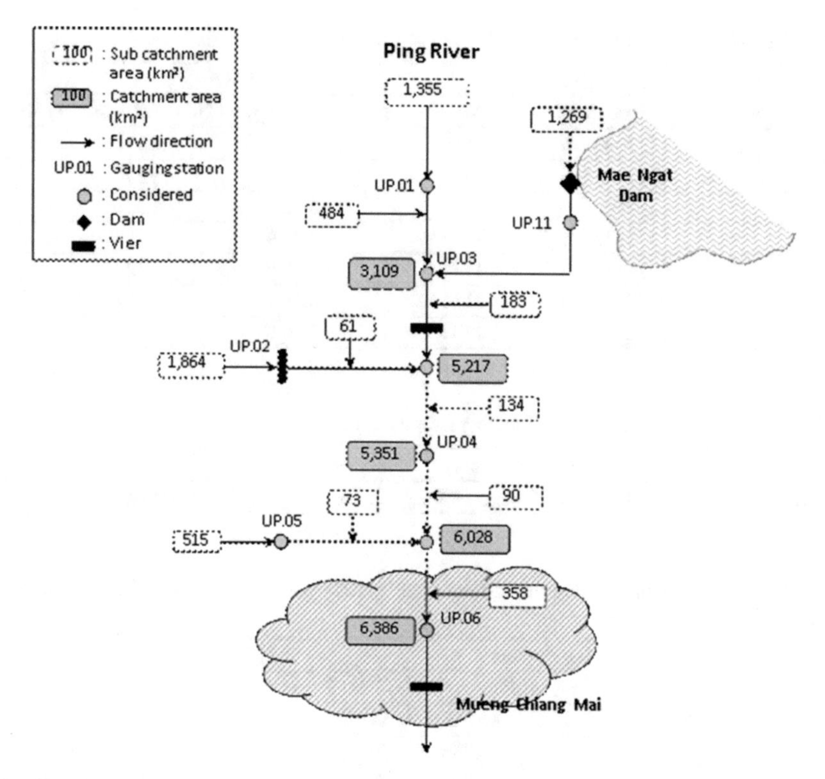

Fig. 10 Diagram of the upper Ping basin river [36]

Table 7 Description of flooding events for Ping river data set [36]

No.	Date	Maximum river level at UP.06 station (cm)	Cause of flood
1	12–18 Aug. 2005	490	Tropical depression
2	9–15 Sep. 2005	381	Low pressure cell
3	20–22 Sep. 2005	471	Tropical depression
4	29 Sep.–3 Oct.2005	493	Tropical depression
5	1–3 Nov. 2005	379	Intertropical convergence
6	25 July–9 Aug. 2006	434	Southwest monsoon
7	25 Aug. –9 Sep. 2006	383	Trough
8	16–20 Sep. 2010	357	River level reached the full supply level
9	27 Sep.–1 Oct. 2010	348	River level reached the full supply level

Table 8 Description of training and blind test sets for Ping river data set

	Training set		Blind test set	
Data group No.	Flood event No.	No. of samples	Flood event No.	No. of samples
1	2, 3, 4, 5, 6, 7	1350	1, 8, 9	351
2	1, 3, 4, 5, 6, 7	1350	2, 8, 9	351
3	1, 2, 4, 5, 6, 7	1350	3, 8, 9	351
4	1, 2, 3, 5, 6, 7	1374	4, 8, 9	327
5	1, 2, 3, 4, 6, 7	1374	5, 8, 9	327
6	1, 2, 3, 4, 5, 7	1134	6, 8, 9	567
7	1, 2, 3, 4, 5, 6	1134	7, 8, 9	567

Table 9 List of input features for the Ping river data set

Model	Feature(time lags:hour)	Target
2 Hrs	$UP04_{(0)}$, $UP04_{(-1)}$, $UP06_{(0)}$, $UP06_{(-1)}$	$UP06_{(+1)}$

Table 10 *MAPE* of the training and validation set from CS-FMLP and CS-MLP for the Ping river data set

		Training set			Validation set		
		CS-FMLP		CS-MLP	CS-FMLP		CS-MLP
Data group	Model	*MAPE*	$MAPE_\varepsilon$	*MAPE*	*MAPE*	$MAPE_\varepsilon$	*MAPE*
1	4-5-1	4.5095	3.4753	4.1956	3.8052	2.7544	3.3844
2	4-4-1	4.5323	3.5797	3.9386	3.9127	2.8912	3.3483
3	4-5-4-1	4.3983	3.4938	4.0336	4.1220	3.1668	3.8913
4	4-5-1	4.0510	3.0466	3.3673	4.0825	3.0963	3.4763
5	**4-4-1**	**4.3179**	**3.3579**	**3.9334**	**3.7992**	**2.7169**	**3.3986**
6	4-5-4-1	4.5701	3.5917	4.2414	4.3966	3.2709	3.8655
7	4-5-3-1	4.6225	3.8059	4.4560	4.1080	3.2321	4.2199

The ***MAPE*** is calculated as

$$\begin{aligned}\mathbf{MAPE} &= \frac{1}{N}\sum_{i=1}^{N}\left|\frac{\mathbf{ActualValue}_i - \mathbf{ForecastValue}_i}{\mathbf{ActualValue}_i}\right| \\ &= \bigcup_{\alpha\in[0,1]}\frac{1}{N}\sum_{i=1}^{N}\left|\frac{\mathbf{ActualValue}_{i_\alpha} - \mathbf{ForecastValue}_{i_\alpha}}{\mathbf{ActualValue}_{i_\alpha}}\right|\end{aligned} \tag{16}$$

In this experiment, we set $[L_b,U_b] = [-1.5,1.5]$, $p_a = 0.2$, $Ne = 20$, $\alpha = 0.01$, MaxGeneration $= 1.5 \times 10^4$, and $\varepsilon = 0.001$. The initial weight is selected using the same scheme as in the previous data sets.

Table 10 shows the best results of CS-FMLP and CS-MLP on the validation set. In each data group, we implement the best model on the blind test data set. The

Table 11 *MAPE* of the blind test set from CS-FMLP and CS-MLP for Ping river data set

		CS-FMLP		CS-MLP
Data group No.	Flood event No.	*MAPE*	$MAPE_\varepsilon$	*MAPE*
1	1	5.5876	5.0770	5.1401
	8	4.5289	4.1225	4.4322
	9	4.3204	3.3180	4.3365
2	2	4.1818	3.3686	3.5130
	8	4.9656	4.6237	4.6283
	9	4.5248	3.6326	3.9552
3	3	4.6281	3.9392	4.1418
	8	5.0402	4.5980	4.3024
	9	4.2043	3.3435	3.5350
4	4	5.3860	5.2194	5.1465
	8	4.8146	4.4097	3.6147
	9	4.5855	3.8830	3.5005
5	**5**	**5.5071**	**4.9615**	**5.3851**
	8	**5.0994**	**4.6970**	**4.8500**
	9	**4.3567**	**3.4540**	**3.9236**
6	6	4.6361	3.2650	4.1715
	8	5.8298	5.5389	7.5722
	9	4.3203	3.2088	3.7429
7	7	4.7170	3.9715	5.1047
	8	7.6020	7.6020	7.7974
	9	5.4573	5.1283	6.4503

results of the blind test data sets are shown in Table 11. The best model is from the fifth data group. The model provides $MAPE = 3.799$ and $MAPE_\varepsilon = 2.717$, whereas *MAPE* of CS-MLP is 3.399. From Table 11, we can see that this model yields $MAPE = 5.099$ and $MAPE_\varepsilon = 4.698$ on the eighth flood event and $MAPE = 4.357$ and $MAPE_\varepsilon = 3.454$ on the ninth flood event. CS-MLP provides *MAPE* of 4.850 and 3.933 on the eighth and ninth flood event, respectively. Please note that these two flood events were from different years from the training data set. The river shape changes every year. The model is still able to predict the water level close to the target water level.

The ***MAPE*** of the blind test data set from each data group model is shown in Fig. 11. However, the ***MAPE*** of the eighth and ninth flood events are from the model of the fifth data group. Figure 12 shows the 7-h predicted fuzzy number and the 7-h target water level. Again, the results of the eighth and ninth flood event are from the model of the fifth data group. This experiment also confirms that the CS-FMLP has an advantage over the CS-MLP in terms of the possibility value of the predicted value. Our method deals with the uncertainty of the data set as shown in this experiment.

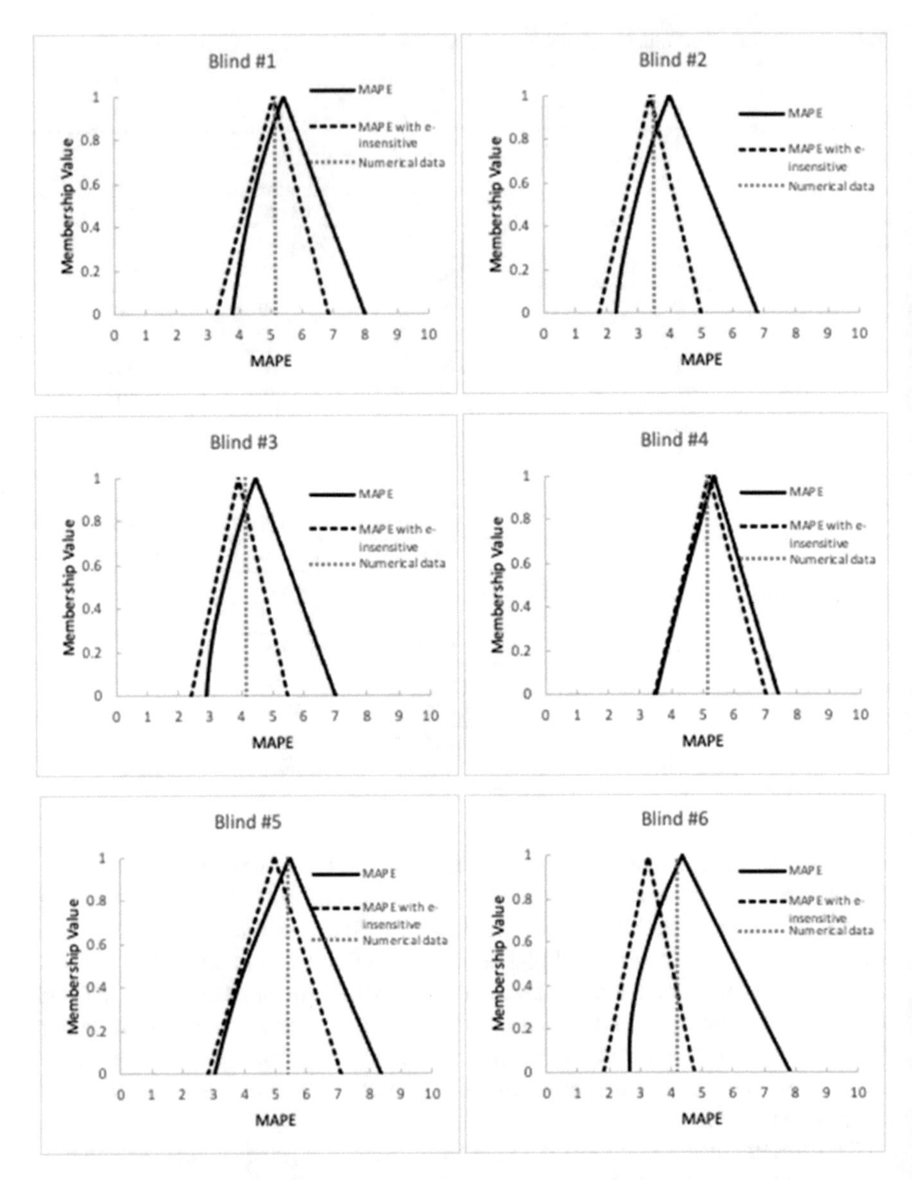

Fig. 11 ***MAPE*** of blind test set from each data group model for Ping river data set

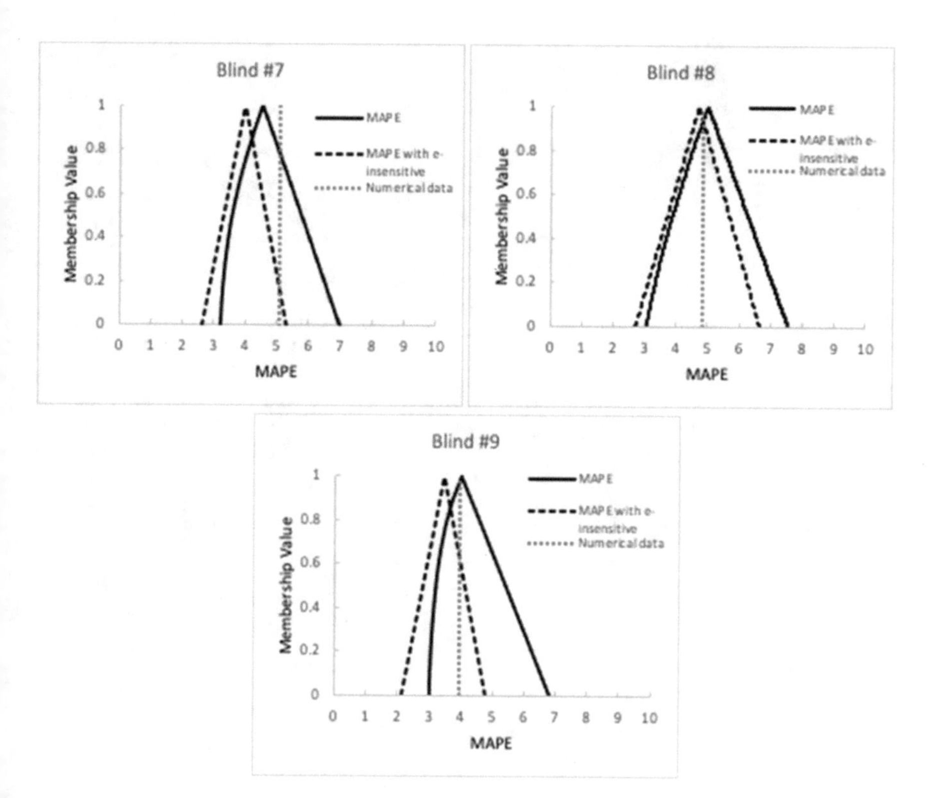

Fig. 11 (continued)

4 Conclusion

In most data sets, there are uncertainties occurring due to the collection method or vagueness in the data sets themselves. Hence, the fuzzy multilayer perceptrons with cuckoo search (CS-FMLP) for regression problems is introduced in this chapter. The CS-FMLP model can deal with fuzzy inputs and fuzzy outputs is provided from the model. The cuckoo search is utilized as an optimization tool for optimal weights and biases finding in the fuzzy multilayer perceptrons. Three real data sets, the yacht hydrodynamics data set, the energy efficiency data set, and the upper Ping river data set, are utilized in the experiment. The multilayer perceptrons with cuckoo search (CS-MLP) is also implemented on the same data sets. The CS-FMLP shows that it can provide better results than the CS-MLP in some cases. In other cases, they are comparable.

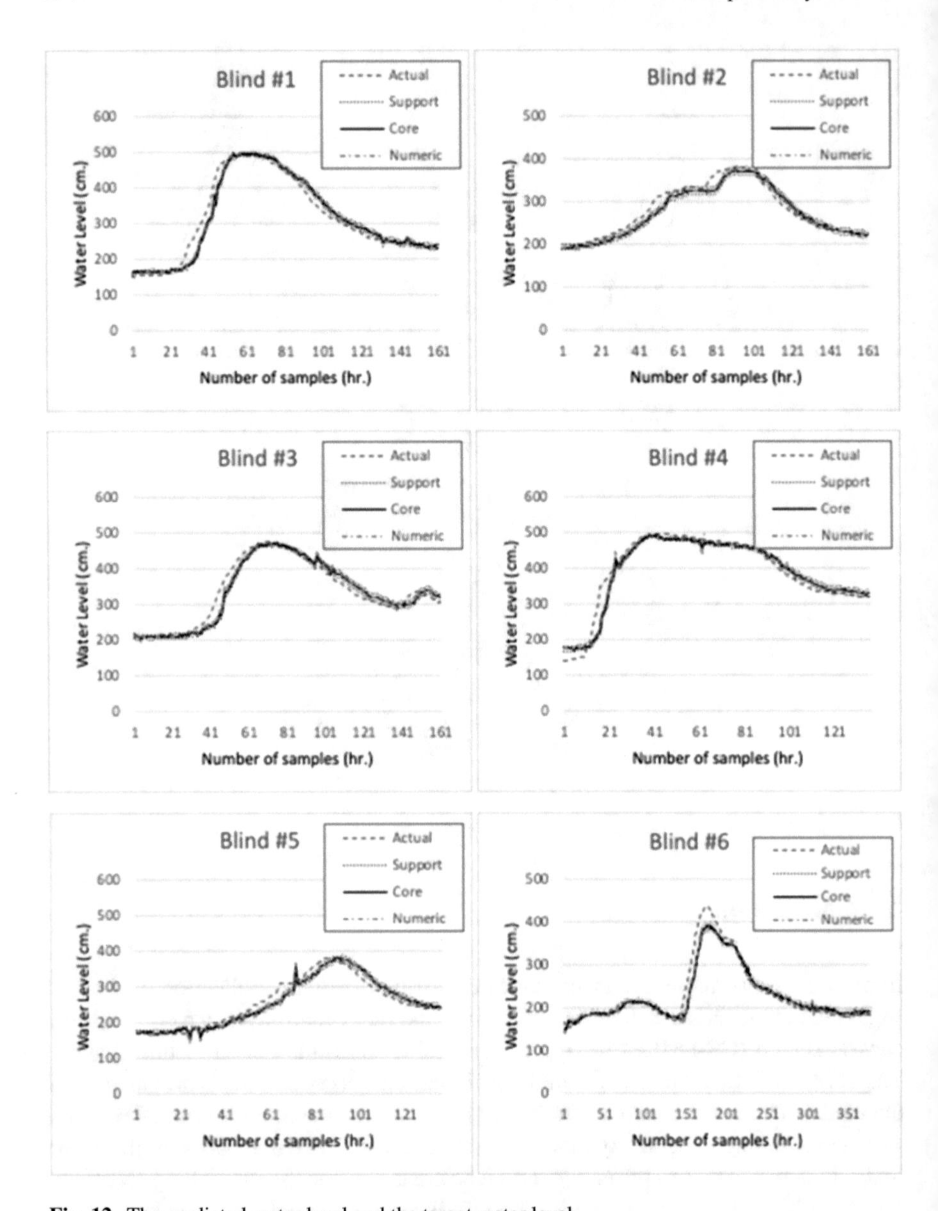

Fig. 12 The predicted water level and the target water level

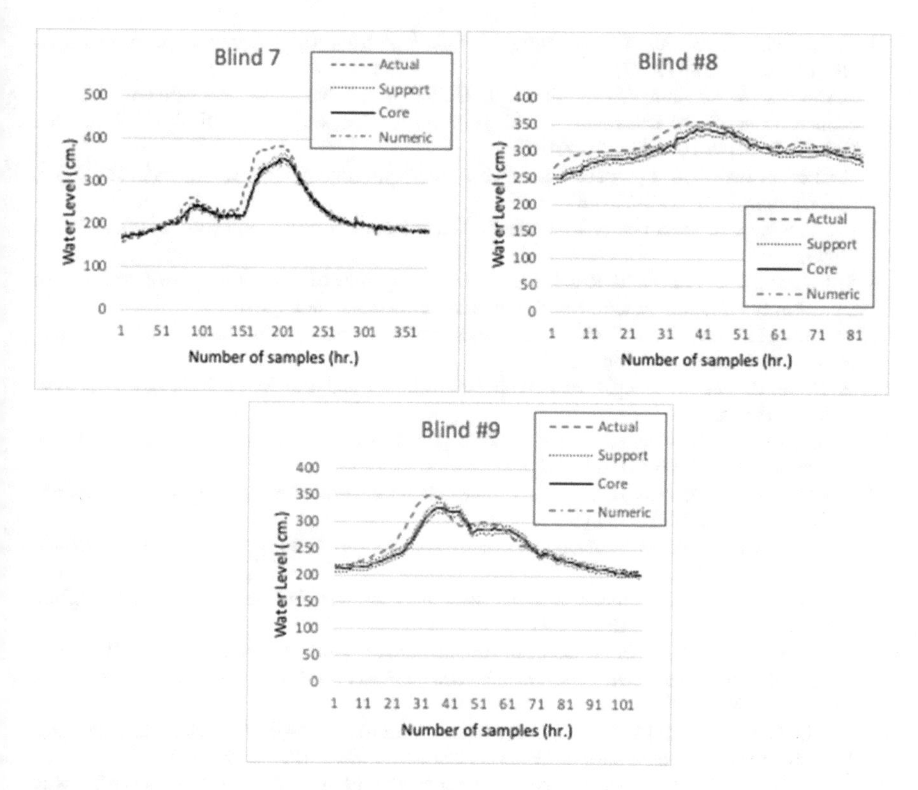

Fig. 12 (continued)

The fuzzy outputs from CS-FMLP can provide the range of the prediction value. If the target value falls inside that range, we can consider it a good prediction point. This is an advantage of the CS-FMLP over the CS-MLP. Although the model in this chapter can deal with fuzzy inputs and fuzzy outputs, the weights are still numeric. In the future, we might need to introduce weight update equation for fuzzy inputs. One might think that if weights are fuzzy numbers, the model might be better coping with data set uncertainties.

References

1. G. Alfonso, A.F. Roldán López de Hierro, C. Roldán, A fuzzy regression model based on finite fuzzy numbers and its application to real-world financial data. J. Comput. Appl. Math. **318**, 47–58 (2017)
2. R.A. Aliev, B. Fazlollahi, R.M. Vahidov, Genetic algorithm-based learning of fuzzy neural networks. Part 1: Feed-forward fuzzy neural networks. Fuzzy Sets Syst. **118**(2), 351–358 (2001)
3. R.A. Aliev, B.G. Guirimov, B. Fazlollahi, R.R. Aliev, Evolutionary algorithm-based learning of fuzzy neural networks. Part 2: Recurrent Fuzzy neural networks. Fuzzy Sets Syst. **160**, 2553–2566 (2009)

4. A. Bargiela, W. Pedrycz, T. Nakashima, Multiple regression with fuzzy data. Fuzzy Sets Syst. **158**(19), 2169–2188 (2007)
5. K.Y. Chan, H.K. Lam, C.K.F. Yiu, T.S. Dillon, A flexible fuzzy regression method for addressing nonlinear uncertainty on aesthetic quality assessments. IEEE Trans. Syst. Man Cybern. Syst. **47**(8), 2363–2377 (2017)
6. T. Chen, A fuzzy Back propagation network for output time prediction in a wafer fab. Appl. Soft Comput. **2/3F**, 211–222 (2003)
7. F. Choobineh, H. Li, An index for Ordering Fuzzy Numbers. Fuzzy Sets Syst. **54**(3), 287–294 (1993)
8. S. Danesh, R. Farnoosh, T. Razzaghnia, Fuzzy nonparametric regression based on adaptive neuro-fuzzy inference system. Neurocomputing **173**, 1450–1460 (2016)
9. W. Dong, F. Wong, Fuzzy weighted averages and implementation of the extension principle. Fuzzy Sets Syst. **21**, 183–199 (1987)
10. W. Dong, H. Shah, F. Wong, Fuzzy computations in risk and decision analysis. Civ. Eng. Syst. **2**, 201–208 (1985)
11. M. Gupta, J. Qi, On Fuzzy Neuron Models. The International Joint Conference on Neural Networks (IJCNN). pp 431–436, (1991)
12. Y. Hayashi, J.J. Buckley, E. Czogala, Fuzzy neural network with fuzzy signals and weights. Int. J. Intell. Syst. **8**, 527–537 (1993)
13. S. Haykin, *Neural Networks: A Comprehensive Foundation* (Prentice Hall, Upper Saddle River, 1999)
14. T. Hong, P. Wang, Fuzzy interaction regression for short term load forecasting. Fuzzy Optim. Decis. Making **13**(1), 91–103 (2014)
15. H. Ishibuchi, R. Fujioka, H. Tanaka, An Architecture of Neural Networks for Input Vectors of Fuzzy Numbers. IEEE International Conference on Fuzzy Systems (FUZZ-IEEE). San Diego. pp. 1293–1300, (1992)
16. H. Ishibuchi, K. Morioka, H. Tanaka, *A Fuzzy Neural Network with Trapezoid Fuzzy Weights*. IEEE International Conference on Fuzzy Systems (FUZZ-IEEE). pp. 228–233, (1994)
17. H. Ishibuchi, K. Kwon, H. Tanaka, A learning algorithm of fuzzy neural networks with triangular fuzzy weights. Fuzzy Sets. Syst. **71**, 277–293 (1995)
18. H. Ishibuchi, K. Morioka, I.B. Turksen, Learning by Fuzzified neural networks. Int. J. Approx. Reason. **13**, 327–358 (1995)
19. O. Kaleva, S. Seikkala, On Fuzzy Metric Spaces. Fuzzy Sets Syst. **12**, 215–229 (1984)
20. G. Klir, B. Yuan, *Fuzzy Sets and Fuzzy Logic: Theory and Application* (Prentice Hall, Upper Saddle River, 1995)
21. P.V. Krishnamraju, J.J Buckley, K.D. Reilly, Y. Hayashi, *Genetic Learning Algorithms for Fuzzy Neural Nets*. IEEE International Conference on Fuzzy Systems (FUZZ-IEEE). pp. 1969–1974, (1994)
22. G. Leng, T.M. McGinnity, G. Prasad, Design for Self-Organizing Fuzzy Neural Networks Based on genetic algorithms. IEEE Trans. Fuzzy Syst. **14**(6), 755–766 (2006)
23. J. Li, W. Zeng, J. Xie, Q. Yin, A new fuzzy regression model based on least absolute deviation. Eng. Appl. Artif. Intell. **52**, 54–64 (2016)
24. M. Mares, *Computation over Fuzzy Quantities* (CRC Press, 1994)
25. F. Megri, A.C. Megri, R. Djabri, An integrated fuzzy support vector regression and the particle swarm optimization algorithm to predict indoor thermal comfort. Indoor Built Environ. **25**(8), 1248–1258 (2016)
26. M. Mizumoto, K. Tanaka, Some properties of fuzzy numbers in advances in fuzzy sets theory and applications, in *Advances in Fuzzy Set Theory and Applications*, ed. by M. M. Gupta, (North-Holland, Amsterdam, 1979), pp. 153–164
27. R. Moore, *Interval Analysis* (Prentice-Hall, Englewood Cliffs, 1966)
28. M. Nii, T. Iwamoto, S. Okajima, Y. Tsuchida, *Hybridization of Standard and Fuzzified Neural Networks from MEMS-Based Human Condition Monitoring Data for Estimating Heart Rate*. International Conference on Machine Learning and Cybernetics (ICMLC). pp. 1–6, (2016)

29. H. Okada, Genetic algorithm with fuzzy genotype values and its application to Neuroevolution. Int. J. Comput. Inf. Sci. Eng. **8**(1), 1–7 (2014)
30. H. Okada, Evolving fuzzy neural networks by particle swarm optimization with fuzzy genotype values. Int. J. Comput. Digit. Syst. **3**(3), 181–187 (2014)
31. S.K. Pal, S. Mitra, Multilayer perceptron, fuzzy sets, and classification. IEEE Trans. Neural Netw **3**(5), 683–696 (1992)
32. S. Phitakwinai, S. Auephanwiriyakul, N. Theera-Umpon, *Fuzzy Multilayer Perceptron with Cuckoo Search. JP Journal of Heat and Mass Transfer*. Special Volume, Issue II: Advances in Mechanical System and ICI-convergence. pp. 257–275, (2018)
33. A.F. Roldán López de Hierro, J. Martínez-Moreno, C. Aguilar-Peña, C. Roldán López de Hierro, Estimation of a fuzzy regression model using fuzzy distances. IEEE Trans. Fuzzy Syst. **24**(2), 344–359 (2016)
34. H. Tanaka, S. Uejima, K. Asai, Linear regression analysis with fuzzy model. IEEE Trans. Syst. Man Cybern. **SMC-12**(6), 903–907 (1982)
35. The UCI, *Repository of Machine Learning Databases and Domain Theories*. Available: http://www.ics.uci.edu/~mlearn/MLRepository.html
36. N. Theera-Umpon, S. Auephanwiriyakul, S. Suteerohnwiroj, J. Pahasa, K. Wantanajittikul, *River Basin Flood Prediction Using Support Vector Machines*. IEEE International Joint Conference on Neural Networks (IJCNN 2008). (Hong Kong, 2008). pp. 3039–3043.
37. J. Wang, X. Yu, P. Li, *Research of Tax Assessment Based on Improved Fuzzy Neural Network* (ICALIP, Shanghai, 2016)
38. H.-C. Wu, Linear regression analysis for fuzzy input and output data using the extension principle. Comput. Math. Appl. **45**(12), 1849–1859 (2003)
39. X.S. Yang, S. Deb, *Cuckoo Search Via Lévy Flights. World Congress on Nature & Biologically Inspired Computing (NaBIC)* (Coimbatore, 2009), pp. 210–214
40. L. Zadeh, Outline of new approach to the analysis of complex systems and decision processes. IEEE Trans. Syst. Man Cybern. Syst. **3**(1) (1973)

Sansanee Auephanwiriyakul (S'98–M'01–SM'06) received the B.Eng. (Hons.) degree in electrical engineering from the Chiang Mai University, Thailand (1993), the M.S. degree in electrical and computer engineering and Ph.D. degree in computer engineering and computer science, both from the University of Missouri, Columbia, in 1996, and 2000, respectively. After receiving her Ph.D. degree, she worked as a post-doctoral fellow at the Computational Intelligence Laboratory, University of Missouri-Columbia.

Dr. Auephanwiriyakul's interest in Computation Intelligence (CI) research area started when she was in her Bachelor degree. She dreamed of building a robot that can automatically move by itself. In that time, the idea of making robot's brain did not come to her mind until she went to University of Missouri-Columbia for her Master and Ph.D. and had a chance to work with the great mentors Krishnapuram, Keller, and Gader and other professors there that she realized that CI might be the answer. She then made a decision of starting the career in this direction.

Now, Dr. Auephanwiriyakul is an Associate Professor in the Department of Computer Engineering and an associate director of the Biomedical Engineering Institute, Chiang Mai University, Thailand. She is a senior member of the Institute of Electrical and Electronics Engineers (IEEE). She is an Associate Editor of the IEEE Transactions on Fuzzy System, IEEE Transactions on Neural Networks and Learning Systems, ECTI Transactions on Computer and Information Technology (ECTI-CIT), and was an Editorial Board of the Neural Computing and Applications, and other various major journals. She was a General Chair of the IEEE International Conference on Computational Intelligence in Bioinformatics and Computational Biology (CIBCB 2016). She was a Technical Program Chair, Organizing Committee in several major conferences including the IEEE International, Conference Fuzzy Systems as well. She is also a chair of the IEEE Computational Society – Webinars and a member of six important IEEE organizations, i.e., Fuzzy System Technical Committee, Bioinformatics and Bioengineering Technical Committee, Data Mining and Big Data Analytics, Distinguished Lecture Program, Women in Computational Intelligence. As a part of these committee members, she has to promote the research, development, education, and understanding of the technology, including the creation of the theory and models as well as their applications.

As a role of researcher, she has written many conference papers that have been published in several major conferences, e.g., the IEEE International Conference Fuzzy Systems and journal papers published in several major journals including IEEE Transactions on Biomedical Engineering. She wrote several textbooks in Thai including Digital Image Processing, as well.

Message:

It is my pleasure to be part of this book. I was also asked to encourage young researchers to pursue their careers in this field. I do not have any thing much to say. But here it is. In many years of working, I have come across many great people. They all remind me of the quote from Forrest Gump (1994) (https://www.imdb.com/title/tt0109830/):

"Life was like a box of chocolate. You never know what you're gonna get".

Another quote from Field of Dream (1989) (https://www.imdb.com/title/tt0097351/) that keeps me going is:

"If you build it, they will come".

So just keep doing what you are doing, it will be your day someday.

Suwannee Phitakwinai received the B.Eng. degree in computer engineering, M.Eng. degree in computer engineering, and Ph.D. degree in computer engineering, all from the Chiang Mai University, Thailand, in 2004, 2008, and 2019, respectively. She started working in computational intelligence with the senior project: "Thai hand writing recognition using HMM". Her master degree thesis topic is also about Thai sign language translation system using FCM, HMM, and SIFT. Her dissertation research was on the fuzzy multilayer perceptron algorithm for fuzzy vector classification and regression.

Her research interests include pattern recognition, neural networks, fuzzy logic, fuzzy set and system, Type-2 fuzzy set, linguistic algorithm, image processing, and computational intelligence applications. She is currently a lecturer in the Department of Computer Engineering, Rajamankala University of Technology Lanna (Tak Campus), Thailand.

Message:

I became interested in the field of computational intelligence since 2004 when I have to find an idea for my senior project. At that time, tablets or smartphones with touchscreen were not widely used and very expensive. Therefore, my advisor gave an idea "How does the computer know what we are writing on the screen?", I think it was a difficult but challenging topic. That has become the beginning of my research interest in Computational Intelligence.

Message for students, "don't think Computational Intelligence is a difficult field but instead a challenging subject. We can do something to change the world. Have fun with it!"

Nipon Theera-Umpon (S'98–M'00–SM'06) received his B.Eng. (Hons.) degree from Chiang Mai University, M.S. degree from University of Southern California, and Ph.D. degree from the University of Missouri-Columbia, all in electrical engineering. He has been with the Department of Electrical Engineering, Chiang Mai University since 1993. He has served as editor, reviewer, general chair, technical chair and committee member for several journals and conferences. He has been bestowed several royal decorations and won several awards. He was associate dean of Engineering and chairman for graduate study in electrical engineering and graduate study in biomedical engineering.

He is presently serving as the director of Biomedical Engineering Institute, Chiang Mai University. He is a member of Thai Robotics Society, Biomedical Engineering Society of Thailand, Council of Engineers in Thailand. He has served as vice president of the Thai Engineering in Medicine and Biology Society and vice president of the Korea Convergence Society. Dr. Theera-Umpon is a senior member of the IEEE, and is a member of IEEE-IES Technical Committee on Human Factors. He has published more than 190 full research papers in international refereed publications. His research interests include Pattern Recognition, Machine Learning, Artificial Intelligence, Digital Image Processing, Neural Networks, Fuzzy Sets and Systems, Big Data Analysis, Data Mining, Medical Signal and Image Processing.

Message:

My Computational Intelligence (C.I.) journey began more than 25 years ago in Columbia, Missouri, where I was with C.I. Lab, University of Missouri-Columbia. The great mentors there, Gader, Keller, and Krishnapuram, taught me how useful and powerful C.I. was. Since then, my life has been filled with C.I. It gives me a great career as a university professor. I also brought hundreds of my academic offspring in C.I. to the world. As mentioned, C.I. is so useful and powerful. It can be applied to solve countless real problems in many fields of application. We, C.I. people, are an important part of the world development and sustainability.

Generalisation in Genetic Programming for Symbolic Regression: Challenges and Future Directions

Qi Chen and Bing Xue

1 Genetic Programming for Symbolic Regression

In recent years, genetic programming for symbolic regression (GPSR), which is a data-driven modelling technique, has become more and more popular in the computer science research community and many multidisciplinary engineering communities. This section introduces how GPSR works including the task of symbolic regression, the components of genetic programming, and how to apply genetic programming to symbolic regression.

1.1 Symbolic Regression

In statistic modelling, regression analysis is the task of identifying the relationship between the input and output variable(s) and expressing this relationship in mathematical/symbolic models for a set of given data. Regression tasks can be denoted as: $Y = f(X) + \alpha$. The p input variables $X = \{x_1, x_2, \ldots, x_p\}$ are also called predictors/independent variables, while the output variable(s) Y is also called dependent/target variable(s). Traditional regression analysis typically finds the (near-)optimal coefficients for a predefined regression model structure, i.e., find the optimal β values for a linear regression model $f(X) = \beta_1 x_1 + \beta_2 x_2 + \cdots + \beta_p x_p$.

The authors are with the Evolutionary Computation Research Group at the School of Engineering and Computer Science, Victoria University of Wellington.

Q. Chen · B. Xue (✉)
Victoria University of Wellington, Wellington, New Zealand
e-mail: qi.chen@ecs.vuw.ac.nz; Bing.Xue@ecs.vuw.ac.nz

A. E. Smith (ed.), *Women in Computational Intelligence*, Women in Engineering and Science, https://doi.org/10.1007/978-3-030-79092-9_13

Symbolic regression is a type of regression analysis that finds the model structure and the coefficients simultaneously during the modelling process. This is a sharp difference from traditional regression analysis. Symbolic regression does not require any assumption of the model structure or the data distribution beforehand. This is particularly useful when there is not much available domain knowledge. Moreover, compared with traditional regression, symbolic regression can effectively reduce bias introduced by analysing the data distribution and predefining the model structure.

1.2 Genetic Programming—An Evolutionary Computation Technique

Genetic programming [37], an evolutionary computation approach with a symbolic representation of solutions, is an attractive technique for symbolic regression. Given a set of data, GP for symbolic regression (GPSR) does not require to specify the form of the solution but builds the models by iteratively combining a set of essential mathematical expressions. Like many other EC techniques, GP follows the principle of Darwinian natural selection and automatically creates computer programs addressing problems at hand [37]. GP starts from a population of randomly generated solutions. Then this population is stochastically updated to a new population simulating evolution using breeding mechanisms and fitness-based selection. At the end of the evolutionary process, the fittest solution is expected to be found. The essential evolutionary process of GP is shown in Fig. 1. A detailed introduction of the basic components of GP is shown in the following subsections.

1.2.1 Representation

In standard GP, computer programs, i.e., candidate solutions, are represented as *parse expressions*; typically, a parse tree represents a candidate solution. Figure 2 shows a parse tree for a solution in GPSR. The nodes/elements in the parse trees are selected from a function set F and a terminal set T, which are predefined. For GPSR, the function set F can include the basic arithmetical functions, e.g., "+", "−", "×", and protected "/", and more advanced functions, such as trigonometric functions and transcendental functions. Each element in F has a fixed number of arguments. For example, function "−" has two arguments and "sine" has one argument. The terminal set T usually consists of independent variables/features and a number of random constants. The selection of the function set and terminal set needs to satisfy the *sufficiency* and *closure* properties [5]. The sufficiency property means these two sets should have enough expressive power to represent a solution for the problem, while the closure property requires each element in F is capable to accept all the possible output(s) returned by any other node in the tree. The introduction

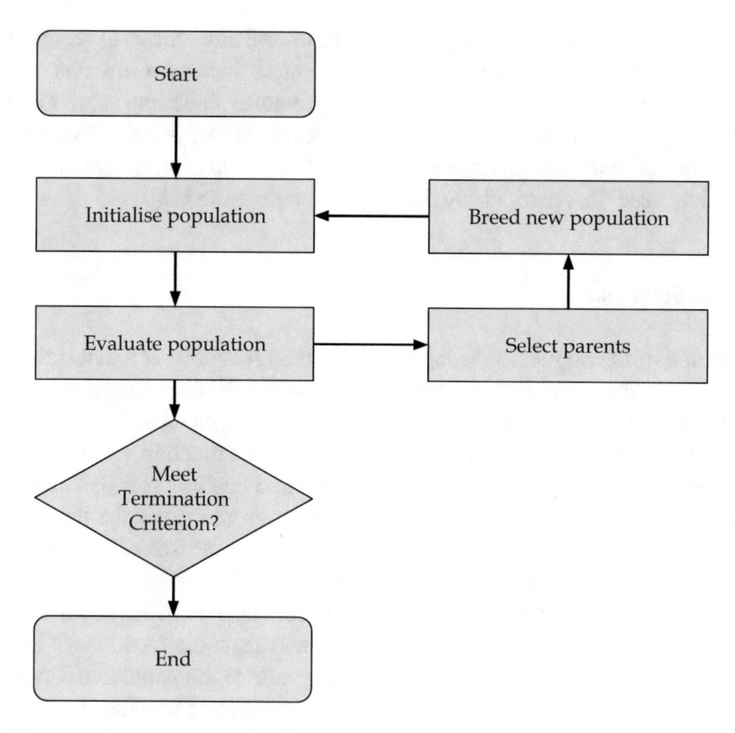

Fig. 1 The evolutionary process of GP

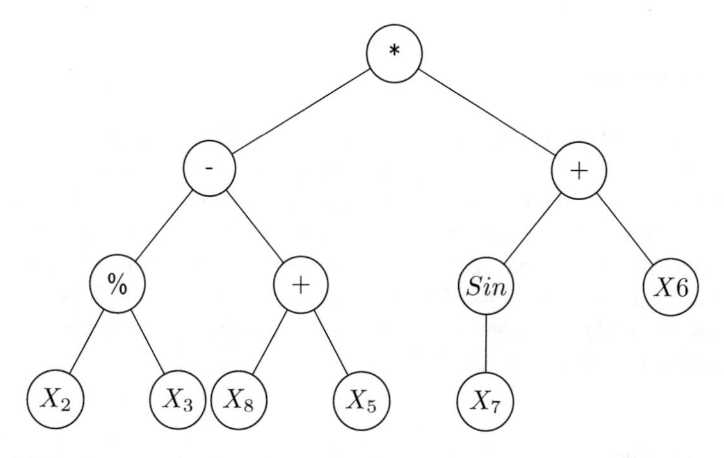

Fig. 2 A GP tree representing the regression model $((x_2/x_3) - (x_8 + x_5)) * (sinx_7 + x_6)$

of protected "/" is to satisfy the closure property in a critical situation of division by 0. In this case, it returns a predefined value (0 or 1). When the terminal set and the function set have been determined, candidate solutions are structured from these sets.

Besides the popular tree-based presentation, several other forms of representation have also been developed for GP. These variants include *Cartesian GP* [45] using a *graphic* structure, GP using *linear* structures including *gene expression programming* [22] and *linear genetic programming* [8]. Among these various ways of representing candidate programs in GP, the tree-based representation is still the most widely used one, particularly for symbolic regression [53].

1.2.2 Initialisation

Initialisation of the population using the predefined function set and terminal set is the first step in the search/evolutionary process of GP. There are several commonly used methods to generate initial individuals in GP. The *grow* method is the simplest one, where a node is selected randomly from either the function set or the terminal set before the maximal tree size or maximal limited depth is reached. If a node is selected from the terminal set, the building of current branch of the tree will stop. The grow method produces trees with various shapes and sizes. The *full* method is another commonly used method for generating GP individuals, which generates trees of the maximal predefined size or depth. Before the maximum depth is reached, the nodes in a GP tree are generated using elements from the function set only. The *ramped-half-and-half* method is introduced to promote the population diversity of the initialisation step. In this case, half of the population is initialised with the full method and the rest are produced by the grow method.

1.2.3 Evaluation

The main driving force of the evolutionary process of GP is seeking individuals with better performance. The evaluation process measures the performance of candidate solutions, which is often a quantitative value defined by a fitness function. The fitness function is calculated on the training set. The fitness measure should be designed to give a fine-grained differentiation between candidate solutions for GP. There are many ways to cast a fitness function. When GP is used for regression problems, generally *mean-squared error*, *root mean-squared error*, and *normalised mean-squared error* can be used.

1.2.4 Selection

After the performance of GP individuals has been measured, the selection operators are used to select individuals to apply genetic operators, which give solutions with above average performing more opportunities. When applying GP for solving symbolic regression, choosing a good selection method is one of the most important decisions. The most commonly used selection method in GP is *tournament selection*. m individuals are first sampled, and the one with the best fitness value (e.g., the lowest regression error for symbolic regression) is selected as a parent for breeding the new population. In tournament selection, the tournament size m leads to different selection pressures. While a large tournament size causes high selection pressure, a small tournament size generally causes low pressure. Many other standard evolutionary selection methods can be used for GP, such as *fitness-proportional selection* and *truncation selection*.

1.2.5 Genetic Operators

Evolution in GP proceeds by transforming parents to offspring by means of genetic operators. The three commonly used GP genetic operators are: *crossover*, *mutation*, and *reproduction*. For *crossover*, two parents are selected using the selection method and a random subtree is selected in each parent. Swapping these two subtrees forms two new trees, which are known as the children and enter the new generation. *Mutation* operates on one parent only. A random subtree of the parent is selected and replaced by a new subtree following the constraints of GP settings. The subtree can contain one node only. *Reproduction* is straightforward. It operates by placing the copy of a selected individual into the population. The commonly used version of reproduction is *elitism* where the top certain percent of GP individuals, ranked according to the fitness value, are copied to the next generation.

1.3 *GP for Symbolic Regression*

The ability of GP in automatically creating various programs including mathematical functions, no requirement for predefined shape and size of solutions, and the expressive powers make it a very suitable approach to symbolic regression.

The key points of tackling a symbolic regression problem by GP are as follows:

- Determine a *representation* scheme for GP produced solutions that is the combination of elements from a terminal set and a function set.
- Design an *evaluation* criterion for estimating the performance of a solution. This usually involves testing a candidate program/solution on a set of data points from the training set.
- Choose a parent *selection* mechanism that determines the way to select a parent or a set of parents.
- Define *genetic operators* for breeding offspring from parents. Standard GP uses two basic variant operators: *subtree crossover* and *subtree mutation*.

GP is a very flexible method, allowing vigorous co-application of all kinds of strategies and meta-strategies. During the past decades, GP has been remarkably successful in solving symbolic regression problems [57, 58, 60]. Despite the many successful applications in GPSR, there are still many open issues, which need to be solved to make GPSR a better regression technique. *Generalisation* is one of the open main issues in GP [50].

2 Generalisation in GPSR

Generalisation is the ability with which the learnt model can achieve good performance on unseen data. Generalisation is one of the most important performance criteria for machine learning algorithms, since it reflects the prediction performance of learned models on unseen data. A major goal of learning algorithms is to find a model that can minimise the expected error $Err = E[L(Y, f(X))]$. Here, $L(Y, f(X))$ refers to the loss function between the target output Y and the output of the model $f(X)$. A set of input X and output Y pairs are drawn from an underlying distribution $P(X, Y) = P(Y|X)P(X)$, where $P(X)$ is the distribution of the input X and the conditional distribution $P(Y|X)$ is based on the input–output relation. The expected generalisation error $Err_T = E[L(Y, f(X))|_T]$ is a quantity related to the expected error [25], which measures the prediction error of the learnt model over a set of unseen/test data for a given training set T. The joint distribution $P(X, Y)$ is needed in order to measure Err, which is typically unknown in most real-world learning tasks. Thus, many learning algorithms rely on the empirical risk minimisation principle, which first computes the errors of a set of candidate models over the training set and then selects the one that obtains the minimum training error among the set of models [25]. The empirical/training error is expected to be a good indicator of the expected test error. However, in many cases, this indicator does not work well, particularly when the number of training samples is too small to represent the real distribution of the data. There often exist a large amount of noise data and/or over-complex models have been learnt.

2.1 Concepts Related to Generalisation

Two key concepts related to generalisation are overfitting and bias–variance decomposition. Existing work on the generalisation of GPSR revolves around these two concepts.

2.1.1 Overfitting

The contrary concept of generalisation is *overfitting*. Overfitting means poor generalisation performance. A more detailed definition of overfitting is given in [46]. According to this definition, a model overfits the training data when there exist some other models with worse training performance but better performance over the entire distribution of instances. GP is prone to overfitting due to many factors.

The downside of the flexible representation of GP is the tendency to learn complex models, which are more likely to overfit the training set. Another important factor is the nature of GP to chase the lowest training error. An ideal learning process is to learn the underlying relationship between the input and output variables while ignoring the noise. Nevertheless, GP has a greedy nature in pursuing models with the lowest error on the training set. Then after the true relationship is learned, GP might continue to fit the noise in the training set to minimise the training error. In this case, GP might generate models that overfit the training set. Furthermore, when the number of available observations/instances is too small to represent the true pattern of the desired function, GP is prone to produce models that overfit the training set.

The high dimensionality of the data is another important issue, which also leads to overfitting in GPSR. The dataset often uses a large number of features, while only a small number of these features are relevant to the target/true model and many of these features are noisy features. Searching a model with unknown structure in a high-dimensional space using a limited number of instance, which indicates a large search space with limited information, runs a high risk of overfitting and poor generalisation.

2.1.2 Bias–Variance Decomposition

A widely accepted wisdom in machine learning is that simpler models can generalise well. A more accurate description is that, to learn a model that achieves a good generalisation on unseen data, it is necessary to obtain a trade-off between achieving an impressive training accuracy and obtaining a reasonably smooth function. The theoretical tool of bias–variance decomposition, i.e. a decomposition of learning error into bias and variance errors, provides insight into this trade-off. This method comes from statistical learning theory and is well known in machine learning [6, 21, 27].

The bias error is a measure of the difference between the expected prediction of the model and the target output value, which is the inherent error of the model. The variance error is taken as the variability of a model prediction for a given data point, which measures how much the predictions for a given data point vary between different approximations of the model. A lower variance error indicates that the models are less sensitive to the training data and thus can potentially generalise well on unseen data.

Figure 3 illustrates this bias–variance trade-off where increasing the model complexity has the effect of reducing bias of the model while increasing its variance error. There exists a "sweet spot" where, with an appropriate model complexity, bias and variance errors are balanced and relatively low, and hence, the generalisation error is minimised. However, how to find the "sweet spot" is not a trivial task and is an attractive research topic for enhancing the generalisation of the learnt models.

Generalisation has been deeply investigated in many other machine learning fields [2, 18]. However, it had not received much attention in GPSR for quite

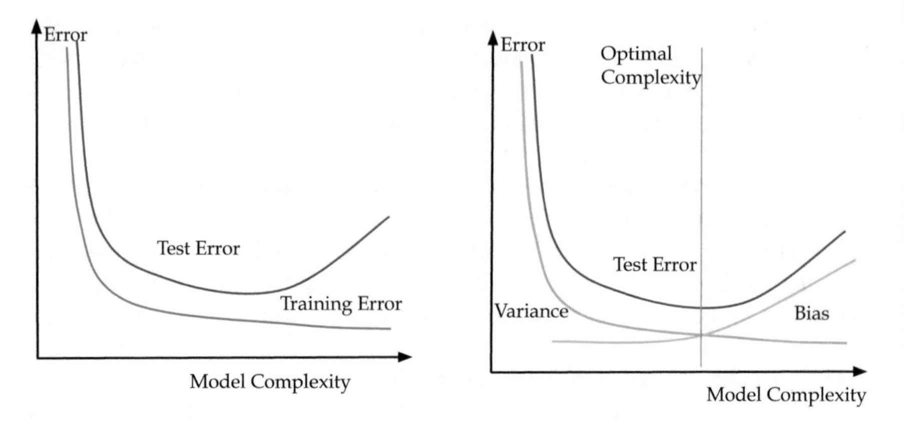

Fig. 3 Bias–variance trade-off

a long time. Before 2000, symbolic regression was mainly considered to be an optimisation problem, which used all the available data for evolving the models and did not examine the generalisation performance of the models on unseen data [3, 29, 30, 33, 69]. In recent years, an increasing number of approaches for enhancing the generalisation ability of GPSR have been proposed [13, 16, 65, 67]. In the following section, a review of these approaches is presented.

3 Enhancing the Generalisation Ability of GPSR

Existing approaches enhance the generalisation of GPSR from many different aspects including improving data quality via data sampling and feature selection [13, 28, 32, 38, 51, 55], utilising the estimated generalisation error [36, 56], proposing new search mechanisms [4, 12, 64], and using an ensemble of models [9, 15]. The categories of these approaches are shown in Fig. 4. A detailed review of these approaches and the challenges are presented in the following subsections.

3.1 Data Sampling and Feature Selection

Data sampling and feature selection methods handle the training data at the instance and variable levels, respectively. These methods aim to reduce overfitting by not exposing the whole training sets to GP individuals and discarding the irrelevant/noisy information.

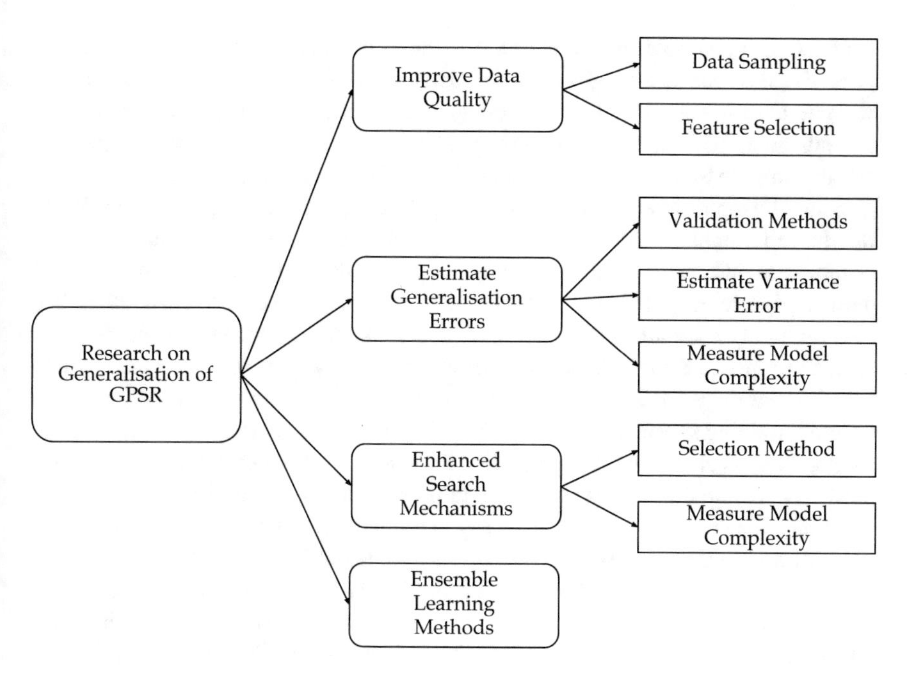

Fig. 4 Categories of research on generalisation

3.1.1 Data Sampling

Data sampling, which is also called instance selection, is a technique to manipulate and select a representative subset of instances to identify the patterns and trends in the original whole set of instances. GPSR with sampling methods typically uses a sample of the training instances dynamically during the evolutionary process and selects GP individuals performing well across different subsets of the training data. These GP individuals usually have low variance and generalise well. The effect of these data sampling methods on improving the generalisation of GPSR has been investigated in previous work [28, 32, 51].

Panait et al. [51] investigated the impact of six sampling methods on the robustness and generalisation of various GP methods including GPSR. These sampling methods are the fixed-random sampling where a set of training instances are sampled beforehand and used during the entire evolutionary process, two dynamical-random sampling methods (including one random set per individual and one random set per generation), two coevolution methods where a population of training sets is evolved along with a population of solutions, and fitness sharing with sampling. The results showed that no single method dominates the others on all problem domains. However, coevolution methods have the best performance for GPSR. The fixed-random method never achieved the top tier performance on the examined problems and should not be used as the first choice. Galves et al. [28] demonstrated that the sampling strategy of using a single random instance and the complete training dataset interleaved at generations leads to better generalisation in GPSR than using a single random instance and the whole set of instances at all generations. Martínez et al. [44] compared the effort of four sampling methods on enhancing GP including reducing overfitting. The four sampling methods are dynamic training subset selection that selects instances according to their associated weight, interval sampling [28], lexicase selection [31], and Keep-Worst Interleaved Sampling, which uses the whole set of instances and the most difficult instances at interval generations. They claimed that these sampling methods are particularly useful for difficult real-world symbolic regression problems, rather than the commonly used artificial benchmark problems in the literature. Kommenda et al. [35] proposed a multi-subpopulation GP method with a sophisticated training sampling strategy for symbolic regression. Each subpopulation is evaluated on a different training subset and a variable training subset that is exchanged between the subpopulations at specific data migration intervals. Their experiment results show a positive effect on preventing overfitting thus enhancing generalisation of GPSR.

These data sampling methods are easy to implement and with potentially notable generalisation gain. However, it is not trivial to determine the sampling frequency and the amount of search effort needed in GPSR. While too much search effort hinders mitigating overfitting, too little search effort might result in GP individuals with poor performance on both the training and the test sets. More importantly, data sampling methods may be susceptible to the noise/outliers in the training data. To solve these challenges, instance sampling based on the difficulty of instances for the learnt models and more uniformly sampling throughout the evolutionary process are promising directions for future research in this direction.

3.1.2 Feature Selection for GPSR

Feature selection is a process of selecting a subset of relevant features/variables and eliminating redundant features from the original large set of features [71]. These features/variables are desired and necessary for describing the dependent variable(s). Feature selection, which reduces the dimensionality of the input space, can decrease the variance error with the cost of obtaining a relatively high bias error. It also allows models to generalise better in many cases due to the elimination of noisy information. Feature selection methods can be divided into three categories: *filter*, *wrapper*, and *embedded methods* [71]. Filter methods typically use information-theory and statistical methods to measure the quality of features individually. Features are selected accordingly. For wrapper methods, during searching for the optimal subset of features, the performance of a learnt model/classifier is utilised to assess the goodness of the subset of features. Filter feature selection methods are independent of any learning algorithm. They are often computationally less expensive and more general than wrapper methods. However, wrapper methods evaluate the feature subsets based on the learning performance, which usually results in better performance for a particular learning algorithm. Embedded methods perform feature selection during the learning process of the model/classifier.

GP has embedded feature selection ability where the features appearing in the better-fitted solutions can be treated as more useful features. The embedded feature selection ability is the by-product of the parent selection pressure and the flexible representation of GP solutions with variable length and shape. However, the embedded feature selection ability of GP is not strong enough when learning from high-dimensional data. More sophisticated feature selection methods are desired.

Xu et al. [70] used random forests for feature selection and GPSR was performed on these selected features and achieved good generalisation performance. Vladislavleva et al. [68] proposed a two-stage GPSR method for predicting the energy output of wind farms. The first stage utilises a multi-objective GP method minimising both the training accuracy and the model complexity for feature selection. Features appearing in the "best" model, which sit at the knee point of the Pareto front, i.e., the point with the smallest distance to the optimal point, are considered. Among these features, the ones that have higher R^2 correlation with the output variable are selected. In the second stage, GPSR learns models using the selected features. These models typically can have good generalisation performance. Dick et al. [20] demonstrated that permutation-based feature importance is more useful in identifying relevant features than the frequency of appearance-based methods in GP programs. Chen et al. [11] divided a GP run into two stages. During the first stage, frequently occurring features are selected from the better-fitted individuals in the population. In the second stage, new individuals are initialized using the selected features to replace the under-average-performing individuals. The two-stage GPSR method brings promising generalisation gains. Later, Chen et al. [13] developed a feature selection method for GPSR, which employed the permutation-based feature importance measure to rank features collected from best-of-run individuals. GPSR learning from the selected features showed a better generalisation gain than many alternative methods.

While feature selection has a promising potential to enhance the generalisation of GPSR when learning from high-dimensional data, the majority of the existing work on feature selection focuses on classification problems. Only a limited number of related researches on symbolic regression have been reported to date. For regression tasks, especially for high-dimensional regression, feature selection is desired to improve the generalisation of GP. More importantly, more attention is needed to explore the inherent feature selection ability of GP to find the subset that enables better generalisation than the original whole set of features, i.e., to develop more advanced GP methods with embedded feature selection ability. Moreover, methods to guarantee that the exploration capability of GP is not beyond the size of the optimum subset of features are promising future directions in feature selection for GPSR.

3.2 *Estimating Generalisation Errors*

3.2.1 Validation Methods

Validation methods measure the error of models on an independent dataset called the validation set and use the assessment as a proxy for the generalisation error of the models. Validation methods can be used to stop an evolutionary run when the potential overfitting is detected. This process is known as *early stopping*. Tuite et al. [61] investigated and compared a number of early stopping heuristics for GP. They found that the Pearson's correlation coefficient between the errors on the training set and the validation set is a good stopping heuristic, stopping the training process when the correlation between the errors falls below a predefined threshold. The method was shown to improve the generalisation of the learnt model. Rivero et al. [56] showed that the use of a validation set on a subset of the population leads to a better generalisation gain than using the validation set on the best individual.

Validation methods can also be used to approximate the generalisation error, thus selecting the best model. Typically, a number of best-of-generation programs sampled throughout the evolutionary process are evaluated on the validation set at the end, with the best one is treated as the best model. Gagné et al. [26] selected the best n solutions on the training set. Among these solutions, the solution with the best performance on the validation set was chosen as the final solution. They found that the validation method usually brings slight benefit to generalisation, while parsimony pressure, i.e., introducing explicit penalisation of larger programs, often results in negative results. However, better generalisation gains can be achieved when combining the two methods.

In a data-rich scenario when there are a sufficient number of data samples, validation methods, which are easy to implement and use, are preferred to reduce overfitting and estimate generalisation error. However, there are two key challenges in using the validation methods. The first one lies with the heuristics that identify the optimal stopping point and select the best candidates to be exposed to the validation set. Another key challenge is on how to ensure the validation set contains instances that are representative of the problem, since the effectiveness of using a validation set highly depends on its instances. Therefore, developing sampling methods that consider the distributions of instances is a promising direction for enhancing the effectiveness of the validation methods on enhancing the generalisation.

3.2.2 Estimating Variance Errors

As mentioned in the previous section, the variance error is a good indicator of the generalisation error. Models with a good trade-off between the bias and the variance errors are more likely to generalise well on unseen data.

Bootstrap techniques have been introduced into GP to estimate the variance error of models [1, 24]. In these methods, the GP population is trained using a list of bootstrap instances. The variance error of GP individuals is assessed according to their error on the bootstrap instances and individuals with a lower variance error are selected. Kowaliw et al. [36] investigated and analysed factors that are important in obtaining a low variance error, e.g., program size, the initialisation of the population, and the function set. The variance errors were approximated by the error between the target outputs and the averaged outputs of the best models obtained from a number of GP runs. They found that a reasonable optimisation of the analysed factors could lead to significant improvements in obtaining the trade-off between bias and variance errors. The change of the function set has the largest gain on the trade-off. Moreover, they found that a larger function set consistently lead to better generalisation. This is in conflict with what has been found in previous research [23, 39], i.e., a simple function set is more likely to evolve solutions with a smaller generalisation error. This indicates that the effect of the function set on the generalisation error might be task-dependent. A larger function set might not lead to better generalisation for GPSR when handling easy tasks.

3.2.3 Model Complexity and Generalisation Error

Another group of methods, which estimate the generalisation error, using various functions, includes the training error, model complexity, and training sample size. The estimated generalisation error is used to rank and select models during the evolutionary process. A key component in these methods is the model complexity measure.

Model complexity and generalisation of GPSR are closely related [2, 17, 18]. Too low model complexity will result in poor generalisation performance due to not fitting the data adequately, while too high model complexity leads to overfitting as the model too closely fits the training instances and the noise in the training data. Many papers have devoted to improving the generalisation of GPSR via controlling model complexity [16, 40]. These contributions can be broadly grouped into two categories: controlling the structural complexity and controlling the functional complexity of GP models/individuals.

Structural Complexity The structural complexity of GP individuals involves the size and shape of the expression tree-based representation. Approaches measuring and controlling structural complexity include counting the number of structural elements such as the number of nodes, leaves, or layers/depth. Then according to the minimal description length and the Occam's razor principle [7], models with a smaller number of nodes or layers are preferred. This can be accomplished through utilising new selection methods [43], new evaluation measures [41], and multi-objective methods [42]. Silva et al. [59, 65] proposed operator equalisation and its dynamic variant to discard over-size models. The methods control the distribution of individual sizes in the population by probabilistically accepting offspring individuals based on their sizes. The probabilities of acceptance are determined by a predetermined target distribution.

Approaches related to controlling the structural complexity are often simple to implement and use. They are typically applicable to a wide range of domains. However, they are relatively ineffective at eliminating overfitting since the structural complexity has a low association with the generalisation capability of GPSR. Many GPSR models have intron, i.e., structurally noneffective sub-programs/sub-models therefore could be complex in structure but smooth in function. Moreover, most GPSR models can be mathematically simplified, which makes the structural complexity decrease significantly, but the generalisation performance remains the same. For this reason research on controlling model complexity in GPSR has progressively shifted from controlling the structural complexity to controlling the functional complexity.

Functional Complexity The functional complexity of a model can be measured by estimating a model's behaviours over a possible input space. Compared with a large number of measures on structural complexity, only a limited amount of research on measuring the functional complexity of GPSR models can be found in the literature. Vladislavleva et al. [67] proposed a behavioural complexity measure for GPSR named order of nonlinearity, which measured the complexity of GP models by the order of the Chebyshev polynomials that approximate these models. Their method did not measure the complexity of GP models directly. Meanwhile, as pointed out in the original work [67], another weakness of this complexity measure is an overestimation of the true minimal degree of Chebyshev approximation of accuracy for unary functions and the approximate nature of the definition for functions of multiple arguments. Even for functions of two arguments, constructing a Chebyshev approximation is performed in terms of tensor products, and it represents a nontrivial computational procedure. For functions of more variables, it is difficult to construct the Chebyshev polynomial approximation for a given accuracy. Vanneschi et al. [65] proposed a complexity measure, which is inspired by the concept of curvature for GP. The proposed measure is only reported for the current best individual on the training set to discover its relationship with bloat and overfitting. Chen et al. [11, 16] introduced the Vapnik–Chervonenkis (VC) dimension [66] to measure the complexity of models for GPSR. An experimental method was introduced to measure the VC dimension of the evolved models and employed structural risk minimisation as the fitness function to favour simpler models. GPSR has shown to achieve a good trade-off between the learning accuracy and the model complexity. Despite the impressive generalisation gains, the method has a potential limitation, i.e., calculating VC dimension is complicated and expensive. This makes it difficult to be widely adopted for GPSR. In their recent work [54], a simpler but effective method based on the Rademacher complexity [34] has been introduced to GPSR to control the functional complexity. The Rademacher complexity of a GP individual is measured by its performance to fit some random outputs. The method has shown positive effect on enhancing generalisation of GPSR.

Approaches related to functional complexity are typically more effective at promoting generalisation due to being more closely linked to overfitting. However, the estimation of functional complexity of a GPSR model is not trivial, which is an obstacle in making it a standard setting to guarantee the generalisation of GPSR.

The use of model complexity for estimating generalisation performance frees the learning from holding out a set of instances for validation and therefore is potentially useful when there is insufficient data to use. These methods have not received enough attention from the GP community. While defining model complexity measures that have a high correlation with the generalisation ability is important, ensuring that the measures are empirically achievable for GPSR is also critical. Therefore, another aspect that needs further investigation is how to measure model complexity in an efficient way. It will pay the way of utilising model complexity to obtain better generalisation for GPSR.

3.3 *Improving Selection and Genetic Operators for Better Generalisation*

One group of works on enhancing generalisation of GP have focused on search mechanisms directly including new selection methods, new genetic operators, and some other search mechanisms [14, 47, 64].

The effect of selection methods on enhancing the generalisation of GPSR was studied in [4, 12, 64]. Azad et al. [4] proposed a new tournament selection scheme that considered both training error and variance of the outputs of GP models. Vanneschi et al. [64] proposed a two-layer selection method to enhance the generalisation of GPSR. The method maintained a list of potentially overfitting individuals. Tournament selection was used on individuals who had a smaller average similarity to these overfitting individuals. Chen et al. [12] introduced a new selection operator to select pairs of parents, which have the largest angle distance between the relative semantics of the parents and the target. It was shown to promote the population diversity and enhance the generalisation of GPSR effectively.

New genetic operators with various additional requirements have been proposed to achieve a better generalisation gain for GPSR via generating compact models, reducing overfittings and obtaining offspring with controllable behaviour. Size-fair crossover [52], which allows exchange of similar-sized subtrees only, was shown to enhance the generalisation to some extent. Nguyen et al. [47] developed a new crossover operator called semantics-aware crossover (SAC) for checking semantic equivalence of subtrees and attempted to increase the semantic diversity. Two subtrees were evaluated on a set of randomly sampled instances. If the output of the two subtrees on this set of sampled instances were close enough (subject to a parameter–semantic sensitivity), then they were considered to be semantically equivalent and allowed to swap. The idea of SAC is novel. However, when compared with standard crossover, it has limited improvement on some test problems. Later, Nguyen et al. [63] proposed the semantic similarity-based crossover (SSC). Different from SAC that checks the semantic equivalence, SSC selected subtrees for crossover by checking semantic similarity. Compared with standard crossover and SAC, SSC was shown to be superior on enhancing the generalisation performance. The idea of SSC was then extended to mutation and led to a semantic similarity-based mutation operator (SSM) [48, 62], which shows much better performance than standard mutation. Chen et al. [12, 14] have proposed a geometric semantic crossover with angle-aware mating, i.e., the perpendicular crossover and a random segment mutation for enhancing the generalisation of GPSR. Chen et al. [10] proposed a mixed search strategy that combined genetic operators with gradient descent based on partial derivatives of the loss function with respect to constants in a GP individual. The new search mechanism was shown to improve generalisation performance of GPSR with a potential limitation of requiring a differentiable fitness function.

3.4 Ensemble Learning

Ensemble learning builds a composite predictor by combining multiple component predictors. The ensemble methods for GPSR utilise the stochastic nature of GP to generate diverse models. Bagging is a technique that works especially well for reducing the variance of models evolved by GP. By using an average model for making predictions, the error contribution due to the high variance can be effectively reduced, and the remainder error contribution is due to bias. Bagging uses bootstrap training samples to train different component models. Predictions are then combined using a simple unweighted average for regression. The component models were evolved sequentially such that each successive model concentrates on training examples that are difficult for the previous models in the sequence. A weight is associated with each training example. During the training of each model in the sequence, weights are increased for those instances that are misclassified by the models and are decreased for other instances.

Chen et al. [15] extended an existing ensemble learning method, AdaBoost [19], for enhancing the generalisation of GP when tackling transfer learning in symbolic regression. An ensemble of GP models, which blended individuals produced by standard syntax-based GP and individuals produced by geometric semantic GP, was proposed in [9]. In addition, they also proposed different pruning criteria based on correlation and entropy to remove similar programs/members and increase/guarantee the diversity of ensemble members, which were shown to be effective for improving generalisation of GPSR while reducing the computational cost.

Ensemble learning methods improve the generalisation of GPSR at the cost of lower interpretability and additional computational cost for training ensemble members. A combination of GP models decreases the interpretability of GPSR, which is a significant drawback. Various combinations of voting and averaging methods can be further explored to maintain the white-box property of GPSR while not preventing the generalisation gain to GPSR. Moreover, the effect of bagging and boosting to enhance the generalisation in GPSR is often problem-dependent, and these two methods produce competitive results. However, comparing to bagging, boosting is more sensitive to noise in the training instances. This is due mainly to the weighted voting mechanism of boosting, which is highly influenced by more accurate ensemble models. One promising direction to improve generalisation performance and computational efficiency in boosting is to select a subset of training instances uniform randomly at each iteration.

4 Conclusions and Future Directions

GPSR aims to discover the underlying data generation process of the given data via searching for the right model that produces the desired outputs. GPSR has gradually overcome the limitations that are commonly appeared in the early literature, and the generalisation of GPSR has received considerably focus in recent years. To make GPSR more widely recognised and a more competitive technique within the larger machine learning community, further research on approaching novel generalisation is fundamental and critical.

We have pointed out some promising future directions when reviewing each group of methods for enhancing the generalisation of GPSR. Meanwhile, some issues related to the generalisation have not been considered much in GPSR. For example, potential of learning the coefficients in symbolic regression models to enhance the generalisation has not been fully explored. Another issue is the undefined mathematical behaviours of the function set, e.g., taking the logarithm of zero or a negative number, and taking the square root of a negative number. Interval arithmetic [33] and analytic quotient [49] have been shown to be able to solve this issue to some extent. However, how they can be used to significantly improve the generalisation of GPSR needs further exploration.

References

1. A. Agapitos, A. Brabazon, M. O'Neill, Controlling overfitting in symbolic regression based on a bias/variance error decomposition, in *Parallel Problem Solving from Nature-PPSN XII* (Springer, Berlin, 2012), pp. 438–447
2. S.-i. Amari, S. Wu, Improving support vector machine classifiers by modifying kernel functions. Neural Netw. **12**(6), 783–789 (1999)
3. D.A. Augusto, H.J. Barbosa, Symbolic regression via genetic programming, in *Proceedings. Vol. 1. Sixth Brazilian Symposium on Neural Networks* (IEEE, Piscataway, 2000), pp. 173–178
4. R.M.A. Azad, C. Ryan, Variance based selection to improve test set performance in genetic programming, in *Proceedings of the 13th Annual Conference on Genetic and Evolutionary Computation (GECCO)* (2011), pp. 1315–1322
5. W. Banzhaf, P. Nordin, R.E. Keller, F.D. Francone, *Genetic Programming—An Introduction: On the Automatic Evolution of Computer Programs and Its Applications* (dpunkt-Verlag and Morgan Kaufmann, San Francisco, 1998)

6. C.M. Bishop et al., *Pattern Recognition and Machine Learning*, vol. 4 (Springer, New York, 2006)
7. A. Blumer, A. Ehrenfeucht, D. Haussler, M.K. Warmuth, Occam's razor. Inf. Process. Lett. **24**(6), 377–380 (1987)
8. M. Brameier, W. Banzhaf, A comparison of linear genetic programming and neural networks in medical data mining. IEEE Trans. Evol. Comput. **5**(1), 17–26 (2001)
9. M. Castelli, I. Gonçalves, L. Manzoni, L. Vanneschi, Pruning techniques for mixed ensembles of genetic programming models, in *Proceedings of the European Conference on Genetic Programming (EuroGP)* (Springer, Berlin, 2018), pp. 52–67
10. Q. Chen, B. Xue, M. Zhang, Generalisation and domain adaptation in GP with gradient descent for symbolic regression, in *2015 IEEE Congress on Evolutionary Computation (CEC)*, May 2015, pp. 1137–1144
11. Q. Chen, B. Xue, L. Shang, M. Zhang, Improving generalisation of genetic programming for symbolic regression with structural risk minimisation, in *Proceedings of the 18th Annual Conference on Genetic and Evolutionary Computation (GECCO)* (ACM, New York, 2016), pp. 709–716
12. Q. Chen, B. Xue, Y. Mei, M. Zhang, Geometric semantic crossover with an angle-aware mating scheme in genetic programming for symbolic regression, in *Proceedings of the European Conference on Genetic Programming (EuroGP)* (Springer, Berlin, 2017), pp. 229–245
13. Q. Chen, M. Zhang, B. Xue, Feature selection to improve generalization of genetic programming for high-dimensional symbolic regression. IEEE Trans. Evol. Comput. **21**(5), 792–806 (2017)
14. Q. Chen, M. Zhang, B. Xue, New geometric semantic operators in genetic programming: perpendicular crossover and random segment mutation, in *Proceedings of the 19th Annual Conference on Genetic and Evolutionary Computation Conference Companion* (2017), pp. 223–224
15. Q. Chen, B. Xue, M. Zhang, Instance based transfer learning for genetic programming for symbolic regression, in *2019 IEEE Congress on Evolutionary Computation (CEC)* (IEEE, Piscataway, 2019), pp. 3006–3013
16. Q. Chen, M. Zhang, B. Xue, Structural risk minimization-driven genetic programming for enhancing generalization in symbolic regression. IEEE Trans. Evol. Comput. **23**(4), 703–717 (2019)
17. Q. Chen, B. Xue, M. Zhang, Rademacher complexity for enhancing the generalization of genetic programming for symbolic regression. IEEE Trans. Cybern. (2020). https://doi.org/10.1109/TCYB.2020.3004361
18. D. Cohn, L. Atlas, R. Ladner, Improving generalization with active learning. Mach. Learn. **15**(2), 201–221 (1994)
19. W. Dai, Q. Yang, G.-R. Xue, Y. Yu, Boosting for transfer learning, in *Proceedings of the 24th International Conference on Machine Learning* (ACM, New York, 2007), pp. 193–200
20. G. Dick, Sensitivity-like analysis for feature selection in genetic programming, in *Proceedings of the 19th Annual Conference on Genetic and Evolutionary Computation (GECCO)* (2017), pp. 401–408
21. P. Domingos, A unified bias-variance decomposition for zero-one and squared loss. AAAI/IAAI **2000**, 564–569 (2000)
22. C. Ferreira, U. Gepsoft, What is gene expression programming (2008)
23. J. Fitzgerald, C. Ryan, On size, complexity and generalisation error in GP, in *Proceedings of the 16th Annual Conference on Genetic and Evolutionary Computation Conference (GECCO)* (2014), pp. 903–910
24. J. Fitzgerald, R. Azad, C. Ryan, A bootstrapping approach to reduce over-fitting in genetic programming, in *Proceedings of the 15th Annual Conference on Genetic and Evolutionary Computation (GECCO)* (2013), pp. 1113–1120
25. J. Friedman, T. Hastie, R. Tibshirani, *The Elements of Statistical Learning*. Springer Series in Statistics, vol. 1 (Springer, New York, 2001)

26. C. Gagné, M. Schoenauer, M. Parizeau, M. Tomassini, Genetic programming, validation sets, and parsimony pressure, in *Proceedings of the European Conference on Genetic Programming (EuroGP)* (Springer, Berlin, 2006), pp. 109–120
27. S. Geman, E. Bienenstock, R. Doursat, Neural networks and the bias/variance dilemma. Neural Netw. **4**(1) (2008)
28. I. Gonçalves, S. Silva, Balancing learning and overfitting in genetic programming with interleaved sampling of training data, in *Proceedings of the European Conference on Genetic Programming (EuroGP)* (Springer, Berlin, 2013), pp. 73–84
29. M. Gulsen, A.E. Smith, A hierarchical genetic algorithm for system identification and curve fitting with a supercomputer implementation, in *Evolutionary Algorithms* (Springer, Berlin, 1999), pp. 111–137
30. M. Gulsen, A. Smith, D. Tate, A genetic algorithm approach to curve fitting. Int. J. Prod. Res. **33**(7), 1911–1923 (1995)
31. T. Helmuth, N.F. McPhee, L. Spector, Lexicase selection for program synthesis: a diversity analysis, in *Genetic Programming Theory and Practice XIII* (Springer, Berlin, 2016), pp. 151–167
32. N.T. Hien, N.X. Hoai, B. McKay, A study on genetic programming with layered learning and incremental sampling, in *2011 IEEE Congress of Evolutionary Computation (CEC)* (IEEE, Piscataway, 2011), pp. 1179–1185
33. M. Keijzer, Improving symbolic regression with interval arithmetic and linear scaling, in *Proceedings of the European Conference on Genetic Programming (EuroGP)* (Springer, Berlin, 2003), pp. 70–82
34. V. Koltchinskii, Rademacher penalties and structural risk minimization. IEEE Trans. Inf. Theory **47**(5), 1902–1914 (2001)
35. M. Kommenda, M. Affenzeller, B. Burlacu, G. Kronberger, S.M. Winkler, Genetic programming with data migration for symbolic regression, in *Proceedings of the 16th Annual Conference on Genetic and Evolutionary Computation (GECCO)* (2014), pp. 1361–1366
36. T. Kowaliw, R. Doursat, Bias-variance decomposition in genetic programming. Open Math. **14**(1), 62–80 (2016)
37. J.R. Koza, *Genetic Programming II, Automatic Discovery of Reusable Subprograms* (MIT Press, Cambridge, 1992)
38. J. Kubalík, E. Derner, R. Babuška, Symbolic regression driven by training data and prior knowledge, in *Proceedings of the 24th Genetic and Evolutionary Computation Conference (GECCO)* (2020), pp. 958–966
39. I. Kuscu, Generalisation and domain specific functions in genetic programming, in *Proceedings of the 2000 Congress on Evolutionary Computation (CEC)*, vol. 2 (IEEE, Piscataway, 2000), pp. 1393–1400
40. N. Le, H.N. Xuan, A. Brabazon, T.P. Thi, Complexity measures in genetic programming learning: a brief review, in *Proceedings of the 2016 IEEE Congress on Evolutionary Computation (CEC)* (IEEE, Piscataway, 2016), pp. 2409–2416
41. S. Luke, L. Panait, Fighting bloat with nonparametric parsimony pressure, in *International Conference on Parallel Problem Solving from Nature (PPSN)* (Springer, Berlin, 2002), pp. 411–421
42. S. Luke, L. Panait, Lexicographic parsimony pressure, in *Proceedings of the 4th Annual Conference on Genetic and Evolutionary Computation (GECCO)* (Morgan Kaufmann, Burlington, 2002), pp. 829–836
43. S. Luke, L. Panait, A comparison of bloat control methods for genetic programming. Evol. Comput. **14**(3), 309–344 (2006)
44. Y. Martínez, E. Naredo, L. Trujillo, P. Legrand, U. López, A comparison of fitness-case sampling methods for genetic programming. J. Exp. Theor. Artif. Intell. **29**(6), 1203–1224 (2017)
45. J.F. Miller, P. Thomson, Cartesian genetic programming, in *Genetic Programming* (Springer, Berlin, 2000), pp. 121–132
46. T.M. Mitchell, *Machine Learning* (McGraw Hill, Burr Ridge, IL, 1997), p. 45

47. Q.U. Nguyen, X.H. Nguyen, M. O'Neill, Semantic aware crossover for genetic programming: the case for real-valued function regression, in *Genetic Programming* (Springer, Berlin, 2009), pp. 292–302
48. Q.U. Nguyen, X.H. Nguyen, M. O'Neill, Examining the landscape of semantic similarity based mutation, in *Proceedings of the 13th Annual Conference on Genetic and Evolutionary Computation (GECCO)* (ACM, New York, 2011), pp. 1363–1370
49. J. Ni, R.H. Drieberg, P.I. Rockett, The use of an analytic quotient operator in genetic programming. IEEE Trans. Evol. Comput. **17**(1), 146–152 (2012)
50. M. O'Neill, L. Vanneschi, S. Gustafson, W. Banzhaf, Open issues in genetic programming. Genet. Program Evolvable Mach. **11**(3–4), 339–363 (2010)
51. L. Panait, S. Luke, Methods for evolving robust programs, in *Proceedings of the 5th Annual Conference on Genetic and Evolutionary Computation (GECCO)* (Springer, Berlin, 2003), pp. 1740–1751
52. G. Paris, D. Robilliard, C. Fonlupt, Exploring overfitting in genetic programming, in *International Conference on Artificial Evolution (Evolution Artificielle)* (Springer, Berlin, 2003), pp. 267–277
53. R. Poli, W.B. Langdon, N.F. McPhee, J.R. Koza, *A Field Guide to Genetic Programming* (2008). http://Lulu.com
54. C. Raymond, Q. Chen, B. Xue, M. Zhang, Genetic programming with rademacher complexity for symbolic regression, in *2019 IEEE Congress on Evolutionary Computation (CEC)* (IEEE, Piscataway, 2019), pp. 2657–2664
55. C. Raymond, Q. Chen, B. Xue, M. Zhang, Adaptive weighted splines: a new representation to genetic programming for symbolic regression, in *Proceedings of the 24th Genetic and Evolutionary Computation Conference (GECCO)* (2020), pp. 1003–1011
56. D. Rivero, E. Fernandez-Blanco, C. Fernandez-Lozano, A. Pazos, Population subset selection for the use of a validation dataset for overfitting control in genetic programming. J. Exp. Theor. Artif. Intell. **32**(2), 243–271 (2020)
57. S.H. Rudy, S.L. Brunton, J.L. Proctor, J.N. Kutz, Data-driven discovery of partial differential equations. Sci. Adv. **3**(4), e1602614 (2017)
58. M. Schmidt, H. Lipson, Distilling free-form natural laws from experimental data. Science **324**(5923), 81–85 (2009)
59. S. Silva, S. Dignum, L. Vanneschi, Operator equalisation for bloat free genetic programming and a survey of bloat control methods. Genet. Program Evolvable Mach. **13**(2), 197–238 (2012)
60. S. Sun, R. Ouyang, B. Zhang, T.-Y. Zhang, Data-driven discovery of formulas by symbolic regression. MRS Bull. **44**(7), 559–564 (2019)
61. C. Tuite, A. Agapitos, M. O'Neill, A. Brabazon, Tackling overfitting in evolutionary-driven financial model induction, in *Natural Computing in Computational Finance* (Springer, Berlin, 2011), pp. 141–161
62. N.Q. Uy, N.X. Hoai, M. O'Neill, Semantics based mutation in genetic programming: the case for real-valued symbolic regression, in *15th International Conference on Soft Computing, Mendel*, vol. 9 (2009), pp. 73–91
63. N.Q. Uy, N.X. Hoai, M. O'Neill, R.I. McKay, E. Galván-López, Semantically-based crossover in genetic programming: application to real-valued symbolic regression. Genet. Program Evolvable Mach. **12**(2), 91–119 (2011)
64. L. Vanneschi, S. Gustafson, Using crossover based similarity measure to improve genetic programming generalization ability, in *Proceedings of the 11th Annual Conference on Genetic and Evolutionary Computation (GECCO)* (2009), pp. 1139–1146
65. L. Vanneschi, M. Castelli, S. Silva, Measuring bloat, overfitting and functional complexity in genetic programming, in *Proceedings of the 12th Annual Conference on Genetic and Evolutionary Computation (GECCO)* (2010), pp. 877–884
66. V. Vapnik, *Estimation of Dependences Based on Empirical Data* (Springer Science & Business Media, Berlin, 2006)

67. E.J. Vladislavleva, G.F. Smits, D. Den Hertog, Order of nonlinearity as a complexity measure for models generated by symbolic regression via pareto genetic programming. IEEE Trans. Evol. Comput. **13**(2), 333–349 (2008)
68. E. Vladislavleva, T. Friedrich, F. Neumann, M. Wagner, Predicting the energy output of wind farms based on weather data: Important variables and their correlation. Renew. Energy **50**, 236–243 (2013)
69. M. Willis, H. Hiden, M. Hinchliffe, B. McKay, G.W. Barton, Systems modelling using genetic programming. Comput. Chem. Eng. **21**, S1161–S1166 (1997)
70. C. Xu, W. Wang, P. Liu, A genetic programming model for real-time crash prediction on freeways. IEEE Trans. Intell. Transp. Syst. **14**(2), 574–586 (2012)
71. B. Xue, M. Zhang, W.N. Browne, X. Yao, A survey on evolutionary computation approaches to feature selection. IEEE Trans. Evol. Comput. **20**(4), 606–626 (2016)

Qi Chen is a Lecturer at Victoria University of Wellington (VUW), New Zealand. She received BE degree in Automation from the University of South China, Hunan, China in 2005 and ME degree in Software Engineering from Beijing Institute of Technology, Beijing, China in 2007, and PhD degree in computer science in 2018 at VUW, New Zealand. Since 2014, she has joined the Evolutionary Computation Research Group at VUW. She worked as a postdoctoral fellow in School of Engineering and Computer Science at VUW from 2018 to 2020. Currently, she is a Lecturer there. Qi's current research mainly focuses on genetic programming for symbolic regression. Her research interests include machine learning, evolutionary computation, feature selection, feature construction, transfer learning, domain adaptation and statistical learning theory. She serves as a reviewer of international conferences, including IEEE Congress on Evolutionary Computation, and international journals, including IEEE Transactions on Evolutionary Computation and IEEE Transactions on Cybernetics.

Message:

It is my honour to participate in this book as a way to encourage women to develop our scientific and engineering abilities and to achieve our full potential. As a women researcher in Computer Science, I can clearly see that women are under-represented in leadership roles in the field. It is vital for the World's future that more women participate fully to address the gender imbalance in Engineering and Computer Science.

Bing Xue (M'10) received BSc degree from the Henan University of Economics and Law, Zhengzhou, China, in 2007, MSc degree in management from Shenzhen University, Shenzhen, China, in 2010, and PhD degree in computer science in 2014 at Victoria University of Wellington (VUW), New Zealand. She is currently a Professor and Program Director of Science in School of Engineering and Computer Science at VUW. She has over 200 papers published in fully refereed international journals and conferences and her research focuses mainly on evolutionary computation, machine learning, classification, symbolic regression, feature selection, evolving deep neural networks, image analysis, transfer learning, multi-objective machine learning. Dr. Xue is currently the Editor of IEEE Computational Intelligence Society (CIS) Newsletter, Chair of IEEE CIS Task Force on Transfer Learning & Transfer Optimization, Vice-Chair of IEEE CIS Evolutionary Computation Technical Committee, Vice-Chair of IEEE Task Force on Evolutionary Feature Selection and Construction, and Vice-Chair of IEEE CIS Task Force on Evolutionary Deep Learning and Applications. She is also served as an Associate Editor of several international journals, such as IEEE Computational Intelligence Magazine, IEEE Transactions of Evolutionary Computation. She is a member of several committees, including Women in Computational Intelligence. She is also a member of IEEE New Zealand Central Section and Secretary of the CI chapter.
Message:
It is my great honour to be part of this great book written mainly by female researchers in computational intelligence. I first started working on Computational Intelligence, particularly evolutionary computation during my Master study in management science and engineering. I was so impressed by how those approaches can solve various complex optimisation problems so that I decided to do further research and started a career in this area. The more research I do in computational intelligence, the more interesting ideas I found. However, I found female researchers are still the minority, which is true for all STEM areas. The difficulties in pursuing a career in these areas might have been overestimated by many females. With my experience in working this area, I can confidently say the opportunities and difficulties are equal for both males and females. The support for women researchers in the computational intelligence community is much more than one normally expected, and female researchers also support each other, where this book is a good example. If you are interested in choosing a career like us in this area, you should go ahead, work hard and be kind, you will be successful.

Neuroevolutionary Models Based on Quantum-Inspired Evolutionary Algorithms

Tatiana Escovedo, Karla Figueiredo, Daniela Szwarcman, and Marley Vellasco

1 Introduction

Neuroevolution is a machine learning technique that uses evolutionary algorithms (EA) to adjust parameters that affect the performance of artificial neural networks, such as topology, learning rate and weights, among others. Each solution of the evolutionary algorithm represents one configuration of these parameters, which are evolved to find the optimal network for the problem.

There are several works related to the use of traditional evolutionary algorithms for parameterisation and configuration of neural networks [1, 9, 12, 19, 28, 43, 48, 51–55, 62, 74, 76]. Paz and Kamath's work in [51] presents the main objectives for using neuroevolution: to specify the weights, training resources and optimal neural network topology.

Most evolutionary systems applied to neural networks (NN) used to include the determination of the networks' optimal synaptic weights, to prevent the local minimum (or maximum) problems when using traditional gradient algorithms [72]. The optimal number of neurons is another critical parameter to ensure proper learning and generalisation performance, which was also among the many applications of evolutionary algorithms for neural networks [51].

T. Escovedo · M. Vellasco (✉)
Pontifical Catholic University of Rio de Janeiro (PUC-Rio), Rio de Janeiro, Brazil
e-mail: tatiana@inf.puc-rio.br; marley@ele.puc-rio.br

K. Figueiredo
State University of Rio de Janeiro (UERJ), Rio de Janeiro, Brazil
e-mail: karlafigueiredo@ime.uerj.br

D. Szwarcman
IBM Research, Rio de Janeiro, Brazil
e-mail: daniela.szw@ibm.com

A. E. Smith (ed.), *Women in Computational Intelligence*, Women in Engineering and Science, https://doi.org/10.1007/978-3-030-79092-9_14

Yao in [74] has also indicated that the volume of data and processing time problems for neural network training can be minimised by data preprocessing to reduce dimensionality with some statistical technique. However, EA can also be considered to evaluate which features are most relevant to the solution of the problem by maximising or minimising some fitness function. This point is highlighted because many authors focus on optimising the topology of a network concerning the number of hidden layers and processors in these layers, but few consider optimising the input layer to select the most relevant variables for the problem [1, 43, 57].

Another strategy to increase generalisation of neural network models is to use an ensemble of classifiers [43, 54, 55, 74]. Many authors propose to use not only the best solution but *n* best individuals from a population of EA, forming an ensemble of classifiers that improves generalisation.

Other models have also considered the evolution of the neural network performance metric, coding, in the chromosome, the type of function used by the neural network model (e.g. deciding between the "sum of squared errors" and the "sum of absolute errors") [52, 62]. Yet another strategy is the definition of the most suitable activation functions for the neurons [1, 41, 70, 77].

More recently, with the significant progress in the performance of deep neural networks for many tasks, the idea of automating the network design has gained the attention of many researchers [37, 79], establishing the field of Neural Architecture Search (NAS) [67]. In this scenario, evolutionary algorithms are one of the search techniques to address the NAS problem [7, 49, 59, 66, 73].

Despite the numerous works involving EA for neural networks, the required computational resources are still a significant issue for many of them, a problem that has worsened in the NAS domain.

Therefore, to apply neuroevolution in real scenarios, models must have superior computational performance and fast convergence. This feature becomes even more relevant in complex environments, such as nonstationary scenarios, since it is necessary to update the model each time new data become available or when some change is detected in data, as well as neural architecture search. The evolutionary process must be fast so as not to compromise the overall performance of the neural network model.

In this scenario, quantum-inspired evolutionary algorithms seem to be an excellent alternative to automatically configure neural network models. Quantum-inspired evolutionary algorithms (QIEA) are a class of evolutionary computation models developed to achieve better performance in computationally intensive problems. This class of algorithms was inspired by quantum computing principles [3, 30, 31] and was designed to obtain better solutions with fewer evaluations when compared to similar algorithms [3, 4].

QIEA have been applied to solve combinatorial, numerical and ordering optimisation problems, based on binary [30, 31], real [3] and ordering representation [63], respectively, providing better results than classical genetic algorithms with less computational effort.

Quantum-inspired evolutionary algorithms with real (QIEA-R) as well as with hybrid (binary and real) representation (QIEA-BR) have been successfully applied to neuroevolution of different neural network models, such as recurrent neural networks [72], Multi-Layer Perceptrons [57], Echo State Neural Networks [18] and even neural network ensembles [22, 23] in a concept drift environment.

More recently, QIEA have been successfully applied to neural architecture search, proving to be a robust and successful approach for automatically developing new topologies for deep neural networks without relying on a priori expert knowledge to reduce the search space [67].

The main objective of this chapter is to provide an overview of this promising area, presenting how QIEA have been applied to the automatic construction of different neural network models, for different applications.

This chapter is divided into four additional sections. Section 2 provides a brief introduction to quantum-inspired evolutionary algorithms, with different chromosome representations. Section 3 describes the neuroevolution of different neural network models, such as Multi-Layer Perceptrons, Echo-State Neural Networks and Convolutional Neural Networks. Section 4 presents some case studies and, finally, Section 5 discusses the conclusion and future work.

2 Quantum-Inspired Evolutionary Algorithms

Some of the advantages of the use of classical evolutionary algorithms are that this class of algorithms does not require rigorous mathematical formulations about the problem to be optimised, besides offering a high degree of parallelism in the search process. This class of evolutionary algorithms has been used successfully to solve complex optimisation problems in a wide range of fields, such as automatic circuit design and equipment, task planning, data mining and software engineering, among many others [30, 31].

However, some problems can have a high computational cost associated with the evaluation of the fitness function, having a significant impact on the evolutionary process. This fact can make optimisation by evolutionary algorithms a slow process for situations where a fast response is desired, as in online optimisation problems. To address this issue, quantum-inspired evolutionary algorithms (QIEA) have been developed. QIEA are a class of estimation distribution algorithms that perform better in combinatorial and numerical optimisation when compared to their homologous canonical genetic algorithms [3, 30, 31]. This class of algorithms is inspired by concepts of quantum physics, in particular in the superposition of states.

Quantum-inspired evolutionary algorithms were initially developed for optimisation problems using binary representation, such as the Quantum-Inspired Evolutionary Algorithm (QIEA-B) [30–33], which uses a chromosome formed by q-bits. A q-bit is a unit vector in a two-dimensional complex vector space for which the orthonormal basis (in Dirac notation) $|0\rangle$, $|1\rangle$ has been fixed. The basis states represent the classical bit values 0 and 1, respectively. A q-bit, in contrast to the classical bit, can be in a superposition of states, that is, the q-bit state ı can be represented as a linear combination of $|0\rangle$ and $|1\rangle$ [31, 58]:

$$|\psi\rangle = \alpha\,|0\rangle + \beta\,|1\rangle \tag{1}$$

Therefore, each q-bit consists of a pair of numbers (α, β), where $|\alpha^2| + |\beta^2| = 1$. The value $|\alpha^2|$ indicates the probability that the q-bit has value 0 when observed, while the value $|\beta^2|$ indicates the probability that the q-bit has value 1 when observed. In QIEA-B, a quantum individual q_i is formed by M q-bits, according to (2):

$$\begin{vmatrix}\alpha_{i1}\\ \beta_{i1}\end{vmatrix}\begin{vmatrix}\alpha_{i2}\\ \beta_{i2}\end{vmatrix}\cdots\begin{vmatrix}\alpha_{iM}\\ \beta_{iM}\end{vmatrix} \tag{2}$$

where α_{ij} and β_{ij} represent the j-th q-bit of chromosome i ($i = \{1, 2, \ldots, N\}$ and $j = \{1, 2, 3, \ldots, M\}$.

This algorithm was then extended to real representation, to better deal with numerical optimisation problems. In these problems, the direct numeric representation is more appropriate, in which real numbers are directly encoded in a chromosome rather than converting binary strings into numbers. With real representation, the memory demand is reduced while the precision is increased [2]. With this motivation, the Quantum-Inspired Evolutionary Algorithm with Real Representation (QIEA-R) was developed [3] (see Sect. 2.1), inspired by the concept of multiple universes of quantum physics. In this scenario, the algorithm allows performing the numeric optimisation process with a smaller number of evaluations, substantially reducing the computational cost.

More recently [68], to cope with the evolution of deep neural networks, a QIEA with categorical representation has been developed. In this case, each gene specifies the probability of a particular category (or function), among a predefined list of available discrete classes.

In the next sections, we briefly describe the main quantum-inspired evolutionary models that were developed for the evolution of different neural network models.

2.1 Quantum-Inspired Evolutionary Algorithm with Real Representation (QIEA-R)

Originally proposed in [Cruz, 2007], this algorithm was used to solve numerical optimisation benchmark problems and the neuroevolution of recurrent neural networks. The results obtained demonstrated the efficiency of this algorithm in the solution to these types of applications.

In QIEA-R, the quantum population $Q(t)$ consists of N quantum individuals q_i ($i = 1, 2, 3, \ldots, N$) which are composed of G quantum genes. Each quantum gene is formed by a probability density function (PDF), which represents the superposition of states and is used to observe the classical gene. Quantum individuals can be represented by:

$$q_i = [g_{i1} = p_{i1}(x), g_{i2} = p_{i2}(x), \ldots, g_{iG} = p_{iG}(x)] \tag{3}$$

where $i = 1, 2, 3, \ldots, N$, $j = 1, 2, 3, \ldots, G$ and p_{ij} functions represent the probability density functions used by the QIEA-R to generate the values for the genes of the classical individuals. In other words, the $p_{ij}(x)$ function represents the probability density of observing a given value for the quantum gene when its overlap is collapsed. The probability density function used by [2] is the square pulse, a uniform function, which can be defined by Eq. 4:

$$p_{ij}(x) = \begin{cases} U_{ij} - L_{ij}, L_{ij} \leq x \leq U_{ij} \\ 0, \text{ otherwise} \end{cases} \tag{4}$$

where L_{ij} is the lower limit and U_{ij} is the upper limit of the interval in which the gene j of the i-th quantum individual can collapse, i.e., assume values when observed.

For the case where $p_{ij}(x)$ is a square pulse, the quantum gene can be represented by storing the position of the centre point of the pulse and its width: μ_{ij} and σ_{ij}, respectively. The QIEA-R also uses a population of quantum individuals, which are observed to generate classical individuals. The updating of the quantum individuals is carried out based on the evaluation of classic individuals: μ_{ij} and σ_{ij} are altered so to bring the pulse to the most promising region of the search space, increasing the probability of observing a particular set of values for the classical gene in the vicinity of the most successful individuals in the classical population.

The pseudocode of the QIEA-R algorithm is shown in Fig. 1.

In Sect. 4.1, we present an application of QIEA-R to evolve voting weights for each classifier member of an ensemble and then determine its final decision. In this application, the chromosome comprises n genes, where n represents the number of ensemble members. Each gene, in turn, specifies the voting weight associated with each classifier. Further details on QIEA-R can be found in [3, 4].

2.2 *Quantum-Inspired Evolutionary Algorithm with Binary-Real Representation (QIEA-BR)*

Many real problems cannot be solved only by numerical decisions or combinatorial decisions. To deal with this kind of problems, algorithms with mixed representation are necessary. More specifically in the field of neural networks, the modelling process may involve combinatorial decisions – like the selection of the most relevant input variables and the number of neurons to be used in the topology – and, simultaneously, numerical decisions, like optimal values for synaptic weights.

```
1    t ← 1
2    create quantum population Q(t) with N individuals with G genes
     each
3    while t ≤ T
4       E(t)← generate classical individuals by observing quantum in-
        dividuals
5            if t=1 then
6                   C(t) ← E(t)
7            else
8                   E(t)← recombination between E(t) and C(t)
9                   Evaluate E(t)
10                  C(t) ← K best individuals from E(t) ∪ C(t)
11           end if
12      Q(t+1) ← updates Q(t) using the N best individuals from C(t)
13           t←t+1
14   end while
```

Fig. 1 The QIEA-R algorithm

To address this kind of problem, [56, 57] proposed an algorithm with quantum inspiration and binary-real representation, called QIEA-BR. This algorithm has a mixed nature because it can be used for simultaneous optimisation of combinatorial and numerical problems. The QIEA-BR algorithm was the first evolutionary algorithm with quantum inspiration and hybrid representation proposed in the literature, which inherits the main characteristics of its precursors, such as global problem-solving ability and probabilistic representation of the search space. This mixed representation results in high population diversity in each quantum individual and the need for fewer individuals in the population to explore the search space.

The QIEA-BR algorithm requires a population of quantum individuals to represent the overlap of possible states that the classical individuals can assume when observed. The quantum population $Q(t)$, at any instant t of the evolutionary process, is formed by a set of N quantum individuals q_i ($i = 1, 2, 3, \ldots, N$). Each quantum individual q_i of this population is formed by L genes g_{ij} ($j = 1, 2, 3, \ldots, L$). The main difference between the QIEA-BR and its predecessors is that part of the L genes is represented by q-bit, like QIEA-B, and another part by real quantum genes (q-real, like QIEA-R). Thus, the representation of a quantum individual i at any time instant t is given by:

$$q_i = \left[(q_i)_b (q_i)_r\right] = \left(\begin{vmatrix} \alpha_{i1} \\ \beta_{i1} \end{vmatrix} \begin{vmatrix} \alpha_{i2} \\ \beta_{i2} \end{vmatrix} \cdots \begin{vmatrix} \alpha_{iM} \\ \beta_{iM} \end{vmatrix} \right)_b \left(\begin{vmatrix} \mu_{i1} \\ \sigma_{i1} \end{vmatrix} \begin{vmatrix} \mu_{i2} \\ \sigma_{i2} \end{vmatrix} \cdots \begin{vmatrix} \mu_{iG} \\ \sigma_{iG} \end{vmatrix} \right)_r \tag{5}$$

```
1   t ← 0
2   create quantum population Q(t) with mixed representation
3   while t ≤ T
4           t ← t+1
5     Generate classical population P(t) with mixed representation
      observing Q(t)
6           Evaluate P(t)
7           if t=1 then
8                 B(t) ← P(t)
9           Else
10      P(t)← classic recombination between P(t) and B(t − 1)
11                Evaluate P(t)
12                B(t) ← best individuals from P(t) ∪ B(t − 1)
13          Updates the binary part of Q(t) using the best individu-
       als of B(t)and a quantum gate
14          Updates the real part of Q(t) using the Best individuals
       of B(t)using quantum crossover
15          End
16  end while
```

Fig. 2 The QIEA-BR algorithm

where index b represents the binary part (q-bit); index r represents the real part (q-real); and α_{ij}, β_{ij}, μ_{ij} and σ_{ij} are the same as before. The complete algorithm is presented in Fig. 2. Further details on QIEA-BR can be found in [56, 57].

2.3 Quantum-Inspired Evolutionary Algorithm with Categorical Representation

Szwarcman et al. [68] extended even further and proposed a new quantum representation to encode categorical spaces with more than two options. In their work, each quantum gene defines a PMF (probability mass function) for the possible choices of a categorical variable. The M options are mapped to integers, so the classical individual $\boldsymbol{p}_i$ with L genes is an array in the form $[g_{i1}, \ldots, g_{iL}]$, with $g_{ij} \in [0, M-1]$. Accordingly, the quantum gene $\boldsymbol{g}_j$ is an array of probabilities [67]:

$$\boldsymbol{g}_j = \left[x_{j1}, \ldots, x_{jM}\right]; \quad x_{jk} \in [0.0, 1.0]; \sum_{k=1}^{M} x_{jk} = 1.0 \tag{6}$$

where x_{jk} is the probability of option k for the variable j.

To generate a classical individual, the proposed algorithm samples from each PMF independently. The decoding process simply maps back the selected integers to the original options [67, 68].

The authors also developed a quantum update heuristic that increases the probability of promising solutions. In other words, the best classical individuals define which function in a node should have its probability increased [67, 68].

In Sect. 3.4, we describe Q-NAS, the QIEA that uses this categorical representation to search for the architecture of deep networks. The reader can refer to [67, 68] for further details on QNAS and the categorical representation.

3 Neuroevolutionary Models Based on QIEA

3.1 Multi-layer Perceptrons

The first example of neuroevolution based on quantum-inspired evolutionary algorithms is the use of QIEA-BR in configuring a Multi-Layer Perceptron for classification problems. The binary part is responsible for three main configurations: selecting the most appropriate input variables; defining which neurons (of a maximum number of neurons) are active in the single hidden layer (1 active neuron, 0 inactive); and specifying the activation function of each neuron in the network (1 for hyperbolic tangent and 0 sigmoid function). The real part, on the other hand, is responsible for determining the values of all synaptic weights. Figure 3 illustrates the information that is encoded in each of the quantum genes, binary or real, of a QIEA-BR chromosome.

In the use of QIEA-BR for modelling MLPs, the evolution of the weights and activation function of an individual neuron in the quantum and classical chromosomes is conditioned to that neuron being active in the corresponding binary

Binary genes

Variables to be selected for the input layer	Neurons to be selected for the hidden layer	Activation functions for the hidden layer	Activation functions for the output layer

Numerical genes

Synaptic weights of the hidden layer, including bias	Synaptic weights of the output layer, including bias

Fig. 3 The QIEA-BR individual structure [56]

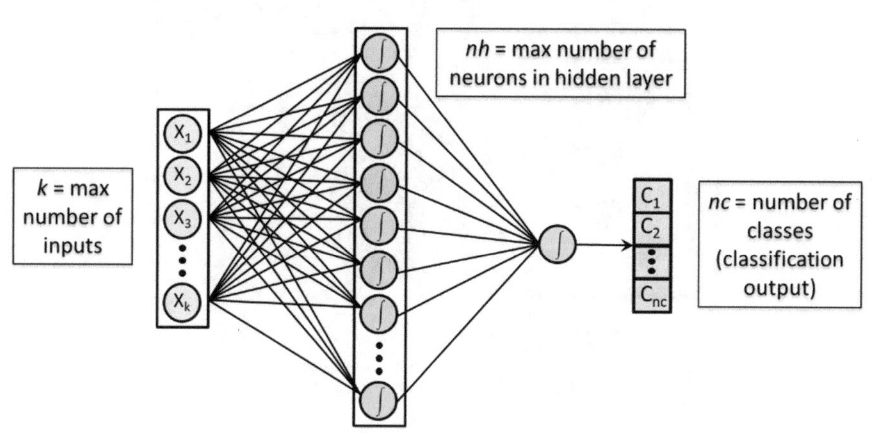

Fig. 4 Neural network created by QIEA-BR

part. In other words, the genes representing the weights and activation functions of inactive neurons remain unchanged by quantum and classical evolutionary process.

A general configuration of a neural network created by QIEA-BR is shown in Fig. 4. The effective number of attributes in the input layer and neurons in the hidden layer are evolved by the QIEA-BR, with the maximum size of inputs equal to the number of available attributes in the dataset (k) and the maximum number of neurons in the hidden layer (nh) configured by the user.

Thus, the total number of genes is given by:

$$\text{num}_{\text{genes}} = (k + 2nh + nc)_b + ((k + 1) \times nh) + ((nh + 1) \times nc)_r \tag{7}$$

where nc is the number of classes in the classification problem. In this case, the evaluation function used is the classification accuracy provided by:

$$\text{Accuracy} = 1 - \frac{1}{n} \sum \left| C_i - \hat{C}_i \right| \tag{8}$$

where C_i is the class of the i-th pattern, while $\hat{C}_i$ is the class predicted by the individual (MLP). When $C_i = \hat{C}_i$ then the result is zero. Otherwise, it is equal to one. Each individual is submitted to this evaluation function, in such a way that the best individuals are those who have greater accuracy.

This specific neuroevolutionary model has been applied to many classification benchmark applications [72], as well as in concept drift environments. The latter is presented in Sect. 4.1.

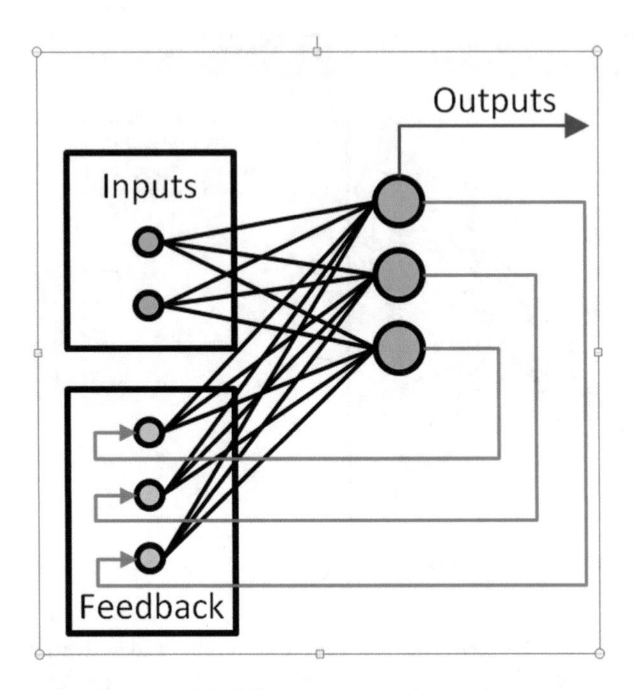

Fig. 5 The general architecture of the fully recurrent neural network

3.2 Fully Recurrent Neural Networks

In addition to models based on MLP architectures, the real-representation quantum-inspired evolutionary algorithm has also been applied to recurrent neural networks with arbitrary topologies [72]. The fundamental issue in the EA process is to determine the best topology and to optimise the weights of a fully recurring neural network (FRNN) with a delay on the processors. This type of structure is suitable for time-series forecasting and control problems, as recurring connections produce a short memory effect.

The basic topology for this network is defined as having one input layer with as many inputs as required by the problem, one hidden layer with an initial number of neurons that can be changed during the evolutionary process, an output layer and recurrent connections. Figure 5 illustrates the general structure of the fully recurrent neural network [2].

Chromosomes, classical or quantum, have genes corresponding to each weight identified in the structure of the recurrent neural network. Thus, the search process performed by the QIEA-R replaces the traditional training based on gradient algorithms. The number of genes required to represent the FRNN is given by the following equation:

$$\text{num}_{\text{genes}} = N_i \times N_p + N_p^2 + N_p \tag{9}$$

where N_i is the number of inputs and N_p is the number of processors. The first product on this equation represents the number of weights required for the connections between the inputs and the hidden layer processors. The squared term represents the number of weights associated with the recurrent connections, and the last term represents the number of biases required (one for each processor).

A new operator (*lesion*) is included in the EA, which is responsible for adding or removing processors from the hidden layer. The lesion operator tries to remove, on every n generations, each neuron one by one, checking the impact on the error performance. If the error is below a certain threshold, the removed neuron is considered unnecessary, and all corresponding weights in quantum and classical individuals are also deleted. If, however, all removed neurons increase the error above the threshold, a new processor is added, with all weights zeroed. By updating the weights during the evolutionary process, the new neuron may have the chance to improve the results, reducing the error obtained by the neural network.

This model has been applied to benchmark control problems [72], as well as multi-agent applications, in a neuro coevolution model [20].

3.3 Echo States Networks

Echo States Networks (ESNs) have been proposed as a model of recurrent neural networks [39]. This type of recurring network is based on the Reservoir Computing framework. As a differential, this network has, as a hidden layer, a large set of randomly interconnected neurons – sparsely connected – called reservoir, which endows the model with dynamic behaviour. Typically, the ESN has nonlinear recurrent connections between output and reservoir neurons and linear connections between the reservoir and output layer. The ESNs exhibit a property called echo state [39] – a dynamic memory – and reservoir states act as a function of input/output history [64].

All weights in the reservoir remain unchanged, whereas only the weights between the reservoir and the output nodes are adjusted through linear regression, which outperforms problems such as slow convergence, local minimum and high computational cost [39].

The dynamic memory of ESN can represent dynamic systems with a fast and straightforward training procedure. These advantages make ESNs an attractive tool for system identification problems. However, these networks have many parameters that need to be correctly tuned so to extract the best performance from the networks: the size and connectivity of the reservoir; the spectral radius of W (reservoir weight matrix); the scale factors of the input and feedback matrices; and the leaky integrator constant. Additionally, although a random reservoir simplifies the training process, this strategy might not be ideal in terms of performance.

There are many works based on evolutionary algorithms (EAs) and other bioinspired optimisation techniques that can handle both issues, acting as ESN

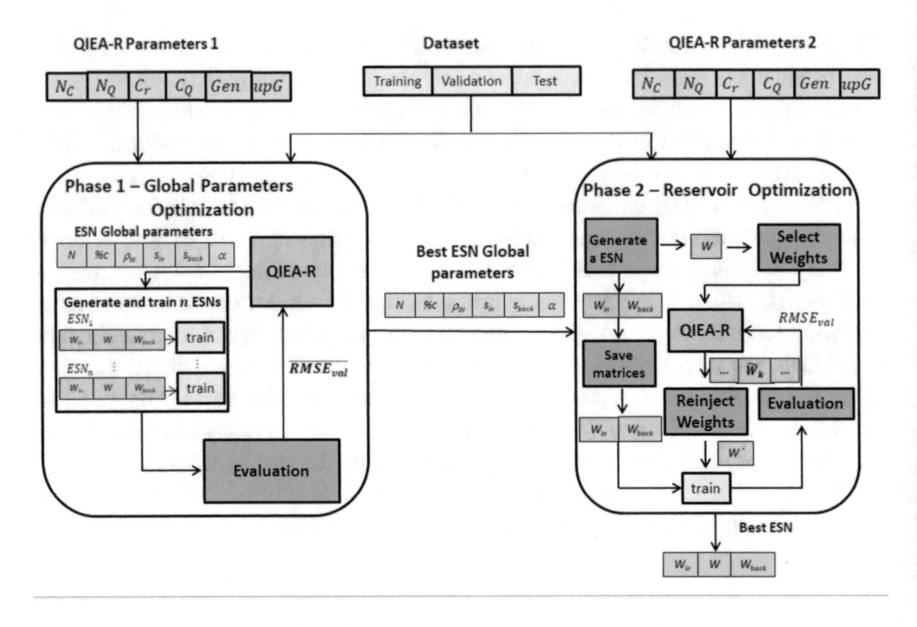

Fig. 6 The proposed method for ESN evolution

N	%c	ρ_w	s_{in}	s_{back}	α

Fig. 7 QIEA-R chromosome (phase 1)

global parameter optimisers [24, 50] or optimising its reservoir weights [8, 15, 25, 38]. However, the high computational cost involved is still an important issue.

Therefore, a quantum-inspired model has been developed to configure Echo State Networks to attain the best performance. Fig. 6 presents the general structure of the proposed method, whose general concept is based on previous evolutionary approaches to ESN optimisation [8, 15, 50]. As can be observed in Fig. 6, the method has two phases: the first phase determines the global parameters and the second one focuses on selecting the most appropriate reservoir weights for the parameters defined in Phase 1.

3.3.1 Phase 1: Global Parameter Optimisation

This phase concentrates in finding the best ESN global parameters by using the QIEA-R. The global parameters which are optimised are reservoir size (N), percentage of connections in the reservoir (%c), spectral radius of $W(\rho_w)$, the scale factor of $W^{\text{in}}(s_{in})$, the scale factor of $W^{\text{back}}(s_{back})$ and the leakage rate (α). More details about these parameters are provided in [18]. The representation of the search model to identify these parameters is the chromosome presented in Fig. 7.

To deal with the stochastic behaviour of the neural network and get robust parameters, for each chromosome from the QIEA-R, n ESNs with this same configuration are generated and trained. After that, each of these ESNs simulates the validation set and the evaluation function (f_{eval}) is returned as the mean of the simulation errors obtained by these n networks. The error metric used as the evaluation function is the RMSE (root mean squared error):

$$f_{\text{eval}} = \overline{\text{RMSE}_{\text{val}}} = \sum_{j=1}^{N} \text{RMSE}_{\text{val}}(j) \Big/ n \tag{10}$$

$$\text{RMSE}_{\text{val}}(j) = {}^{1}\Big/_{N_y} \sum_{i=1}^{N_y} \sqrt{\sum_{k=1}^{S_v} \left(\left(\hat{y}_i(k) - y_i(k) \right)^2 \Big/ S_v \right.} \tag{11}$$

where N_y is the number of output variables; $\hat{y}_i$ and y_i are, respectively, the simulated value from network j and the actual value of output i at time step k; and S_v is the validation set size.

After finishing this process, the best global parameters that will be used in phase 2 are presented by the QIEA-R.

3.3.2 Phase 2: Reservoir Optimisation

In phase 2, a new QIEA-R model is used to estimate the best reservoir (weights) for the configuration of global parameters previously defined in phase 1. Only the weights of W with values different from zero are optimised to maintain the percentage of connectivity (%c), determined in phase 1, and to reduce the search space. The matrices W^{in} and W^{back} remain fixed in the process and, before an evaluation, matrix W is rescaled to the spectral radius ρ_w optimised in phase 1. Figure 8 presents the complete ESN evolutionary process.

This model has been applied to many system identification benchmark problems [17]. Two of them are presented in Sect. 4.2.

3.4 Convolutional Neural Networks

Deep convolutional networks have recently become a dominant theme in machine learning research. They are behind the striking advance in the performance of machine learning models in tasks such as image recognition [37, 79]. One noticeable advantage is that convolutional networks provide a way to learn feature extractors, which eliminates the necessity for feature engineering. However, the success of such networks has eventually created a demand for architecture engineering, shifting the

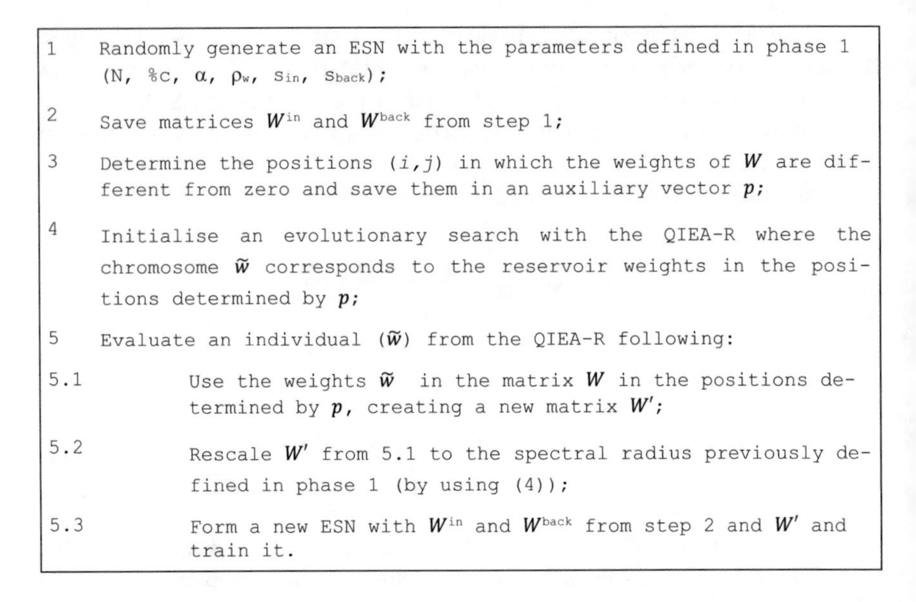

1 Randomly generate an ESN with the parameters defined in phase 1 (N, %C, α, ρ_w, s_{in}, s_{back});

2 Save matrices W^{in} and W^{back} from step 1;

3 Determine the positions (i,j) in which the weights of W are different from zero and save them in an auxiliary vector p;

4 Initialise an evolutionary search with the QIEA-R where the chromosome $\tilde{w}$ corresponds to the reservoir weights in the positions determined by p;

5 Evaluate an individual ($\tilde{w}$) from the QIEA-R following:

5.1 Use the weights $\tilde{w}$ in the matrix W in the positions determined by p, creating a new matrix W';

5.2 Rescale W' from 5.1 to the spectral radius previously defined in phase 1 (by using (4));

5.3 Form a new ESN with W^{in} and W^{back} from step 2 and W' and train it.

Fig. 8 The ESN evolutionary algorithm

paradigm of feature design to network design. The field of Neural Architecture Search (NAS) emerged with the idea of automating the configuration of deep networks, as it is a process that requires expert knowledge and significant time [37].

Neuroevolution research has presented many successful algorithms to automate the design of smaller networks, adjusting their topology and weights [5, 65, 72]. These approaches, on the other hand, did not directly scale to the context of deep architectures, as they optimise connections at the neuron level. NAS research is not restricted to evolutionary methods and considers the layers of the networks instead of individual neurons [66].

Recently, Szwarcman et al. [68] proposed a quantum-inspired algorithm, named Q-NAS, to search for deep networks to execute a predefined task. The quantum individuals are composed of two parts: the first one encodes some numerical hyperparameters related to training, and the second encodes the structure of the network. The authors highlight that Q-NAS evaluates the classical individuals as one unique solution containing both the network and hyperparameters.

The numerical representation is a modified version of the idea introduced in Sect. 2.1. Q-NAS adopts the uniform distribution as the PDF to represent the numerical space, but it defines the quantum genes with the lower and upper limits of the function. More specifically, if a hyperparameter h_i to be evolved can assume values in the range $[l_i, u_i]$, then the quantum gene is defined as the pair (l_i, u_i). The authors claim that this representation simplified their quantum update procedure that corrects a problem in the idea presented in [72].

The second part of the chromosome is the core of Q-NAS since it encodes the network architecture. Following the NAS general idea of considering network layers

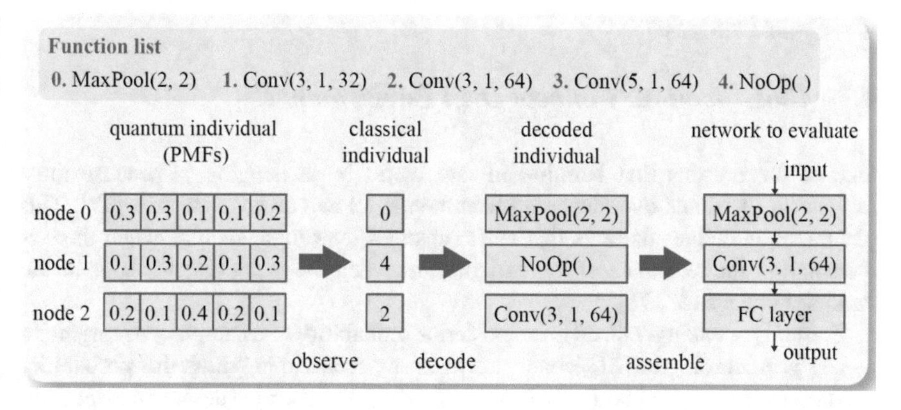

Fig. 9 Generation procedure. The list at the top defines the available functions. Conv(k, s, f) is a convolution layer with kernel size $k \times k$, stride s and f filters. To observe the quantum individual, the PMFs are sampled independently for each node. The decoding procedure maps integers to function names. The final network includes a fully connected classifier layer

instead of neurons, the topologies are represented as a sequence of L nodes, each one with an associated function. The functions can be either a single layer (e.g. a convolutional or pooling layer) or a more complex block of layers. The search space is the same for all nodes, and it is determined by a list of predefined layer functions that the algorithm can select. The authors suggest including a no-operation function in this list, so it is possible to represent variable-length networks.

Observe that the search space for the nodes is categorical, but not binary. As seen in Sect. 2.3, the authors developed a new quantum representation in which a quantum gene defines a PMF for the available options. Therefore, each quantum gene determines the probabilities of each function in a network node (see Fig. 9). The user also specifies the initial PMF, which is the same for all nodes [68].

Q-NAS has execution steps similar to other quantum-inspired evolutionary algorithms. Essentially, the steps in the evolution loop are (1) generate classical individuals, (2) evaluate and rank them, (3) select the best ones to be stored and (4) update the quantum individuals according to the best classical ones. It is essential to mention that the evaluation step involves training the candidate networks for a small number of epochs and assigning the validation score as the fitness of the individual. When the evolution process ends, the final network is retrained from scratch for more epochs using the also evolved hyperparameters. The reader can refer to [67] for further details.

The Q-NAS model has been evaluated in CIFAR-10 and CIFAR-100 image classification benchmarks. A summary of the results obtained is presented in Sect. 4.3.

4 Case Studies

4.1 Classification: Concept Drift Environment

One of the reasons that learning in real world is challenging is because most real-world problems experience a phenomenon known as concept drift [27]. This phenomenon defines datasets that suffer changes over time, such as when there is a change in the relevance of the variables, or when the mean and variance of the variables change [42, 71].

Formally speaking, consider the posterior probability of a sample x belonging to a class y. According to [21], concept drift is any scenario in which this probability changes over time, that is: $P_{t+1}(y|x) \neq P_t(y|x)$). We can also define concept drift, in a supervised learning scenario, when the relationship between the input data and the target variable changes over time [47]. An environment from which this kind of data is obtained is considered a nonstationary environment. When concepts often evolve, the system may be unable to adapt to the new information, hence dramatically deteriorating its performance [45, 78].

Learning from incremental and dynamic data extracted from a nonstationary environment is challenging. However, in the context of neural networks, the problem becomes even more complicated, since most of the existing models must be retrained when a new data block is available, using the whole set of patterns learned until then. Therefore, a classifier in a concept drift environment must, ideally, be able to [60]:

- Track and detect any changes in the underlying data distribution;
- Learn with new data without the need to present the whole dataset again for the classifier;
- Adjust its parameters to incorporate the detected changes on data;
- Forget what has been learned when that knowledge is no longer useful for classifying new instances.

Many approaches have been devised to accomplish some or all of the abilities mentioned above. One of the older and simpler approaches is a sliding window (not always continuous) on the input data used to train the classifier with the data delimited by this window [36]. Another method is to detect deviations and, if they occur, to adjust the classifier [13]. Some models, in turn, use rule-based classifiers, like [6, 10, 11, 60]. A more successful and widely used approach is to use a group of different classifiers (ensemble) to cope with changes in the environment. Several different ensemble models have been proposed in the literature, including recent approaches like [26, 44, 75], and may or may not weigh each of its members. Most models using weighted classifier ensembles determine the weights for each classifier using a set of heuristics related to classifier performance in the most recent data received [42].

4.1.1 Classification in Concept Drift Scenarios Using Quantum-Inspired Neuroevolution

Although several algorithms have already been proposed in the literature for classification in concept drift scenarios – many even using ensembles – neuroevolution has still been little explored. The use of EA in an ensemble environment is also able to dynamically adjust the entire ensemble, a task that would be very arduous if performed manually, due to the complexity involved.

Because of the architecture complexity, the neuroevolutionary models based on classifier ensembles must have good computational performance and fast convergence, to be able to be applied in real scenarios. This feature becomes even more relevant in nonstationary environments, since it is necessary to update the ensemble each time new data become available or when some change is detected in data. This step must be fast so as not to compromise the overall performance of the model, making this scenario suitable for the quantum-inspired evolutionary algorithms, presented in Sect. 2.

Quantum-inspired evolutionary algorithms can be used to model the neural network ensemble and to determine the voting weights for each ensemble member. Thus, each time a new block of data arrives, the ensemble can be optimised, improving its classification performance for the new data.

Therefore, a new quantum-inspired neuroevolution ensemble model, called NEVE: NEuroevolutionary model for learning in Nonstationary Environments, has been proposed in [23]. NEVE is a self-adaptive and flexible model with good accuracy and is suitable for learning in nonstationary environments that also can use an active detection approach (DetectA, detailed in [22]), being an important differential compared to the existing approaches in the literature. NEVE has the following characteristics:

- Can be used with the concept drift detection mechanism DetectA [22], with the ability to detect changes proactively or reactively, allowing the reaction and adjustment of the model whenever necessary;
- Performs the automatic generation of new classifiers for the ensemble, most suitable for the new input data, using the quantum-inspired evolutionary algorithm for numerical and binary optimisation (QIEA-BR) [57];
- Automatically determines the voting weights of each ensemble member, using the quantum-inspired evolutionary algorithm for numerical optimisation (QIEA-R) [3, 72].

Four different variations of NEVE were implemented: ND-NEVE (without detection), RD-NEVE (with reactive detection), PDGL-NEVE (with proactive detection and Group Label approach) and PDPMS-NEVE (with proactive detection and Pattern Mean Shift approach). These variations differ from each other in the way they detect and treat drifts and will be detailed in the next section.

NEVE is based on an ensemble Multi-Layer Perceptron (MLP), where each neural network member is trained and has its parameters (topology, weights, among others) optimised by QIEA-BR algorithm (see Sect. 2). NEVE is composed of three

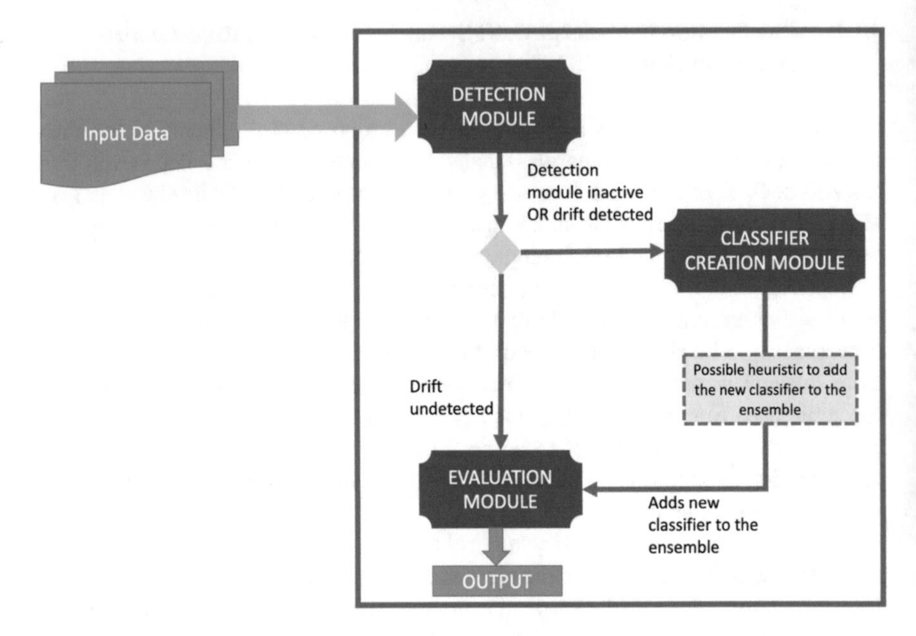

Fig. 10 The modular structure of NEVE

main modules, detailed below and illustrated in Fig. 10: *Drift Detection*; *Classifier Creation*; and *Evaluation and combination weights*.

Drift Detection Module

This module is optional, and, if activated, for each new input data block received, the detection module checks if any drift has occurred. The model works with data blocks of configurable size. If it is necessary or desired to work with individual data inputs, the block can be set to size to 1. It should be mentioned, however, that the strategy of working with one instance at a time is not the most suitable for this model, as it may compromise its computational performance. Two methods of detection were proposed: proactive and reactive detection methods, resulting in four different approaches implemented for this drift detection module:

No detection: simply classifies the new data block, without warring the presence or absence of drift.

Reactive detection: waits until the real data block labels are available to check if a drift has occurred in relation to the previous data block.

Proactive detection (Group Label approach): for each new data block received, a clustering algorithm is performed using the centroids of the last labelled data block as initial centroids. Based on the results of the clustering algorithm, the detection mechanism checks if a drift has occurred. If so, a new MLP is created and trained with the new block, using the class labels suggested by the clustering algorithm.

Proactive detection (Pattern Mean Shift approach): it is similar to the Group Label approach, with the difference that when a drift is detected, instead of creating a new MLP with the new data block, the old data block is used to train the MLP and the drift is "removed" from the new data block. While in the Group Label approach the new MLP is adjusted to the new data, in Pattern Mean Shift approach, the new data is adjusted to the old MLP.

Classifier Creation Module

This module is responsible for creating a new classifier, which may or may not be added to the ensemble, depending on its maximum size defined by the user. The decision to build or not a new neural network is linked to the drift detection mechanism. If a new neural network is created, the new classifier is added to the ensemble if space is available or by replacing an older classifier of worse accuracy. This approach gives the ensemble the ability to learn the new data without having to analyse the old data, as well as allow to forget the data that is no longer needed. In short, this module determines the complete configuration of the new MLP network ensemble member using the QIEA-BR algorithm (presented in Sect. 2). QIEA-BR selects the most relevant input variables, specifies the number of neurons in the hidden layer (respecting the maximum limit configured by the user), and determines the weights and activation functions of each neuron. The number of output neurons is equal to the number of classes in the application.

Evaluation Module

This module determines the final response of the classifier ensemble by combining the results presented by the classifier members. The QIEA-R algorithm is used to determine the most suitable voting weight for each classifier dynamically. The optimisation of weights allows the model to quickly adapt to sudden data changes by assigning higher weights to the classifiers best suited to the current concepts that govern the data. Three possible voting methods were implemented:

Linear Combination: It uses the QIEA-R algorithm to generate a voting weight for each classifier, which is multiplied by the output of each ensemble member (between 0 and 1), on a weighted average. The result of this weighted average is used to determine the ensemble response. If the problem has only two classes, the output is assigned to class 0 if the result is less than 0.5 and to class 1 otherwise; in case of problems with multiple categories, the class will be the one that presents the output with the highest value.

Weighted Majority Voting: As in the previous case, it uses the QIEA-R algorithm to generate a voting weight for each classifier. However, the outputs of the neurons from each ensemble network are first rounded (for values 0 or 1) and then multiplied by the corresponding classifier weight, thus forming a weighted average.

Simple Majority Voting: The output of each ensemble member is rounded to one of the possible classes, and the final ensemble output is the most chosen class among all classifiers. In this case, there is no need to determine voting weights.

Considering the four different approaches implemented for this drift detection module, we implemented four possible variations of NEVE (more details are presented in [23]):

ND-NEVE, without detection
RD-NEVE, with reactive detection
PDGL-NEVE, with proactive detection and the Group Label approach
PDPMS- NEVE, with proactive detection and the Pattern Mean Shift approach

4.1.2 Experiments

Several experiments have been performed, with artificial and real datasets, to validate and compare the performance of the proposed model with other existing models for learning in nonstationary environments. The idea was to verify how the detection model affects the performance and accuracy of NEVE and which model variation and configurations achieved the best results. We varied the voting method, the maximum number of neurons in the hidden layer and the maximum size of the ensemble.

It was found that the ND-NEVE, RD-NEVE and PDPMS-NEVE approaches produce the best results in terms of accuracy and computational performance. It was also observed that the linear combination is the best voting method in terms of accuracy, and simple majority voting the best in terms of computational performance. The unlimited ensemble strategy has worse accuracy and computational performance than limited ensembles, with no significant difference between the five and ten networks. Compared with other consolidated models of the literature, the accuracy of NEVE was found to be superior in most cases. It appeared that the ND-NEVE and RD-NEVE approaches provide uniformly superior results in terms of accuracy. However, the addition of the detection method has resulted in substantial gains in some cases. This fact reinforces that the neuroevolutionary ensemble approach is an appropriate choice for situations in which datasets are subject to sudden behavioural changes.

4.2 System Identification

Another case study of neuroevolutionary models based on quantum-inspired evolutionary algorithms is the application of evolved Echo-State Neural Networks for System Identification.

The identification of nonlinear systems is derived from the equation:

$$\hat{y}(k) = f\,(\varphi(k)) \tag{12}$$

where k is the discrete-time step, $\hat{y}$ is the model output, $f(.)$ is a nonlinear function and $\varphi(k)$ is the regression vector. The regression vector $\varphi(k)$ can include past information about process inputs, model or system outputs and prediction errors.

The nonlinear system identification problem can be summarised in finding an appropriate representation for $f(.)$, for an available input/output data, determining its structure and estimating its parameters through an algorithm that minimises the error between model and system outputs. Additionally, it is critical to choose the maximum lags in input, output and noise signal, the regression terms (to be included in $\varphi(k)$) for model validation.

The model presented in Sect. 3.3, used to evaluate the performance of evolved Echo-State Networks for system identification, has been applied to many benchmark problems [17], with impressive results. The results obtained with the flexible robotic arm [16] benchmark are presented below.

Identifying a flexible robotic arm installed in an electric motor is the first benchmark problem (dataset is available at DaISy repository [16]). The sample was obtained from the test made from the reaction torque of structure (input) to the acceleration of the flexible arm (output) with an estimated sampling period of 1.0 s. A total of 1024 samples were generated and are separated for training, validation and test with 312, 200 and 512, respectively.

To obtain the f_{eval} (Eq. 10), 10 ESNs ($n = 10$) were used for each experiment, with a total of ten trials. The activation functions used were tan$h$ and linear for reservoir and output processors, respectively. Matrices $\mathbf{W}_\mathbf{0}^{\mathbf{in}}$ and $\mathbf{W}_\mathbf{0}^{\mathbf{back}}$ were randomly generated in the interval $[-1,1]$ while $\mathbf{W_0}$ between the limits $[-0.5, 0.5]$. Another significant advantage of using ESN is that training of output layer weight is performed by a pseudoinverse. The washout time was set up with 30 for the robot arm and with 100 for the cascade tanks. Table 1 shows the variable range, and Table 2 shows the parameters used in phase 1 and phase 2 of QIEA-R.

4.2.1 Robot Arm Results

Two tests were performed: an evolved ESN without Phase 2 of the evolutionary process (see Sect. 3.3), where the reservoir is randomly selected, and another considering the complete model, including Phase 2, that is, passing through the optimiser module. In the first case, the ESN was obtained by selecting the network

Table 1 QIEA-R variable range (phase 1 and 2)

Phase	Variable	Range	
		Robot arm	Cascaded tanks
1	N	[20,150]	[40,200]
	$\%c$	[0.10,1]	[0.10,1]
	ρ_W	[0.10,0.99]	[0.10,0.99]
	s_{in}	[0.0001,1]	[0.0001,1]
	s_{back}	[0,1]	[0,1]
	α	[0.10,1]	[0.10,1]
2	$\tilde{w}$	[−0.5,0.5]	[−0.5,0.5]

Table 2 QIEA-R parameter setup

Dataset	Phase	N_c	N_Q	C_r	Gen	upG	C_Q
Robot arm	1	60	20	0.6	100	5	0.2
	2	40	10	0.3	1000	5	0.8
Cascaded tanks	1	30	10	0.9	100	5	0.6
	2	30	5	0.6	800	5	0.4

Table 3 Results for the robot arm benchmark on the test set

Method	RMSE			
	mean	min	max	deviation
Random ESN	0.046168	0.032082	0.086773	0.016241
QIEA-R/ESN	0.024391	0.16050	0.032857	0.004588

with the least $RMSE_{val}$ from a set ten ESNs randomly generated with the global parameters determined by phase 1. The results are presented in Table 3.

As can be verified in Table 3, the complete ESN model, with global parameters and reservoir adjusted by the evolutionary process, provided results 1.89 times better than the average of 10 RMSEs and almost 2.0 times greater than the best solution found considering the ten experiments when compared to Random ESN.

Figure 11 shows the output signal of the best ESN founded by the proposed method.

The comparison among the values obtained with the proposed method and those presented in other works is shown in Table 4. The metrics are fit(%) (Eq. 13) and MSE (Eq. 14):

$$\text{fit}\,(\%) = 100\left(1 - \frac{||\hat{y} - y||}{||y - \overline{y}||}\right) \tag{13}$$

$$\text{MSE} = \sum_{k=1}^{N_d} \big(y(k) - \hat{y}(k)\big)^2 / N_d \tag{14}$$

where $||.||$ is the Euclidian distance, $\hat{y}$ is the mean value of system output sequence and N_d is the data size.

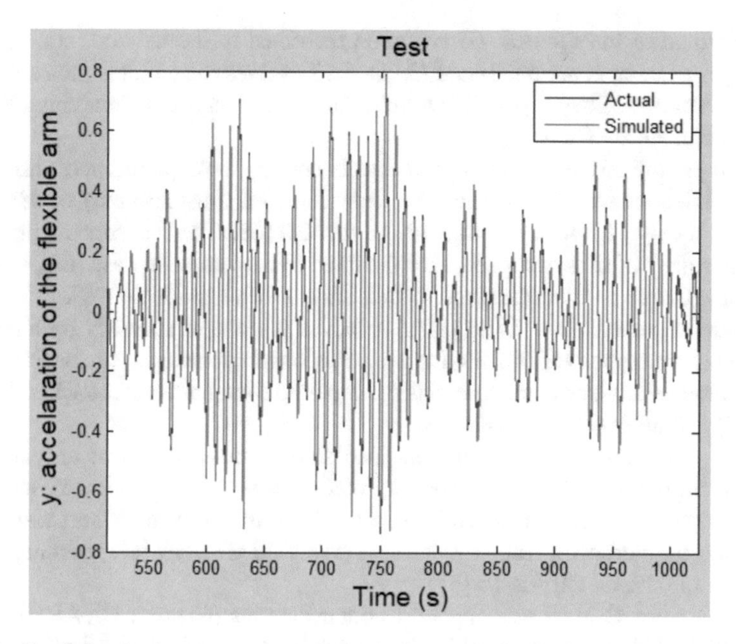

Fig. 11 Simulation for the robot arm system using the best QIEA-R/ESN solution (red, simulated; blue, actual) – test set

Table 4 Comparison among different methods for the test set of robot arm benchmark

Method	Metric
	Fit (maximise)
QIEA-R/ESN (mean)	91.0
SMOPE-GSA	92.9
	MSE (minimise)
QIEA-R/ESN (best)	0.0003
zVF-DT	0.0042
zOVF-DT	0.0042
VF-DT CR	0.0588
VF-DT IT	0.0695

Although the mean fit value in ten experiments is higher than 90.0%, the proposed model doesn't outperform the mean result obtained by the SMOPE-GSA method [40]. However, in terms of MSE, Table 4 shows that the best QIEA-R/ESN solution overcomes all other results of the four methods used in [61].

4.3 Deep Learning with Neural Architecture Search (Q-NAS)

To validate the algorithm, the authors conducted experiments applying Q-NAS to an image classification task, i.e., the benchmark dataset CIFAR-10 [46]. Two scenarios were defined to run Q-NAS: (i) evolution restricted to the network structure (all the hyperparameters are fixed) and (ii) Q-NAS evolves the architectures and some hyperparameters. They repeated five runs for each scenario, maintaining all other Q-NAS parameters fixed.

Their preliminary results are promising concerning the exploration of simple and more efficient convolutional networks. They reported a test accuracy of 88% for a network of only 11 layers, running Q-NAS for 52 h [68]. The best network presented a test accuracy of more than 89%, for the fixed hyperparameter case [68], which is at the level of some hand-designed models such as Maxout (90%) [29].

In a more recent work [67], Q-NAS achieved 93.85% of accuracy for a residual network search space and 93.70% for a regular convolutional space in the CIFAR-10 task. These results overcome the accuracy reported for architectures such as ResNet [34, 35] and other NAS works.

To improve efficiency, the authors also studied the addition of an early-stop mechanism [67, 69]. The early-stopping method was able to reduce the evolution time by more than 60% in several test cases [67]. In one of their best results, the execution time dropped from 249 to 54 h (using 20 GPUs) while reaching a final accuracy of 93% on CIFAR-10 [67].

Additionally, Q-NAS was applied to other datasets [67]. For CIFAR-100, they obtained 74.23% of accuracy, which is comparable to a ResNet with 164 layers [34, 35]. In a case study with a real dataset, Q-NAS was able to generate a network in 40 h using 5 GPUs that surpassed the state-of-the-art model specially created for this task.

Finally, Q-NAS was applied to a real case study related to seismic image classification [67]. The architectures designed by Q-NAS were able to outperform a hand-designed model (Danet-3 [14]) especially developed for this task. For the seismic dataset, Q-NAS best network (with 0:82 million parameters) reached an accuracy of 98:57%, while Danet-3, which has more than 14 million weights, achieved 97:45%.

5 Conclusions and Future Work

This chapter provided an overview of quantum-inspired evolutionary algorithms for the automatic design of different neural network models. Quantum-inspired evolutionary algorithms are a class of EA that provides faster convergence to (near) optimal solutions. This feature is essential for problems with high computational cost to evaluate the fitness function, which is the case for neuroevolution.

We have presented the proposed quantum-inspired evolutionary algorithms for neuroevolution and discussed their application in modelling different neural network architectures, such as Multi-Layer Perceptron, Recurrent Neural Networks and Deep Convolutional Neural Networks.

All neuroevolutionary models were tested in various benchmark applications, as well as in real data problems. The results demonstrated that the different quantum representations allowed to evolve both the neural network architecture and its hyperparameters, with excellent performance. In many cases, the performance obtained was superior to hand-made architectures, indicating that automatic configuration of neural networks is feasible in a wide range of problems: classification in a concept drift environment, control, system identification and, more recently, image classification.

Although neural networks have resurfaced in the 1980s, the recent increase in machine learning research is mostly associated with deep neural networks, which have been predominantly designed by human experts. The manual neural network engineering process can be time-consuming and prone to errors [37], problems that are aggravated in the deep learning environment. Deep neural networks are highly sensitive to hyperparameters and design decisions, which makes neural architecture search a very relevant area that will continue to receive a lot of attention from the research community.

Researchers have been attempting to improve the efficiency of NAS algorithms, trying to simplify the problem by reducing the search space. However, these approaches incorporate a significant human bias in the search process, depending on previous knowledge from the user. Advances in NAS research are still needed to address the efficiency problem without introducing such bias in the search. The proposed Q-NAS model is an attempt to provide considerable flexibility to the user, letting her/him choose how complex the search space should be.

Authors propose future investigations for the Q-NAS model, including a more comprehensive study on the evolution of hyperparameters, to identify which ones should be evolved. Also, they intend to apply Q-NAS to other problems that are suitable for neural networks to solve, such as regression tasks.

Acknowledgements The authors would like to thank the Brazilian Agencies CNPQ (Conselho Nacional de Pesquisa e Desenvolvimento) and FAPERJ (Fundação de Amparo à Pesquisa do Estado do Rio de Janeiro) for their financial support in the development of the research projects presented in this chapter.

References

1. A. Abraham, Meta-learning evolutionary artificial neural networks. Neurocomputing **56**, 1–38 (2004)
2. A.V. Abs da Cruz, Algoritmos evolutivos com inspiração quântica para otimização de problemas com representação numérica. PhD thesis, Pontifical Catholic University of Rio de Janeiro, Brazil, (2007) (in Portuguese)
3. A.V. Abs da Cruz, M.M.B.R. Vellasco, M.A.C. Pacheco, Quantum-inspired evolutionary algorithm for numerical optimisation, in *Book Series in Computational Intelligence, Vol. 75 – Hybrid Evolutionary Algorithms*, (Springer, Berlin/Heidelberg, 2007), pp. 19–37
4. A.V. Abs da Cruz, M.M.B.R. Vellasco, M.A.C. Pacheco, Quantum-inspired evolutionary algorithms applied to numerical optimisation problems, in *IEEE Congress on Evolutionary Computation (IEEE CEC 2010)*, (2010), pp. 3899–3904

5. P.J. Angeline, G.M. Saunders, J.B. Pollack, An evolutionary algorithm that constructs recurrent neural networks. IEEE Trans. Neural Netw. **5**(1), 54–65 (1994)
6. P.P. Angelov, X. Zhou, Evolving fuzzy-rule-based classifiers from data streams. IEEE Trans. Fuzzy Syst. **16**(6), 1462–1475 (2008)
7. F. Assunção, N. Lourenço, P. Machado, B. Ribeiro, DENSER: deep evolutionary network structured representation. Genet. Program Evolvable Mach. **20**(1), 5–35 (2019)
8. S. Basterrech, E. Alba, V. Snasel, An experimental analysis of the echo state network initialization using the particle swarm optimization, in *2014 Sixth World Congress on Nature and Biologically Inspired Computing (NaBIC 2014)*, (2014), pp. 214–219
9. A. Blanco et al., A real-coded genetic algorithm for training recurrent neural networks. Neural Netw. **14**(1), 93–105 (2001)
10. A. Cano, B. Krawczyk, Learning classification rules with differential evolution for high-speed data stream mining on GPU s, in *2018 IEEE Congress on Evolutionary Computation (IEEE CEC 2018)*, (2018)
11. A. Cano, B. Krawczyk, Evolving rule-based classifiers with genetic programming on GPUs for drifting data streams. Pattern Recogn. **87**, 248–268 (2019)
12. G. Capi, K. Doya, Evolution of recurrent neural controllers using an extended parallel genetic algorithm. Robot. Auton. Syst. **52**(2), 148–159 (2005)
13. V. Carvalho, W. Cohen, Single-pass online learning: performance, voting schemes and online feature selection, in *Proceedings of the 12th ACM SIGKDD International Conference on Knowledge Discovery and Data Mining (KDD'06)*, (2006), pp. 548–553
14. D. Chevitarese, D. Szwarcman, E. Brazil, and B. Zadrozny, "Efficient Classification of Seismic Textures", 2018 International Joint Conference on Neural Networks (IJCNN 2018), 2018
15. N. Chouikhi, B. Ammar, N. Rokbani, A.M. Alimi, PSO-based analysis of Echo State Network parameters for time series forecasting. Appl. Soft Comput. **55**, 211–225 (2017)
16. B. L. R. De Moor (ed.), *DaISy: Database for the Identification of Systems*, Technical Report 97–70, http://homes.esat.kuleuven.be/~smc/daisy/ (Department of Electrical Engineering, ESAT/STADIUS, KU Leuven, Belgium, 2018)
17. P.R.M. de Paiva, Modelos neuroevolucionários com Echo State Networks aplicados à Identificação de Sistemas, MSc dissertation, Pontifical Catholic University of Rio de Janeiro, Brazil, 2018 (in Portuguese)
18. P.R.M. de Paiva, M. Vellasco, J. Amaral, Quantum-inspired optimisation of echo state networks applied to system identification, in *2018 IEEE Congress on Evolutionary Computation (IEEE CEC 2018)*, (2018), pp. 2089–2096
19. M. Delgado, M.C. Pegalajar, A multiobjective genetic algorithm for obtaining the optimal size of a recurrent neural network for grammatical inference. Pattern Recogn. **38**(9), 1444–1456 (2005)
20. E.D.M. Dias, M.M.B.R. Vellasco, A.V.A. Cruz, Quantum-inspired neuro coevolution model applied to coordination problems. Expert Syst. Appl. (2020). https://doi.org/10.1016/j.eswa.2020.114133
21. R. Elwell, R. Polikar, Incremental learning of concept drift in nonstationary environments. IEEE Trans. Neural Netw. **22**(10), 1517–1531 (2011)
22. T. Escovedo, A. Koshiyama, A. Abs da Cruz, M. Vellasco, DetectA: abrupt concept drift detection in nonstationary environments. Appl. Soft Comput. **62**, 119–133 (2018)
23. T. Escovedo, A. Abs Da Cruz, M. Vellasco, A. Koshiyama, Neuroevolutionary learning in nonstationary environments. Appl. Intell. **50**, 1590–1608 (2020)
24. A.A. Ferreira, T.B. Ludermir, Genetic algorithm for reservoir computing optimization, in *2009 International Joint Conference on Neural Networks, Atlanta, GA*, (2009), pp. 811–815
25. A.A. Ferreira, T.B. Ludermir, R.R.B. De Aquino, An approach to reservoir computing design and training. Expert Syst. Appl. **40**(10), 4172–4182 (2013)
26. R.S. Ferreira, G. Zimbrão, L.G.M. Alvim, AMANDA: semi-supervised density-based adaptive model for nonstationary data with extreme verification latency. Inf. Sci. **488**, 219–237 (2019)
27. J. Gama, I. Žliobaite, A. Bifet, M. Pechenizkiy, A. Bouchachia, A survey on concept drift adaptation. ACM Comput. Surv. **46**(4), Article 44 (2014)

28. F. Gomez et al., Accelerated neural evolution through cooperatively coevolved synapses. J. Mach. Learn. Res. **9**, 937–965 (2008)
29. I.J. Goodfellow, D. Warde-Farley, M. Mirza, A. Courville, Y. Bengio, Maxout networks. Proc. Mach. Learn. Res. **28**(3), 1319–1327 (2013)
30. K. Han, J. Kim, Genetic quantum algorithm and its application to combinatorial optimisation problem, in *Proceedings of the 2000 Congress on Evolutionary Computation (IEEE CEC 2000)*, vol. 2, (2000), pp. 1354–1360
31. K. Han, J. Kim, Quantum-inspired evolutionary algorithm for a class of combinatorial optimisation. IEEE Trans. Evol. Comput. **6**(6), 580–593 (2002)
32. K. Han, J. Kim, On setting the parameters of QEA for practical applications: some guidelines based on empirical evidence, in *Genetic and Evolutionary Computation Conference (GECCO 2003)*, (2003), pp. 427–428
33. K. Han, J. Kim, Quantum-inspired evolutionary algorithms with a new termination criterion, He gate, and two-phase scheme. IEEE Trans. Evol. Comput. **8**(2), 156–169 (2004)
34. K. He, X. Zhang, S. Ren, J. Sun, Deep residual learning for image recognition, in *2016 IEEE Conference on Computer Vision and Pattern Recognition (CVPR), Las Vegas, NV*, (2016a), pp. 770–778
35. K. He, X. Zhang, S. Ren, J. Sun, Identity mappings in deep residual networks, in *European Conference on Computer Vision (ECCV 2016)*, (2016b), pp. 630–645
36. G. Hulten, L. Spencer, P. Domingos, Mining time-changing data streams, in *Proceedings of the Seventh ACM SIGKDD International Conference on Knowledge Discovery and Data Mining (KDD'01)*, (2001), pp. 97–106
37. F. Hutter, L. Kotthoff, J. Vanschoren (eds.), *Automated Machine Learning: Methods, Systems, Challenges* (The Springer Series on Challenges in Machine Learning, 2019)
38. K. Ishu, T. Van Der Zant, V. Becanovic, P. Ploger, Identification of motion with echo state network, in *Oceans'04 MTS/IEEE Techno-Ocean '04, Kobe*, vol. 3, (2004), pp. 1205–1210
39. H. Jaeger, The "echo state": approach to analysing and training recurrent neural networks, in *GMD Report*, vol. 148, (2001)
40. H. Jaeger, Simple toolbox for ESNs, http://reservoircomputing.org/software (2009)
41. W. Jia, D. Zhao, T. Shen, C. Su, C. Hu, A new optimized GA-RBF neural network algorithm. Comput. Intell. Neurosci. **2014**, Article 982045 (2014)
42. M.T. Karnick, M. Ahiskali, M. Muhlbaier, R. Polikar, Learning concept drift in nonstationary environments using an ensemble of classifiers based approach, in *2008 International Joint Conference on Neural Networks (IJCNN 2008)*, (2008), pp. 3455–3462
43. K.-J. Kim, S.-B. Cho, Evolutionary ensemble of diverse artificial neural networks using speciation. Neurocomputing **71**(7–9), 1604–1618 (2008)
44. B. Krawczyk, A. Cano, Online ensemble learning with abstaining classifiers for drifting and noisy data streams. Appl. Soft Comput. **68**, 677–692 (2018)
45. B. Krawczyk, L.L. Minku, J. Gama, J. Stefanowski, M. Woźniak, Ensemble learning for data stream analysis: a survey. Inf. Fusion **37**, 132–156 (2017)
46. A. Krizhevsky, *Learning Multiple Layers of Features from Tiny Images*, Technical Report TR-2009 (University of Toronto, 2009)
47. L.I. Kuncheva, Classifier ensemble for changing environments, in *Multiple Classifier Systems, Lecture Notes in Computer Science*, vol. 3077, (Springer, Berlin/Heidelberg, 2004)
48. E. Lacerda, A.C.L.F. Carvalho, A.P. Braga, Evolutionary radial basis functions for credit assessment. Appl. Intell. **22**, 167–181 (2005)
49. H. Liu, K. Simonyan, O. Vinyals, C. Fernando, K. Kavukcuoglu, Hierarchical representations for efficient architecture search, in *International Conference on Learning Representations (ICLR 2018)*, (2018)
50. G. Martins, M. Vellasco, R. Schirru, P. Vellasco, Closed-loop identification of nuclear steam generator water level using ESN network tuned by genetic algorithm, in *Engineering Applications of Neural Networks (EANN 2015), Communications in Computer and Information Science*, vol. 517, (Springer, 2015)

51. E. Paz, C. Kamath, An empirical comparison of combinations of evolutionary algorithms and neural networks for classification problems. IEEE Trans. Syst. Man Cybern. B Cybern. **35**(5), 915–927 (2005)
52. N. Pedrajas, D.O. Boyer, A cooperative constructive method for neural networks for pattern recognition. Pattern Recogn. **40**(1), 80–98 (2007)
53. N. Pedrajas, C. H-Martínez, J. Muñoz-Perez, Multiobjective cooperative coevolution of artificial neural networks (multiobjective cooperative networks). Neural Netw. **15**(10), 1259–1278 (2002)
54. N. Pedrajas, C. Hervas-Martinez, J. Munoz-Perez, COVNET: a cooperative coevolutionary model for evolving artificial neural networks. IEEE Trans. Neural Netw. **14**(3), 575–596 (2003)
55. N. Pedrajas, C. Hervas-Martinez, D. Ortiz-Boyer, Cooperative coevolution of artificial neural network ensembles for pattern classification. IEEE Trans. Evol. Comput. **9**(3), 271–302 (2005)
56. A. Pinho, Algoritmo evolucionário com inspiração quântica e representação mista aplicado a Neuroevolução. Master's dissertation, Pontifical Catholic University of Rio de Janeiro, Brazil, (2010) (in Portuguese)
57. A. Pinho, M. Vellasco, A. Abs da Cruz, A new model for credit approval problems: a quantum-inspired neuro-evolutionary algorithm with binary-real representation, in *2009 World Congress on Nature & Biologically Inspired Computing (NaBIC)*, (2009), pp. 445–450
58. M. Platel, S. Schliebs, N. Kasabov, Quantum-inspired evolutionary algorithm: a multimodel EDA. IEEE Trans. Evol. Comput. **13**(6), 1218–1232 (2009)
59. E. Real, S. Moore, A. Selle, S. Saxena, Y.L. Suematsu, J. Tan, Q.V. Le, A. Kurakin, Large-scale evolution of image classifiers, in *Proceedings of the 34th International Conference on Machine Learning*, vol. 70, (2017), pp. 2902–2911
60. J. Schlimmer, R. Granger, Incremental learning from noisy data. Mach. Learn. **1**, 317–354 (1986)
61. R. Schumacher, G.H.C. Oliveira, Uma nova abordagem vector fitting para identificação de sistemas com dados no domínio do tempo. XII Simpósio Brasileiro de Automação Inteligente, Brazil, 283–288 (2015) (in Portuguese)
62. R.S. Sexton, R.E. Dorsey, Reliable classification using neural networks: a genetic algorithm and backpropagation comparison. Decis. Support. Syst. **30**(1), 11–22 (2000)
63. L. Silveira, R. Tanscheit, M. Vellasco, Quantum inspired evolutionary algorithm for ordering problems. Expert Syst. Appl. **67**, 71–83 (2017)
64. M. Skowronski, J. Harris, Automatic speech recognition using a predictive echo state network classifier. Neural Netw. **20**(3), 414–423 (2007)
65. R. Stanley, O. Kenneth, Miikkulainen, Evolving neural networks through augmenting topologies. Evol. Comput. **10**(2), 99–127 (2002)
66. M. Suganuma, S. Shirakawa, T. Nagao, A genetic programming approach to designing convolutional neural network architectures, in *Proceedings of the Twenty-Seventh International Joint Conference on Artificial Intelligence (IJCAI-18)*, (2018), pp. 5369–5373
67. D. Szwarcman, Quantum-inspired neural architecture search. PhD thesis, Pontifical Catholic University of Rio de Janeiro, Brazil, 2020
68. D. Szwarcman, D. Civitarese, M. Vellasco, Quantum-inspired neural architecture search, in *2019 International Joint Conference on Neural Networks (IJCNN 2019)*, (2019a)
69. D. Szwarcman, D. Civitarese, M. Vellasco, Q-nas revisited: exploring evolution fitness to improve efficiency, in *2019 8th Brazilian Conference on Intelligent Systems (BRACIS)*, (2019b), pp. 509–514
70. D.L. Tong, R. Mintram, Genetic algorithm-neural network (GANN): a study of neural network activation functions and depth of genetic algorithm search applied to feature selection. Int. J. Mach. Learn. Cybern. **1**, 75–87 (2010)
71. A. Tsymbal, The problem of concept drift: definitions and related work, in *Technical Report, Trinity College Dublin, ICD-CS-2004-15*, (2004)
72. M.M.B.R. Vellasco, A.V. Abs da Cruz, A.G. Pinho, Quantum-inspired evolutionary algorithms applied to neural network modeling, in *IEEE World Congress on Computational Intelligence (IEEE WCCI 2010), Plenary and Invited Lectures*, ed. by J. Aranda, S. Xambó, (2010), pp. 125–150

73. L. Xie, A. Yuille, Genetic CNN, in *2017 IEEE International Conference on Computer Vision (ICCV)*, (2017), pp. 1388–1397
74. X. Yao, Evolving artificial neural networks. Proc. IEEE **87**(9), 1423–1447 (1999)
75. R. Ye, Q. Dai, A novel greedy randomised dynamic ensemble selection algorithm. Neural. Process. Lett. **47**, 565–599 (2018)
76. L. Zhan et al., ANN-GA approach of credit scoring for mobile customers, in *IEEE Conference on Cybernetics and Intelligent Systems*, (2004), pp. 1148–1153
77. L.M. Zhang, Genetic deep neural networks using different activation functions for financial data mining, in *2015 IEEE International Conference on Big Data (Big Data)*, (2015), pp. 2849–2851
78. B. Zhang, L. Xue, W. Wang, S. Qin, D. Wang, Model updating mechanism of concept drift detection in data stream based on classifier pool. EURASIP J. Wirel. Commun. Netw. (2016)
79. B. Zoph, Q.V. Le, Neural architecture search with reinforcement learning. https://arxiv.org/abs/1611.01578. (2016)

Tatiana Escovedo received the BSc and MSc degrees in Computer Science from the Pontifical Catholic University of Rio de Janeiro (PUC-Rio), Brazil, in 2005 and 2007, respectively, and the PhD degree in Electrical Engineering from the Pontifical Catholic University of Rio de Janeiro (PUC-Rio) in 2015. Dr Escovedo currently works as a Data Scientist at Petrobras, in Rio de Janeiro, Brazil, and as an Assistant Professor and Course Coordinator at PUC-Rio. She is the author of several books and papers in the fields of Software Engineering and Data Science. Her research interests include (but are not limited to) Data Science, Artificial Intelligence, Software Engineering, Machine Learning and Business Intelligence.

"I decided to study Computer Science because I loved Math when I was a child, and I always loved to solve puzzles. When I started programming, I was really excited in the beginning, but when things started to get hard, I was worried if I would be able to keep going on advanced fields. I had a lot of help, excellent teachers and a lot of dedication. That episode has taught me that I could learn everything I wanted, even rocket science, with hard work. Now that I am a part-time Professor, I am always looking for new and non-traditional resources to help students on their learning journey.

I am passionate about teaching and learning, and I decided to dive into the Artificial Intelligence and Computational Intelligence fields. I love to work on raw data and to transform it into insights to support business decisions. I especially appreciate when people think that this is a kind of superpower! I am also a data-driven evangelist because everybody needs to know that data is the new oil and if we don't invest a lot of time working on data issues, we won't have information and knowledge".

Karla Figueiredo received the BSc degree in Electrical Engineering from the Federal University of Rio de Janeiro (UFRJ), Brazil, in 1990; the MSc degree in Electrical Engineering from Pontifical Catholic University of Rio de Janeiro (PUC-Rio), Brazil, in 1994; and the PhD in Computer Systems from the Pontifical Catholic University of Rio de Janeiro (PUC-Rio), Brazil, in 2003.

She is currently the head of the Applied Computational Intelligence and Robotics Lab (LIRA) at PUC-Rio and, since 2016, also an assistant professor at the State University of Rio de Janeiro (UERJ). She has participated and coordinated more than 30 projects in Artificial Intelligence over the last 25 years. She is the author of 10 book chapters and more than 50 scientific articles in the area of soft computing and machine learning. She has supervised more than 20 PhD theses and MSc dissertations. Her research interests are related to Artificial Intelligence methods and applications, including Data Science, Data Mining, Deep Learning, Neural Networks, Fuzzy Logic, Evolutionary Computation, Hybrid Intelligent Systems and Robotics, applied to pattern classification, time series forecasting, optimisation, data mining and decision support systems.

"During my master's degree, I worked in computer companies as a systems analyst for three years, which encouraged me to switch from electrical engineering to computer systems. Despite having studied engineering, I have always been interested in biology and medicine. I believe that the machine learning techniques inspired by biology that attracted me to Artificial intelligence. Then, after four years of studying and working with AI at PUC-Rio, I decided to pursue an academic career, through a doctorate, but without losing interest in applying this knowledge to real problems".

Daniela Szwarcman received her B.Sc. and M.Sc. degrees in electrical engineering from Pontifical Catholic University of Rio de Janeiro (PUC-Rio), Rio de Janeiro, Brazil, in 2013 and 2016, respectively. She recently received her PhD in electrical engineering at PUC-Rio.

In 2017, she joined IBM Research as an intern, and now she works as a research scientist. Her research interests include neural networks, neural architecture search, AutoML, and evolutionary computation.

Daniela has been interested in computers since she was a kid. However, at the end of high school, she was very curious about computer hardware and electronics. Then, she decided to enter the undergraduate course in Electrical Engineering - Electronics. After college, Daniela decided she wanted to continue working with electronics, and she joined a semiconductor laboratory in her university. Then, she started the Master's degree course in Electrical Engineering, focusing on infrared photodetectors. At the end of the course, Daniela was sure she wanted to work on a research lab, with real-world applications. At this time, deep learning was starting to grow, and a lot of research opportunities were emerging. Inspired by the excitement around deep learning, Daniela applied to a PhD in Electrical Engineering to work with neural networks, and later she would focus on Neural Architecture Search. She has been working with neural networks and machine learning ever since.

Marley Maria Bernardes Rebuzzi Vellasco received the BSc and MSc degrees in Electrical Engineering from the Pontifical Catholic University of Rio de Janeiro (PUC-Rio), Brazil, in 1984 and 1987, respectively, and the PhD degree in Computer Science from the University College London (UCL) in 1992. She is the founder and the Head of the Computational Intelligence and Robotics Laboratory (LIRA) at PUC-Rio. Marley has served on the Board of Governors of the International Neural Network Society and is currently Vice-President for Conferences of the IEEE Computational Intelligence Society. She is an Action Editor of the Neural Networks Journal and Associate Editor of IEEE Transactions on Fuzzy Systems, IEEE Transactions on Neural Networks and Learning Systems and IEEE Systems Journal. She is the author of four books and more than 450 scientific papers in the area of soft computing and machine learning. She has supervised more than 40 PhD Thesis and 90 MSc Dissertations and has coordinated more than 50 research projects with industries, some of them resulting in Technology Innovation prizes.

Her research interests are related to computational Intelligence methods and applications. Her main research area is related to Hybrid Intelligent Systems, developing Neuro-Evolutionary, Fuzzy-Evolutionary and Neural Architecture Search models based on Quantum-Inspired Evolutionary Algorithms.

Marley has always been interested in applied sciences. Before entering High School, she had already decided to follow an engineering degree. During her undergraduate course, Marley grew her interest in scientific research, deciding to pursue MSc and PhD degrees to explore an academic career. After graduating, she decided to do her Master's degree in Electrical Engineering, but with a Dissertation more related to Computer Science. The MSc degree was the first step to the final goal, which was to do her PhD abroad. While applying to some universities, she came to know the area of Artificial Neural Networks, which made her very excited. She got her PhD in Computer Science focusing on the development of VLSI chips for Neurocomputers. After she returned to PUC-Rio in 1992, now as a Lecturer, she decided to broaden her area from Neural Networks to other techniques associated with Computational Intelligence. She studied Fuzzy Logic and Evolutionary Computation, and founded one of the first research labs in this area in Rio de Janeiro, Brazil, which she heads still today. She has been teaching and developing computational intelligence models and applications since then.

Weightless Neural Models: An Overview

Teresa B. Ludermir

1 Introduction

Research in the field of Artificial Neural Networks (ANNs) has developed rapidly in recent years. There have been a great number of successful applications of neural networks, where computers are replacing manual and mental labor. For example, computers can now recognize very well the contents of images. In fact, they have approached human performance on some benchmarks over the last five years.

Miniaturization of computers components is taking us from classical to quantum physics. Further reduction in computer component size eventually will lead to the development of computer systems whose components will be on such a small scale that quantum physics intrinsic properties must be taken into account. The expression quantum computation and a first formal model of a quantum computer were first employed in the eighties. With the discovery of a quantum algorithm for factoring exponentially faster than any known classical algorithm in 1997, quantum computing began to attract industry investments for the development of a quantum computer and the design of novel quantum algorithms. Some advantages of quantum neural models over classical models are the exponential gain in memory capacity [45], quantum neurons can solve non-linearly separable problems [48], and a non-linear quantum learning algorithm with linear time over the number of examples in the data set is presented in [37]. However, these quantum neural models cannot be viewed as a direct generalization of a classical neural network.

The high speed of the learning process in WNNs, due to the existence of mutual independence between nodes when entries are changed, is very attractive. The WiSARD machine built in the 1980s was the first ANN machine that successfully

T. B. Ludermir (✉)
Centro de Informática, Universidade Federal de Pernambuco, Recife, Brazil
e-mail: tbl@cin.ufpe.br

A. E. Smith (ed.), *Women in Computational Intelligence*, Women in Engineering and Science, https://doi.org/10.1007/978-3-030-79092-9_15

identified human faces [5]. The effectiveness of WNN models for real-world applications has been demonstrated in a number of experimental studies [5, 8–10, 13, 14, 26, 28, 39, 41, 43, 46]. A recent study presented an extensive experimental evaluation of WiSARD's classification capability in comparison to methods from the state of the art [14]. The WiSARD Classifier was very close in performance to the best methods available in most popular machine learning libraries with the possibility of online training. An approach to data stream clustering based on WiSARD is presented in [9]. The model has useful characteristics such as inherent incremental learning capability and patent functioning speed. An experimental evaluation showed that the proposed system had excellent performance according to multiple quality standards. An iconic training algorithm was developed for WNN investigating the possibility of Machine Consciousness. The iconic training algorithm associates the trained patterns with themselves. The idea is to make a many-to-some mapping of the input in which each node samples the patterns that occur at the input [3, 4].

The purpose of this paper is a description of weightless models, including quantum weightless models [15, 16, 19, 20, 42]. Previous papers published literature reviews on weightless models [6, 31]. However, none of them focused on quantum models. Besides that, those reviews were published more than 10 years ago. The main objective of this review is to identify and discuss what has been published recently on WNNs and the trends and the main challenges of this field.

The remainder of this paper is divided into three sections. Section 2 introduces definitions and learning algorithms of classical weightless models. In Sect. 3, quantum weightless models are formally described. Section 4 presents some final remarks and concludes this overview.

2 Weightless Neural Networks

The model of a neuron employed in the great majority of work regarding artificial neural networks is related to variations of the McCulloch-Pitts neuron [32], which will be called the *weighted-sum-and-threshold neuron*, or *weighted neuron* for short. A typical weighted neuron specifies a linear weighted sum of the inputs, followed by some non-linear transfer function [2, 31]. On the other hand, in this paper, the ANNs analyzed are based on artificial nodes that often have binary inputs and outputs, and no adjustable weights between nodes. The functions computed by the neurons are stored in look-up tables that can be implemented using commercially available Random Access Memories (RAMs).

The process of learning for these systems usually consists of changing the contents of look-up table entries, which results in highly flexible and fast learning algorithms. These systems and the nodes that they are composed of will be described respectively as *Weightless Neural Networks* and *weightless nodes* [2, 31]. They differ from other models, such as the weighted neural networks, whose training

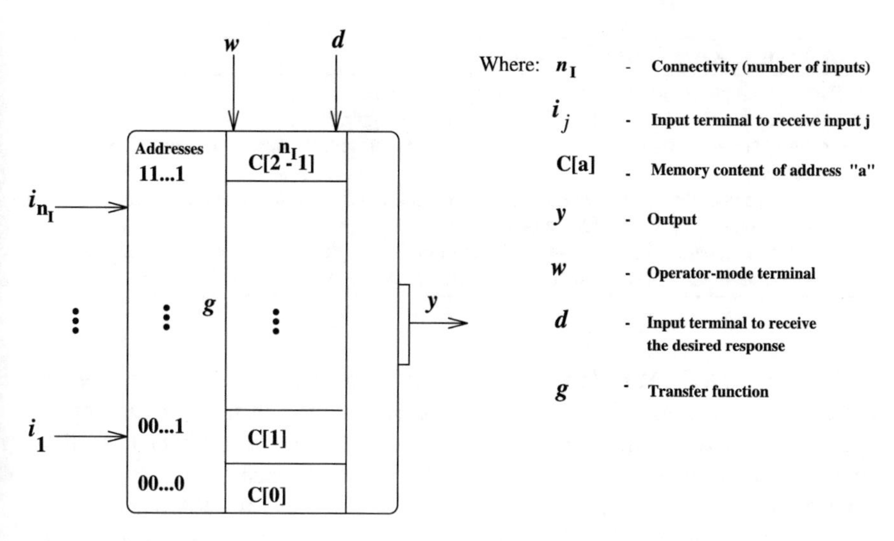

Fig. 1 A RAM node

is accomplished by means of adjustments of weights. In the literature, the terms "RAM-based" and "N-tuple based" have been used to refer to WNNs [28, 40].

The rest of this section describes the models RAM, PLN, MPLN, and pRAM.

The RAM Model

Definition 1 A *RAM neuron* (Fig. 1) is an artificial neuron with the following sub-definitions and restrictions:

- $I = \{0, 1\}^{n_I}$ is *the set of inputs for the node*, where n_I is *the number of input terminals*.
- $A = \{0, 1\}^{n_I}$ is *the set of addresses or locations of the node*. For each address $\mathbf{a} \in A$, there is a cell $C[\mathbf{a}]$, which stores the contents or learned information (local memory) in the form of a bit. A binary signal $\mathbf{i} \in I$ on the input terminals will access only one of these locations, that is, the one for which $\mathbf{a} = \mathbf{i}$. The location $\mathbf{a}$ accessed is called the *activated location*.
- $y \in S$ is *the output of the node*, where the set S is either $[0, 1]$ or $\{0, 1\}$.
- $d \in S$ is *the teaching terminal*, which provides the desired response.
- $w = \{0, 1\}$ is *the operator-mode terminal*, which indicates if the neuron is in the learning or recalling phase.
- $g : I \to S$ is *the transfer function*, which computes y from the k-bit stored in the memory location determined by the input terminal, that is, $y = g(C[\mathbf{a} = \mathbf{i}])$.

The bit, $C[\mathbf{a}]$, stored at the activated memory $\mathbf{a} = \mathbf{i}$ represents the output of a RAM node, that is, $y = C[\mathbf{a}]$. Thus, g is the identity function. In other words, the Boolean function performed by the neuron is determined exactly by the contents of the RAM.

Algorithm 1: RAM learning algorithm

1 Present an input (training) pattern to the input terminals
2 Select the RAM nodes that should learn (α nodes are going to learn, where α is a parameter—learning rate—set before training) and present the desired output to the vector d
3 Set to 1 the operator-mode terminals, w, of the nodes that are going to learn. Repeat by going back to Step 1 for all N training patterns
4 The algorithm halts when the error of the solution is acceptable (this parameter is also set before training). Otherwise, repeat the whole procedure by going back to Step 1

Learning in a RAM node takes place simply by writing into the corresponding look-up table entries. Such a process is much simpler than the adjustment of weights. The RAM node, as previously defined, can compute all binary functions of its input, whereas the weighted-sum-and-threshold nodes can only compute linearly separable functions. There is no generalization in the RAM node itself. Nevertheless, there is generalization in networks composed of RAM nodes [1]. Generalization can be introduced, for instance, by considering the Hamming distance from training patterns or masks.

Definition 2 A *Weightless Neural Network* (WNN) or *RAM-based neural network* is a neural network whose neurons are RAM nodes.

A simple learning algorithm for RAM networks is summarized in Algorithm 1:

The PLN Model

The Probabilistic Logic Node (PLN) differs from the RAM node in that a 2-bit is now stored at the addressed memory location. The content of this location is turned into the probability of firing (i.e. generating 1) at the overall output of the node. That is, a PLN consists of a RAM node augmented with a probabilistic output generator [29]. Hence, like in a RAM node, the n_I binary inputs to a PLN form an address to one of the 2^{n_I} addressable locations $\mathbf{a} \in A$. The classic RAM nodes then output the stored value directly. In contrast, in a PLN, the content at this address is passed through a transfer function which converts it into a binary node output. Such a content could be either 0s, 1s, or u's. The undefined state u implies the node flipping its output between 0 and 1 with equal probability. The use of a third logic value, u (undefined), makes possible the use of an "unknown" state in the operation of WNNs architectures. This value is stored in all the memory contents before the learning phase, indicating the ignorance of the network before it was trained.

Definition 3 A PLN neural network is a WNN in which the neurons are 2-bit nodes (PLNs). The output of the PLN is given by:

$$y = \begin{cases} 0 & \text{if } C[\mathbf{a}] = 0 \\ 1 & \text{if } C[\mathbf{a}] = 1 \\ \mathsf{random}(0, 1) & \text{if } C[\mathbf{a}] = u \end{cases} \quad (1)$$

where $C[\mathbf{a}]$ is the content in the address position associated with the input pattern $\mathbf{i}$ (i.e. $\mathbf{a} = \mathbf{i}$) and random(0,1) is a random function that generates zeros and ones with the same probability.

Training PLN neural networks becomes a process of replacing u's with 0s and 1s so that the network consistently produces the correct output pattern in response to training input patterns. At the beginning of training, all stored values in all nodes are set to u, and thus the net's behavior is completely unbiased. A generic gradient descent learning algorithm, proposed by [35], uses several presentations of the training set to teach PLN neural networks by using reward and punish phases.

A simple learning algorithm for multi-layer PLN networks is summarized in Algorithm 2:

Algorithm 2: PLN learning algorithm

```
1  All memory contents are set to u
2  while some stopping criterion is not met do
3  |   One of the N training patterns p is presented to the network
4  |   learn ← FALSE
5  |   for t=1 to η do
6  |   |   The network is allowed to produce the output for the pattern p
7  |   |   if s is equal to the desired output for the pattern p then
8  |   |   |   learn ← TRUE
9  |   |   |   break
10 |   |   end
11 |   end
12 |   if learn then
13 |   |   all the addressed memory content are made to assume their current output values,
   |   |     making those with u become definitely 0 or 1, accordingly
14 |   else
15 |   |   all the addressed memory content are made to assume the u value
16 |   end
17 end
```

Despite its simplicity, a WNN has good generalization capacity and computation power. The computation power (computability) of PLN networks was studied in [30], where a new recognition method was proposed. The computability of a PLN network is identical to the computability of a probabilistic automaton [38]. In [22] another study of computation power is presented, with a single-layer sequential weightless neural networks, such a class of WNN is an important representation of the research on temporal processing in WNNs.

Other Extensions of the Classical RAM Node Model

The development of the PLN led to the definition of the m-state PLN or MPLN [34, 35]. The main difference between the MPLN and the PLN is that the first allows a wider, but still discrete, range of probabilities to be stored at each memory content unit. One result of extending the PLN to MPLN is that the node locations can now store output probabilities which are more finely graded than in the PLN. An MPLN, for instance, could output 1 with 15% probability under a certain input. The reinforcement learning procedure for PLNs was extended to take this into account and was applied to model delay learning in invertebrates [33].

Definition 4 An MPLN neural network is a WNN where the nodes are m-bit value nodes, for $m > 2$, where the activation function g, $g : I \rightarrow \{0, 1\}$, is a probabilistic function.

In [44], Taylor proposed a model of noisy neurons that presents equations which incorporate and formalize many known properties of biological neurons. The evolution of such a model was shown to be equivalent to that of networks of noisy (*probabilistic*) RAM nodes or pRAM nodes [23]. The pRAM node is also an extension of the PLN like the MPLN, but in which continuous probabilities can be stored, that is, values in the range [0, 1]. This kind of node outputs 1 (spike) with frequency related to the memory content being addressed. The pRAM node was further extended, so as to be able to map continuous inputs to binary outputs.

Definition 5 A pRAM neural network is a neural network whose neurons are pRAM nodes.

In general, pRAM nodes have been used in pyramids [39] and trained by means of reinforcement procedures [24]. The convergence of the reinforcement procedure for pRAM neural networks was investigated and the generalization properties of these networks were addressed in [12]. An important aspect about pRAM nodes and their reinforcement training algorithm is the fact that they are hardware implementable, and chips are commercially available [11].

For more details about the models introduced in this section or on other types of weightless nodes, such as the General RAM (GRAM) node and the Goal-Seeking Node (GSN), the reader is referred to [2, 7, 31].

3 Quantum Weightless Networks

A mathematical quantization of a Random Access Memory was proposed in [15] starting from its matrix representation. This section presents definitions of the following quantum weightless neurons: qRAM, qPLN, and qMPLN. The quantum weightless models described in this section are direct realizable in quantum circuits and have a natural adaptation of the classical learning algorithms and physical feasibility of quantum learning.

The quantum information unit is the quantum bit or "qubit." Nik Weaver in the Preface of his book *Mathematical Quantization* [47] explains the quantization procedure in a simple way with just a phrase which says it all: "The fundamental idea of mathematical quantization is *sets are replaced with Hilbert spaces*".

The quantization of Boolean circuit logic can be done by embedding the classical bits {0, 1} in a convenient Hilbert space. The classical bits {0, 1} are represented as the (orthonormal) basis of a Complex Hilbert space. These basis elements are called the computational-basis states. Linear combinations of the basis span the whole space whose elements, called states, are said to be in *superposition*. Any basis can be used, but in Quantum Computing it is common to use the most conventional to define the model of one qubit: $|0\rangle, |1\rangle$ are a pair of orthonormal basis vectors representing each classical bit, or "cbit", as column vector, $|0\rangle = \begin{bmatrix} 1 & 0 \end{bmatrix}^T$ and $|1\rangle = \begin{bmatrix} 0 & 1 \end{bmatrix}^T$. A general state of the system can be written as: $|\psi\rangle = \alpha|0\rangle + \beta|1\rangle$, where α and β are complex numbers and $|\alpha|^2 + |\beta|^2 = 1$. The model of one qubit uses tensor products to multiply qubits. It is necessary to say how the tensor behaves on the basis: $|i\rangle \otimes |j\rangle = |i\rangle|j\rangle = |ij\rangle$, where $i, j \in \{0, 1\}$. Quantum Mechanics Principles [36] restrain the kind of permissible operations. Operations on qubits are carried out only by unitary operators (i.e. matrices U such that $UU^\dagger = U^\dagger U = I_n$, where I_n is the identity $n \times n$ matrix and $U^\dagger$ is *conjugate transpose* also called the *Hermitian adjoint* of the matrix U). Quantum algorithms on n bits are represented by unitary operators U over the 2^n-dimension complex Hilbert space: $|\psi\rangle \rightarrow U|\psi\rangle$. Thus all operators on qubits are reversible. The only exception is a special class of operations called measurement which is how information is retrieved from a quantum system. In a sense, it is a destructive operation that loses the information about the superposition of states. After measuring a general state $|\psi\rangle = \alpha|0\rangle + \beta|1\rangle$ it collapses (projects) into either the state $|0\rangle$ or the state $|1\rangle$, with probability $|\alpha|^2$ or $|\beta|^2$, respectively.

Quantum operator **U** over n qubits is a unitary complex matrix of order $2^n \times 2^n$. For example, some operators over 1 qubit are: Identity **I** (which does nothing), NOT **X** (flip operator, which behaves as the classical NOT on the computation basis) and Hadamard **H** (which generates superposition of state =), described in Eqs. (2) and (3) in matrix form and operator form, respectively. The combination of these unitary operators forms a quantum circuit.

$$\mathbf{I} = \begin{bmatrix} 1 & 0 \\ 0 & 1 \end{bmatrix} \begin{matrix} \mathbf{I}|0\rangle = |0\rangle \\ \mathbf{I}|1\rangle = |1\rangle \end{matrix} \quad \mathbf{X} = \begin{bmatrix} 0 & 1 \\ 1 & 0 \end{bmatrix} \begin{matrix} \mathbf{X}|0\rangle = |1\rangle \\ \mathbf{X}|1\rangle = |0\rangle \end{matrix} \tag{2}$$

$$\mathbf{H} = \frac{1}{\sqrt{2}} \begin{bmatrix} 1 & 1 \\ 1 & -1 \end{bmatrix} \begin{matrix} \mathbf{H}|0\rangle = 1/\sqrt{2}(|0\rangle + |1\rangle) \\ \mathbf{H}|1\rangle = 1/\sqrt{2}(|0\rangle - |1\rangle) \end{matrix} \tag{3}$$

Quantum operators are represented as quantum circuits with corresponding quantum gates. Figure 2 shows an n-qubit controlled gate U, where U is an arbitrary unitary operator, whose action on the target qubit (bottom-most) is applied or not depending on the $n - 1$ (topmost) control qubits [36]. The output is checked by measurement gates.

Following the definitions of classical WNNs in Sect. 2 and the basic principles of quantum computing, it is possible now to introduce quantum weightless models.

The Quantum RAM Model

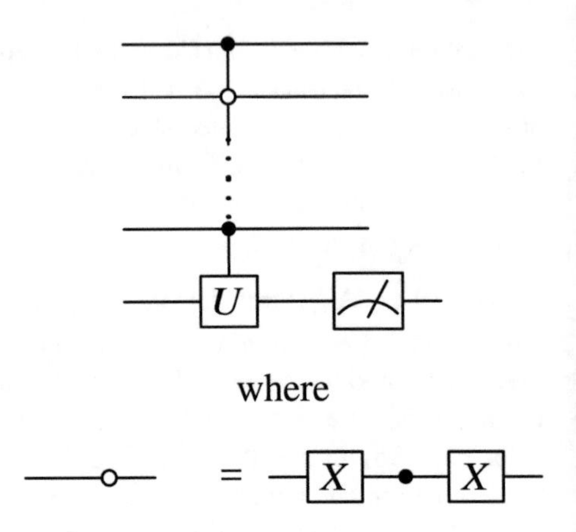

Fig. 2 A quantum circuit

Initially the qRAM node which stores just one qubit is defined. This qubit will not be directly stored in a quantum register as in the classical case. It will use a selector (parameter) and an operator, which applied to the selector will produce the desired qubit. This form of quantization of weightless neural networks was proposed in [15, 16] using matrix A defined in Eq. (4):

$$\mathsf{A} = \begin{pmatrix} 1\,0\,0\,0 \\ 0\,1\,0\,0 \\ 0\,0\,0\,1 \\ 0\,0\,1\,0 \end{pmatrix} \tag{4}$$

The matrix A defines a quantum operator over two qubits that simply flip the second qubit if the first is in the state $|1\rangle$ and it does nothing if the first state is in the state $|0\rangle$. The reader familiar with Quantum Computing terminology will notice that matrix A is well known as the controlled not or c-NOT gate.

Definition 6 A *Quantization of the RAM-based Neural Network*: *qRAM* node with n inputs is represented by the operator N described in Eq. (5). The inputs, selectors, and outputs of N are organized in three quantum registers $|i\rangle$ with n qubits, $|s\rangle$ with 2^n qubits and $|o\rangle$ with 1 qubit. The quantum state $|i\rangle$ describes the qRAM input, and quantum state $|s\rangle|o\rangle$ describes the qRAM state.

$$\mathsf{N} = \sum_{i=0}^{2^n-1} |i\rangle_n \langle i|_n A_{s_i,o} \tag{5}$$

In Fig. 3, a quantum circuit of a qRAM node with two inputs $|\psi\rangle$ and $|\varphi\rangle$ is described. There are four selectors $|s_i\rangle$ and four operators $\mathsf{A}_{s_i,o}$, each one equal to the A operator defined in Eq. (4), where the first qubit is the selector $|s_i\rangle$ and the second is the state in the output register $|o\rangle$. Learning, in qRAM, is achieved by adapting the values of the selectors according to the training set. The main advantage of qRAM over classical weightless neural networks is its capacity to receive inputs

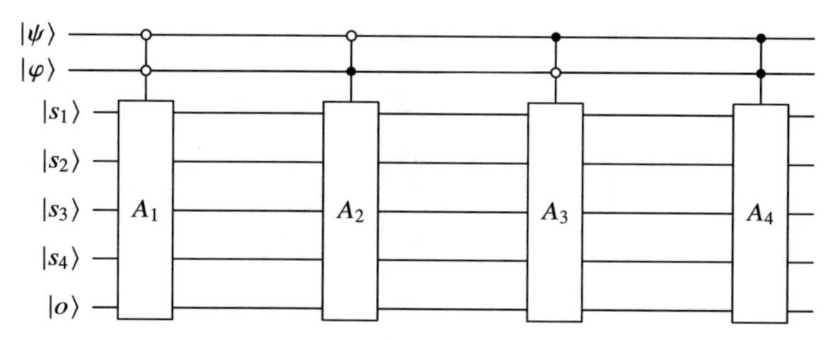

Fig. 3 A qRAM node

Algorithm 3: qRAM—Superposition based learning algorithm

1. Initialize all the qubits in register **s** with the quantum state $\mathsf{H}|0\rangle$
2. Initialize the registers **p**, **o** , **d** with the quantum state $|p\rangle = \sum_{i=1}^{p} |p_i, 0, d_i\rangle$
3. $|\psi\rangle = \mathsf{N}|\psi\rangle$, where N is a quantum operator representing the action of the neuron
4. Use a quantum oracle to change the phase of the states where registers **p**, **o**, **d** $= \sum_{i=1}^{p} |p_i, d_i, d_i\rangle$
5. Apply the operator inverse to the neuron in the state $|\psi\rangle$, $|\psi\rangle = \mathsf{N}^{-1}|\psi\rangle$, to disentangle the state in register s
6. Apply the inversion about the mean operation (as in Grover's algorithm) in register **s**
7. Repeat steps 3, 4, 5 and 6 $T = \frac{\pi}{4}\sqrt{n}$ times, where n is the number of selectors of the networks
8. Measure register **s** to obtain the desired parameters

and selectors in superposition. When the selectors are in the computational basis, the qRAM node acts exactly as a RAM node.

Algorithm 3, proposed by da Silva et al. [42], makes use of superposition and quantum properties to train qRAMs. It is based on Grover's algorithm [25], that amplifies amplitudes of the qubits given an input $|\psi\rangle$ and its desired states $|d\rangle$.

The Quantum PLN Model

de Oliveira et al. [16] shows that by applying a Hadamard operator in the output qubit, the qRAM can simulate a PLN Node. The qPLN Node is proposed with more features beyond the PLN Node.

Definition 7 A *Quantum Probabilistic Logic*: *qPLN* . The values stored in a PLN Node 0, 1, and u are, respectively, represented by the qubits $|0\rangle$, $|1\rangle$, and $H|0\rangle = |u\rangle$. The probabilistic output generator of the PLN is represented as a measurement of the corresponding qubit.

There is a relationship between outputs of a PLN and that of a qPLN that associates i with $|i\rangle$, where $i = 0, 1$ or u. The random properties of the PLN are guaranteed to be implemented by measurement of $\mathbf{H}|0\rangle = |u\rangle$ from quantum mechanics principles. In Fig. 4, a representation of a qPLN node, with one input, is described.

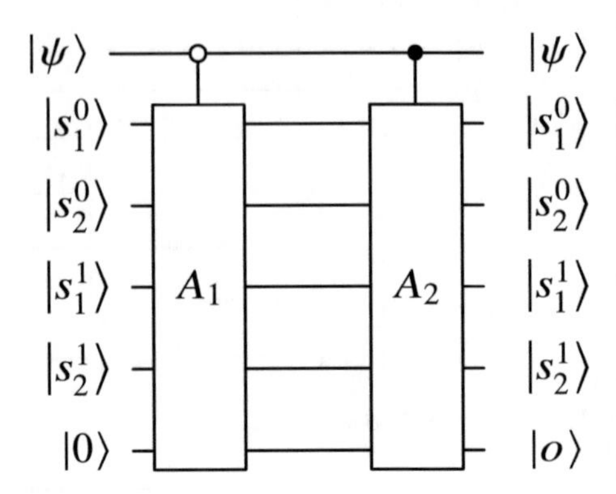

Fig. 4 A qRAM of one input

Algorithm 4: Naive qPLN learning algorithm

1 All selectors $|s\rangle$ are set to $|u\rangle$
2 **while** *some stopping criterion is not met* **do**
3 One of the N training patterns $|p\rangle$ is presented to the network
4 Set $|c\rangle = |1\rangle$ and $|o\rangle = 0$
5 Let $d(p)$ be the desired output of pattern p. Set $|d\rangle = |d(p)\rangle$
6 Apply the quantum circuit described in Fig. 5 **if** $|c\rangle = |0\rangle$ **then**
7 all the addressed memory content are made to assume their current output values, making those with $|u\rangle$ become definitely $|0\rangle$ or $|1\rangle$, accordingly
8 **else**
9 all the addressed memory content are made to assume the $|u\rangle$ value
10 **end**
11 **end**

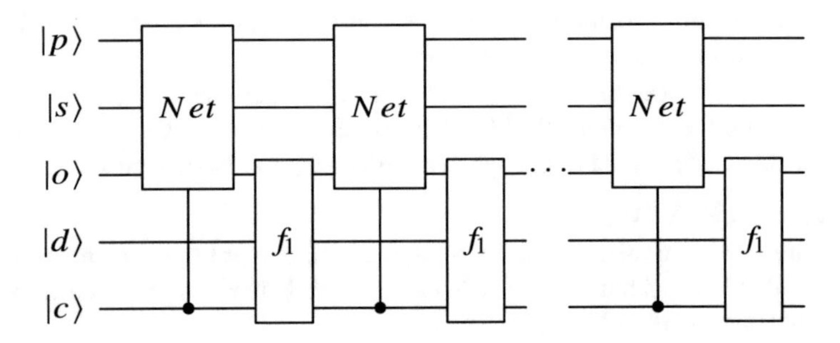

Fig. 5 A PLN learning iteration

With a small modification of the original PLN training algorithm it is possible to train the qPLN. For this, simply change items 1, 4, and 5: All selectors are set to produce $|u\rangle$; if 7 is true all addressed A gates have their selectors changed to produce their current output values; otherwise all addressed A gates are made to produce $|u\rangle$ value. Return to step 2. Observe that Step 7 in the algorithm can be made in superposition thus dramatically speeding up the learning process. A simple learning algorithm for qPLN networks is summarized in Algorithm 4.

Definition 8 A *Quantum Multi-valued PLN: qMPLN*. In the classical case the difference between PLN and MPLN is that each MPLN node stores the probability of the node having the output 1 (out of a discrete set of values, in general bigger than just $0, 1, u$). The same happens for qMPLN. In order to accomplish that matrices U_p are defined:

$$U_p = \begin{pmatrix} \sqrt{1-p} & -\sqrt{p} \\ \sqrt{p} & \sqrt{1-p} \end{pmatrix} \tag{6}$$

With this notation it is routine to check that $U_0|0\rangle = |0\rangle, U_1|0\rangle = |1\rangle$ and $U_{\frac{1}{2}}|0\rangle = |u\rangle$ are the corresponding values in a qPLN node. A general qMPLN node has p varying in the (complex) interval $[0, 1]$ which is useful for the quantization of the pRAM networks. It is a simple exercise to check that all matrices of this form are unitary [27]. The universal matrix A for qMPLN has form similar to the qPLN but with the diagonal blocks being the U_p.

Dynamical qubit extraction methods were proposed to characterize the behavior of qRAM [18, 19]. The methods allow complex values manipulations. By employing an open quantum systems approach, one quantum information extraction method in quantum RAM node dynamics, where complex values are iterated, was proposed in [21]. Experimentally, it was possible to show bifurcation and chaos emergence by varying the parameters of the methods.

4 Conclusions and Future Directions

This paper presented the results of research in weightless models: classic and quantum.

Recent publications reported the use of WNN in the development of commercial solutions to important and difficult problems, such as automated video surveillance, robotics, acceleration of 3D video animation, automated text categorization, clustering, and others.

Inspired by quantum computing, quantum weightless neural models were defined. Section 3 described a definition of a quantum WNN and its learning algorithms. The quantum WNNs and learning algorithms described have simplicity, computational capacity, and quantum features—such as quantum parallelism, using quantum superpositions. Previously proposed quantum neural networks present some problems, for instance in [48] where the quantum neuron can enter into a non-unitary configuration. A matrix representation able to depict classical (weighted, RAM-based, PLN, etc.) and quantum (weighted, weightless) neural models, as particular cases, was suggested [17]. WNNs, for simplicity, allow direct quantization. qRAM neural networks can simulate classical weightless neural networks models and can be trained with a classical or quantum algorithm. Reference [20] showed how to perform artificial neural network architecture selection through the learning procedure of a qRAM weightless neural network. The strategy described is the first algorithm proposed to perform neural network architecture selection in a quantum computer.

The study of small pieces of quantum memories as neurons of a neural network may possibly allow the application of quantum computing using only small quantum processors. Using neurons with low dimensionality, simulation of this kind of quantum neural network will be possible in classical computers.

The advent of quantum computation dramatically changes the way that computers are programmed and opens the possibility for the solution of some problems more efficiently. Quantum algorithms that improve upon known classical algorithms use the concept of quantum parallelism.

As in the neural network field in general, there are many directions for research in weightless systems. Some challenges and future directions are considered below:

1. Sample complexity. Most neural configurations need many examples to learn a particular task. Neural systems need many more examples than a human needs to learn anything.
2. Truly cognitive and conscious human beings. High-level concepts and knowledge about the world are provided by humans to the neural network. Truly cognitive and conscious human beings in machines are not yet achieved. Artificial Intelligence (AI) systems work well in a set of tasks but when they fail to work, they fail in ways that are very different from humans; this reveals somehow they are taking advantage of fairly superficial clues rather than capturing the kind of high-level abstractions that humans use.
3. Social implication of Artificial Intelligence. AI will have profound impact on our society. So, it is important to ask: How are we going to use it? Immense positive uses may come along with negative ones such as military use or rapid disruptive changes in the job market. As far as threat to jobs goes, it is necessary to re-think our education system and to re-skill a huge number of workers.
4. Social value applications. To lessen AI damage it is important to invest in social value added applications, like health, to improve treatment with personalized medicine, to provide medical advice to the multitude of people who do not have access to it right now, and to provide all kinds of services like legal services, through natural language understanding applications.
5. Artificial Intelligence is different from human intelligence. Machines are not going to make humans redundant through the self-perpetuation of machine learning techniques.
6. Artificial Intelligence regulation. There is no question that AI needs to be regulated. AI is too important not to. The big question is how to approach regulation. Regulation can consider safety, explainability, fairness, accountability, and balancing potential harms, especially in high-risk areas, with social opportunities.

References

1. I. Aleksander, Emergent intelligent properties of progressively structured pattern recognition nets. Pattern Recogn. Lett. **1**, 375–384 (1983)
2. I. Aleksander, H. Morton, *An Introduction to Neural Computing*, 2nd edn. (Chapman and Hall, London, UK, 1995)
3. I. Aleksander, H. Morton, *Aristotle's Laptop - The Discovery of our Informational Mind*, vol. 1 of *Series on Machine Consciousness* (World Scientific, 2012)
4. I. Aleksander, H. Morton, Learning state prediction using a weightless neural explorer, in *22th European Symposium on Artificial Neural Networks, ESANN 2014*, pp. 505–510 (2014)
5. I. Aleksander, M.V. Thomas, P.A. Bowden, WiSARD: a radical step forward in image recognition. Sensor Review **4**(3), 120–124 (1984)
6. I. Aleksander, M. de Gregorio, F.M.G. França, P.M.V. Lima, H. Morton, A brief introduction to weightless neural systems, in *ESANN 2009, 17th European Symposium on Artificial Neural Networks*, pp. 299–305 (2009)
7. J. Austin, RAM-based neural networks: A short history, in *RAM-Based Neural Networks*, ed. by J. Austin (World Scientific, UK, 1998), pp. 3–17
8. D.O. Cardoso, J. Gama, F.M.G. França, Weightless neural networks for open set recognition. Machine Learning **106**(9-10), 1547–1567 (2017)
9. D.O. Cardoso, F.M.G. França, J. Gama, WCDS: A two-phase weightless neural system for data stream clustering. New Gener. Comput. **35**(4), 391–416 (2017)
10. S.S. Christensen, A.W. Andersen, T.M. Jorgensen, C. Liisberg, Visual guidance of a pig evisceration robot using neural networks. Pattern Recogn. Lett. **17**(4), 345–355 (1996)
11. T.G. Clarkson, C.K. Ng, D. Gorse, J.G. Taylor, Learning probabilistic RAM nets using VLSI structures. IEEE Trans. Comput. **41**(12), 1552–1561 (1992)
12. T.G. Clarkson, Y. Guan, J.G. Taylor, Generalization in probabilistic RAM nets. IEEE Trans. Comput. **4**(2), 360–363 (1993)
13. M. de Gregorio, M. Giordano, Background estimation by weightless neural networks. Pattern Recogn. Lett. **96**, 55–65 (2017)
14. M. de Gregorio, M. Giordano, An experimental evaluation of weightless neural networks for multi-class classification. Appl. Soft Comput. **72**, 338–354 (2018)
15. W.R. de Oliveira, Quantum RAM based neural networks, in ed. by M. Verleysen, *ESANN'09: Advances in Computational Intelligence and Learning*, pp. 331–336 (2009). ISBN 2-930307-09-9
16. W.R. de Oliveira, A.J. da Silva, T.B. Ludermir, A. Leonel, W.R. Galindo, J.C. Pereira, Quantum logical neural networks, in *Brazilian Symposium on Neural Networks*, pp. 147–152 (2008)
17. W. de Oliveira, A.J. da Silva, T.B. Ludermir, Vector space weightless neural networks, in *European Symposium on Artificial Neural Networks 2014*, pp. 535–540 (2014)
18. F.M. de Paula Neto, T.B. Ludermir, W.R. de Oliveira, A.J. da Silva, Fitting parameters on quantum weightless neuron dynamics, in *2015 Brazilian Conference on Intelligent Systems, BRACIS 2015, Natal, Brazil, November 4–7, 2015* (IEEE Computer Society, 2015), pp. 169–174
19. F.M. de Paula Neto, W.R. de Oliveira, A.J. da Silva, T.B. Ludermir, Chaos in quantum weightless neuron node dynamics. Neurocomputing **183**, 23–38 (2016)
20. A.J. da Silva, W.R. de Oliveira, T.B. Ludermir, Weightless neural network parameters and architecture selection in a quantum computer. Neurocomputing **183**, 13–22 (2016)
21. F.M. de Paula Neto, W.R. de Oliveira, T.B. Ludermir, A.J. da Silva, Chaos in a quantum neuron: An open system approach. Neurocomputing **246**, 3–11 (2017)
22. M.C.P. de Souto, T.B. Ludermir, W.R. de Oliveira, Equivalence between ram-based neural networks and probabilistic automata. IEEE Trans. Neural Netw. **16**(4), 996–999 (2005)
23. D. Gorse, J.G. Taylor, On the equivalence and properties of noisy neural networks and probabilistic RAM nets. Phys. Lett. A **131**(6), 326–332 (1988)

24. D. Gorse, J.G. Taylor, Reinforcement training strategies for probabilistic RAMs, in *International Symposium on Neural Networks and Neurocomputing (NEURONET90)*, ed. by M. Novak, E. Pelikan, pp. 180–184 (1990)
25. L.K. Grover, Quantum mechanics helps in searching for a needle in a haystack. Phys. Rev. Lett. **79**, 325–328 (1997)
26. L. Hepplewhite, T.J. Stonham, N-tuple texture recognition and the zero crossing sketch. Electronics Letters **33**(1), 45–46 (1997)
27. K. Hoffman, R. Kunze, *Linear Algebra* (Prentic-Hall, 1971)
28. T.M. Jorgensen, Classification of handwritten digits using a RAM neural net architecture. Int. J. Neural Syst. **8**(1), 17–25 (1997)
29. W.K. Kan, I. Aleksander, A probabilistic logic neuron network for associative learning, in *Proc. of the IEEE International Conference on Neural Networks*, vol. II, pp. 541–548, San Diego, California (June 1987)
30. T.B. Ludermir, Computability of logical neural networks. J. Intell. Syst. **2**(1), 261–290 (1992)
31. T.B. Ludermir, A. de Carvalho, A.P. Braga, M.C.P. de Souto, Weightless neural models: A review of current and past works. Neural Comput. Surv. **2**, 41–61 (1999)
32. W.S. McCulloch, W. Pitts, A logical calculus of the ideas immanent in nervous activity. Bull. Math. Biophys. **5**, 115–137 (1943)
33. C. Myers, *Delay Learning in Artificial Neural Networks* (Chapman & Hall, 1992)
34. C. Myers, I. Aleksander, Learning algorithms for probabilistic logic nodes, in *Abstracts of I Annual INNS Meeting*, p. 205, Boston (1988)
35. C. Myers, I. Aleksander, Output functions for probabilistic logic nodes, in *Proc. IEE International Conference on Artificial Neural Networks*, pp. 310–314, UK (1989)
36. M.A. Nielsen, I.I.L. Chuang, *Quantum Computation and Quantum Information* (Cambridge University Press, 2000)
37. M. Panella, G. Martinelli, Neural networks with quantum architecture and quantum learning. Int. J. Circuit Theory Appl. **39**(1), 61–77 (2011)
38. M.O. Rabin, Probabilistic automata. Inf. Control **6**(3), 230–245 (1963)
39. S. Ramanan, R.S. Petersen, T.G. Clarkson, J.G. Taylor, pRAM nets for detection of small targets in sequence of infra-red images. Neural Networks **8**(7-8), 1227–1237 (1995)
40. R. Rohwer, M. Morciniec, A theoretical and experimental account of n-tuple classifier performance. Neural Computation **8**(3), 629–642 (1996)
41. R. Rohwer, M. Morciniec, The theoretical and experimental status of the n-tuple classifier. Neural Networks **11**(1), 1–14 (1998)
42. A.J. Silva, W.R. de Oliveira, T.B. Ludermir, Classical and superposed learning for quantum weightless neural networks. Neurocomputing **75**, 52–60 (2012)
43. M. Staffa, M. Berardinelli, G. Acampora, M. Giordano, M. de Gregorio, F. Ficuciello, A weightless neural network as a classifier to translate EEG signals into robotic hand commands, in *27th IEEE International Symposium on Robot and Human Interactive Communication* (IEEE, 2018), pp. 487–490
44. J.G. Taylor, Spontaneous behaviour in neural networks. J. Theor. Biol. **36**, 513–528 (1972)
45. C. A. Trugenberger, Quantum pattern recognition. Quantum Inf. Process. **1**, 471–493 (2002)
46. Y.S. Wang, B.J. Griffiths, B.A. Wilkie, A novel system for coloured object recognition. Comput. Ind. **32**(1), 69–77 (1996)
47. N. Weaver, *Mathematical Quantization*. Studies in Advanced Mathematics (Chapman & Hall/CRC, Boca Raton, FL, 2001)
48. R. Zhou, Q. Ding, Quantum M-P neural network. Int. J. Theor. Phys. **46**(12), 3209–3215 (2007)

Teresa B. Ludermir received her B.Sc. in Computer Science from Universidade Federal de Pernambuco (Brazil) in 1983. She decided to study computer science because she loved mathematics and used to help her friends with difficulties with numbers.

During her B.Sc., the career of University professor and researcher had captivated her. She thought that the best job in the world is studying and teaching what she likes best. So she went to improve her qualification to be a professor. She finished a master in Computing in 1986 and decided to study Artificial Neural Networks, effervescent worldwide at the time. Then she went to do a Ph.D. at Imperial College London with Igor Aleksander, one of the leading researchers in the field. With an education and expertise acquired, in addition to contact with a number of professionals in the field, after being a lecturer at Kings College London for two years, she returned to Brazil in order to bring and foster this field of research. At the time, 1992, Artificial Neural Network research in Brazil was almost non-existent. So she wrote books, in Portuguese, gave seminars and short courses to transform that reality. She organized the Brazilian symposium on neural networks (today it is called Brazilian symposium on computational intelligence). Through those initiatives, and creating a mass of trained and active professionals, her long-term goals were achieved.

Teresa has been the recipient of many awards including The National Order of Scientific Merit (is an honor bestowed upon Brazilian and foreign personalities recognized for their scientific and technical contributions to the cause and development of science in Brazil).

During the last 30 years, Teresa has published over 400 articles in scientific journals and conferences and three books in Neural Networks (in Portuguese). Her research interests include weightless Neural Networks, hybrid neural systems, and machine learning. She is IEEE and INNS senior members and a Research fellow level 1A (the higher level) of the National Research Council of Brazil (CNPq). She is a member of the Board of Governors of INNS, IEEE CIS Summer School Sub-Committee and IEEE Frank Rosenblatt Award and was of the Neural Networks Technical Committee of CIS/IEEE. She has experience in journal editorial boards: IEEE Transactions on Neural Networks and Learning Systems, Neural Network, International Journal of Computational Intelligence and Applications (Editor in chief from 2004 to 2012), and others. She was IJCNN 2018 PC chair and served in the international program committee for over 50 conferences (such as IJCNN, ICANN, HIS, HAIS, BRACIS).

Message:

I appreciate to be part of this book project written only by women in Computational Intelligence field. Women and girls remain excluded from full participation in science. Worldwide, women still represent only 28% of engineering graduates and 40% of computer science and information technology graduates. A UNESCO survey points to greater disparities in highly qualified areas, such as artificial intelligence, where only 22% of professionals are women. So in highly qualified areas such as Computational Intelligence we are only 22% and only 12% of the members of national science academies are women. We need more women in sciences to reduce the gender disparities and I am sure we, women, are capable of doing a very good job in science in academic career and industry. If you want to work in science do not give up. You have the necessary scientific skills and with perseverance you will be a good professional in science.

Part IV
Optimization

Challenges Applying Dynamic Multi-objective Optimisation Algorithms to Real-World Problems

Mardé Helbig

1 Introduction

Many real-world optimisation problems have multiple dynamic objectives in conflict with one another, as well as constraints that should not be violated. This type of problems is referred to as constrained dynamic multi-objective optimisation problems (DMOPs).

Traditionally, a multi-objective optimisation problem (MOP) was solved by converting the MOP to a single-objective optimisation problem (SOP), through the weighted sum approach [52]. However, problems with this approach are as follows [9, 39]:

- This approach requires *a priori* knowledge of the decision-maker's preference, i.e. knowing the decision-maker's preference with regard to the various objectives before running the algorithm. However, if the preferences are unknown, the optimal weight combination becomes an optimisation problem in itself.
- Each run of the algorithm results in a single solution.
- The MOP must be convex, i.e. all constraints and objectives have to be convex.

Multi-objective computational intelligence algorithms solve MOPs without using weights to convert the problem to a SOP, but rather by finding a set of optimal trade-off solutions. These algorithms are based on nature and are population-based, where each entity of the algorithm presents a possible solution in the search space. Therefore, each run of the algorithm produces multiple solutions. The algorithms can also find solutions in all areas of the search space, without being restricted as in the case of the weighted-sum approach.

M. Helbig (✉)
School of ICT, Griffith University, Southport, QLD, Australia
e-mail: m.helbig@griffith.edu.au

A. E. Smith (ed.), *Women in Computational Intelligence*, Women in Engineering and Science, https://doi.org/10.1007/978-3-030-79092-9_16

Most of the research in the field of dynamic multi-objective optimisation (DMOO) focuses on unconstrained DMOPs. Furthermore, many of the benchmark functions (BFs) that are used to evaluate the newly developed dynamic multi-objective algorithms (DMOAs) are not really representative of real-world DMOO problems (RWPs). This leads to certain aspects of research in the field being neglected, since they are typically not required when solving the BFs.

Aspects that relate to BFs include decision variable values that are not necessarily continuous, dealing with constraints (either static or dynamic in nature), changes that occur in a non-cyclic fashion and with different severity over time and decision variable values that are not always known (uncertainty).

Other aspects include performance measures that are used when the optimal solution for the RWP is unknown, selecting a DMOA to solve the RWP and decision-making.

This chapter provides the following contributions:

- A taxonomy of DMOO RWPs is presented.
- From the taxonomy, typical characteristics of an RWP are highlighted.
- Aspects of the RWP characteristics that are not currently addressed in the BFs are discussed.
- Challenges that should be addressed within the field of DMOO to solve RWPs are highlighted and discussed.

The rest of the chapter's layout is as follows: Sect. 2 provides background information on multi-objective optimisation (MOO) and DMOO that is required for the rest of the chapter. RWPs are discussed in Sect. 3, firstly by categorising DMOO RWPs into a taxonomy and then highlighting typical characteristics of DMOO RWPs. Challenges that should be addressed in the field of DMOO to solve RWPs are discussed in Sect. 4. Finally, Sect. 5 concludes the chapter.

2 Background

This section presents the definitions and concepts that are required for the rest of the chapter. MOPs and DMOPs are discussed in Sects. 2.1 and 2.2, respectively. Section 2.3 discusses the multi-objective algorithms (MOAs), both evolutionary computation and swarm intelligence algorithms, referred to in this chapter.

2.1 Multi-objective Optimisation Problems

A MOP is an optimisation problem with more than one objective, where typically at least two objective functions are in conflict with each other. Therefore, improving one objective leads to a reduction in quality of at least one other objective.

The concept of vector domination is used to compare the quality of two solutions. A solution, $\mathbf{x}_1$, dominates another solution, $\mathbf{x}_2$, indicated by $\mathbf{x}_1 \prec \mathbf{x}_2$, if and only if

- at least one of the objective function values of $\mathbf{x}_1$ is better than the corresponding objective value of $\mathbf{x}_2$, and
- each of the objective function values of $\mathbf{x}_1$ is equal to or better than the corresponding objective function value of $\mathbf{x}_2$.

When solving a MOP, it is important to determine whether a solution is Pareto-optimal or not. A solution $\mathbf{x}_1$ is Pareto-optimal if no other feasible solution dominates it, i.e. $\forall \mathbf{x}' \in F, \nexists \mathbf{x}' : \mathbf{x}' \prec \mathbf{x}_1$, where F is the feasible space that contains all solutions that do not violate the constraints.

The main goal of a MOA is to find the set of Pareto-optimal solutions, referred to as the Pareto-optimal set (POS) in the decision space. Mathematically, the POS, PS, is defined as

$$PS = \{\mathbf{x}^* \in F | \nexists \mathbf{x}' \in F : \mathbf{x}' \prec \mathbf{x}^*\} \tag{1}$$

The corresponding set of Pareto-optimal solutions in the objective space is referred to as the Pareto-optimal front (POF). The POF, PF, is defined as follows:

$$PF = \{\mathbf{f} = (f_1(\mathbf{x}^*), f_2(\mathbf{x}^*), \ldots, f_k(\mathbf{x}^*))\}, \forall \mathbf{x}^* \in PS \tag{2}$$

A solution set is deemed to be of a good quality if it is close to the POF (accurate) and has a good spread of solutions along the entire POF.

2.2 *Dynamic Multi-objective Optimisation Problems*

A DMOP is MOP where the objective functions and/or constraints change over time. Most DMOO BFs configure the environmental changes through a parameter t set as follows:

$$t = \frac{1}{n_t} \left\lfloor \frac{\tau}{\tau_t} \right\rfloor \tag{3}$$

where n_t controls the severity of the change, τ_t controls the frequency of the change and τ is an iteration counter. An example BF, FDA5 [16, 28], is defined as

$$FDA5 = \begin{cases} Minimise: f(\mathbf{x}, t) = (f_1(\mathbf{x}, g(\mathbf{x}_{II}, t)), \ldots, f_k(\mathbf{x}, g(\mathbf{x}_{II}, t))) \\ f_1(\mathbf{x}, g, t) = (1 + g(\mathbf{x}_{II}, t)) \prod_{i=1}^{k-1} cos\left(\frac{y_i \pi}{2}\right) \\ f_a(\mathbf{x}, g, t) = (1 + g(\mathbf{x}_{II}, t)) \prod_{i=1}^{k-a} cos\left(\frac{y_i \pi}{2}\right) \\ \qquad sin\left(\frac{y_{k-a+1}\pi}{2}\right), \forall a = 2, \ldots, k-1 \\ f_k(\mathbf{x}, g, t) = (1 + g(\mathbf{x}_{II}, t)) sin\left(\frac{y_1}{\pi}\right) \\ where: \\ g(\mathbf{x}_{II}, t) = G(t) + \sum_{x_i \in \mathbf{x}_{II}} (x_i - G(t))^2 \\ G(t) = |sin(0.5\pi t|, t = \frac{1}{n_t}\left\lfloor \frac{\tau}{\tau_t} \right\rfloor \\ y_i = x_i^{F(t)}, \forall i = 1, \ldots, (k-1) \\ F(t) = 1 + 100 sin^4(0.5\pi t) \\ \mathbf{x}_{II} = (x_k, \ldots, x_n); x_i \in [0, 1], \forall i = 1, \ldots n \end{cases} \tag{4}$$

where k is the number of objective functions. Both FDA5's POF and POS change over time. Its POF is non-convex, and the spread of solutions in the POF changes over time. For three objective functions, its POS is $x_i = G(t), \forall x_i \in \mathbf{x}_{II}$, and its POF is $f_1^2 + f_2^2 + f_3^2 = (1 + G(t))^2$ [23, 28]. A POF found by a DMOA for one environment is presented in Fig. 1.

Farina et al. [16] categorised DMOPs into four types, where a DMOP is referred to as

- Type I if the POS changes, while the POF remains static.
- Type II if both the POS and the POF change.
- Type III if the POF changes, while the POS remains static.
- Type IV if both the POS and the POF remain static, but the environment changes.

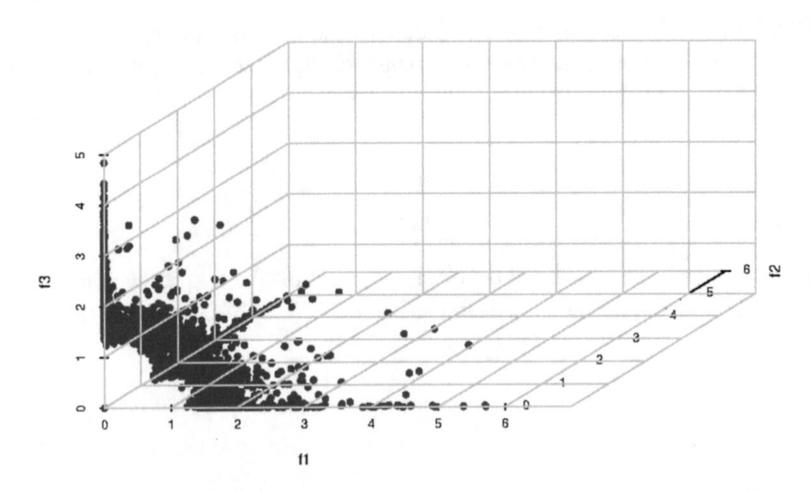

Fig. 1 POF of FDA5 for one environment

Algorithm 1: MOEA/D Algorithm

1. (a) create and initialise a population
   (b) create weight vectors, that are uniformly spread
   (c) calculate the T closest weight vectors for each weight vector
   (d) initialise a vector z containing the best value for each objective
2. while stopping condition has not been reached
3. for each objective do
4. produce offspring
5. apply repair heuristic to the offspring
6. update z
7. update the neighbouring solutions
8. update the archive

2.3 Multi-objective Algorithms

This section discusses the main types of MOAs referred to in this chapter. Section 2.3.1 presents evolutionary computation algorithms, and swarm intelligence algorithms are discussed in Sect. 2.3.2.

2.3.1 Multi-objective Evolutionary Computation Algorithms

Multi-objective Evolutionary Algorithm Based on Decomposition
The multi-objective evolutionary algorithm based on decomposition (MOEA/D) [53] decomposes a MOP into a number of scalar optimisation problems, where each sub-problem is optimised by only considering information from its neighbouring sub-problems. The various steps of MOEA/D are presented in Algorithm 1.

Euclidean distance is used to measure the distance between two weight vectors. In steps 2–8, the T neighbouring sub-problems are used in the calculations. For reproduction (step 4), two random solutions are selected from the neighbouring sub-problems, where each solution is the best solution for that specific sub-problem. Offspring are then created using genetic operators on these selected solutions. Once the offspring have been created, a problem-specific heuristic is used to repair offspring violating any constraints and/or optimising the offspring further (step 5).

Non-dominated Sorting Genetic Algorithm II
The non-dominated sorting genetic algorithm II (NSGA-II), introduced by Deb et al. [10, 11], is a multi-objective genetic algorithm that uses non-domination and is an improved version of the non-dominated sorting genetic algorithm (NSGA) developed by Srinivas and Deb [46]. The steps of NSGA-II are illustrated in Fig. 2.

The first step in Fig. 2 is to produce offspring. Before the algorithm starts, an initial parent population, P_0, is randomly created. P_0 is then sorted based on non-domination, where each individual is assigned a fitness based on its non-domination

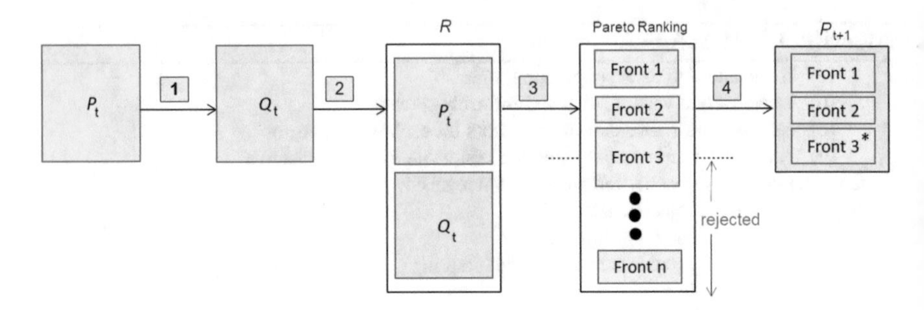

Fig. 2 Steps of NSGA-II

level. Binary tournament selection is performed to select parents for cross-over to produce n_p offspring. Mutation is then performed on the offspring.

After the offspring have been created, the parents and offspring are combined to create a new population, R, of size $2n_p$. This is the second step in Fig. 2. Pareto-ranking is performed on R, i.e. R is sorted according to the individuals' level of non-domination, as explained below. This sorting process is the third step in Fig. 2.

In order to speed up the sorting process, Deb et al. introduced the fast non-dominated search procedure [10, 11]. The first individual of R is placed in a new population P', and each individual, $\mathbf{x}_i \in R$, is then compared against all individuals in P'. If $\mathbf{x}_i$ dominates any individuals in P', the dominated individuals are removed from P'. If $\mathbf{x}_i$ is dominated by any individual in R, $\mathbf{x}_i$ is not placed in P'. However, if $\mathbf{x}_i$ is non-dominated with regard to all individuals in R, $\mathbf{x}_i$ is placed in R. After all $\mathbf{x}_i \in R$ have been compared against the individuals in P', the individuals in P' form the first front with the best rank. To determine the second front, all individuals of the first front are removed from R and not considered, and the search process is repeated. This whole process is repeated until all fronts have been found, i.e. R is empty.

Once all the individuals in R have been assigned a front, a new population, P_{t+1}, is selected for the next generation using crowded sorting [10, 11] to ensure diverse solutions. NSGA-II does not make use of an archive but preserves elitism through its selection mechanism, where the best μ individuals from both the parents and the offspring are selected for the next generation.

When solving DMOPs, diversity is injected into the population after a change has occurred by randomly selecting a certain percentage of the population and then either [13]

- replacing these individuals with randomly created individuals or
- replacing these individuals with mutated solutions of randomly selected existing solutions.

Algorithm 2: MOPSO Algorithm

```
1.  create and initialise a swarm
2.  while stopping condition has not been reached
3.      for each particle in swarm do
4.          calculate new velocity
5.          calculate new position
6.          manage boundary constraint violations
7.      update archive
8.      update the particles' allocation to hypercubes
9.      for each particle in swarm do
10.         update pbest
```

2.3.2 Multi-objective Swarm Intelligence Algorithms

Multi-objective Particle Swarm Optimisation

The multi-objective particle swarm optimisation (MOPSO) algorithm was proposed by Coello Coello and Salazar Lechuga [7] as one of the first particle swarm optimisation (PSO) algorithms extended for MOO. Algorithm 2 lists the various steps of the MOPSO algorithm.

The first step of the MOPSO algorithm initialises each particle's position, velocity and personal best (pbest) and sets the swarm size, neighbourhood size and the control parameters. The particles' initial positions are initialised in such a way that they are uniformly spread over the search space. The particles' velocities are initialised to zero, and their pbests are set to their current positions.

MOPSO uses the velocity equation of PSO to update the velocity of the particles. However, the global best (gbest) is a global guide selected from the archive. MOPSO updates each particle's pbest, referred to as the local guide, as follows: the new position of the particle is compared to its pbest, taking all objective functions into account, and

- if the new position dominates the current pbest, the new position is selected as the pbest, otherwise
- if the new position is non-dominated with regard to the current pbest, the new pbest is randomly selected between the particle's position and the current pbest.

In contrast to NSGA-II, MOPSO uses an archive to preserve elitism. During the search, the particles are evaluated, and the positions of the particles that are non-dominated are stored in the archive. The search space that has been explored so far is divided into hypercubes, and all particles are placed in a hypercube based on their positions in the objective space.

The original version of MOPSO struggles to converge to the POF in the presence of many local POFs [8]. To overcome this problem, Coello et al. [8] introduced an updated version of MOPSO that uses a mutation operator to increase the swarm's exploration ability. Initially, the mutation operator is applied to all particles, but then the number of particles being mutated decreases rapidly as the number of iterations increases.

Lechuga [33] extended MOPSO for DMOO [7]. To detect changes in the environment, a specified number of particles, called sentry particles [5], are randomly selected and re-evaluated after the algorithm has performed the specific iteration, but before the next iteration starts. If the sentry particle's fitness value differs after re-evaluation with more than a specified value, the swarm is notified that a change in the environment has occurred. Once a change has been detected, one of the following approaches is used to react to the change:

- the pbest of the particle is set to its current position if the current position dominates the pbest, or
- the pbest of the particle is set to its current position.

Max–Min Ant System

The max–min ant system (MMAS) [47, 48] is an ant colony optimisation algorithm based on the foraging behaviour of ants. In all ant colony optimisation (ACO) algorithms, the pheromone trail of the ants is represented by real numbers associated with solution attributes used to update the pheromone trail during the search/run of the algorithm. The pheromone trails can be thought of as adaptive memory of previously found solutions in the search space that are updated after each iteration. Local search is used to improve solutions. In max-min ant system (MMAS), only one ant is allowed to update the pheromone trails, and the selected ant lays down a constant amount of pheromone. However, an evaporation constant is used on the trail intensity variables to define how fast/slow the information gathered in previous iterations is forgotten, i.e. how fast/slow the pheromone trail evaporates. To balance the exploration of new search areas with the exploitation of found solutions, MMAS restricts the pheromone trail strength to a specified interval. The main steps of MMAS are presented in Algorithm 3.

To speed up the running time of ACO, the algorithm was extended for parallel processing, where the ants are divided into groups and each group is assigned to a processor. Information is exchanged between processors to communicate and update the pheromone trails [6]. This algorithm is referred to as the adaptive parallel ant colony (PACO) algorithm.

Algorithm 3: MMAS Algorithm

```
1.  initialise the peromone trails and parameters
2.  while stopping condition has not been reached
3.      for each ant do
4.          construct a solution
5.          improve solution with local search
6.          update pheromone trail
```

Algorithm 4: BBO Migration Operator

```
1.  while stopping condition has not been reached
2.      for each habitat H_i
3.          perform selection on H_i based on immigration rate λ_i
4.          if H_i selected, for each habitat H_j
5.              perform selection on habitat H_j based on emigration rate μ_j
6.              if H_j selected
7.                  randomly select a SIV from H_j
8.                  copy SIV to H_i
```

Non-dominated Sorting Biogeography-Based Optimisation

The biogeography-based optimisation (BBO) algorithm [37, 45] is a population-based algorithm based on the distribution of animals and plants in different habitats over time and space. Mathematical models of biogeography [37] describe migration, speciation and extinction of species in various islands, where an island refers to any habitat that is geographically isolated from other habitats. Habitats are characterised by factors represented as suitability index variables (SIVs) to indicate the habitat suitability index (HSI). Each habitat experiences both immigration (arrival of new species) and emigration (species leaving).

Two operators are used, namely migration and mutation. The migration operator's steps are presented in Algorithm 4.

When BBO is applied to MOPs, non-dominated sorting is added to deal with the multiple conflicting objectives. The multi-objective version of BBO is referred to as the non-dominated sorting biogeography-based optimisation (NSBBO) algorithm [45].

Vector-Evaluated Particle Swarm Optimisation

The vector evaluated particle swarm optimisation (VEPSO) algorithm, inspired by the vector evaluated genetic algorithm (VEGA) [44], was introduced by Parsopoulos et al. [42]. It consists of two layers, namely a top layer that manages the sub-swarms and a lower layer that contains the sub-swarms. At the top layer, the algorithm manages the sharing of knowledge between the sub-swarms, and at the lower layer, each sub-swarm optimises only its assigned objective function.

The search process of VEPSO is driven through local and global guides. The local guides, or pbests, contain information about the particles' own experience with regard to a single objective (assigned to their sub-swarm). On the other hand, the global guides, which are the global bests (or gbests), contain information obtained by a pre-defined neighbourhood of particles with regard to another objective. A knowledge sharing topology determines which gbest is used. Two knowledge sharing topologies, namely ring [42] and random [18], are presented in Fig. 3. With the ring topology, each sub-swarm uses the gbest of its neighbouring sub-swarm. With the random topology, each sub-swarm uses the gbest randomly selected between its own gbest and the gbest of its neighbouring sub-swarm.

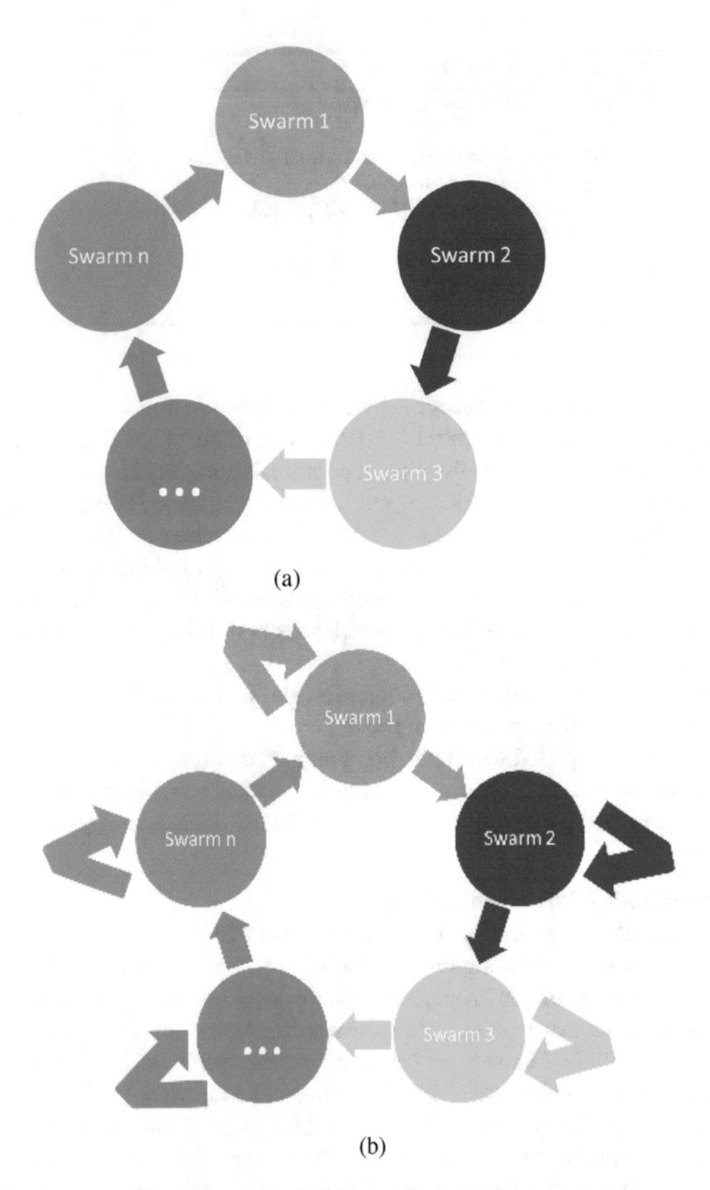

Fig. 3 Topologies of VEPSO algorithm. (**a**) Ring topology. (**b**) Random topology

Similar to VEPSO, the search process of dynamic VEPSO (DVEPSO) [18] is driven through local and global guides. In contrast to VEPSO, DVEPSO's guide update approaches use Pareto-dominance information.

If a change has been detected by one or more of the sub-swarms, DVEPSO responds to the change to ensure tracking of the changing POF by either

- re-evaluating all particles in the sub-swarm or
- re-initialising a percentage of the particles in the sub-swarm.

Re-evaluating the particles ensures that all previously obtained information is preserved. However, the particles already converged towards the POF, and therefore the diversity of the sub-swarm has to be increased to ensure exploration of a new environment. If re-evaluation is used, additional approaches should be used to increase the sub-swarm's diversity. Re-initialisation introduces diversity by re-initialising a certain percentage of the sub-swarm's particles and preserves previously obtained information from the particles that are not re-initialised. However, it may occur that particles with optimal positions in the new environment are re-initialised, and thereby the information is lost.

Greeff and Engelbrecht [17] proposed that the responses can be applied to either all sub-swarms or only the sub-swarm(s) whose objective function has changed. Applying the response to all sub-swarms increases the diversity of all sub-swarms and thereby increases the exploration of the sub-swarms. If a sub-swarm's objective function did not change and re-initialisation is used, a percentage of previously obtained information is removed. However, the increasing diversity may lead to exploration of the search space that was not explored before.

3 Real-World Problems

This section discusses DMOO RWPs. A taxonomy is presented in Sect. 3.1 of real-world DMOPs that were solved with DMOAs. The characteristics of these RWPs are highlighted in Sect. 3.2.

3.1 Taxonomy of Real-World Dynamic Multi-objective Optimisation Problems

This section categorises various RWPs according to the following aspects: application, the objectives and constraints, DMOAs evaluated on the problem, performance measures used to evaluate the performance of the DMOAs and the dynamic elements of the problem. In Table 1, # refers to 'number of'.

The DMOAs referred to in Table 1 are as follows:

- AI-NSGA-II [32]: NSGA-II using adaptive immigrants.
- DC-NSGA-II [49]: a dynamic constrained NSGA-II, adapted for dynamic environments, and the Pareto-dominance comparison. The algorithm has been adapted and the Pareto-dominance comparison used by the algorithm has been adapted to deal with constraint violations.

Table 1 Taxonomy of real-world dynamic multi-objective optimisation problems

Application	Objectives and constraints	Performance measures	DMOAs	Dynamic elements
Dynamic workflow scheduling for the cloud [31]	6 objectives: minimise makespan, cost, energy and degree of imbalance; maximise reliability and utilisation	Hypervolume, Schott's spacing, #non-dominated solutions	NN-DNSGA-II, DNSGA-II-A, DNSGA-II-B, DNSGA-II-HM, DNSGA-II-RI, DMOPSO	Resource failures (hardware/software), #objectives, uncertainties
Dynamic travelling salesman (combining travelling salesman and knapsack) [34]	3 objectives: minimise cost function per objective, can be scaled		Parallel implementation	#cities, cost between cities, #objectives, change intervals
Dynamic travelling thief [29]	2 objectives: travelling salesman and knapsack	Hypervolume, spread	NSGA-II [4] extended for dynamic problem	Dynamic city location, dynamic item location, dynamic item values (weight/profit/capacity)
Economic emission load dispatch to charge electric vehicles [36]	2 objectives: minimise economic load dispatch and emission load dispatch; 4 constraints: power balance (equality), power capacity, ramp rate limits (inequality)	Average power cost	Dy-NSBBO [36]	Objectives
Scheduling smart home appliances [49]	3 objectives: minimise energy consumption, energy cost and dissatisfied requests; 9 constraints: both equality and inequality constraints	Hypervolume	DC-NSGA-II [49]	#appliances, #tenants, energy policy/pricing
Hydro-thermal power scheduling [13]	2 objectives: minimise emission and cost; 6 constraints: equality (including quadratic)	Hypervolume ratio	DNSGA-II-A, DNSGA-II-B	Demand

(continued)

Table 1 (continued)

Application	Objectives and constraints	Performance measures	DMOAs	Dynamic elements
Energy consumption and productivity improvement in batch electrodialysis [43]	2 objectives: maximise the concentration of the concentrate tank and minimise energy consumption; 2 constraints: inequality and equality	Hypervolume, spacing and diversity	NSGA-II	Objectives
Greenhouse control [54]	4 objectives: maximise crop; minimise indoor heat, indoor CO_2 and constraint violations; equality constraints	Convergent ratio, average density, coverage scope and coverage rate	Multi-objective immune algorithm [54], NSGA-II, SPEA2	Density of CO_2
Railway junction rescheduling [15]	2 objectives: minimising timetable variation and additional energy expenditure (want to eventually remove expenditure objective)	Hypervolume, generational distance	DM-PACO-ST, DM-PACO-R, DM-MMAS-SC, DM-MMAS-ST, DM-MMAS-NC, DM-MMAS-NT, NSGA-II, FCFS	#trains, frequency of trains changing objective function values
Raw ore allocation in mineral processing [14]	5 objectives: maximise concentrate yield, concentrate grade and metal recovery and minimise concentrate cost; 7 inequality constraints	Coverage ratio, hypervolume	NSGA-II with random immigrants and gradient-based local search [14], NSGA-II with immigrants [13], DSS, DVEPSO, MOEA/D-KF	Capacity of ball mill grinding cell

(continued)

Table 1 (continued)

Application	Objectives and constraints	Performance measures	DMOAs	Dynamic elements
Social network clustering [32]	2 objectives: maximise min–max cut and global silhouette	Binary ϵ indicator	AI-NSGA-II, HI-NSGA-II, NSGA-II, DNSGA-II-RI, EI-NSGA-II	Data
Route Planning [21]	3 objectives: minimise length, travel time and ease of driving; constraints: hard (traffic regulations) and soft		Single objective genetic algorithm, multi-objective genetic algorithm, Dijkstra algorithm	Traffic
Control of time-varying unstable plants [30]	3 objectives: minimise rising time, shooting time and settling time; 3 constraints: inequality		Dynamic multi-objective algorithm based on membrane computing	Objectives
Identifying vulnerable zones based on water quality [22]	2 objectives: minimise cost and maximise treasure; constraints: equality and inequality	Generational distance, inverted generational distance, hypervolume	Parity-Q deep Q network (PQDQN) algorithm [22]	Treasure resulting in changes in POS and/or POF
Scheduling in welding industry [35]	3 objectives: minimise the makespan, the total penalty of machine load and instability; 16 constraints: equality and inequality	Generational distance, inverted generational distance, spread	Hybrid multi-objective grey wolf optimiser (HMOGWO) [35], MOGWO, NSGA-II, SPEA2	Machine breakdown, job with poor quality and job release delay
Financial markets—currency exchange [2]	2 objectives: minimise total transaction cost and maximise net profit	Net profit, winning trades, coefficient of variation	DVEPSO, MOPSO with crowding distance, DNSGA-II	Currency exchange rates

- Dy-NSBBO [36]: a dynamic non-dominated multi-objective biogeography-based optimisation algorithm that uses migrants and mutation.
- DM-MMAS-NC [15]: ACO algorithm where the solutions in the archive is repaired and pheromone trails are cleared after a change occurred.
- DM-MMAS-NT [15]: ACO where the archive is repaired after a change occurred, but the pheromone values are retained.
- DM-MMAS-SC [15]: ACO where the pheromone matrix is re-initialised and the archive is cleared after a change in the environment occurred.
- DM-MMAS-ST [15]: ACO where after a change in the environment occurred, an evaporation variable is used on the pheromone values and the archive is cleared.
- DMOPSO [8]: a multi-objective particle swarm optimisation algorithm that uses an adaptive grid and mutation and re-initialises a percentage of the population when a change is detected in the environment.
- DM-PACO-R [15, 20]: ACO with built-in memory that at each decision point randomly selects a pheromone matrix.
- DM-PACO-ST [15, 20]: ACO with built-in memory that uses an average-weight-rank method.
- DNSGA-II-A [12]: dynamic non-dominated sorting genetic algorithm II (DNSGA-II) that re-initialises a percentage of the population when a change in environment has been detected.
- DNSGA-II-B [12]: DNSGA-II that replaces a percentage of the population with solutions created through mutation of existing solutions.
- DNSGA-II-HM [50]: DNSGA-II that introduces diversity into the population by increasing the mutation rate for a certain number of generations, after which the mutation rate is decreased again.
- DNSGA-II-RI [19, 31]: a version of DNSGA-II that, instead of trying to detect changes in the environment and then increasing diversity, re-initialises a percentage of the population at each generation.
- DSS [51]: a prediction-based multi-objective algorithm that uses a direct prediction strategy.
- DVEPSO [26]: a multi-population particle swarm optimisation algorithm where each sub-swarm optimises only one objective and then shares knowledge between the various sub-swarms.
- EI-NSGA-II [32]: NSGA-II using elitism-based immigrants.
- FCFS [15]: a heuristic often used by train dispatchers to recover the timetable after perturbations.
- HI-NSGA-II: NSGA-II using a hybrid immigrant scheme consisting of both random and elitism-based immigrants.
- MOEA/D-KF [40]: a multi-objective evolutionary algorithm that uses Kalman filtering prediction.
- NN-DNSGA-II [31]: a prediction-based DNSGA-II that integrates an artificial neural network with NSGA-II.

3.2 Characteristics of Real-World Dynamic Multi-objective Optimisation Problems

From Table 1, the following characteristics of DMOO RWPs can be highlighted:

- Some problems are unconstrained, but with boundary constraints for the variables. However, the majority of the problems have both equality and inequality constraints. In addition, some constraints are hard constraints and cannot be violated, where as other constraints are soft constraints where violation of the constraints should only be minimised.
- The performance measures used to evaluate the performance of DMOAs vary. However, the hypervolume, spread and the generational distance (or inverted generational distance) are frequently used.
- The DMOAs used to solve the RWPs vary greatly and even more so the algorithms that they are evaluated against.
- The dynamic elements of the problems affect both the objectives and the constraints. Furthermore, the number of objectives can change over time.
- Some problems also have uncertainty with regard to decision variable values.
- Some problems have continuous variables and other problems discrete variables.

4 Challenges of Solving Real-World Problems

This section discusses the various challenges experienced when applying DMOAs to RWPs. The variable types, i.e. discrete-valued or continuous-valued, and the challenges associated with these types are discussed in Sect. 4.1. Issues that have to be addressed with regard to benchmark functions and performance measures are discussed in Sects. 4.2 and 4.3, respectively. The difficulty of selecting an algorithm to solve a specific RWP is presented in Sect. 4.4. Section 4.5 highlights the issues with regard to decision-making.

4.1 Discrete- Versus Continuous-Valued Variables

RWPs can have discrete-valued or continuous-valued variables or both. However, typically, an algorithm will be adapted specifically for a certain type of variable. Most research in DMOO focuses on continuous optimisation problems. Most of the research on discrete optimisation problems is applied research, where a specific algorithm has been developed to solve an RWP. It can for example be that an algorithm that performs really well on continuous problems does not perform as well when solving discrete problems, even if certain modifications are made. Furthermore, certain operators have to be adjusted when solving discrete problems. Therefore, more research is required to understand the impact of the various modifications of these operators on the performance of the algorithms and to investigate whether algorithms that were developed for continuous DMOPs can efficiently solve discrete problems when the required operators are modified.

4.2 Benchmarks

Helbig and Engelbrecht [27] proposed ideal characteristics that a BF suite should have. However, the characteristics of RWPs, highlighted in Sect. 3.2, identified the following aspects that should still be addressed to ensure that DMOO BFs are more representative of RWPs: constraints, frequency and severity of change, as well as uncertainty.

Constraints
Most DMOO BFs are unconstrained, with only boundary constraints. However, many RWPs do require constraints to be considered (refer to Table 1). Some BFs with constraints have been proposed [3], but most research focuses on solving unconstrained BFs. Therefore, more BFs should be developed that contain various types of constraints, namely

- equality and inequality constraints,
- static and dynamic constraints and
- soft and hard constraints.

Frequency and Severity of Change
Most BFs also have cyclic changes. However, when solving RWPs, changes happen at random and do not necessarily follow a specific pattern, such as cyclic changes. Furthermore, not all RWPs have the same severity of change occurring every time. Therefore, BFs should be developed that contain

- changes that occur at different frequencies, which are more random and not cyclic and
- severity of changes that differ and that are not constant throughout the run.

Uncertainty
In addition, some RWPs also have uncertainty around certain variables' values. Therefore, BFs should be developed that incorporate uncertainty.

4.3 Performance Measures

When solving RWPs, the hypervolume is a natural choice to measure a DMOA's performance, since the calculation of the hypervolume does not require prior knowledge of the POF. However, depending on the type of problem and how the POF changes over time, the hypervolume values can be misleading when a DMOA loses track of the changing POF [25]. Therefore, a visual check of the found POFs that are being compared should be conducted.

4.4 Algorithm Selection

Selecting a DMOA to solve a specific RWP is a daunting task. Many DMOAs were developed. However, no comprehensive comparison of these algorithms has been conducted that provide a good understanding of the types of problems that certain DMOAs perform well on and which types of problems they are struggling to solve. Furthermore, even if such a study would be conducted, the following issues would have to be addressed:

- With reference to Sect. 4.2, which BFs (that are representative of RWPs) should the DMOAs be evaluated on? In addition, which severity and frequency of changes should be used?
- Which performance measures should be used to evaluate the performance of the DMOAs? Furthermore, how should the performance measure values be used for the comparison? Using the average and standard deviation values will not necessarily represent how the DMOAs performed in the various environments. One possibility is to plot the measure values over time, but when many BFs are used with different combinations of frequency and severity of changes, then this approach is not feasible. Another approach has been proposed by Helbig and Engelbrecht, referred to as the wins-and-losses approach for dynamic environments [24]. The wins-and-losses approach enables the quantification of the performance of DMOAs across various benchmarks, specific categories such as environment types or DMOP types. Furthermore, it indicates how much one DMOA is statistically significantly better than another DMOA by taking the DMOAs tracking ability into account by not only using average values but also looking at the DMOA's performance in the various environments.
- How is a fair comparison between the DMOAs ensured and what constitutes a fair comparison? Should the population size be set equal amongst the DMOAs or the fitness evaluations? What if one DMOA's complexity is much more than the other DMOAs? What if some DMOAs make use of an archive (or even multiple archives) and other DMOAs do not? Should these aspects even be taken into consideration to ensure a fair comparison?

When the results are obtained from the study, it is not sufficient to only report on the performance of the DMOAs in general. An analysis should be conducted on:

- Which type of BFs did each DMOA perform really well on or struggle with and why? But to answer this question, how should the BFs be categorised for the discussion? For example, should the BFs be categorised according to the DMOP type defined by Farina et al. [16] or according to other BF characteristics such as types of constraints?

- Which type of change did each DMOA perform really well on or struggle with and why? However, this leads to the question of which types of changes should be considered? For example, should the changes be categorised according to frequency of changes, severity of changes or a combination of these two aspects or perhaps according to whether the objectives and/or constraints changed over time?

When this analysis of the results is conducted, it may not always be trivial to know why a DMOA performed well or struggled with a specific BF. Sometimes, in order to understand the behaviour of a specific DMOA, visualisation of the entities (individuals, particles etc.) over time, and especially after changes occurred, can be a useful tool. However, the question then becomes: what exactly should be visualised to deepen the understanding of the DMOA's behaviour? Should the movement of the entities over time with regard to the POS or POF be visualised or the change in performance measure values over time or all of these?

4.5 Decision-Maker

An important aspect of solving RWPs is decision-making. However, decision-making has not been addressed sufficiently yet in the field of DMOO. When solving static MOP, typically, the role of the decision-maker in the process is [9]

- interactive during the search, by providing input that guides the search process, and/or
- selecting the best solution after the search.

However, when solving dynamic RWPs, if the decision-maker provides input during the search, should the input be extrapolated after a change in the environment occurs to automatically adapt the input for the new environment? Or should the decision-maker guide the search by providing input after every change that occurs in the environment? Approaches that enable a decision-maker to guide the search in DMOO have been proposed [1, 38, 41]. However, more research is required.

A similar question occurs about selecting the best solution from the POS. Should the decision-maker select the best solution for each environment? i.e. after a change in the environment occurs and the optimal solutions are found for that specific environment, the decision-maker selects the best solution, and then this process repeats for each change that occurs. Or should the decision-maker's selection of a solution be extrapolated and used after a change occurs to automatically select a solution for the following environments? Furthermore, which visualisation techniques can be used to assist the decision-maker with selecting a solution from the POS?

5 Conclusion

Most research in the field of dynamic multi-objective optimisation (DMOO) focuses on unconstrained dynamic multi-objective optimisation problems (DMOPs). However, most real-world DMOO problems (RWPs) have constraints. Therefore, benchmark functions should be created that are representative of RWPs.

This chapter presented a taxonomy of DMOO RWPs, identifying the area of application, the objectives and constraints, dynamic multi-objective algorithms (DMOAs) that were evaluated on RWPs and performance measures that were used, as well as the dynamic elements of RWPs. This taxonomy highlighted aspects that are lacking in current benchmark functions (BFs).

These aspects or challenges were each discussed in more detail. The challenges included discrete problems versus continuous problems, BFs used to evaluate the performance of newly developed DMOAs, performance measures used to evaluate the quality of the found solutions, the selection of a DMOA to solve the RWP and decision-making.

BFs should be developed that are more representative of RWPs and an in-depth evaluation of DMOAs should be conducted on this set of BFs.

References

1. R.A. Adekoya, Driving dynamic multi-objective optimizations constrained by decision makers' preferences. Master's thesis, University of Pretoria, 2019
2. F. Atiah, M. Helbig, Effects of decision models on dynamic multi-objective optimization algorithms for financial markets, in *Proceedings of IEEE Congress on Evolutionary Computation (CEC)* (2019), pp. 762–770
3. R. Azzouz et al., Handling time-varying constraints and objectives in dynamic evolutionary multi-objective optimization. Swarm Evol. Comput. **39**, 222–248 (2018)
4. J. Blank, K. Deb, S. Mostaghim, Solving the bi-objective traveling thief problem with multi-objective evolutionary algorithms, in *Evolutionary Multi-Criterion Optimization*, ed. by H. Trautmann, G. Rudolph et al. (Springer, Basel, 2017), pp. 46–60
5. A. Carlisle, G. Dozler, Tracking changing extrema with adaptive particle swarm optimizer, in *Proceedings of the Biannual World Automation Congress, Orlando* (2002), pp. 265–270
6. L. Chen, H.Y. Sun, S. Wang, Parallel implementation of ant colony optimization on MPP, in *Proceedings of the International Conference on Machine Learning and Cybernetics (ICMLC)*, vol. 2 (2008), pp. 981–986
7. C.C. Coello, M. Lechuga, MOPSO: a proposal for multiple objective particle swarm optimization. Proc. Congr. Evol. Comput. **2**, 1051–1056 (2002)
8. C.C. Coello, G. Pulido, M. Lechuga, Handling multiple objectives with particle swarm optimization. IEEE Trans. Evol. Comput. **8**(3), 256–279 (2004)
9. K. Deb, *Multi-objective Optimization Using Evolutionary Algorithms*. Wiley-Interscience Series in Systems and Optimization (Wiley, Chichester, 2001)
10. K. Deb, S. Agarwal, A. Pratap, T. Meyarivan, A fast and elitist multiobjective genetic algorithm: NSGA-II. Tech. Rep. 200001, Indian Institute of Technology Kanpur, Kanpur Genetic Algorithms Laboratory (KanGAL), Kanpur, 2000
11. K. Deb, A. Pratap, S. Agarwal, T. Meyarivan, A fast and elitist multiobjective genetic algorithm: NSGA-II. IEEE Trans. Evol. Comput. **6**(2), 182–197 (2002)

12. K. Deb, N. Rao, S. Karthik, Dynamic multi-objective optimization and decision-making using modified NSGA-II, in *Proceedings of the International Conference on Evolutionary Multi-Criterion Optimization*. Lecture Notes in Computer Science, vol. 4403 (Springer, Berlin, 2007), pp. 803–817
13. K. Deb, U. Bhaskara Rao, S. Karthik, Dynamic multi-objective optimization and decision-making using modified NSGA-II: a case study on hydro-thermal power scheduling, in *Evolutionary Multi-Criterion Optimization*, ed. by S. Obayashi, K. Deb, C. Poloni et al. (Springer, Berlin, 2007), pp. 803–817
14. J. Ding et al., Dynamic evolutionary multiobjective optimization for raw ore allocation in mineral processing. IEEE Trans. Emerg. Top. Comput. Intell. **3**(1), 36–48 (2019)
15. J. Eaton, S. Yang, M. Gongora, Ant colony optimization for simulated dynamic multi-objective railway junction rescheduling. IEEE Trans. Intell. Transp. Syst. **18**(11), 2980–2992 (2017)
16. M. Farina, K. Deb, P. Amato, Dynamic multiobjective optimization problems: test cases, approximations, and applications. IEEE Trans. Evol. Comput. **8**(5), 425–442 (2004)
17. M. Greeff, A. Engelbrecht, Solving dynamic multi-objective problems with vector evaluated particle swarm optimisation, in *Proceedings of World Congress on Computational Intelligence (WCCI): Congress on Evolutionary Computation, Hong Kong* (2008), pp. 2917–2924
18. M. Greeff, A. Engelbrecht, Dynamic multi-objective optimisation using PSO, in *Multi-Objective Swarm Intelligent Systems*, ed. by N. Nedjah, L. dos Santos Coelho, L. de Macedo Mourelle. Studies in Computational Intelligence, vol. 261 (Springer, Berlin, 2010), pp. 105–123
19. J. Grefenstette, Genetic algorithms for changing environments, in *Parallel Problem Solving from Nature, Honolulu, Hawaii* (1992), pp. 139–146
20. M. Guntsch, M. Middendorf, Pheromone modification strategies for ant algorithms applied to dynamic TSP, in *Applications of Evolutionary Computing*, ed. by E. Boers (Springer, Berlin, 2001), pp. 213–222
21. H.K.K. Hara, Hybrid genetic algorithm for dynamic multi-objective route planning with predicted traffic in a real-world road network, in *Proceedings of the Annual Conference on Genetic and Evolutionary Computation 2008, Atlanta, GA* (2008), pp. 657–664
22. M. Hasan et al., Dynamic multi-objective optimisation using deep reinforcement learning: benchmark, algorithm and an application to identify vulnerable zones based on water quality. Eng. Appl. Artif. Intell. **86**, 107–135 (2019)
23. M. Helbig, Solving dynamic multi-objective optimisation problems using vector evaluated particle swarm optimisation. Ph.D. thesis, University of Pretoria, 2012
24. M. Helbig, A. Engelbrecht, Analysing the performance of dynamic multi-objective optimisation algorithms, in *Proceedings of the IEEE Congress on Evolutionary Computation, Cancún* (2013), pp. 1531–1539
25. M. Helbig, A. Engelbrecht, Issues with performance measures for dynamic multi-objective optimisation, in *Proceedings of the Symposium on Computational Intelligence in Dynamic and Uncertain environments, Singapore* (2013), pp. 17–24
26. M. Helbig, A. Engelbrecht, Dynamic multi-objective optimization using PSO, in *Metaheuristics for Dynamic Optimization*, ed. by E. Alba, A. Nakib, P. Siarry (Springer, Berlin, 2013), pp. 147–188
27. M. Helbig, A. Engelbrecht, Benchmarks for dynamic multi-objective optimisation algorithms. ACM Comput. Surv. **46**(3), 1–39 (2014)
28. M. Helbig, A. Engelbrecht, Benchmark functions for CEC 2015 special session and competition on dynamic multi-objective optimization. Tech. rep., University of Pretoria, 2015
29. D. Herring, M. Kirley, X. Yao, Dynamic multi-objective optimization of the travelling thief problem. Tech. rep., Cornell University, 2020
30. L. Huang, I. Suh, A. Abraham, Dynamic multi-objective optimization based on membrane computing for control of time-varying unstable plants. Inf. Sci. **181**(11), 2370–2391 (2011)
31. G. Ismayilov, H. Topcuoglu, Neural network based multi-objective evolutionary algorithm for dynamic workflow scheduling in cloud computing. Future Gener. Comput. Syst. **102**, 307–322 (2020)

32. K. Kim, R. McKay, B.R. Moon, Multiobjective evolutionary algorithms for dynamic social network clustering, in *Proceedings of the Annual Conference on Genetic and Evolutionary Computation, Portland, OR* (2010), pp. 1179–1186
33. M. Lechuga, Multi-objective optimisation using sharing in swarm optimisation algorithms. Ph.D. thesis, University of Birmingham, 2009. http://etheses.bham.ac.uk/303/
34. W. Li, M. Feng, A parallel procedure for dynamic multi-objective TSP, in *Proceedings of the International Symposium on Parallel and Distributed Processing with Applications, Leganes* (2012), pp. 1–8
35. C. Lu et al., A hybrid multi-objective grey wolf optimizer for dynamic scheduling in a real-world welding industry. Eng. Appl. Artif. Intell. **57**, 61–79 (2017)
36. H. Ma et al., Multi-objective biogeography-based optimization for dynamic economic emission load dispatch considering plug-in electric vehicles charging. Energy **135**, 101–111 (2017)
37. R. Macarthur, E. Wilson, *The Theory of Island Biogeography* (Princeton University Press, Princeton, 2001)
38. T. Macias-Escobar et al., Plane separation: a method to solve dynamic multi-objective optimization problems with incorporated preferences. Future Gener. Comput. Syst. **110**, 864–875 (2020)
39. R. Marler, J. Arora, The weighted sum method for multi-objective optimization: new insights. Struct. Multidiscipl. Optim. **41**, 853–862 (2010)
40. A. Muruganantham, K. Tan, P. Vadakkepat, Evolutionary dynamic multiobjective optimization via Kalman filter prediction. IEEE Trans. Cybern. **46**(12), 2862–2873 (2016)
41. A. Nebro et al., InDM2: interactive dynamic multi-objective decision making using evolutionary algorithms. Swarm Evol. Comput. **40**, 184–195 (2018)
42. K. Parsopoulos, M. Vrahatis, Recent approaches to global optimization problems through particle swarm optimization. Nat. Comput. **1**(2–3), 235–306 (2002)
43. F. Rohman, N. Aziz, Performance metrics analysis of dynamic multi-objective optimization for energy consumption and productivity improvement in batch electrodialysis. Chem. Eng. Commun. **0**(0), 1–13 (2019)
44. J. Schaffer, Multiple objective optimization with vector evaluated genetic algorithms, in *Proceedings of the International Conference on Genetic Algorithms* (L. Erlbaum Associates Inc., Hillsdale, 1985), pp. 93–100
45. S. Singh, E. Mittal, G. Sachdeva, NSBBO for gain-impedance optimization of Yagi-Uda antenna design, in *Proceedings of the IEEE World Congress on Information and Communication Technologies* (2012), pp. 856–860
46. N. Srinivas, K. Deb, Muiltiobjective optimization using nondominated sorting in genetic algorithms. Evol. Comput. **2**(3), 221–248 (1994)
47. T. Stútzle, An ant approach to the flow shop problem, in *Proceedings of the European Congress on Intelligent Techniques & Soft Computing (EUFIT), Verlag* (1997), pp. 1560–1564
48. T. Stützle, H. Hoos, Improvements on the ant-system: introducing the MAX-MIN ant system, in *Artificial Neural Nets and Genetic Algorithms* (Springer, Vienna, 1998), pp. 245–249
49. W. Trabelsi et al., Leveraging evolutionary algorithms for dynamic multi-objective optimization scheduling of multi-tenant smart home appliances, in *Proceedings of the IEEE Congress on Evolutionary Computation, Vancouver* (2016), pp. 3533–3540
50. F. Vavak, K. Jukes, T. Fogarty, Adaptive combustion balancing in multiple burner boiler using a genetic algorithm with variable range of local search, in *Proceedings of the International Conference on Genetic Algorithms, East Lansing, MI* (1999), pp. 719–726
51. Y. Wang, B. Li, Multi-strategy ensemble evolutionary algorithm for dynamic multi-objective optimization. Memetic Comput. **2**(1), 3–24 (2009)
52. L. Zadeh, Optimality and non-scalar-valued performance criteria. IEEE Trans. Automat. Contr. **8**(1), 59–60 (1963)
53. Q. Zhang, H. Li, MOEA/D: a multiobjective evolutionary algorithm based on decomposition. IEEE Trans. Evol. Comput. **11**(6), 712–731 (2007)
54. Z. Zhang, Multiobjective optimization immune algorithm in dynamic environments and its application to greenhouse control. Appl. Soft Comput. **8**(2), 959–971 (2008)

Mardé Helbig is a Senior Lecturer at Griffith University, Australia. She obtained her PhD at the University of Pretoria, South Africa. Before joining Griffith University, she worked at the Council for Scientific and Industrial Research (CSIR) and the University of Pretoria in South Africa.

She started working in the field of computational intelligence (CI) with her PhD studies, where she investigated solving optimisation problems with multiple often conflicting goals, where the goals and/or constraints change over time. These problems are called dynamic multi-objective optimisation problems (DMOPs). She selected this topic due to her love for mathematics and being curious about the advantages of using nature-inspired population-based algorithms to solve DMOPs, since DMOPs occur in many situations in the real world. Her current research interests are solving DMOPs using CI algorithms (evolutionary algorithms and swarm intelligence algorithms), decision making and visualisation.

She was the main organiser of special sessions on dynamic multi-objective optimisation (DMOO) at numerous conferences and a competition on DMOO at CEC 2015. She presented a tutorial on DMOO at the IEEE Symposium Series on Computational Intelligence (SSCI) 2015, the International Conference on Swarm Intelligence (ICSI) 2016 and the World Congress on Computational Intelligence (WCCI2018). She was invited as a keynote speaker on DMOO at the International Conference on Soft Computing and Machine Learning (ISCMI) 2016 and the International Conference on Mechanical and Intelligent Manufacturing Technologies (ICMIMT) 2017–2019 and presented the first Memorial Lecture of Prof Zadeh at ISCMI 2017. She was the local organiser of SSCI 2015 and the co-chair of the ACO and SI track of the Genetic and Evolutionary Computation Conference (GECCO) in 2020. She is the co-chair of the ACO and SI track of GECCO 2021 and the special session chair of SSCI 2021. She is a regular reviewer for the top conferences and journals in the field. In addition, she is the chair of the IEEE Computational Intelligence Society (CIS) Chapters sub-committee and a subcommittee member of the IEEE CIS Women in Computational Intelligence and the IEEE CIS Young Professionals.

In 2016, she was awarded a Y1 rating by the National Research Foundation of South Africa, indicating that she is an emerging researcher who already enjoys international recognition for her research on DMOO. In 2017, she was selected as a member of the South African Young Academy of Science (SAYAS) and in 2017 and 2018 as an executive committee member of SAYAS. In 2019, she was awarded the TW Kambule-NSTF Emerging Researcher award for her research on DMOO.

Computational Intelligence Methodologies for Multi-objective Optimization and Decision-Making in Autonomous Systems

Sanaz Mostaghim

1 Introduction

Multi-criteria decision-making is a difficult task since it involves multiple conflicting objectives resulting in a set of optimal alternatives. Choosing an optimal alternative means losing the others. Usually, it is the duty of a (human) decision-maker to decide which alternative to take using personal preferences. The science of multi-criteria decision-making has attracted the attention of many scientists over the last 100 years. Starting from late 1890, Vilfredo Pareto worked on the individual's choices and the application of utility functions for defining an order for preferences in decision-making [33]. He is known for the Pareto-optimality and Pareto-efficiency concepts. Generally, we define a multi-objective optimization problem as a problem which has several conflicting objective functions to be optimized at the same time. The solution of this problem is a set of so-called Pareto-optimal solutions (called trade-offs in engineering) from which we have to select one (i.e., make a decision).

In the last decades, special decision support systems have been developed to help the decision-makers to take the best possible alternative. Very often in critical applications such as in a cancer treatment plan, the role of a decision support system goes beyond providing the best alternative. It gives more confidence to the decision-maker about her/his decision. Decision support systems are already fully integrated in design toolboxes for engineers. Toolboxes in MATLAB or other commercial software such as modeFRONTIER offer tools for finding and selecting the optimal alternative for engineering design. It will be a long list and out of the scope of this chapter to name all of these applications (some are listed in [8, 16, 27, 28, 30]).

S. Mostaghim (✉)
Faculty of Computer Science, Otto von Guericke University Magdeburg, Magdeburg, Germany
e-mail: sanaz.mostaghim@ovgu.de

A. E. Smith (ed.), *Women in Computational Intelligence*, Women in Engineering and Science, https://doi.org/10.1007/978-3-030-79092-9_17

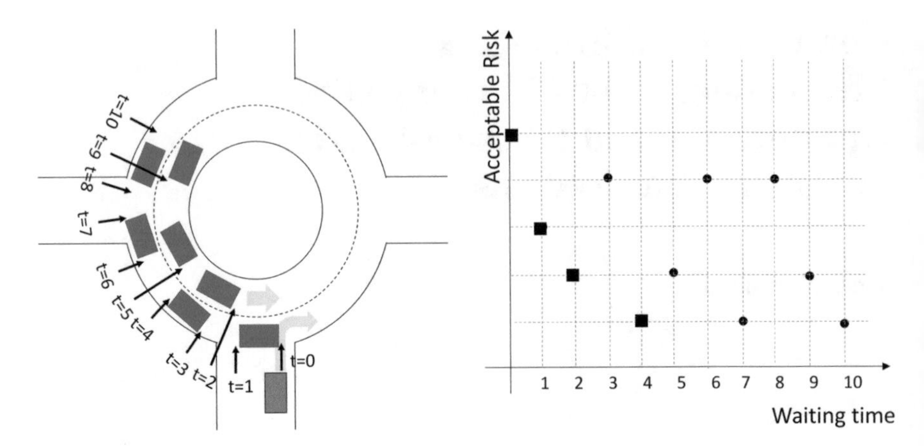

Fig. 1 Example of an autonomous driving car deciding to enter a roundabout and confronted with two conflicting objectives: (1) minimize waiting time and (2) minimize risk. The right figure shows the alternative actions in the objective space

Conflicting objectives appear in many navigation and exploration scenarios in which an autonomous robot or a self-driving car is confronted with. Imagine an autonomous driving car at a crowded roundabout. It should decide: "Should I wait long and avoid risk to enter or should I enter the roundabout and take a bit of acceptable risk". Figure 1 shows this example with two objectives and some of possible actions.

Other exemplary scenarios are as follows: an autonomous flying robot: "should I fly forward and consume energy, or should I rather stay in hovering mode and save energy", an autonomous driving robot in rescue scenario: "Should I take a route with many intersections but shorter distance to point of interest, or should I go for a longer route and less intersections", and finally an autonomous flying robot: "should I fly through a tunnel and get to the point of interest faster, or should I avoid collisions with the walls and take a longer route". In these scenarios, a wrong decision can end up with drastic consequences and can lead to an unsuccessful termination of the mission.

For many years, scientists have been working on decision-making algorithms. However, in most existing approaches, either we assume that the optimal alternatives are given (or we have enough time to calculate them) and/or the decision policy (preferences) to take a possible alternative is predefined by the designer (such as a set of weights). This is not the case in many real-world applications, where the autonomous systems are confronted with unforeseen conflicting objectives during a mission. The goal of this chapter is to give an overview about the multi-criteria decision-making algorithms, corresponding challenges, and the existing methodologies for autonomous systems particularly in navigation and exploration scenarios.

2 Decision-Making in Various Disciplines

The field of operations research has intensively investigated multi-objective problems and the multi-criteria decision-making (MCDM) algorithms in different domains from politics to industrial applications to individual use cases such as in medical treatments [16, 27, 28]. Preferences in multi-criteria decision-making are usually defined in various forms and are intensively studied. A preference is simply a policy that a decision-maker uses to decide for an alternative. Scalarization functions such as a set of weights defining the importance of each of the objectives are the mostly used form of preferences [16, 28]. Other approaches concern defining various utility functions such as preference ordering [1]. Value functions can be used to model the preferences. In this case, the decision-maker is asked to make pairwise comparisons of certain preselected alternatives. This information is then used to build the model [11]. Very often a decision-maker is uncertain about her/his preferences. In this case, we can incorporate the preferences in the form of a desirability function which is a nonlinear function over the range of an objective. The preference is based on a probability function, indicating the most probable range which is preferred by the human decision-maker [17, 49]. Uncertainties in decision-making are usually modelled using quantitative and qualitative approaches [7, 16]. In all of these, the preferences or models are set by (or for) a human decision-maker.

The field of neuroscience has a long list of literature about understanding decision-making of humans and animals. A very recent work addresses several models for finding the optimal policy for multi-alternative decisions [47]. They study the reaction time in neurons as a function of the number of alternatives. It is known according to Hick's law that the more alternatives we offer to a decision-maker, it is more difficult for her/him to come up with an optimal decision [19].

The field of machine learning offers several topics for preference representation (conditional preference nets, logical representations, and fuzzy constraints), reasoning with preferences (decision theory, constraint satisfaction, and non-monotonic reasoning), and preference acquisition (preference elicitation and preference learning) [12]. In preference learning methods, the preferences are learned based on large data sets and samples (rankings) such as in recommender systems.

In the field of robotics, the two topics of planning and decision-making are essential components for achieving autonomy [48]. Very often the decision-making refers to a range of algorithms from path to motion to task planning for taking actions that help robots understand the world around them. Most of the methodologies about decision-making in navigation scenarios in robotics [20] vary from A* to recent advances in Reinforcement Learning (RL) [46]. They are intensively used for task and path planning. These (path, task, or motion) planners highly rely on one objective and do not consider the conflict between several objectives. In a very recent survey in [44], the authors provide an overview about emerging trends and challenges in the field of intelligent and autonomous cars. They categorize the existing approaches into game-theoretic, probabilistic, and learning-based approaches. Game-theoretic and probabilistic methods are developed for predicting the behaviour of other systems (e.g., the other cars or pedestrians on the road or sidewalk). Decision-making is addressed in the learning-based approaches. Among these, Inverse Reinforcement Learning (IRL) also known as inverse optimal control has been identified as a prominent approach. In IRL, an unknown reward function is learned from expert (human) demonstrations (e.g., [43]). Even if many of the problems contain conflicting objectives, the exiting approaches use scalarization techniques such as a weighted sum approach and reduce the conflicting objectives into one single objective. Both RL and IRL (even with the combination of deep learning, such as deep reinforcement learning methods [29]) rely on one objective. These mechanisms simply follow a greedy approach and select the action (alternative) which gives the maximum reward. The drawback of a greedy approach is that very often an optimal action at a certain time-step can lead to degraded actions in later time-steps. This is of great importance for systems that are in long missions such as an agent in a computer game or a mobile robot or an autonomous car. Only recently, there have been a few works on multi-criteria decision-making in robotics [2, 31], which will be discussed later.

Concerning time critical decision-making, modern RL-based approaches, such as Monte Carlo Tree Search (MCTS) [14], combine the greedy term with an exploration mechanism, use a tree policy, and integrate a Monte Carlo approach for estimation of future states. In this way, they can be efficiently used for time critical applications, where the agent has to make quick actions (decisions). The prominent application of MCTS is the DeepMind's Alpha-Go Zero [45]. Incorporating conflicting objectives in RL has been subject to study, e.g., [41, 42]. The existing approaches rely on scalarization functions and turn the objectives into one reward function [13]. We have also investigated a new variation of MCTS with a scalarization function [36, 37].

Furthermore, learning under conflicting objectives has been studied in the new research field of multi-objective learning [22, 23, 34]. In most of these approaches, special mechanisms are developed to find the Pareto-optimal alternatives, but decision-making is not addressed.

3 Multi-objective Optimization and Decision-Making

As already mentioned, in multi-objective optimization we are confronted with several conflicting objectives $f_i(\mathbf{x})$, $i = 1, \ldots, m$, which are to be optimized (without loss of generality, we take minimization):

$$\begin{aligned} &\text{Minimize } \mathbf{f}(\mathbf{x}) = (f_1(\mathbf{x}), \ldots, f_m(\mathbf{x})) \\ &\text{Subject to } \mathbf{x} \in S \end{aligned} \tag{1}$$

where $\mathbf{x}$ corresponds to a decision variable in n-dimensional search space S. The solution of this problem is a set of so-called Pareto-optimal solutions denoted by P. For each $\mathbf{x} \in P$, there is no other $\mathbf{y} \in S$, which dominates $\mathbf{x}$ (denoted by $\mathbf{y} \prec \mathbf{x}$ referring to $\mathbf{y}$ dominates $\mathbf{x}$):

$$\begin{aligned} \mathbf{y} \prec \mathbf{x} : & f_i(\mathbf{y}) \leq f_i(\mathbf{x}), \forall i = 1, \cdots m \text{ and} \\ & f_j(\mathbf{y}) \leq f_j(\mathbf{x}), \exists j \end{aligned} \tag{2}$$

In this way, the solutions in P are all optimal and indifferent from one another. We usually represent these solutions in the so-called objective space (representing the objective values). An example of such objective space is shown in Fig. 1 (right). The image of these solutions in the objective space is called the Pareto-front (Pareto Frontier). The goal of multi-objective optimization algorithms is to find multiple Pareto-optimal solutions which can provide a good representation of the Pareto-front. In Fig. 1 (right), Pareto-optimal solutions are indicated by ■. The ultimate goal of multi-criteria decision-making is to select one of these Pareto-optimal solutions.

Usually, there are two ways of dealing with multi-criteria decision-making: either the preferences of the decision-maker are given a priori or a posteriori to the optimization process [16, 28]. A priori methods such as scalarization algorithms provide only one optimal solution which matches to the given preference, while a posteriori methods find a set of alternatives from which the decision-maker can select one.

Weighted Sum Approach The prominent scalarization method is the weighted sum approach, which is widely used in engineering [28]. In this a priori approach, the decision-maker predefines her/his preferences in form of a weight vector $(w_1, \cdots, w_m)$, where $\sum_{i=1}^{m} w_i = 1$. The multi-objective problem is then transformed to a single-objective problem:

$$\begin{aligned} &\text{Minimize } f(\mathbf{x}) = w_1 f_1(\mathbf{x}) + \cdots + w_m f_m(\mathbf{x})) \\ &\text{Subject to } \mathbf{x} \in S \end{aligned} \tag{3}$$

This mechanism can be additionally used for a posteriori decision-making to select an optimal alternative from a set of Pareto-optimal solutions. In this case, given a set P and a set of weight vectors, we take the solution $\mathbf{x}^*$ so that

$$\mathbf{x}^* = arg \min_{\mathbf{x} \in P}(w_1 f_1(\mathbf{x}) + \cdots + w_m f_m(\mathbf{x})) \tag{4}$$

ϵ-Constraint Approach This is another a priori method, which transforms the multi-objective problem into a single-objective problem by selecting one objective to be optimized. The other objectives are taken as constraints:

$$\begin{aligned} &\text{Minimize } f_i(\mathbf{x}) \\ &\text{Subject to } \mathbf{x} \in S \text{ and} \\ &\qquad f_j(\mathbf{x}) < \epsilon_j, \forall j = 1, \cdots, m \text{ and } j \neq i \end{aligned} \tag{5}$$

The value of ϵ_j can either be set to one value for all of the objectives that are considered as constraints or it can be different. These values have to be defined by the decision-maker a prior to the optimization.

Similar to the weighted sum approach, we can use this technique for an a posteriori decision-making to select one solution from the set P. The solution that fulfils the constraints $(f_j(\mathbf{x}) < \epsilon_j, \forall j = 1, \ldots, m \text{ and } j \neq i)$ and additionally delivers the minimum value $\mathbf{x}^* = arg\, \min_{\mathbf{x} \in P} f_i(\mathbf{x})$ is selected.

4 Multi-objective Optimization and Decision-Making in Autonomous Systems

The main challenge in the context of decision-making for autonomous systems is that there is usually not much time for optimization before taking an action during a mission. While the a priori methods are very fast, they usually find one Pareto-optimal solution according to predefined preferences of the decision-maker. The problem in most of the cases is that the predefined preferences cannot fully capture unforeseen situations. Therefore, there is a need to re-optimize and find alternatives. This will be very time consuming and cannot be used within a limited time frame during the mission. Other a posteriori approaches such as evolutionary multi-objective optimization (EMO) algorithms can deliver a set of alternatives in one run, due to their parallel nature [8]. In the past 20 years, EMO algorithms have been improved in terms of the performance and there is a long list of applications that successfully use them [24]. Nevertheless, multi-objective optimization during a mission brings new challenges for these algorithms. One major aspect is that at each time-step, we need to restart the optimization algorithms to find a set of appropriate alternatives for that time-step. This is a computational challenge if we want to have a smooth movement of the agents. In the following, we have identified two forms of decision-making during a mission for two different timescales:

1. Decision-making during the mission: time to decide is about a few seconds or minutes.
2. Decision-making in time critical missions: time to decide is in a few milliseconds.

In the following, we introduce challenges for both of these cases and propose methodologies to deal with them.

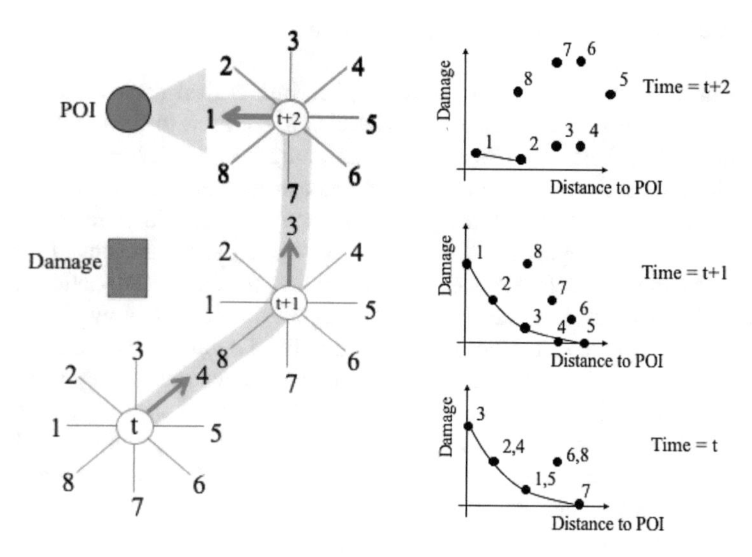

Fig. 2 A simple example of an agent in three time-steps, t, $t+1$, and $t+2$. The two conflicting objectives are to minimize both the distance to the POI and the amount of damage

4.1 Decision-Making During the Mission

Consider the following navigation scenario of an autonomous system as shown in Fig. 2. Here, we observe the agent in three time-steps, t, $t+1$, and $t+2$. At each time-step, it has to decide among the 8 routes (directions). The two conflicting objectives are to minimize the distance to the POI, and to minimize the damage caused by an obstacle. In the illustration, you do not see the optimization process for finding the optimal routes. Here, we depict the multi-criteria decision-making by the agent (arrows show the taken alternatives over time). For each of the 8 directions, the agent has to measure the corresponding objective values. Figure 2 (right) shows the objective space for the different time-steps. We note that the shape of the Pareto-front can change. Change of the shape of a Pareto-front is known in the context of dynamic optimization. However, the challenge at this stage concerns the fact that the shape and therefore the complexity of the problem can change over time.

The multi-criteria decision-making at this stage can be based on various modelling of preferences such as weighted sum, ϵ-constraint, or a desirability function [49]. In order to enable the autonomous system to change its policy, we need to develop adaptive mechanisms. In [2, 31, 36], we adapt the value of the weight vector to the current situation. The main idea is to enable the agent to change this vector over time. The agent needs to observe the current status for each of the objectives and has to change the weight vector so that the objective that is in a critical status gets a larger weight value. The same procedure can be used for other decision-making policies. In such a critical mission, three major challenges for optimization algorithms are limited computational time, dynamic problems, and multi-modality.

4.1.1 Computational Time for Optimization

Even if at this stage, we are not dealing with time critical decision-making and have a few seconds or minutes to take a decision, we need to deal with a limited computational budget. We can take the advantage of the nature of navigation and exploration tasks in which certain actions can remain constant over a certain number of time-steps. For example, while moving from one point to the next, unless unexpected events happen, the direction for navigation stays unchanged for a few seconds or minutes. Therefore, we can work on the so-called macro-actions [38]. A macro-action is defined as one action that is repeated for a certain number of time-steps. In this way, we can allocate enough time for the optimization algorithms to find optimal alternatives. In addition to macro-actions, we can use algorithms with low computational budget. There is a large number of related works in the literature of multi-objective optimization which deal with limited computational budget (in terms of the function evaluations, i.e., computing time). The most prominent work is the ParEGO algorithm [25]. It uses a specific initialization procedure and learns a Gaussian process model of the search landscape, which is updated after every function evaluation. Other approaches rely on surrogate models [3].

4.1.2 Dynamic Multi-objective Optimization

Problems with changing Pareto-front in terms of shape and position are called dynamic multi-objective optimization problems. Solving these problems is in fact a very difficult task, since the changes are unpredictable. Change detection mechanisms are the core of existing approaches [5]. Recent approaches [18, 52] address adaptation to the varying complexity of the problem. This means that at some time-steps, it is very difficult to find a near-optimal solution, due to the complexity of the problem. But in other time-steps due to a new shape of the front, it is very easy to find possible alternatives. One of such problems with varying complexity over time is the so-called distance minimization problem [21, 32].

This problem is of great relevance to navigation and exploration applications for mobile robots. The goal of such problems is to find a set of optimal alternatives that have minimum distances to m objective points in a given space. Depending on the distance function L1 or L2 norms (i.e., Manhattan or Euclidean distance functions), the difficulty of the problem changes. The complexity additionally depends on the positions of the objective points [5].

Figure 3 shows an example for three objective points O1, O2, and O3 for given predefined positions using Euclidean (left) and Manhattan (middle and right) metrics. Using Manhattan metric increases the complexity of the problem, since the Pareto-front shown by P contains two lines. Finding solutions in a continuous space which exactly lie on a line is almost impossible, given a very limited time budget. Varying the positions of the objective points will additionally change the shape and therefore the complexity of the problem as shown in Fig. 3 middle and right.

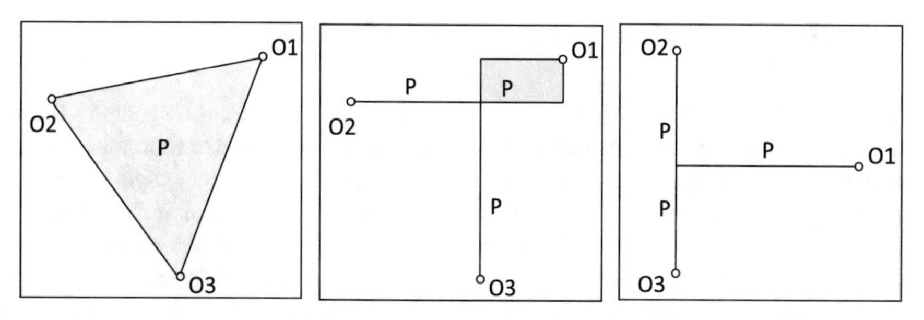

Fig. 3 Distance minimization problem with three conflicting objectives and the corresponding Pareto-front (P) with two different distance metrics: Euclidean (left) and Manhattan (middle and right)

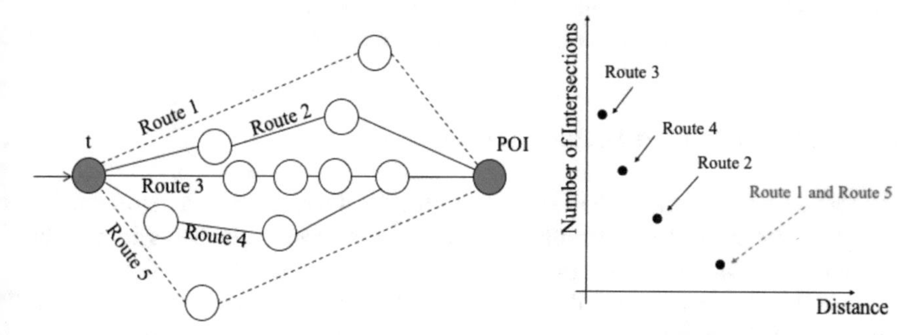

Fig. 4 Example of a multi-modal multi-objective Problem: minimizing both the distance to POI and the number of intersections. Alternatives are shown in the objective space (right). Both Routes 1 and 5 (multi-modal alternatives) map to the same objective values

4.1.3 Multi-modality in Optimization

One challenge in navigation and exploration applications is to find all possible alternatives in a given time. Here, we can be confronted with the so-called multi-modal problem [26]. Such problems have several alternatives which yield the exact same objective values. It is of great importance for decision-making to obtain all such alternative solutions. Suppose the following example in a navigation scenario. The goal is that the agent minimizes the distance to get to a point of interest (POI) while minimizing the number of intersections along the way (Fig. 4). These two goals are in conflict with each other as can be observed in the figure. The two multi-modal solutions are shown as Route 1 and Route 5 with dashed lines. State-of-the-art EMO methodologies such as NSGA-II [9], NSGA-III [10], MOEA/D [50], and many other approaches unfortunately fail to find all these alternatives, since they operate in the objective space and remove the possible duplicates by using a crowding distance operator (objective space is shown in Fig. 4). This means that only either Route 1 or Route 5 can be found. The existing approaches for multi-modal multi-objective optimization are designed to cope with this problem [26].

4.2 Decision-Making in Time Critical Missions

The task of decision-making gets more difficult if we consider the agents in time critical missions. We assume that such missions have a very limited time for finding and taking an action (a few milliseconds in computer games or a flying robot). Adapting preferences (i.e., finding the optimal policy) can help exploit the potential for perfect navigation and exploration and will give the agent the ability to act differently at the same decision-making situations but with different settings, in terms of the energy level or environmental dynamics such as wind. For instance, an agent with not much time and energy (fuel) can act in a more aggressive way to at least reach a certain point of the mission than an agent that can take the advantage of available resources and take safe actions. Given the limited time for optimization and decision-making, there are two major challenges such as real-time decision-making algorithms and reduction of the number of alternatives.

4.2.1 Limited Time for Decision-Making

Time as a constraint limits the exploration time for the optimization algorithms to a large extent, even if we use macro-actions or surrogate methods (as in already mentioned for non-time critical missions). Here, we assume that such time critical missions have a very limited time for finding and taking an action. In "real-time games" or "video game playing", this critical time is set to less than 40 milliseconds. Novel approaches based on Monte Carlo Tree Search (MCTS) can help to deal with time limitations [14]. Such methods take the advantage of the given time to explore the possible actions in a tree policy and then use the concept of Monte Carlo Reinforcement Learning [36] to estimate the expected return from that point.

Figure 5 (left) shows the concept for the tree policy. $Q_t(a)$ is the so-called action value function. The agent usually selects the action which gives the maximum value function. The exploration term involves the bandit-based methods to help get rid of the local optima shown as $U_t(a)$. Other methods such as Rolling Horizon Evolutionary Algorithms (RHEAs) [35] use the available time to explore possible actions using Evolutionary Algorithms (EAs) [15]. Since EAs are among the anytime algorithms [51], we can stop the optimization process at any time and have some good (near-optimal) alternatives. In all these variations, the difficulty arises when we have multiple conflicting objectives as shown in Fig. 5 (right). In [36], a new variation of MCTS called Multi-Objective MCTS (MO-MCTS) is proposed, which is designed to deal with multi-objective problems containing several conflicting objectives. They use a so-called hypervolume metric, which is an indicator to evaluate the quality of a set of Pareto-optimal solutions [8], and integrate it into the MCTS equations. The actions are then taken according to a predefined set of preferences [37]. The results show that MO-MCTS approaches can significantly outperform single-objective algorithms in specific real-time scenarios.

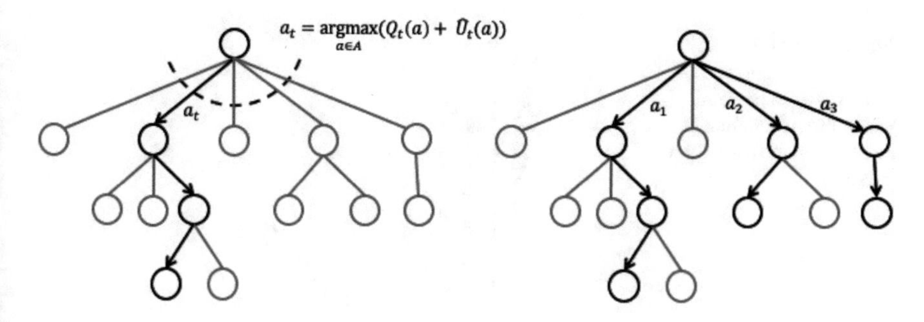

Fig. 5 An example of tree policy in MCTS. Left: single objective. Right: multi-objective with 3 alternative optimal actions

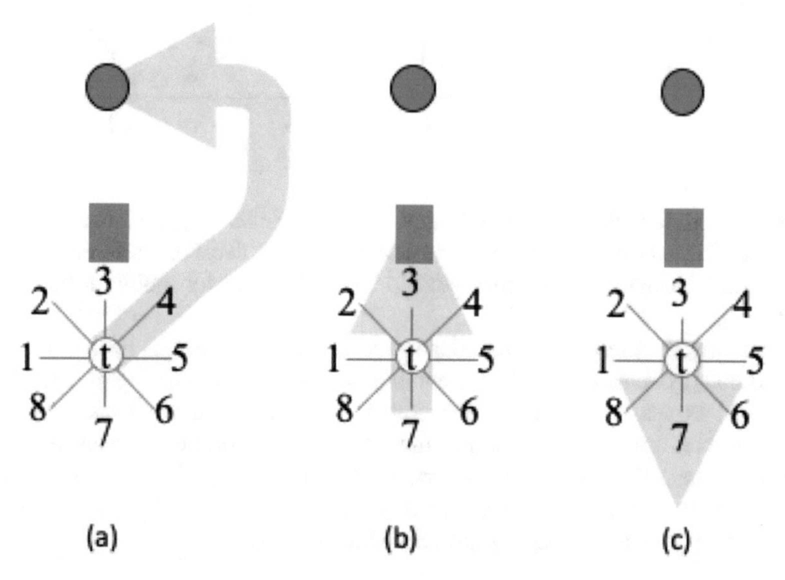

Fig. 6 Example of three human-like behaviours in decision-making: (**a**) weight vector (0.5, 0.5) as in Fig. 2, (**b**) "Aggression behaviour" with weight vector (1, 0), and (**c**) "Fear behaviour" with weight vector (0, 1)

This kind of real-time multi-criteria decision-making can lead to several behaviours for the same agent. Let us assume a simple example shown in Fig. 2. Taking a simple weighted sum approach [28] of weights (0.5, 0.5) for the two objectives to minimize $(f1, f2)$ = (distance to POI, amount of damage) will lead to taking actions as shown in the figure. Figure 6 shows two further scenarios in (b) and (c). If we take a weight vector of (1, 0), the mission will terminate in time-step $t + 1$, since the agent goes towards the maximum damage. On the other hand, if we take (0,1) as the weight vector, the agent will not be able to reach the POI. These three simple settings of weight vectors for decision-making show that we can

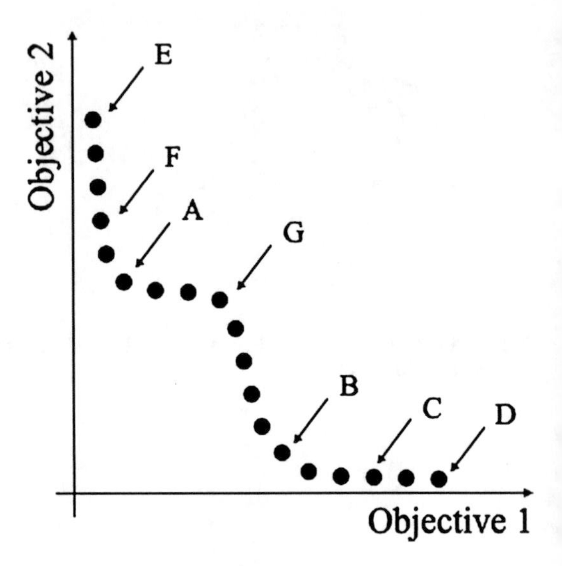

Fig. 7 Identifying the alternative solutions E, A, G, B, and D as knee points which provide sufficient information to make a good decision

implicitly produce three different human-like behaviours (aggression, neutral, and fear), similar to human behaviours when selecting their decision preferences. These human-like behaviours have been studied by Braitenberg for mobile robots in the 1980s [4].

Changing the preferences (similar to the above example) has been studied for autonomous systems. In [2, 31], the authors study an adaptive setting of preferences (modelled as weight vectors) in the presence of conflicting objectives: minimize both fuel consumption and mission time. The motivation comes from the aerial robotics, where the robots have a very limited energy capacity. They have found out that adaptive settings of preferences (weights) can be very helpful to reach an even better performance than for predefined fixed settings.

4.2.2 Reducing the Number of Alternatives

In order to make a reasonably good decision in a very limited time, we need to reduce the number of alternatives to a small set. According to Hick's law [19], it is very difficult for a human decision-maker to make a reasonable decision from a large number of alternatives. In the following, we aim to reduce the number of alternatives to identify only the most "important" ones called knee points. In Fig. 7, we explain the idea behind the methodology. This figure shows an example of a set of Pareto-optimal alternatives when minimizing both Objectives 1 and 2. While all of these solutions are optimal and indifferent from one another, some of them (such as A, B, D, E, and G) seem to be very important for the decision-making, as they either indicate the extremes of the front (e.g., D and E) or are located in knee areas (e.g., A, G, and B). The two solutions E and D define the extreme points of the Pareto-front, which deliver information about each single-objective function. When comparing C with D, we realize that D is better than C in terms of Objective 2, but the proportion that C is better than D in terms of Objective 1 is much larger. In this way, we can prefer the solution C to D by only sacrificing a bit for Objective 2. In fact, the same reasoning holds, when comparing B and C, and the alternative B can be preferred over C. Similarly, we can conclude the same for A and G. The alternatives A, B, and G are known to be the so-called knee points of the Pareto-front. There are different ways of finding these knee points [6]. While clustering methods can help to identify such solutions, fast algorithms such as the cone-domination concept [6] will have no additional computational evaluations and can identify the knee points. The difficulty arises, when we increase the number of objectives. In this case, we produce many alternatives and the parameter settings of cone-domination will be very challenging. In [39, 40], we used the concept of cone-domination and introduced an adaptive parameter setting so that the knee points can be efficiently identified. It is shown that by increasing the number of objectives to 7, using cone-domination can reduce the number of alternatives from 50 to 5.

5 Conclusions

This chapter addresses the problem of multi-objective decision-making for autonomous systems. This topic is very relevant for the application of autonomous systems in critical missions, where the intervention of the human operator is not always possible. We have defined the problem and provided the corresponding challenges for navigation and exploration scenarios. Autonomous decision-making requires autonomy in selecting an optimal alternative without the intervention of the human controller. This can raise many ethical and social questions. However, the major focus of this chapter concerns the algorithms and computational intelligence methodologies for optimization and decision-making during the mission at run-time.

References

1. S. Angilella, S. Greco, F. Lamantia, B. Matarazzo, Assessing non-additive utility for multicriteria decision aid. Eur. J. Oper. Res. **158**(3), 734–744 (2004)
2. P. Bartashevich, D. Koerte, S. Mostaghim, Energy-saving decision making for aerial swarms: PSO-based navigation in vector fields, in *IEEE Swarm Intelligence Symposium* (2017)
3. K. Bhattacharjee, H. Singh, T. Ray, J. Branke, Multiple surrogate assisted multiobjective optimization using improved pre-selection, in *IEEE Congress on Evolutionary Computation* (IEEE, 2016)
4. V. Braitenberg, *Vehicles: Experiments in Synthetic Psychology* (MIT Press, 1986)
5. J. Branke, *Evolutionary Optimization in Dynamic Environments* (Springer, 2002)

6. J. Branke, K. Deb, H. Dierolf, M. Osswald, Finding knees in multi-objective optimization, in *Parallel Problem Solving from Nature - PPSN VIII*, ed. by X. Yao, E.K. Burke, J.A. Lozano, J. Smith, J.J. Merelo-Guervós, J.A. Bullinaria, J.E. Rowe, P. Tiňo, A. Kabán, H.-P. Schwefel (Springer, Berlin, Heidelberg, 2004), pp. 722–731
7. N. Bulling, A survey of multi-agent decision making. Künstliche Intelligenz **28**(3), 147–158 (2014)
8. K. Deb, *Multi-Objective Optimization Using Evolutionary Algorithms* (Wiley, 2001)
9. K. Deb, A. Pratap, S. Agarwal, T. Meyarivan, A fast and elitist multiobjective genetic algorithm: NSGA-II. IEEE Trans. Evol. Comput. **6**(2), 182–197 (2002)
10. K. Deb, H. Jain, An evolutionary many-objective optimization algorithm using reference-point-based nondominated sorting approach, part i: Solving problems with box constraints. IEEE Trans. Evol. Comput. **18**(4), 577–601 (2014)
11. J. Figueira, S. Greco, R. Slowinski, Building a set of additive value functions representing a reference preorder and intensities of preference: Grip method. Eur. J. Oper. Res. **195**(2), 460–486 (2009)
12. J. Furnkranz, E. Hullermeier, *Preference Learning* (Springer, 2011)
13. J. Fürnkranz, E. Hüllermeier, W. Cheng, S.-H. Park, Preference-based reinforcement learning: a formal framework and a policy iteration algorithm. Machine Learning **89**(1), 123–156 (2012)
14. S. Gelly, Y. Wang, R. Munos, O. Teytaud, Modification of UCT with patterns in Monte-Carlo go, Tech. Rep. 2, Inst. Nat. Rech. Inform. Auto. (INRIA), 2006
15. D.E. Goldberg, *Genetic Algorithms in Search, Optimization and Machine Learning* (Addison-Wesley Longman Publishing, 1989)
16. S. Greco, M. Ehrgott, J. Figueira, *Multiple Criteria Decision Analysis: State of the Art Surveys*, Series Volume 233 (Springer, 2016)
17. J. Harrington, The desirability function. Ind. Qual. Control **21**(10), 494–498 (1965)
18. M. Helbig, H. Zille, M. Javadi, S. Mostaghim, Performance of dynamic algorithms on the dynamic distance minimization problem, in *ACM Genetic and Evolutionary Computation Conference (GECCO) Companion* (ACM, 2019)
19. W.E. Hick, On the rate of gain of information. Q. J. Exp. Psychol. **4**(1), 11–26 (1952)
20. L. Hofer, *Decision-making algorithms for autonomous robots*. Ph.D. thesis, Robotics Universite de Bordeaux, 2017
21. H. Ishibuchi, M. Yamane, N. Akedo, Y. Nojima, Many-objective and many-variable test problems for visual examination of multiobjective search, in *IEEE Congress on Evolutionary Computation* (IEEE, 2013)
22. Y. Jin, *Multi-Objective Machine Learning*. Springer Lecture Notes in Computer Science, vol. 16 (2006)
23. Y. Jin, B. Sendhoff, Pareto-based multiobjective machine learning: An overview and case studies. IEEE Trans. Syst. Man Cybern. C (Appl. Rev.) **38**(3), 397–415 (2008)
24. K. Klamroth, J. Knowles, G. Rudolph, M. Wiecek, Personalized multiobjective optimization: An analytics perspective, in *Dagstuhl Seminar 18031* (2018)
25. J. Knowles, Parego: A hybrid algorithm with on-line landscape approximation for expensive multiobjective optimization problems. IEEE Trans. Evol. Comput. **10**(1), 50–66 (2006)
26. J.J. Liang, C.T. Yue, B.Y. Qu, Multimodal multi-objective optimization: A preliminary study, in *2016 IEEE Congress on Evolutionary Computation*, pp. 2454–2461 (2016)
27. A. Lotov, V.A. Bushenkov, G. Kamenev, *Interactive Decision Maps, Approximation and Visualization of Pareto Frontier* (Springer, 2004)
28. K. Miettinen, *Nonlinear Multiobjective Optimization* (Springer, 1998)
29. V. Mnih, K. Kavukcuoglu, D. Silver, A.A. Rusu, J. Veness, M.G. Bellemare, A. Graves, M. Riedmiller, A.K. Fidjeland, G. Ostrovski, S. Petersen, C. Beattie, A. Sadik, I. Antonoglou, H. King, D. Kumaran, D. Wierstra, S. Legg, D. Hassabis, Human-level control through deep reinforcement learning. Nature **518**(7540), 529–533 (2015)
30. I. Moser, S. Mostaghim, The automotive deployment problem: A practical application for constrained multiobjective evolutionary optimisation, in *Proceedings of the IEEE Congress on Evolutionary Computation, CEC* (IEEE, 2010), pp. 1–8

31. S. Mostaghim, C. Steup, F. Witt, Energy aware particle swarm optimization as search mechanism for aerial micro-robots, in *IEEE Swarm Intelligence Symposium* (2016)
32. S. Mostaghim, C. Steup, H. Zille, Multi-objective distance minimization problems and its application in technical systems. at - Automatisierungstechnik **66**(11), 964–974 (2018)
33. V. Pareto, *Cours d'Économie Politique Professé*. Ph.D. thesis, l'Université de Lausanne, 1986
34. S. Parisi, M. Pirotta, M. Restelli, Multi-objective reinforcement learning through continuous Pareto manifold approximation. ACM J. Artif. Intell. Res. **57**(1), 187–227 (2016)
35. D. Perez, S. Samothrakis, S. Lucas, P. Rohlfshagen, Rolling horizon evolution versus tree search for navigation in single-player real-time games, in *ACM Genetic and Evolutionary Computation Conference (GECCO) Companion* (ACM, 2013)
36. D. Perez, S. Mostaghim, S. Samothrakis, S. Lucas, Multi-objective Monte Carlo tree search for real-time games. IEEE Trans. Comput. Intell. AI Games **7**(4), 347–360 (2015)
37. D. Perez-Liebana, S. Mostaghim, S. Lucas, Multi-objective tree search approaches for general video game playing, in *IEEE Congress on Evolutionary Computation* (2016)
38. D. Perez-Liebana, M. Stephenson, R.D. Gaina, J. Renz, S.M. Lucas, Introducing real world physics and macro-actions to general video game AI, in *IEEE Conference on Computational Intelligence and Games (CIG)* (IEEE, 2017)
39. C. Ramirez-Atencia, S. Mostaghim, D. Camacho, A knee point based evolutionary multi-objective optimization for mission planning problems, in *Proceedings of the Genetic and Evolutionary Computation Conference*, GECCO '17 (Association for Computing Machinery, New York, NY, USA, 2017), pp. 1216–1223
40. C. Ramirez-Atencia, S. Mostaghim, D. Camacho, sKPNSGA-II: Knee point based MOEA with self-adaptive angle for mission planning problems (2020). arXiv:2002.08867
41. D.M. Roijers, S. Whiteson, *Multi-Objective Decision Making* (Morgan & Claypool Publishers, 2017)
42. D.M. Roijers, P. Vamplew, S. Whiteson, R. Dazeley, A survey of multi-objective sequential decision-making. J. Artif. Int. Res. **48**, 67–113 (2013)
43. D. Sadigh, S. Sastry, S. Seshia, A. Dragan, Planning for autonomous cars that leverage effects on human actions. Robot. Sci. Syst. XII (2016)
44. W. Schwarting, J. Alonso-Mora, D. Rus, Planning and decision-making for autonomous vehicles. Annu. Rev. Control Robot. Auton. Syst. **1**(1), 187–210 (2018)
45. D. Silver, J. Schrittwieser, K. Simonyan, I. Antonoglou, A. Huang, A. Guez, T. Hubert, L. Baker, M. Lai, A. Bolton, Y. Chen, T. Lillicrap, F. Hui, L. Sifre, G. van den Driessche, T. Graepel, D. Hassabis, Mastering the game of go without human knowledge. Nature **550**, 354–359 (2017)
46. R.S. Sutton, A.G. Barto, *Introduction to Reinforcement Learning* (MIT Press, 1998)
47. S. Tajima, J. Drugowitsch, N. Patel, A. Pouget, Optimal policy for multi-alternative decisions. Nature Neuroscience **22**(9), 1503–1511 (2019)
48. S. Thrun, W. Burgard, D. Fox, *Probabilistic Robotics* (MIT Press, 2005)
49. H. Trautmann, C. Weihs, On the distribution of the desirability index using Harrington's desirability function. Metrika **63**(2), 207–213 (2006)
50. Q. Zhang, H. Li, Moea/d: A multiobjective evolutionary algorithm based on decomposition. IEEE Trans. Evol. Comput. **11**(6), 712–731 (2007)
51. S. Zilberstein, Using anytime algorithms in intelligent systems. AI Magazine **17**(3), 73–83 (1996)
52. H. Zille, A. Kottenhahn, S. Mostaghim, Dynamic distance minimization problems for dynamic multi-objective optimization, in *IEEE Congress on Evolutionary Computation* (2017)

Sanaz Mostaghim is a full professor of computer science and the head of the chair of Computational Intelligence at the Otto von Guericke University Magdeburg, Germany. She holds a Ph.D. degree (2004) in electrical engineering from the University of Paderborn, Germany and has worked as a postdoctoral fellow at ETH Zurich in Switzerland (2004–2006) and as a lecturer at Karlsruhe Institute of Technology (KIT), Germany (2006–2013). Sanaz received her habilitation degree (the highest academic degree in Germany) in applied computer science from KIT in 2012.

Her research interests are in the area of evolutionary multi-objective optimization and decision-making, swarm intelligence and their applications in robotics and science.

She is the deputy chair of the executive board of Informatics Germany, member of the advisory board on digitalization at the ministry of digitalization and economy in state Saxony-Anhalt, Germany and the head of the RoboCup team of the University of Magdeburg.

Sanaz is an active member of IEEE Computational Intelligence Society (IEEE CIS) and serves as the vice president for member activities (2021–2022). She was the member of the administrative committee of IEEE CIS (2015–2020).

She is an associate editor of IEEE Transactions on Artificial Intelligence, IEEE Transactions on Evolutionary Computation and member of the editorial boards of several international journals.

Message:

With a great pleasure, I participated in this book project. In this way I intend to encourage women in the field of engineering and computer science to pursue their goals. I have been always fascinated by the field of engineering and as a child could easily follow the logic in the field of mathematics more than other fields such as medicine or law. Studying electrical engineering and then pursuing my academic career in the field of computational intelligence was a very natural decision for me. However, the more I pave this career, the more I see the difficulties. Unfortunately, women are still underrepresented in these fields and are confronted with many challenges at work and along their academic career. Being the chair of Women in Computational Intelligence at the IEEE CIS (2016–2017), I have heard many stories from women across the world about the difficulties of being a woman in the field. Lack of acceptance from the male colleagues is one of the major difficulties making this career path even more challenging. Nevertheless, this book is an example of a group of women in the field of computational intelligence who intend to provide their support. Academic career path is difficult for both men and women, however, due to the lack of role models, support, and very often cultural barriers, many women feel discouraged to pave this career path. Here is a humble advice from me: with dedication, patience and perseverance you can reach your goals. Never give up. Set your goals and do not let others destroy your dreams.

A Framework-Based Approach for Flexible Evaluation of Swarm-Intelligent Algorithms

Eva Kühn and Vesna Šešum-Čavić

1 Introduction

Swarm-inspired algorithms are a very promising research area and have a broad spectrum where they can be used. In order to raise awareness and confidence in them, a method for the fair and flexible evaluation of swarm-inspired algorithms is needed.

1.1 Relevance of Swarm-Inspired Algorithms

Combinatorial optimization tasks belong to the most commonly arising computational tasks necessary and crucial in many practical applications [3]. Many real-world optimization scenarios are challenging from the computational point of view and have been shown to be NP-hard. Therefore, it is essential to develop adequate metaheuristics that obtain near-optimal solutions in a relatively short time and also to perform a systematic, theoretical study of such metaheuristics. This encompasses ways to measure, analyze, compare, and improve their performance. Interesting types of metaheuristics are those inspired by some mechanism or phenomenon from nature [39], in particular swarm intelligence.

These types of algorithms support the optimization and robustness of highly dynamic distributed systems where autonomous agents interact without central control. High complexity of distributed systems implies the necessity of finding

E. Kühn · V. Šešum-Čavić (✉)
TU Wien, Faculty of Informatics, Compilers and Languages Group, Institute of Information Systems Engineering, Wien, Austria
e-mail: eva@complang.tuwien.ac.at; vesna@complang.tuwien.ac.at

A. E. Smith (ed.), *Women in Computational Intelligence*, Women in Engineering and Science, https://doi.org/10.1007/978-3-030-79092-9_18

new approaches of creating, developing, and maintaining systems. The behavior of such complex systems is typically unpredictable, yet exhibits various forms of adaptation and self-organization. According to [22], there is a constant necessity for self-organizing mechanisms in distributed systems. Swarm-inspired metaheuristics address these issues. The design of good metaheuristic algorithms is a very active area of research where one continues to find new methods and techniques that are increasingly important in tackling large real-world optimization problems.

1.2 General Applicability

Basically, swarm-inspired algorithms are not bounded to any particular application domain but support and address more generic patterns like optimization, search, etc. However, when approaching a particular use case, the usually applied methodology is to implement a highly specific solution that integrates the algorithm. A good example for that is the load balancing pattern, where one can find an enormous amount of research dedicated to different aspects of this problem (load balancing of highly adaptive and irregular parallel applications, for parallel database systems, Distributed Hash Table (DHT)-based networks, etc.). For each of these specific requirements, usually a new approach is taken and a new solution is constructed. That leads to the problem of a fair comparison and evaluation of different approaches that use different swarm-inspired algorithms with a large number of parameters.

Moreover, it is difficult to derive a recommendation whether an algorithm that suits one problem (e.g., adaptive load balancing) would also be useful for another problem domain (e.g., distributed information retrieval).

1.3 Need for Fair and Flexible Evaluation Methodology

The motivation for research in this chapter stems from our work on different complex distributed use cases, where the advantages of different kinds of swarm algorithms (ants, bees, slime mold, etc.) should be proven [7]. The objective is to generalize highly specialized solutions in the area of swarm-inspired algorithms, so that a broad class of applications can profit from these algorithms and to establish guidelines when to apply which algorithm.

Therefore, the major research questions we target are: How is a set of suitable algorithms determined? What are the conditions under which algorithms need to be compared, such as fairness, independence on hardware/platform, topology etc.? Which swarm algorithm is the best for which use case? Is an algorithm applicable to only one problem or to a more general class of problems? Can results be reproduced?

1.4 Requirements on the Evaluation Methodology

Nowadays, there exists no recognized method that encompasses all of the following requirements:

Provisioning of a General Framework We differentiate between the complete solution for a domain specific use case (aka scenario) and the algorithm(s) integrated/used by the solution. As there is no "one-fits-all" solution, in order to find the best algorithm for a use case, a generalized framework is needed that allows the testing and tuning of different algorithms for a specific use case and environment. By framework we understand a general, reusable architectural solution that contributes to solving the communication and coordination generics found in a certain use case. For example, for a use case that requires data collected from many sensors to be evenly distributed to many processing nodes, these generics can be abstracted to the problem of load balancing. The motivation is that load balancing is an issue that occurs in many other scenarios. In other words, a framework can be seen as the implementation of a high level communication and coordination pattern that represents a general, reusable architectural solution to a certain scenario. Note that a framework should be the result of passing through the following steps: (I) analyze the problem and (II) abstract it to a generic case. A framework does not per se solve the entire use case but serves as a necessary base for the spectrum of algorithms used. It abstracts the general communication and coordination requirements found in a use case in the form of flexible software architecture.

Composability of the Architecture In order to achieve a generalized framework, it is worthwhile to decompose it into subcomponents that in turn implement subpatterns. A design pattern [23] describes a recurring and reusable solution in software design. In the load balancing pattern, there exist subpatterns for routing, task allocation, etc. The generality of a framework can be measured by its architecture flexibility. The architecture must be flexible, so that neither new requirements on specific algorithms nor other assumptions on the network infrastructure become "architecture breakers." In other words, the flexible exchange of components and algorithms as well as combinations of different components within the framework shall be possible. This can be achieved by means of composition.

Autonomy and Self-Organizing Properties Especially use cases where swarm-inspired algorithms unfold their benefits typically show high dynamics and complex coordination requirements. This increased complexity, diversity of requirements, and dynamically changing configurations force identifying new solutions based on self-organization, autonomic computing, and mobile agents. Intelligent algorithms require agents as they are advantageous in situations that are characterized by high dynamics, unforeseeable events, and heterogeneity.

Support of Arbitrary Configurations The objective is to stay problem and domain independent and to allow component composition towards arbitrary network topologies. A general framework must be able to cope with all these demands at the same time and offer means to abstract hardware and network heterogeneities.

Benchmarking in Different Environments Benchmarking has the following goals: fine tuning of parameters in order to obtain the best set of parameters; comparison of different algorithms (and combinations of them) in different use cases under different environmental settings, collection, and interpretation of the results; and recommendation of which algorithm suits the best for a given use case.

Possibility of Reconstructing the Solution Each finding shall be made transparent, so that the scientific and developer community can easily reconstruct it.

1.5 Related Work

The positioning of this chapter in the context of related work is as follows: We target a new methodology for the development of frameworks for a systematic, fair, and flexible comparison of swarm-inspired algorithms that possesses all the properties enumerated above (Sect. 1.4).

There are many research works on using swarm-inspired algorithms for dedicated use cases regarding highly specialized solutions (examples for some of them are: [9, 12, 28, 37, 62, 65]). These works are very promising and we believe that swarm-inspired algorithms can contribute to the improvement and optimization of diverse IT problems, especially in highly dynamic situations where automatic adaptation and self-organizing properties are required. The problem is that there is no general methodology for how to decide when to apply which swarm-algorithm, or in which context a possible coexistence of algorithms would lead to the best results. Also, the works implementing specialized solutions are carried out on different platforms, and usually the source code is not provided, so that: (1) a reconstruction of the results for the own problem is basically impossible, and (2) the different approaches simply cannot be compared to each other.

The existing related work does not fulfill the requirements of a general framework.

Only separated properties such as flexible architecture or a component-oriented approach are addressed. The software engineering community has proposed a number of development methodologies to provide flexibility where requirements can be met by component-orientation [16, 66], exchanging software components [40, 45], support of arbitrary configurations [46] and agent-based software architectures [46], and autonomy and self-organizing properties [13–15, 41].

With regard to the comparison of swarm-inspired algorithms, there exist only specific use cases and general overviews of different sets of swarm-inspired algorithms. A generally accepted development process for a framework that possesses all the identified properties does not exist.

The contribution of the chapter is a method that describes the steps from algorithm design to a systematic evaluation of algorithms in different settings. The provision of use cases and frameworks for evaluation is a part of the method. The chapter is structured as follows: Sect. 2 describes the proposed methodology. Section 3 discusses projects where this methodology was already successfully applied to use cases. The evaluation is presented in Sect. 4. Section 5 summarizes the results.

2 Proposed Methodological Approach

The construction of a general framework is the important part of our approach that allows for exchanging different algorithms through plugging. Different algorithms shall be tested in combination at different levels. The goal is to ease the selection of the best algorithm(s) for a certain problem scenario. Blackboard-based communication offers a high degree of decoupling (time, reference, space) and supports agents' autonomy. Our research focuses on a new conception of a self-organizing coordination infrastructure that suggests a combination of coordination spaces, self-organization, adaptive algorithms, and multi-agent technologies. Clearly, also conventional ("unintelligent") algorithms with fewer requirements on dynamics can be evaluated with this method. The proposed methodology comprises software engineering, interdisciplinary, and mathematical methods. Figure 1 shows the interplay of the proposed methods. It is not necessary to perform each of these steps for any problem. For instance, once a middleware has been defined, it can be reused for further iterations.

2.1 Mathematical Methods

M1: Mathematical Models After the analysis of swarm-mechanisms in nature, the behaviors that exist in a respective swarm-mechanism and govern natural phenomena must be described using mathematical representations as the abstraction of the reality, i.e., by using stochastic modeling and discrete mathematics tools (in particular, graph theory and discrete probability theory). It is assumed that we need nonlinear dynamic models. A constructed model is evaluated and assessed through the combination of several statistical analyses. The following techniques for model evaluation are suggested to be used: analysis of linear regression, analysis of fitting errors, concordance correlation coefficient, diverse evaluation measurements, mean square error of prediction, nonparametric analysis, and data distribution comparison [62].

M2: Mathematical Proofs The algorithms themselves must be analyzed and their behavior explained and theoretically proven (whether the algorithm converges, what kind of convergence exists, what speed of convergence is present, etc.)

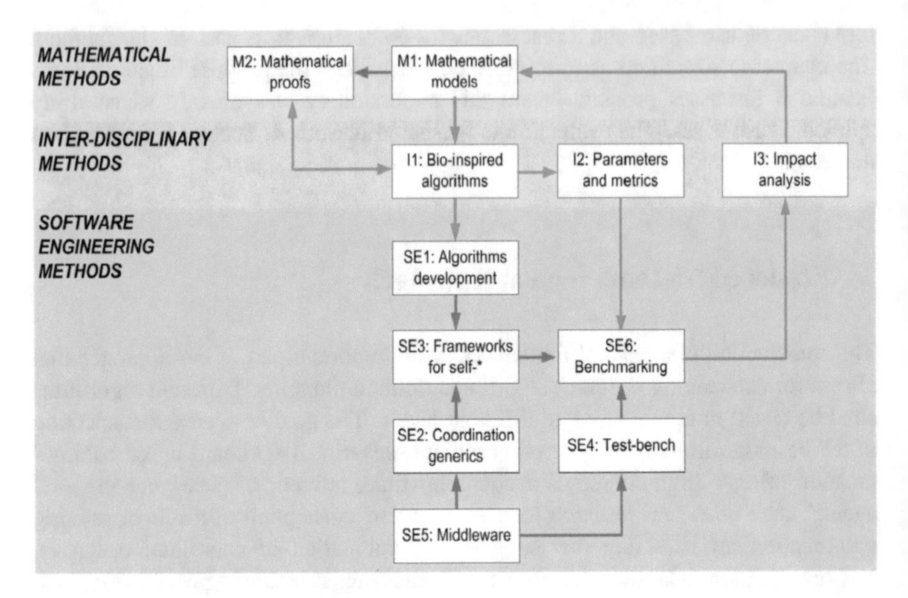

Fig. 1 Interplay of mathematical (M), software engineering (SE), and interdisciplinary (I) methods. Note: "Self-*" means different aspects of self-organizing properties such as self-healing, self-repairing, self-configuring, etc.

2.2 *Interdisciplinary Methods*

I1: Swarm-Inspired Algorithms Swarm-inspired algorithms appear as a consequence of used mathematical models. For each investigated swarm-mechanism in nature, one must analyze how it can be mapped to IT terminology. In mapping, software agents play the role of a particular swarm and perform self-organized actions characteristic for the respective swarm-colony. As an example, bees in nature, searching for nectar, can be understood as software agents searching for work to be done [47]; or ants in nature placing pheromones on their way are software agents picking up quality information about solutions found so far [49].

I2: Parameters and Metrics The so-called best parameter settings must be derived for each algorithm for each considered problem scenario, as all swarm-inspired algorithms are characterized by a huge number of different environmental parameters influencing the behavior of artificial swarms (e.g., number of ants or the invocation policy for load balancing). Evaluation criteria and specification of suitable metrics for scalability, performance, quality of solution, etc. must be identified.

I3: Impact Analysis A feedback loop to improve the mathematical models based on the benchmark results is suggested.

2.3 Software Engineering Methods

SE1: Algorithms The newly invented algorithms must be implemented based on the mathematical and algorithmic description (see above). In addition, other known (intelligent and/or unintelligent) algorithms shall be implemented that can also be applied to the problem at hand for comparison purposes.

SE2: Coordination Generics Reusable generic coordination mechanisms must be extracted for each use case as "patterns" which can be understood as universal blue prints (e.g., master/worker, routing, split/join, etc.). They must be implemented as reusable components. Their designs shall be inspired by multi-agent technologies [59]. Agents must be self-responsible, up and running, implement a certain reactive and continuous behavior, and may dynamically join and leave. This avoids a central coordinator, supports peer-to-peer collaboration, and thus is a necessary prerequisite for scalable, adaptive, and self-organized systems.

SE3: Frameworks for Self-Organization A particular challenge is that for each use case scenario a generic framework with self-organizing properties as a composition of the components described above must be designed and developed. The framework allows the exchange of different, swarm-inspired as well as other algorithms simply through "plugging", and must support many different network topology settings through configuration. The plugging approach is achieved by a component-based design, where each agent represents a certain exchangeable policy or behavior. This allows fair comparison of different algorithms in a neutral framework. It is possible to compare different combinations of algorithms within one problem scenario. The composition of entire frameworks into even larger ones shall also be supported (e.g., integration of load clustering and load balancing).

SE4: Test-Bench A test environment for automatically running and interpreting of benchmarks in all desired framework configurations and algorithms combinations must be developed. We suggest the test to be run on a cluster of computers and/or in a cloud in order to obtain significant results within a larger setting.

SE5: Middleware In order to achieve SE2 and SE4, a suitable middleware architecture must be selected that supports (a) good decoupling of all components (in order to make them easily exchangeable), (b) dynamic joining and leaving of processes, (c) agent-based communication patterns, (d) composition of pattern-based solutions, (e) interoperability between platforms, and (f) good abstraction of distribution (e.g., in form of overlay networks). Our suggestion is to use tuple-space-based middleware, which provides these desired properties [32].

SE6: Benchmarking Finally, the benchmarks need to be performed and the results evaluated using the parameters and metrics of I2. This includes trying out different algorithms (in pure form, hybrid form, and combinations of different algorithms) and different topology configurations. The results can continuously be used to improve the models, algorithms, and their parameter settings (see I3).

3 Experiences with Swarm-Inspired Algorithms and Self-Organizing Use Cases

Our work on self-organization so far focused on the following use cases: load balancing, load clustering, peer clustering, information placement, and retrieval in heterogeneous networks and distributed routing [56].

Load Balancing can be described as finding the best possible workload (re)distribution and addresses ways to transfer excessive load from busy nodes to idle nodes [30]. [47] explains for the first time how bee intelligence can be mapped to load balancing. [30] presents the load balancing coordination pattern SILBA (Self Initiative Load Balancing Agents) that assumes autonomous agents and decentralized control. It can be composed towards arbitrary network topologies, foresees exchangeable policies for load balancing and allows the combination of different algorithms at different network levels simultaneously – in sub-networks and between sub-networks [50, 51]. Self-organization is achieved by the application of swarm-based algorithms: the bee algorithm is modified and adapted for the load balancing problem as well as ant algorithms (MinMax Ant System and AntNet).

Load Clustering [31] tries to make further optimizations of the load distribution based on the content of the load items. [31] is a prototype for a clustering coordination pattern termed SILCA (Self Initiative Load Clustering Agents) for the problem of load clustering, developed and tested with a few known algorithms, demonstrating that load clustering is also a promising candidate for a self-organizing use case. The plugging of combinations of different algorithms as well as arbitrary network topologies is possible. Among different clustering and classifying algorithms (K-Means, Fuzzy C-Means, Genetic K-Means, Hierarchical Clustering, K-Nearest Neighbor, Decision Trees), SILCA demonstrates a usage of ant intelligence in clustering. SILCA works with numbered clustering algorithms succesfully.

[48, 52, 53, 55] deal with data placement and retrieval in the internet. Unstructured peer-to-peer (P2P) overlay network technologies are combined with swarm intelligence implemented with the help of a coordination middleware. It is proven that a good query capability with good scalability can be achieved by using swarm-based algorithms. Further, we extended this use case to streaming in fully decentralized P2P networks. P2P applications (that support the streaming delivery method) rely on hybrid approaches, and therefore, are not fully decentralized. This use case includes the possibility to combine user collaboration on both streaming and application level. Besides ant intelligence and bee intelligence [48, 52], the lookup mechanism in this use case includes for the first time usage of a slime mold intelligence [53] and bark beetle intelligence [55]. In [55], a new swarm-inspired algorithm based on the collective feeding of bark beetles, BB-P2P, is presented and applied to the problem of search and lookup in unstructured P2P networks. Additionally, the already existing Physarum Polycephalum algorithm is adapted for search in unstructured P2P networks. These algorithms are implemented, evaluated and compared with Gnutella flooding, k-Walker, AntNet and Dd-slime mold (SM-P2P) algorithms.

[54] presents modeling of the lifecycle of cellular slime molds (Dictyostelium discoideum) and bee-behavior based on the foraging mechanism of honey bees in order to create fully distributed routing algorithms, SMNet and BeeNet, for unstructured P2P networks. A modeling and adaptation of Dd-type slime mold intelligence is done for the first time for routing in unstructured P2P networks. Bee intelligence was already applied to the routing problem in general [64]. However, in [54], another type of mapping and adaptation is proposed.

4 Evaluation

From the experiences presented in Sect. 3, we derived the single steps of the proposed methodology. The overview, where each step has been proven, is given in Table 1. In the continuation of this section, a recommendation for each particular use case is proposed. Further, the basic characteristics of the swarm-inspired algorithms used are presented including theoretical considerations.

4.1 Algorithm Recommendation for Use Cases

Based on the results obtained in the selected use cases, the proposed methodology leads to a recommendation [57] of which algorithm (or type of algorithms) is the best for which use case including dimensions of the treated problem. This is also connected to the issue of configurable parameter recommendations. A systematic approach for the fine tuning of parameters, comparison of a large set of algorithms, obtained results on different topologies etc. leads to the precise specifications about the (type of) algorithm, the set of parameters used, and the dimension of the problem

Table 1 Derivation of the proposed method from previous works

Step	Proof of concept
M1	[47, 50–55]
M2	[51, 53–55]
I1	[47–55]
I2	[32, 50, 53–55]
I3	[30, 31]
SE1	[32, 48–55]
SE2	[31, 47, 48, 50, 51, 55]
SE3	[30, 31, 48, 51]
SE4	[30, 31, 51, 53, 55]
SE5	[32]
SE6	[30, 31, 48–55]

Table 2 Basic characteristics of used swarm-based algorithms

Algorithm	Bio-mechanism	Mechanism of exploration	Mechanism of exploitation
Ant algorithm	Ant foraging behavior, brood sorting, nest building	Pheromone trail evaporation	Using the learned pheromone trails
Bee algorithm	Bee foraging behavior	Random search of scout bees	Neighborhood search in good food sources
Slime Mold (Dd) algorithm	Cellular Dd slime mold /amoebae foraging	Dispersal	Vegetative state
Slime Mold (Physarum) algorithm	Physarum acellular slime mold / amoebae foraging	Variable level of flow through edges – new (source, sink) pairs	Level of conductivity of some tubes – selected tubes grow bigger
Bark beetle algorithm	Feeding aggregation of bark beetles	Anti-attractant pheromone	Higher level of pheromone radius

Table 3 Modelling and communication models in the used swarm-based algorithms

Algorithm	Modelling	Communication
Ant algorithms	Differential equation modelling	Stigmeric, broadcast-like
Bee algorithms	Differential equation modelling	Stigmeric, broadcast-like
Slime Mold (Dd) algorithm	Differential equation modelling // cellular automation + agent-based modelling	Stigmeric
Slime Mold (Physarum) algorithm	Differential equation modelling // fluid dynamics	Stigmeric
Bark beetle algorithm	Differential equation modelling	Stigmergic

that provide the best results. Before presenting use cases, we provide the basic characteristics of used swarm-inspired algorithms (Table 2).

All of the numbered algorithms have some kind of mathematical modeling and use different communication models (Table 3).

4.1.1 Load Balancing

- The concurrent multi-level load balancing strategy is investigated: a complex scenario is assumed where a network can consist of several sub-nets; in the considered case, there were two levels on which load balancing is realized concurrently – between several subnets and within each subnet; load balancing is performed on different levels through different algorithms (both swarm-based

and conventional). The success of a particular combination depends on a network topology.
- The performance metric used was the absolute execution time.
- The following algorithms and their combinations were used: MinMax Ant System (MinMaxAS) [10], AntNet [10], Bee Algorithm [63], adapted Genetic Algorithm (GA) [67], Sender and Round Robin.
- In the chain topology, both combinations of BeeAlgorithm/Sender[1] and MinMaxAS/MinMaxAS were equal good.
- In the ring topology, both combinations of BeeAlgorithm/Sender and MinMaxAS/RoundRobin were equal good.
- In the star topology, both combinations of BeeAlgorithm/BeeAlgorithm and GA/AntNet were equal good.
- In the full topology, a combination of RoundRobin/BeeAlgorithm was the best.
- In almost each topology (except the star topology), the best combination is made by one swarm-intelligent and one conventional algorithm.
- Bee algorithms play a significant role in nearly every topology, as the best obtained results in each topology are based on bee algorithms either used inside subnets or used between subnets or both.

4.1.2 Load Clustering

- An approach similar to the one explained in 4.1.1 is taken: the concurrent multi-level load clustering strategy is investigated.
- The performance metric used was the absolute execution time.
- From the group of clustering algorithms, Hierarchical Clustering obtained the best results.
- The following algorithms and their combinations were used: Ant K-Means [58], Ant-Miner [43], cAnt Miner [42], K-Means [25], Fuzzy C-Means [11], Genetic K-Means [29], Hierarchical Clustering [17, 18], K-Nearest Neighbor [36], and Decision Trees [5].
- From the group of classification algorithms, the Ant-Miner algorithm was the best because of its fast rule generation phase and a good and distinct classification.
- In a small network with only one client that supplies the load, the combination of the Hierarchical algorithm with any other, except the Genetic K-Means algorithm, leads to a good execution time. The best result was delivered by the combination of Hierarchical/Fuzzy C-Means algorithm.
- For large and more complex networks, an intelligent approach with an appropriate similarity function will help.

[1]A combination of Algorithm1/Algorithms2 refers to the concurrent application of Algorithm1 inside subnets and Algorithm 2 between subnets.

4.1.3 Information Placement and Retrieval

- In case of data placement and retrieval in the internet, the following algorithms are used: MinMax Ant System [10], AntNet [10], Bee Algorithm [63], and Gnutella flooding mechanism [2].
- Data placement is placement in two different ways: randomly and by applying some specific strategy as "brood sorting" [6].
- The performance metric used was the absolute execution time.
- The random strategy was generally better than brood strategy. The brood-based ant algorithms obtained better results than the random-based ant algorithms only on small instances. However, the results of the random-based ant algorithms are better with increasing dimensions.
- The Random/AntNet[2] combination supports dynamic processes and is better than the Random/MinMaxAS;
- The bee algorithm obtained the best results especially on large instances; this algorithm informs the "starting place" of the search directly and therefore got the better results.

4.1.4 P2P Streaming

- The previous use case (4.1.3) is extended, i.e., the underlying searching mechanism for finding and retrieving information is used in case of P2P Streaming.
- The following algorithms are applied: AntNet [10], Slime Mold (Dd) Algorithm [37], Physarum Slime Mold [33], k-walker [34], Bark Beetle Algorithm [55], and Gnutella flooding mechanism [1].
- Metrics used were absolute time, average message per node, and success rate.
- The algorithms are evaluated in an environment with two replication strategies: small replication with replicas equal to 2% of the network size and high replication with replicas equal to 16% of the network size.
- In the case of high replication, AntNet shows better performance than Slime Mold in terms of absolute time and success rate.
- In case of small replication, the success rate of Slime Mold is better than the one of AntNet. Slime Mold can be used in cases of very small replication, as it is scalable, and at the same time shows better success rate than AntNet.
- Regarding absolute time:
  - For a small replication rate (2%) and all network sizes, Physarum Slime Mold outperforms the other algorithms.
  - For a bigger replication rate (16%) and all network sizes, the Bark Beetle Algorithm outperforms the other algorithms.

[2] A combination Algorithm1/Algorithms2 refers to the following: Algorithm1 is used for data placement, whereas Algorithm2 is used in information retrieval.

- Regarding average message per node:
  - For all replication rates and all network sizes, Bark Beetle Algorithm outperforms the other algorithms.
- Regarding success rate:
  - For a small replication rate (2%) and small network sizes (50 and 100 nodes), the Bark Beetle Algorithm has comparable success rate to Gnutella.
  - For a bigger replication rate (16%) and all network sizes, the Bark Beetle Algorithm has 100% success rate.

4.1.5 Routing in P2P

- For routing in P2P, the following algorithms were used: AntNet [10], BeeNet [54], BeeHive [64], Slime Mold (Dd) Algorithm [37], Physarum Slime Mold [33], k-walker [34], and Gnutella flooding mechanism [1].
- Metrics used were data packet delivery ratio, average data packet delay, average data packet hop count, and routing overhead messages.
- Slime Mold (Dd) algorithm outperformed all other benchmarked routing algorithms regarding the average delivery delay of data packets with increasing amounts of network nodes and data packet traffic.
- The Bee routing algorithm took the overall second place right after SMNet.
- Regarding the data packet delivery ratio metric, all algorithms show a perfect delivery ratio except the k-Random Walker algorithm.
- Regarding average data packet hop count, Gnutella shows the best results regarding that metric at all network sizes and all traffic levels.
- Regarding routing overhead messages, BeeNet shows the lowest amount of routing overhead messages on all network sizes and also is resilient to an increment of the data packet network traffic.

4.2 Theoretical Considerations

The asymptotic behavior of swarm-inspired algorithms is mainly partially researched. For certain metaheuristics, theoretical work regarding convergence has been partially established with some encouraging results, whereas for many others, no theoretical background exists. In the following subsection, a short overview of available state-of-the-art regarding asymptotic behavior of the swarm-inspired algorithms used is presented.

4.2.1 Convergence

Ant Algorithms In [10], a possibility to prove asymptotic convergence for a particular subset of Ant Colony Optimization (ACO) algorithms is stated. Asymptotic convergence in value is proved for Ant Colony System (ACS) and Min-MaxAS, while convergence in solution is proved for Graph-Based Ant System and $ACO_{bs,\tau min(\theta)}$. [19, 20] investigate the geometric pheromone decrement on not reinforced arcs affected by a constant evaporation factor and state that it is too fast (i.e., implies the premature convergence to suboptimal solutions). Introducing a fixed lower pheromone bound implies random search. A compromise is to allow pheromone trails to tend to zero slower than geometrically by decreasing evaporation factors or by "slowly" decreasing lower pheromone bounds. [44] provides convergence results for a specific type of ant algorithm applied to a specific use case: an Ant Routing (ARA) Algorithm for wireline, packet switched communication networks that are acyclic.

Bee Algorithms [8, 26] investigate the convergence of the Bee Colony Optimization (BCO) algorithm and prove that the current best solution converges to one of the optimal solutions (with the probability of one) as the number of iterations increases.

Slime Molds Algorithms Some theoretical work that considers the behavior of the Physarum-based algorithm partially exists, whereas there are still no results in the scope of the Dd-based algorithm. [27] analyzes a mathematical model of the Physarum growth dynamics. They show how to encode general linear programming (LP) problems as instances of the Physarum and prove that under the growth dynamics, the Physarum is guaranteed to converge to the optimal solution of the LP. [61] analyses the convergence speed and proves that many of the Physarum-inspired algorithms suffer from a low convergence speed. [4] also provides both a proof of the convergence property and the algorithm's correctness.

Bark Beetle As this is a "new-born" algorithm [55], theoretical investigations are still in inception.

4.3 *Advantages of Space-Base Coordination Middleware*

Through the use of space-based middleware [32], our methodology provides the desired properties of general frameworks, composability, autonomy, arbitrary configurations, and easy benchmarking of different algorithms and parameters, whereas self-organization is achieved through swarm-intelligent mechanisms.

With space-based middleware, applications communicate via accessing data in shared spaces. This offers a high decoupling between communicating software components with regard to time, reference, and space. In contrast to message queues, the data in the space are accessed by more flexible coordination laws than solely in FIFO order. Depending on the used space technology, this includes template matching, key- or label-based access, SQL-like query languages, etc.

The advantage is that highly concurrent and complex communication and interaction patterns are well supported. Moreover, it is possible to read data in the space multiple times. An event-driven style is supported: agents can subscribe to space changes and are automatically notified if the required event occurs. Software agents that access the space can be mobile, dynamically join, and leave the system. The distribution and replication of data according to flexible strategies is hidden in the middleware. This makes applications highly independent of changing requirements that concern distribution and data placement issues and leads to flexible and general solutions. The composition of pattern-based components is achieved by letting agents belong to different pattern access spaces that are shared among patterns.

5 Conclusion

In this chapter, we proposed a new methodology for the evaluation of both swarm-intelligent and conventional algorithms, which are based on the following pillars:

- Finding a high-level abstraction of the problem's communication and interaction generics in the form of composable, agent-based coordination patterns
- Development of generic and flexible components based on these patterns
- Development of different swarm-inspired algorithms
- Identification of configuration and evaluation parameters
- Composition of frameworks from components and algorithms in a way that the algorithms can be flexibly exchanged
- Development of test benches.
- Systematic evaluation of different configurations of algorithms, topologies, and parameters

As already mentioned, this methodology can be also applied for the evaluation of conventional algorithms. However, the main benefit of the proposed methodology can be seen in the process of evaluation of swarm-based metaheuristics, because it is crucial for a fair comparison of swarm-based metaheuristics taking in consideration that results depending on the sensitivity analysis and fine tuning of parameters. In general, metaheuristics have large amounts of parameters that should be properly tuned by using some of the available methodologies for automatic tuning, e.g., meta-evolution [21], sequential parameter optimization [2], estimation of distribution [38], racing [35], sharpening [60], or adaptive capping [24].

The open source coordination middleware has proven beneficial for implementations leading to flexible patterns. It was also helpful for the development of test beds. The abstraction into generic patterns was done successfully for load balancing, load clustering, and information placement and retrieval. We succeeded in developing a stable framework for each case where the algorithm can simply be exchanged through "plugging" and leaving the entire scenario implementation untouched (which makes the comparison fair). Also we succeeded in deploying different topologies simply by configurations of the coordination patterns. The result is a reproducible argument for recommendations of which algorithm, or combination of algorithms, best fits the use case at hand. As a proof of concept, we presented several use cases successfully treated by this methodology and showed the conclusions drawn from it. This methodology will contribute to selecting the right algorithm and fostering increasing usage of swarm-inspired algorithms in IT scenarios that require aspects of self-organization.

In future work, a community platform shall be established where all implementations will be provided as open source. Also, we intend to investigate more use cases and algorithms in order to improve the recommendation system.

References

1. S. Androutsellis-Theotokis, D. Spinellis, A survey of peer-to-peer content distribution technologies. ACM Comp. Surv. **36**, 335–371 (2004)
2. T. Bartz-Beielstein, C. Lasarczyk, M. Preuss, Sequential parameter optimization. IEEE Congr. Evolut. Comput. **1**(1), 773–780 (2005)
3. L. Bianchi, M. Dorigo, L.M. Gambardella, W.J. Gutjahr, A survey on metaheuristics for stochastic combinatorial optimization. Nat. Comput. **8**(2), 239–287 (2009)
4. V. Bonifaci, K. Mehlhorn, G. Varma, Physarum can compute shortest paths. J. Theor. Biol. **309**, 121–133 (2012)
5. L. Breiman, J. Friedman, C.J. Stone, R.A. Olshen, *Classification and Regression Trees*, 1st edn. (Chapman and Hall/CRC, 1984)
6. M. Casadei, R. Menezes, M. Viroli, R. Tolksdorf, A Self-organizing Approach to Tuple Distribution in Large-Scale Tuple-Space Systems. International Workshop on Self-Organizing Systems, IWSOS'07, 146–160, 2007
7. M. Chiarandini, L. Paquete, M. Preuss, et al. Experiments on Metaheuristics: Methodological Overview and Open Issues. Tech. Report DMF-2007-03-003, the Danish Mathematical Society, 2007
8. T. Davidovic, D. Teodorovic, M. Selmic, Bee Colony optimization part I: The algorithm overview. YU J. Oper. Res. **25**(1), 33–56 (2015)
9. S.K. Dhurandher, S. Misra, P. Pruthi, S. Singhal, S. Aggarwal, I. Woungang, Using bee algorithm for peer-to-peer file searching in mobile ad hoc networks. J. Netw. Comp. Appl. **34**(5), 1498–1508 (2011)
10. M. Dorigo, T. Stützle, *Ant Colony Optimization* (MIT Press, Cambridge, 2004)
11. J.C. Dunn, A fuzzy relative of the ISODATA process and its use in detecting compact well-separated clusters. J. Cybernetics **3**(3), 32–57 (1973)
12. S. Farzi, Efficient job scheduling in grid computing with modified artificial fish swarm algorithm. Int. J. Comp. Theory Eng. **1**(1), 1793–8201 (2009)
13. D. Garlan, R. Monroe, D. Wile, Acme: Architectural description of component- based systems. Foundations of Component-based Systems, 47–67, 2000
14. D. Garlan, B. Schmerl, S. Cheng, Software architecture-based self-adaptation. Autonomic Computing and Networking, 31–55, 2009
15. I. Georgiadis, J. Magee, J. Kramer, Self-organising software architectures for distributed systems. 1st Workshop on Self-healing Systems, 33–38, 2002
16. M. Goedicke, G. Neumann, U. Zdun, Design and implementation constructs for the development of flexible, component-oriented software architectures. Generative and Component-Based Software Engineering, 114–130, 2001

17. K.C. Gowda, G. Krishna, Agglomerative clustering using the concept of mutual nearest neighborhood. Pattern Recogn. **10**(2), 105–112 (1978)
18. K.C. Gowda, T.V. Ravi, Divisive clustering of symbolic objects using the concepts of both similarity and dissimilarity. Pattern Recogn. **28**(8), 1277–1282 (1995)
19. W. Gutjahr, A converging ACO algorithm for stochastic combinatorial optimization, 2nd Int. Symp. Stochastic Algorithms: Foundations and Applications, 10–25, 2003
20. W. Gutjahr, ACO algorithms with guaranteed convergence to the optimal solution. Inf. Process. Lett. **3**, 145–153 (2002)
21. N. Hansen, *The CMA Evolution Strategy: A Comparing Review* (Springer, Berlin, 2006), pp. 75–102
22. K. Herrmann, Mesh mdl – a middleware for self-organization in ad hoc networks. 23rd International Conference on Distributed Computing Systems, IEEE ICDCSW, 2003
23. G. Hohpe, B. Woolf, *Enterprise Integration Patterns: Designing, Building, and Deploying Messaging Solutions* (Addison-Wesley, Boston, 2003)
24. F. Hutter, H. Hoos, K. Leyton-Brown, T. Stützle, ParamILS: an automatic algorithm configuration framework. J. Artif. Intell. Res. **36**, 267–306 (2009)
25. A.K. Jain, M.N. Murty, P.J. Flynn, Data clustering: a review. ACM Comput. Surv. **31**, 264–323 (1999)
26. T. Jakšić-Krüger, T. Davidović, D. Teodorović, et al., The bee colony optimization algorithm and its convergence. Int. J. Bio-Inspir. Comput. **8**(5), 340–354 (2016)
27. A. Johannson, J. Zou, A Slime Mold Solver for Linear Programming Problems, How the World Computes: Turing Centenary Conference and 8th Conference on Computability in Europe, 344–354, 2012
28. K.O. Jones, G. Boizante, Comparison of firefly algorithm optimisation, particle swarm optimisation and differential evolution. 12th International Conference on Computer Systems and Technologies, CompSysTech, 55–62, 2011
29. K. Krishna, M. Narasimha-Murty, Genetic K-means algorithm. IEEE Trans. Syst. Man Cybern. Part B (Cybernetics) **29**(3), 433–439 (1999)
30. E. Kühn, V. Šešum-Čavić, A Space-Based Generic Pattern for Self-Initiative Load Balancing Agents. 10th International Workshop on Engineering Societies in the Agents World, 17–32, 2009
31. E. Kühn, A. Marek, T. Scheller, V. Šešum-Čavić, M. Vögler, S. Craß, A Space-Based Generic Pattern for Self-Initiative Load Clustering Agents. 14th International Conference on Coordination Models and Languages, 230–244, 2012
32. E. Kühn, S. Craß, G. Joskowicz, A. Marek, T. Scheller, Peer-Based Programming Model for Coordination Patterns. 15th International Conference on Coordination Models and Languages, 121–135, 2013
33. K. Li, C. Torres, K. Thomas, L. Rossi, C.-C. Shen, Slime mold inspired routing protocols for wireless sensor networks. Swarm Intell. **5**, 183–223 (2011)
34. Q. Lv, P. Cao, E. Cohen, K. Li, S. Schenker, Search and replication in unstructured peer-to-peer networks. 16th ACM Int. Conf. on Supercomputing, 84–95, 2002
35. M. Lopez-Ibanez, J. Dubois-Lacoste, T. Stützle, M. Birattari, *The irace package, iterated race for automatic algorithm configuration*, Tech. Report TR/IRIDIA/2011-004, IRIDIA (Université Libre de Bruxelles, Belgium, 2011)
36. F. Mhamdi, M. Elloumi, A new survey on knowledge discovery and data mining. RCIS, 42–432, 2008
37. D. Monismith, B. Mayfield, Slime mold as a model for numerical optimization. Swarm Intelligence Symposium, IEEE SIS, 1–8, 2008
38. V. Nannen, A.E. Eiben, Relevance estimation and value calibration of evolutionary algorithm parameters. 20th International Conference on Artificial Intelligence, 975–980, 2007
39. F. Neumann, C. Witt, *Bio-inspired Computation in Combinatorial Optimization: Algorithms and Their Computational Complexity* (Springer, Berlin, 2010)
40. B. Nuseibeh, Weaving together requirements and architectures. Computer **34**(3), 115–119 (2001)

41. P. Oreizy, M. Gorlick, R. Taylor, D. Heimhigner, G. Johnson, N. Medvidovic, A. Quilici, D. Rosenblum, A. Wolf, An architecture-based approach to self- adaptive software. Intell. Syst. Appl. **14**(3), 54–62 (1999)
42. F.E.B. Otero, A.A. Freitas, C.G. Johnson, Handling continuous attributes in ant colony classification algorithms. IEEE CIDM, 225–231, 2009
43. R. Parpinelli, H. Lopes, A. Freitas, Data mining with an ant Colony optimization algorithm. IEEE Trans. Evolut. Comput. Ant Colony Algorithms **6**(4), 321–332 (2002)
44. P. Purkayastha, J.S. Baras, Convergence results for ant routing algorithms via stochastic approximation. ACM Trans. Auton. Adapt. Syst. **8**(1)., Art. 3, 34 pages (2013)
45. A. Repenning, A. Ioannidou, M. Payton, W. Ye, J. Roschelle, Using components for rapid distributed software development. Software **18**(2), 38–45 (2001)
46. C. Schmitt, A. Freitag, G. Carle, Comada: An adaptive framework with graphical support for configuration, management, and data handling tasks for wireless sensor networks. 9th International Conference on Network and Service Management, 68–73, 2013
47. V. Šešum-Čavić, E. Kühn, Instantiation of a generic model for load balancing with intelligent algorithms. 3rd International Workshop on Self-Organizing Systems, 311–317, 2008
48. V. Šešum-Čavić, E. Kühn, A Swarm Intelligence Appliance to the Construction of an Intelligent Peer-to-Peer Overlay Network. International Conference on Complex, Intelligent and Software Intensive Systems, 1028–1035, 2010
49. V. Šešum-Čavić, E. Kühn, Applying Swarm Intelligence Algorithms for Dynamic Load Balancing to a Cloud Based Call Center. 4th International Conference on Self- Adaptive and Self-Organizing Systems, 255–256, 2010
50. V. Šešum-Čavić, E. Kühn, Comparing Configurable Parameters of Swarm Intelligence Algorithms for Dynamic Load Balancing. Self-Adaptive Networking Workshop, 4th International Conference on Self-Adaptive and Self-Organizing Systems, 42–49, 2010
51. V. Šešum-Čavić, E. Kühn, Self-organized load balancing through Swarm intelligence. Next generation data technologies for collective computational intelligence, Chap. 8, in *Studies in Computational Intelligence*, (Springer, Berlin, 2011), pp. 195–224
52. V. Šešum-Čavić, E. Kühn. Algorithms and Framework for Comparison of Bee-Intelligence Based Peer-to-Peer Lookup. 4th International Conference on Advances in Swarm Intelligence, 404–413, 2013
53. V. Šešum-Čavić, E. Kühn, D. Kanev, Bio-inspired search algorithms for unstructured P2P overlay networks, in *Swarm and Evolutionary Computation*, vol. 29, (Elsevier, Amsterdam, 2016), pp. 73–93
54. V. Šešum-Čavić, E. Kühn, S. Zischka, Swarm-inspired routing algorithms for unstructured P2P networks. Int. J. Swarm Intell. Res. **9**(3)., Article 2 (2018)
55. V. Šešum-Čavić, E. Kühn, L. Fleischhacker, Efficient search and lookup in unstructured P2P overlay networks inspired by swarm intelligence. IEEE Trans. Emerg. Topics Comput. Intell. **4**(3), 351–368 (2020)
56. V. Šešum-Čavić, Swarm Intelligence in Distributed Systems Use-Cases, Keynote Lecture, 11th International Joint Conference on Computational Intelligence IJCCI, 9–15, 2019
57. J. Silberholz, B. Golden, Comparison of metaheuristics, in *Handbook of Metaheuristics*, (Springer, Cham, 2010), pp. 625–640
58. P.S. Shelokar, V.K. Jayaraman, B.D. Kulkarni, An ant colony approach for clustering. Anal. Chim. Acta **509**(1), 187–195 (2004)
59. Y. Shoham, K. Leyton-Brown, *Multiagent Systems Algorithmic, Game-Theoretic, and Logical Foundations* (Cambridge University Press, Cambridge, 2009)
60. S.K. Smit, A.E. Eiben, Comparing parameter tuning methods for evolutionary algorithms. IEEE Congr. Evolut. Comput. **399–406** (2009)
61. D. Straszak, N.K. Vishnoi, IRLS and Slime Mold: Equivalence and Convergence, CoRR abs/1601.02712, 2016
62. L. Tedeschi, Assessment of the adequacy of mathematical models. Agric. Syst. **89**(2–3), 225–247 (2006)

63. D. Teodorovic, P. Lucic, G. Markovic, M. Dell'Orco, Bee Colony Optimization: Principles and Applications, 8th Seminar on Neural Network Applications in Electrical Engineering, 151–156, 2006
64. H.F. Wedde, M. Farooq, Y. Zhang, BeeHive: an efficient fault-tolerant routing algorithm inspired by honey bee behavior, in *Ant Colony Optimization and Swarm Intelligence*, (Springer, Berlin, 2004), pp. 83–94
65. L.P. Wong, M.Y.H. Low, C.S. Chong, A bee colony optimization algorithm for traveling salesman problem. 2nd Asia International Conference on Modelling & Simulation, IEEE AMS, 818–823, 2008
66. P. Zoeteweij, F. Arbab, A component-based parallel constraint solver. Coord. Models Lang., 44–68, series 0302-9743 (2004)
67. A.Y. Zomaya, Y.H. Teh, Observations on using genetic algorithms for dynamic load-balancing. IEEE Trans. Parallel Distrib. Syst. **12**(9), 899–911 (2001)

Eva Maria Kühn holds a tenured professor position at Faculty of Informatics at TU Wien. She is the recipient of the Heinz-Zemanek Research Award for her PhD on "multi-database systems" and the Kurt-Gödel Research Grant from the Austrian Government for a sabbatical at the Indiana Center for Databases at Purdue University, USA. She holds international publications and teaching and research projects in the areas of software development for concurrent and distributed systems, models, languages, and patterns for coordination, space-based computing, coordination middleware, peer-to-peer systems and algorithms, blockchain and distributed ledger technologies, self-organizing bio-inspired systems, usability measurement of programming APIs, and asynchronous transaction models. Project coordinator of many nationally (FWF, FFG) and internationally (EU Commission) funded research projects and projects with industry. She is also member of the governing board of the Austrian and European UNIX systems user group, member of the ISO working group for the standardization of Prolog, member of the senate of the Christian Doppler Forschungsgesellschaft (CDG), member of the science and research council of the Federal State of Salzburg. She also holds international software patents for research work on a new "coordination system" and management and leadership experience as chief technological officer (CTO) of an Austrian spin-of company for software development and has supervised more than 150 master's and doctoral theses. She has more than 120 peer-reviewed international publications and has served as conference chair, program committee member, and organizer of international conferences.

Message:

It was a great pleasure to participate in the book project, and many thanks to Alice E. Smith for this initiative and for perfectly managing it! It is a great sign that women can deliver great research and that women recognize the work of female researchers and support each other, which is unfortunately not always the case in reality. Why? Because women are still severely underrepresented in computer science, and I had to watch quite often that they would rather play in relevant men's networks in order to be able to assert their position instead of supporting other women against these alliances. However, this book is an important step in the right direction.

Vesna Šešum-Ĉavić is Associate Professor at the Faculty of Computer Sciences, Belgrade University Union and Senior Scientist and University Lecturer in Computational Intelligence, Institute of Information Systems Engineering, Compilers and Languages Group, Vienna University of Technology, Austria. She is a graduated mathematician (Dipl.Math) and Magistar of Computer Science (Mag.) from University of Belgrade. She received a doctoral degree in Computer Science from Vienna University of Technology (Dr. techn.). Her research interests cover swarm intelligence, evolutionary computation, network optimization, p2p systems, theory and design of algorithms, combinatorial optimization, complex systems, self-organization, and multi-agent systems. She combines her double background – mathematics and computer sciences, using them in the scope of computational intelligence, specifically nature-inspired computational methodologies, and applying to complex real-world problems from the domain of distributed systems. Computational Intelligence was her main choice as it is a multidisciplinary, interesting area with numerous applications in real life.

Vesna was the Chair of IEEE Women in Computational Intelligence in 2019 and member of the IEEE Women in Engineering Committee. She is a member of the IEEE Women in Engineering Committee and the IEEE CIS Member Activities Committee. She has served as conference chair, program committee member, and keynote speaker of international conferences. Currently, she serves as Associated Editor of IEEE Open Journal of Intelligent Transportation Systems and IEEE Transactions on Emerging Topics in Computational Intelligence.

Message:

I am pleased to contribute to this remarkable book, which provides a motivation and support for young female scientists to continue their research carriers, and also encourage female students to follow their dreams. Unfortunately, some environments still keep encouraging the gender gap. Being aware of all obstacles that a female researcher could face, my message would be:

Work on yourself, strive to your goals, follow your dreams, be persistent and self-confident. With the right combination of tech skills, social awareness, and inspiration, anything is possible and any career aspiration is within reach.

An Improved Bat Algorithm with a Dimension-Based Best Bat for Numerical Optimization

Lingyun Zhou and Alice E Smith

1 Introduction

Many numerical optimization problems in engineering and science exhibit complex characteristics, such as discontinuity, non-convexity, non-differentiability, multi-modality, and multiple constraints. It is difficult for traditional methods to solve these problems. However, swarm intelligence algorithms inspired by the collective behavior of nature can often solve these complex continuous problems efficiently due to the swarm paradigm's robustness, effectiveness, and implementation simplicity. Good examples are ant colony optimization (ACO) [7], particle swarm optimization (PSO) [11], the firefly algorithm (FA) [16], and the bat algorithm (BA) [18].

The bat algorithm (BA) is a recent swarm intelligence algorithm and uses the echolocation behavior of microbats as its natural inspiration. The BA combines and leverages the strengths of other successful swarm algorithms, particularly PSO and Harmony Search [17]. Some comparative studies have shown that BA algorithms are superior to other swarm intelligence algorithms, including particle swarm optimization, over several numerical benchmarks and real applications [1, 5, 10, 12, 15]. Existing application fields of BA are available from recent review papers including [6, 9, 19].

In the BA, there is a population of bats and each bat represents a feasible solution in the search space. At each iteration, the bat with the global best fitness among the population, called the current global best bat, is preserved, and at the next iteration,

L. Zhou
College of Computer Science, South-Central University for Nationalities, Wuhan, China
e-mail: zhouly@mail.scuec.edu.cn

A. E. Smith (✉)
Department of Industrial & Systems Engineering, Auburn University, Auburn, AL, USA
e-mail: smithae@auburn.edu

A. E. Smith (ed.), *Women in Computational Intelligence*, Women in Engineering and Science, https://doi.org/10.1007/978-3-030-79092-9_19

all of the bats tend to fly toward it. Most of the BA variants only use the current best bat to guide the search direction. A few BA variants utilize the experience of both the current global best bat and another randomly selected bat. However, neither of these two methods typically offers a better value in each and every dimension. By using these existing approaches, some favorable information of the other bats in some certain dimensions is ignored, which may result in a slow convergence rate and low accuracy.

Therefore, in this chapter, we propose a novel BA variant, with a dimension-based best bat, named DBBA. At each iteration after all the bats update their positions, the current global best bat is identified. Then, favorable search information of the population in different dimensions is integrated with the global best bat information. This generates a dimension-based best bat that makes full use of the search experience of all of the bats and guides the population to search promising regions at the next iteration over all dimensions of the problem.

The rest of the chapter is organized as the following. In Sect. 2, we provide a brief introduction to the BA and some BA variants. In Sect. 3, we propose our approach. In Sect. 4, we describe the experimental setup, including the comparison algorithms, and analyze the results. Finally, in Sect. 5, we have a summary and future research directions.

2 Related Works

2.1 The Bat Algorithm

The bat algorithm (BA) is a type of meta-heuristic algorithm and carries out the search process by using a population of artificial bats which represents feasible solutions in the search space. It utilizes the concept of swarm intelligence where the agents (microbats, in this case) share information among the swarm and the behavior of an agent is shaped by both its own knowledge and the swarm's collective knowledge.

The BA is designed based on the natural inspiration of microbat colonies as articulated and described in [18]. (1) In nature, bats use echolocation, a type of sonar, to sense distance and also discern the difference between food/prey and background barriers; (2) Bats fly with velocity v_i at position x_i with frequency f_i and loudness A_i to search for a target. They can adjust the frequency of their emitted pulses during echolocation and also change the rate of pulse emission r in [0, 1], depending on the nearness to their target. (3) It is assumed in the BA algorithm that the loudness A_i varies from a maximum to a minimum value.

In a D-dimensional search space, the size of the population is denoted N and the population is defined as $X = (x_1, x_2, ..., x_N)$. Each bat is an individual agent associated with a velocity and a position. Let v_i^t and x_i^t be the velocity and the position of the bat i at time step t, respectively, where i is an integer between 1 and N. The velocity and position update equations can be expressed by

$$f_i = f_{min} + (f_{max} - f_{min})\beta, \tag{1}$$

$$v_i^t = v_i^{t-1} + (x_i^{t-1} - x_*) f_i, \tag{2}$$

$$x_i^t = x_i^{t-1} + v_i^t, \tag{3}$$

where β is a random variable that is uniformly distributed within the range [0, 1]. f_i denotes the frequency of the ith bat which helps define the pace and range of its movement. f_{max} is the maximum of the frequency and f_{min} is the minimum. x_* is the current global best bat, which has the best fitness among all of the population. Note that the bats update their locations each iteration weighted by the current best bat location according to (2) and (3).

To improve optimization ability in the BA, local search is performed on a bat by comparing a random number with a specified threshold. The local search mechanism is defined by

$$x_i = x_* + \epsilon A^t, \tag{4}$$

where $\epsilon \in [-1, 1]$ is a uniform random number, while A^t is the average loudness of the population at the tth time step (i.e., iteration).

A bat's properties in each time step are generated by adjusting frequency, loudness, and pulse emission rate. A bat is synonymous with a solution to the optimization problem and also includes the bat's properties at that time step. Whether a new solution (an updated bat) is accepted or not depends on its quality. Loudness and emission rates are altered only if the bat has improved in quality since the last time step. The loudness A_i of the ith bat updates as the iterations proceed and is determined by

$$A_i^{t+1} = \alpha A_i^t, r_i^{t+1} = r_i^0[1 - e^{-\gamma t}], \tag{5}$$

where α and γ are parameters, which are usually set to constants. A_i^t is the loudness of the ith bat at the tth time step. r_i^{t+1} is the pulse emission rate of the ith bat at the time step $t + 1$, and r_i^0 is its initial pulse emission rate, which is a uniform random value between [0, 1]. Initially, each bat in the population has different values of loudness and pulse emission rate.

The pseudo-code of the classic BA is described in Algorithm 5. In the BA, it is clear that the current global best bat x_* affects the movement direction of each individual in the population.

Algorithm 5: The BA

1: Input the objective functions $f(x)$ and the size of population N;
2: Initialize the parameters f_{min}, f_{max}, α, γ, pulse rate r, frequency f, loudness A;
3: Generate N bats in the search space and calculate their fitness;
4: Record the number of function evaluations $FEs = N$;
5: Find the current global best bat x_*;
6: **while** $FEs < FE_{max}$ **do**
7: **for** $i = 1:N$ **do**
8: Adjust frequency f_i according to Eq. (1);
9: Update velocities v_i and locations x_i according to Eqs. (2) and (3);
10: **if** $(rand > r_i)$ **then**
11: Generate a local solution according to Eq. (4);
12: **end if**
13: Evaluate the new solution and $FEs = FEs + 1$;
14: **if** $(rand < A_i)$ and $(f(x_i) < f(x_*))$ **then**
15: Accept the new solution;
16: Increase r_i and Reduce A_i according to Eq. (5);
17: **end if**
18: Update the current best solution x_*.
19: **end for**
20: **end while**
21: Output the global optimum;

2.2 *Some BA Variants*

The BA has proven to be a powerful optimizer due to its natural inspirations and has successfully been applied to a wide range of engineering and science applications [6, 19]. Despite such progress, some studies reported that its performance may diminish when it searches complex problem space. To overcome this deficiency, several strategies have been proposed. For example, Arora et al. introduced a random inertial weight, akin to the one used in PSO, to the velocity equation [3]. Subsequently, Ramli et al. modified the inertia weight based on the distance between an individual's best position and the global best bat position to increase the exploitation capability [13]. They also proposed a dynamic dimensional size, which performs updating only on a randomly selected subset of dimensions where this random number changes randomly with iteration. Their method was tested on some IEEE CEC 2006 benchmark functions, and experimental results showed that their method surpassed some BA variants.

On the other hand, some other studies have hybridized BA with other optimization methods to take advantage of different approaches. Haruna et al. employed an elite opposition-based learning in the BA for population initialization to diversify the solution search space [8]. The work of Ram et al. uses opposition-based learning (OBL), which is embedded in both initialization and the iteration process [12]. Al-Betar et al. proposed that the strategy of an island model adapted for a bat-inspired algorithm could improve its capability in controlling diversity. Their results showed that the proposed approach could obtain better performance compared with the BA and some recent firefly algorithm (FA) variants [1]. Wang et al. pointed out that embedding existing optimization methods in the BA can improve performance only at certain times. They suggested combining several optimization methods that are chosen randomly. Experimental results showed that the performance of their approach was superior to that of comparison algorithms [15]. However, this approach requires many parameters.

Another work of Al-Betar et al. pointed out that the global best selection scheme may not be that useful because worthier bat location guidance may be found elsewhere [2]. They assessed six natural selection mechanisms derived from other evolutionary algorithms to select the "best" bat. Their experimental results demonstrated that a bat algorithm with a tournament selection scheme (termed TBA) is one of the most successful versions. This work motivated us to look for an alternative scheme by using both the current global best bat and valuable information from the other bats instead of the global best selection scheme.

There are many other BA variants in the literature; however, most of them rely on the bat with the current best fitness among the population to guide its search direction, which may result in slow convergence rate and low accuracy when searching in complex problem spaces. To overcome this problem, Chakri et al. presented a directional echolocation. In their strategy, a bat updates its position by using the experience of the current global best bat and a randomly selected bat [5]. Their algorithm obtained better performance on most of the functions of the IEEE CEC 2006 benchmark compared with ten other swarm intelligence algorithms. Yildizdan extended the new position determination strategy of Chakri et al. by using the difference between a random dimension of a randomly selected neighboring bat and that of an existing bat [20]. The bat neighborhood can be defined in many ways such as Euclidean distance. The bats formed this way can have more useful information in different dimensions; however, they cannot contain all the useful information of the population in every dimension. To achieve valuable search information of the population in every dimension, a more thorough approach is needed.

There is a modified firefly algorithm (FA) proposed by Verma et al., in which each firefly moves to a virtual firefly calculated by a dimension-based updating [14]. The core idea is to change each dimension value of the virtual firefly at a time with the corresponding dimension value of the other fireflies, one by one, and then if its fitness value improves, the virtual firefly is updated. Their experimental results showed that this strategy has a significant positive influence on the FA algorithm.

Based on the gaps identified by the papers mentioned above and the ideas of moving beyond using only the best bat, we form a dimension-based best bat by making use of the search experience of the current best bat and the other bats. Then, we embed the dimension-based best bat into the BA framework to result in a novel algorithm that we term the dimension-based bat algorithm (DBBA).

3 The Proposed DBBA

This section first puts forward a learning strategy of a dimension-based best bat for BA. Then, this concept is integrated to the BA to create the DBBA algorithm.

3.1 Dimension-Based Best Bat

As mentioned in the earlier sections, the current global best bat location is relied upon to generate the new bats' locations. However, the individual with the current best fitness selected from the population does not necessarily have the best value in every dimension, and the other individuals may have better values in certain dimensions. This can be explained in Fig. 1.

In Fig. 1, there is an objective function $f(x) = x^2 + y^2$ with global minimum position of [0, 0]. Suppose that there is a population of three individuals $x_1(1, 3)$, $x_2(2, 2)$, and $x_3(3, 1)$. Obviously, their fitness values are 10, 8, and 10, respectively. So, x_2 is the current global best individual among the population. But we can see that x_1 has a better value than x_2 in the first dimension and x_3 has a better value in the second dimension. If we combine them to form a new position [1, 1] whose first dimension value comes from x_1 and second dimension value comes from x_3, we can get a better individual with a fitness of 2, better than x_2. In this way, we can get a dimension-based best solution that makes the bat fly faster toward the global optimum [0, 0].

The procedure of obtaining the dimension-based best bat $DbBest$ is described in Algorithm 6. Essentially, the procedure inserts the value of each bat in the population for each dimension, one by one, into $DbBest$. It is a greedy local search using the $DbBest$ as a base with the other bats providing the local search components. If a better objective value is found, this change is kept in. Therefore, all of the useful information can be mined from the current population, which can then guide the population to search more promising regions to find the global optimum at the next generation. A potential downside of this approach is possible dependence

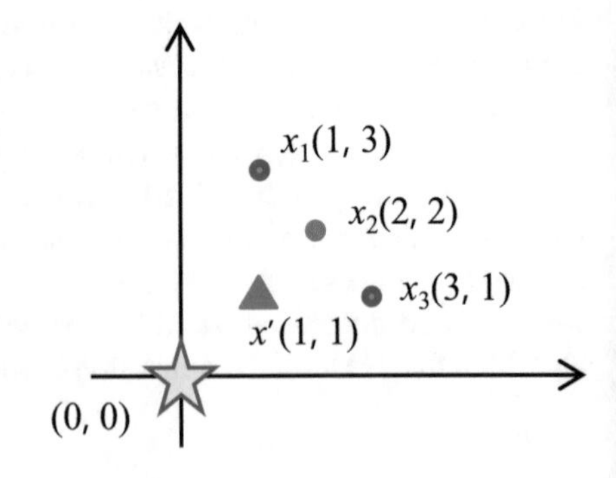

Fig. 1 An example of forming a dimension-based best bat. x' is the dimension-based best bat

Algorithm 6: The DBB

```
1: Input the position of the current global best bat Best, its fitness BestFit, and the
   positions of all of the bats x
2: DbBest = Best;
3: DbBestFit = BestFit;
4: for j = 1:D do
5:    temp = DbBest;
6:    for i = 1:N do
7:       temp(j) = x(i, j);
8:       Evaluate temp and save its fitness to tempFit;
9:       if (tempFit <DbBestFit) then
10:         DbBest = temp;
11:         DbBestFit = tempFit;
12:      end if
13:   end for
14: end for
15: Output the position of the dimension-based best DbBest and its fitness DbBestFit
```

on the ordering of how the dimensions/variables are considered. We explore the sensitivity of the approach to that discussed later in the chapter.

Let D be the dimension of the objective function and N be the population size. It can be analyzed from Algorithm 6 that the total time complexity of the DBB is $O(ND)$, quite reasonable for solving even large dimensional problems with large population sizes.

3.2 The Framework of DBBA

We incorporate the DBB into BA to form a new algorithm named DBBA. In DBBA, the initialization process is the same as that of BA. In each iteration of DBBA, besides performing DBB, there are two features to make it different from that of the standard BA. Both essentially substitute the dimension-based best bat for the current global best bat. One is the velocity updating of bats defined as follows:

$$v_i^t = v_i^{t-1} + (x_i^t - DbBest) f_i. \tag{6}$$

The other is the local search part that can been seen as

$$x_i = DbBest + \epsilon A^t. \tag{7}$$

In the local search part, the $DbBest$ is used, which means that the local search process is near the dimension-based best bat, not the current global best bat. As discussed before, the dimension-based best bat is the same or better than the current global best bat, so the local search is conducted near more promising regions.

Algorithm 7: DBBA

1: Input the objective function $f(x)$ and the size of population N;
2: Initialize the parameters f_{min}, f_{max}, α, γ, pulse rate r, frequency f, loudness A;
3: Randomly generate N bats in the search space and calculate their fitness;
4: Record the number of function evaluations $FEs = N$;
5: Find the current global best bat x_*;
6: $DbBest = x_*$;
7: **while** $FEs < FE_{max}$ **do**
8: **for** $i = 1{:}N$ **do**
9: Adjust frequency f_i according to Eq. (1);
10: Update velocities v_i and locations x_i according to Eq. (6) and Eq. (3), respectively;
11: **if** $(rand > r_i)$ **then**
12: Generate a local solution according to Eq. (7);
13: **end if**
14: Evaluate the new solution and $FEs = FEs + 1$;
15: **if** $(rand < A_i)$ and $(f(x_i) < f(x_{DbBest}))$ **then**
16: Accept the new solutions;
17: Increase r_i and reduce A_i according to Eq. (5);
18: **end if**
19: Update the current best solution x_*.
20: **end for**
21: perform DBB;
22: $FEs = FEs + N * D$;
23: **end while**
24: Output the best found solution;

Let D to be the dimension of the objective function. N is the population size, and T is the maximum number of iterations, which is the stopping criterion of the algorithm. Steps 1–6 comprise the initialization phase, and its time complexity is $O(ND)$. Steps 7–23 comprise the iterative evolution stage, where step 21 is the DBB operator with the time complexity $O(ND)$ as explained earlier, and steps 8–20 are the bat movement operation with the time complexity $O(ND)$. Since a loop over N is conducted in the iterative stage, the time complexity of this stage is $O(TND)$. In summary, the total time complexity of DDBA is $O(ND)+O(TND)$. Omitting the lower order terms, the time complexity of DDBA is $O(TND)$, which is the same as that of the standard BA.

We postulate that the DBBA mechanism can achieve more promising results in terms of the quality of the solutions than ordinary BA. Experimental results in Sect. 4 support this hypothesis.

4 Experimental Results and Discussion

4.1 Benchmark Functions and Simulation Environment

To validate the performance of our approach, the IEEE CEC 2017 benchmark suite is used and these are all minimization problems. Besides including the ability to shift and rotate, this test suite has several advanced features such as some new basic problems, composing test problems by extracting features dimension-wise from several problems, a graded level of linkages, and rotated trap problems. These all make the search spaces more complex [4]. The test suite has 29 functions including unimodal functions f_1 and f_3 (f_2 has been excluded [4]), simple multimodal functions f_4–f_{10}, hybrid functions f_{11}–f_{20}, and composition functions f_{21}–f_{30}. The range of these functions is $[-100, 100]^D$, and the dimension of the functions is set to 30. The experiments were conducted using MATLAB 2019 on a computer with Intel(R) Core(TM) i7-2600 CPU @3.40GHz, 8 GB RAM.

The same maximum number of function evaluations is used as the terminating criterion for all of the algorithms to ensure fair comparisons. The population size is set to 40. The initial population is generated randomly in the search space. To gauge variability, all problem instances are run 30 times, varying only by random number seed. For all the algorithms considered, the common parameters are set to the values shown in Table 1, and the other (algorithm specific) parameters are set as introduced in their original papers.

For all the algorithms, x_{best} and x_{opt} are used to indicate the known optimum solution and the best solution obtained once the search stops after the termination criterion. The fitness error value $\left|f(x_{best}) - f(x_{opt})\right|$ is used to assess the performance of each algorithm. The best values are shown in boldface.

4.2 Comparison of DBBA with a Standard BA

In this subsection, to assess the performance of the proposed DBBA, a comparative study is conducted. The terminating criterion is 3E+5 function evaluations for each algorithm. The mean error, denoted as Mean, of the 30 runs, the corresponding standard deviation, denoted as Std, and the average time of each run, denoted as Time, are recorded in Table 2.

For a statistical evaluation of the experimental results, the non-parametric Wilcoxon rank sum test at a 5% significance level is used to measure the significance of the performance difference between the Mean of BA and that of DBBA. The comparisons between DBBA and BA are summarized in Table 3, which depicts that DBBA wins on win functions, loses on $lose$ functions, and ties on tie functions. The $p\text{-}values$ are recorded as well. From Tables 2 and 3, it can be seen that DBBA gives better results in terms of the mean and the standard deviation of error on the majority

Table 1 Parameter settings used for the algorithms

Parameter	Value
f_{min}	0
f_{max}	1
α	0.95
γ	0.95

Table 2 Mean, Std, and Time (s) for BA and DBBA

Func.	BA			DBBA		
	Mean	Std	Time	Mean	Std	Time
f_1	1.61E+05	6.30E+04	3.86	**2.91E+03**	**3.55E+03**	**1.75**
f_3	**8.42E-02**	**8.87E-002**	4.18	4.38E+04	1.78E+04	**1.76**
f_4	**8.13E+0**	**1.67E+01**	4.14	5.19E+01	2.99E+01	**1.70**
f_5	3.60E+02	7.75E+01	4.56	**1.56E+02**	**4.35E+01**	**2.11**
f_6	7.49E+01	1.02E+01	6.01	**7.85E+00**	**3.92E+0**	**3.70**
f_7	1.11E+03	2.62E+02	4.70	**1.98E+02**	**5.14E+01**	**2.25**
f_8	2.73E+02	7.67E+01	4.57	**1.49E+02**	**4.45E+01**	**2.23**
f_9	8.70E+03	2.95E+03	4.69	**4.86E+03**	**1.72E+03**	**2.20**
f_{10}	4.63E+03	**5.26E+02**	4.94	**3.31E+03**	6.42E+02	**2.52**
f_{11}	1.73E+02	7.32E+01	4.37	**8.19E+01**	**4.08E+01**	**1.97**
f_{12}	**4.25E+05**	**3.00E+05**	4.72	1.98E+06	1.85E+06	**2.25**
f_{13}	1.73E+05	7.16E+04	4.54	**2.58E+04**	**2.58E+04**	**2.08**
f_{14}	**2.50E+03**	**2.31E+03**	5.03	8.08E+04	8.79E+04	**2.61**
f_{15}	6.80E+04	5.12E+04	4.37	**1.21E+04**	**1.39E+04**	**1.91**
f_{16}	1.69E+03	**3.68E+02**	4.71	**1.30E+03**	3.85E+02	**2.18**
f_{17}	1.20E+03	3.35E+02	6.21	**7.79E+02**	**2.98E+02**	**3.75**
f_{18}	**1.14E+05**	**6.12E+04**	4.67	9.70E+05	8.04E+05	**2.15**
f_{19}	2.00E+05	1.57E+05	12.59	**9.85E+03**	**1.25E+04**	**10.19**
f_{20}	9.15E+02	**2.04E+02**	6.43	**7.64E+02**	2.45E+02	**4.00**
f_{21}	5.31E+02	6.89E+01	6.43	**3.52E+02**	**3.99E+01**	**3.93**
f_{22}	5.08E+03	**8.34E+02**	6.83	**3.41E+03**	1.45E+03	**4.45**
f_{23}	1.01E+03	1.31E+02	6.25	**4.79E+02**	**3.75E+01**	**4.81**
f_{24}	1.13E+03	1.54E+02	6.69	**6.93E+02**	**1.07E+02**	**5.30**
f_{25}	4.09E+02	2.85E+01	5.81	**3.86E+02**	**9.08E+0**	**4.40**
f_{26}	5.99E+03	2.60E+03	7.08	**2.93E+03**	**6.52E+02**	**5.70**
f_{27}	1.00E+03	3.31E+02	7.79	**5.42E+02**	**1.69E+01**	**6.43**
f_{28}	**3.25E+02**	5.21E+01	6.79	4.48E+02	**3.86E+01**	**5.51**
f_{29}	2.04E+03	4.91E+02	6.64	**1.13E+03**	**2.46E+02**	**5.29**
f_{30}	2.09E+05	9.72E+04	12.19	**8.31E+03**	**5.67E+03**	**11.63**

of test functions, which confirms the effectiveness of DBBA. DBBA performs better than BA on 23 functions. Only on functions f_3, f_4, f_{12}, f_{14}, f_{18}, and f_{28} are the results of DBBA slightly worse than those of BA. From Table 3, we can see that the p-values are less than the selected significance level 5%, which suggests that the results of DBBA are significantly better than those of BA. From Table 2, we also find the notable result that the Means of DBBA on function f_{19} and function f_{30} are more than twenty times better than those of BA. These two functions are a hybrid function and a composition, respectively, which are the most complicated functions among the IEEE CEC 2017 benchmark functions. This can signify that DBBA can perform better in complex search space.

Table 3 The Wilcoxon rank sum test for the Means of DBBA and BA

win	*lose*	*tie*	*p-value*
23	6	0	0.017897

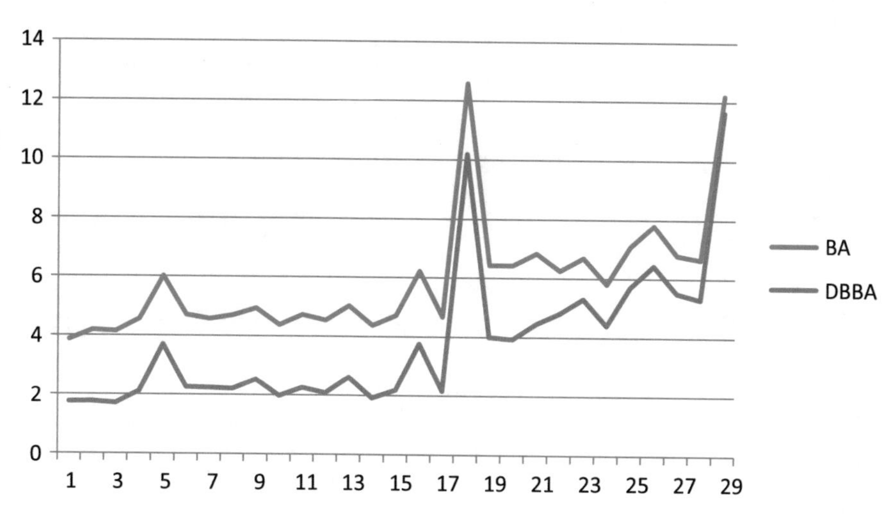

Fig. 2 Run time of BA and DBBA for each function

The average times of each run for each function are plotted in Fig. 2. The vertical axis is the average time over 30 runs and the unit is CPU seconds. The horizontal axis is the number (i.e., identifier) of each function. It is clear that DBBA needs less time than BA over all functions. Compared with BA, the run time of DBBA is reduced up to 58.94%, at least by 4.59%, and by an average of 40.05%. Through the theoretical analysis in Sect. 3, we know that the time complexity of these two algorithms is equal. DBBA still runs in less total time than BA because the number of function evaluations for DBBA is more than that for BA at each iteration. So for a termination criterion of a maximum number of function evaluations, the number of iterations of DBBA is less than that of BA. This means that DBBA performs fewer movement operations of the bats. Therefore, DBB enables the DBBA algorithm to progress faster during each iteration, and the run time of DBBA is less than that of BA.

The convergence curves in terms of the best solution obtained by each algorithm over a single typical run are depicted in Fig. 3. Due to the space limitations, only sample curves of one unimodal function f_1, one multimodal function f_5, one hybrid function f_{11}, and one composition function f_{23} are illustrated. The horizontal axis is the number of function evaluations, and the vertical axis is the fitness error (that is, the amount over the function optimal). This figure indicates that DBBA generally outperforms BA in achieving both faster convergence speed and higher convergence accuracy.

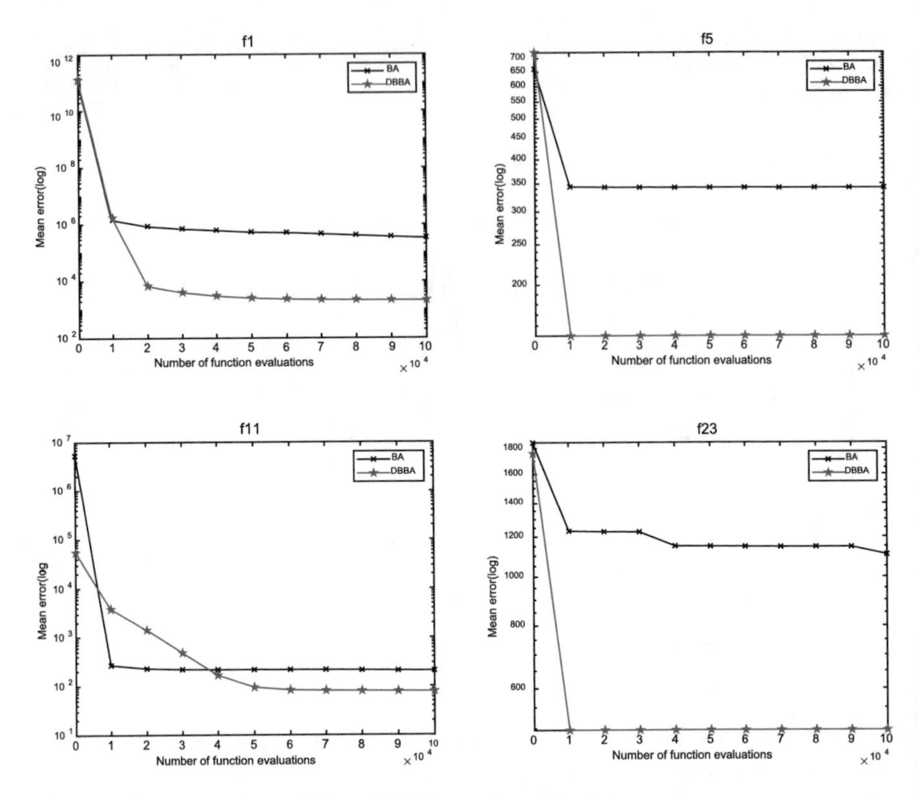

Fig. 3 Convergence curves of BA and DBBA on selected functions

In general, DBBA has the same time complexity of that of BA while achieving more accurate solutions and converging in fewer CPU seconds and fewer iterations. This evidence supports the effectiveness of the DBB approach.

4.3 Comparison of the DBBA with Some Related Algorithms

In this subsection, the DBBA is further compared with several recent BA variants and another swarm-based meta-heuristic algorithm, the firefly algorithm, as detailed below. The parameters of the algorithms are set as the values suggested in the original papers.

(1) OBA is the opposition-based bat algorithm [12]. Its unique parameter is $Jr = 0.4$, which represents the opposition-based generation jumping rate. The opposition learning strategy a candidate solution and its mirror image in parameter space to diversify the search with the aim of improving convergence.

(2) TBA is a bat-inspired algorithm with a tournament selection mechanism [2]. Its unique parameter is t=0.8, which represents the tournament size.
(3) EBA is a bat algorithm using elite opposition-based learning [8].
(4) ODFA is the opposition- and dimension-based firefly algorithm [14].

The reason for choosing these algorithms for comparison is that they are similar to our proposed algorithm DBBA in terms of improvement ideas. OBA and EBA also attempt to leverage favorable information from the search experience of population. The TBA algorithm uses the well-known tournament selection mechanism to select a bat instead of the x_*, reducing the dependence on the globally best bat. ODFA is another meta-heuristic algorithm that uses the firefly algorithm framework along with an opposition-based initial population and a dimension-based updating solution in the framework of the firefly algorithm. The results achieved by OBA, TBA, EBA, ODFA, and DBBA are reported in Table 4.

It is clear in Table 4 that DBBA gives better results in terms of the mean error and the standard deviation for the majority of test functions. Further comparison results between DBBA and the other algorithms are summarized in Table 5, showing by function where DBBA wins (win), loses ($lose$), and ties (tie). The statistical p-values are recorded as well.

Compared with these three BA variants, DBBA achieves the best values on 22 out of the 29 functions. OBA performs better than DBBA only for function f_{28}. EBA is better than DBBA on functions f_3, f_4, and f_{12}. Both OBA and EBA use opposition-based learning to discover useful information from the opposite space of individuals. However, they do not integrate favorable information derived from different individuals, as DBBA does. TBA surpasses DBBA for functions f_{14}, f_{18}, and f_{20}. The performance of TBA is related to the tournament size t, which is set to 0.8 according to the original paper. For the IEEE CEC 2017 benchmark functions, which have more complex search spaces, a higher value of t might be preferred. Compared with the meta-heuristic ODFA that utilizes opposition-based learning and dimension-based updating in the framework of the standard firefly algorithm, DBBA surpasses it on 28 functions. Only on function f_{20} does ODFA achieve better results than those of DBBA.

The convergence curves in terms of the best solution obtained by the BA variants and ODFA over one run are depicted in Fig. 4. Due to the tight space limitation, only sample curves of one unimodal function, one simple multimodal function, two hybrid functions, and two composition functions are illustrated. The horizontal axis is the number of function evaluations, and the vertical axis is the fitness error (that is, the amount over the objective function value). It is apparent that DBBA outperforms the other involved BA variants and ODFA in terms of final solution and convergence speed on most functions. For the convergence progress on the majority of the functions, DBBA improves the best solution steadily, which means it has less tendency to stall out during search.

To compare the total performance of all of the algorithms involved in our experiments, we use SPSS 19, a statistical package, to conduct the non-parametric Friedman test (analogous to ANOVA) on the experimental results. Table 6 presents

Table 4 Mean ± Std for DBBA and the comparative algorithms

Func.	OBA	TBA	EBA	ODFA	DBBA
f_1	8.67E+05± 3.27E+05	3.99E+05± 1.77E+05	4.42E+05± 1.82E+05	4.83e+09± 2.40e+09	**2.91E+03± 3.55E+03**
f_3	3.06E-1± 2.63E-01	3.06E-01± 2.14E-01	**2.54E-1±2.31e-01**	7.78e+04± 2.39e+04	4.38E+04± 1.78E+04
f_4	4.48E+01± 3.41E+01	7.11E+00± 1.41E+01	**5.39E+00±1.27E+01**	7.53e+02± 4.40e+02	5.19E+01± 2.99E+01
f_5	2.83E+02± 5.64E+01	2.88E+02± 7.49E+01	2.78E+02± 5.17E+01	3.73e+02± 4.45e+01	**1.56E+02±4.35E+01**
f_6	6.94E+01± 1.07E+01	6.45E+01± 6.93E+00	6.43E+01± 1.02E+01	6.85e+01± 1.35e+01	**7.85E+00±3.92E+00**
f_7	6.63E+02± 1.08E+02	5.74E+02± 8.52E+01	5.98E+02± 7.95E+01	7.52e+02± 1.60e+02	**1.98E+02±5.14E+01**
f_8	2.30E+02± 5.03E+01	2.43E+02± 5.77E+01	2.30E+02± 5.25E+01	3.18e+02± 3.27e+01	**1.49E+02±4.45E+01**
f_9	6.16E+03± 1.56E+03	6.57E+03± 1.76E+03	6.53E+03± 1.58E+03	1.14e+04± 2.31e+03	**4.86E+03±1.72E+03**
f_{10}	4.44E+03± 6.14E+02	4.71E+03± 6.73E+02	4.64E+03± 8.25E+02	6.45e+03± 5.53e+02	**3.31E+03±6.42E+02**
f_{11}	1.46E+02± 5.05E+01	1.57E+02± 4.91E+01	1.69E+02± 5.90E+01	1.35e+03± 6.37e+02	**8.19E+01±4.08E+01**
f_{12}	2.83E+06± 2.58E+06	6.98E+05± 4.72E+05	**5.66E+05±5.94E+05**	4.20e+08± 3.13e+08	1.98E+06± 1.85E+06
f_{13}	1.79E+05± 7.99E+04	1.90E+05± 5.11E+04	1.95E+05± 6.94E+04	7.33e+07± 3.62e+07	**2.58E+04±2.58E+04**
f_{14}	2.93E+03± 3.21E+03	**2.99E+03±2.44E+03**	3.59E+03± 3.98E+03	5.88e+05± 5.10e+05	8.08E+04± 8.79E+04
f_{15}	7.96E+04± 7.85E+04	6.55E+04± 3.58E+04	6.99E+04± 3.12E+04	1.31e+07± 6.94e+06	**1.21E+04±1.39E+04**
f_{16}	1.84E+03± 4.04E+02	1.59E+03± 4.27E+02	1.70E+03± 5.43E+02	2.55e+03± 4.94e+02	**1.30E+03±3.85E+02**
f_{17}	1.10E+03± 3.65E+02	9.53E+02± 2.96E+02	9.07E+02± 2.81E+02	1.15e+03± 2.59e+02	**7.79E+02±2.98E+02**
f_{18}	1.44E+05± 9.49E+04	**9.58E+04±3.28E+04**	1.09E+05± 4.99E+04	7.83e+06± 6.54e+06	9.70E+05± 8.04E+05
f_{19}	1.04E+06± 3.87E+05	3.62E+05± 1.54E+05	3.09E+05± 9.47E+04	5.39e+07± 4.45e+07	**9.85E+03±1.25E+04**
f_{20}	6.22E+02± 1.87E+02	**5.57E+02±1.51E+02**	6.13E+02± 1.84E+02	7.33e+02± 1.92e+02	7.64E+02± 2.45E+02
f_{21}	5.02E+02± 5.67E+01	4.59E+02± 5.45E+01	4.81E+02± 6.87E+01	5.30e+02± 6.21e+01	**3.52E+02±3.99E+01**
f_{22}	4.49E+03± 2.08E+03	4.14E+03± 1.91E+03	4.21E+03± 1.69E+03	4.24e+03± 3.03e+03	**3.41E+03±1.45E+03**
f_{23}	9.97E+02± 1.40E+02	1.01E+03± 1.77E+02	9.79E+02± 1.63E+02	8.72e+02± 1.10e+02	**4.79E+02±3.75E+01**
f_{24}	1.18E+03± 1.04E+02	1.08E+03± 1.47E+02	1.06E+03± 1.60E+02	9.01e+02± 8.60e+01	**6.93E+02±1.07E+02**
f_{25}	4.00E+02± 2.61E+01	4.11E+02± 2.94E+01	4.07E+02± 2.74E+01	8.41e+02± 2.08e+02	**3.86E+02±9.08E+00**
f_{26}	6.57E+03± 1.77E+03	5.76E+03± 1.54E+03	5.74E+03± 2.00E+03	5.48e+03± 1.59e+03	**2.93E+03±6.52E+02**
f_{27}	1.06E+03± 3.06E+02	9.82E+02± 3.02E+02	1.08E+03± 3.05E+02	9.77e+02± 1.99e+02	**5.42E+02±1.69E+01**
f_{28}	**3.32E+02±4.94E+01**	3.37E+02± 5.84E+01	3.35E+02± 5.01E+01	1.06e+03± 2.72e+02	4.48E+02± 3.86E+01
f_{29}	1.76E+03± 3.60E+02	1.85E+03± 3.51E+02	1.86E+03± 3.97E+02	2.02e+03± 4.40e+02	**1.13E+03±2.46E+02**
f_{30}	1.14E+06± 1.02E+06	2.47E+05± 1.97E+05	1.99E+05± 1.52E+05	6.11e+07± 4.72e+07	**8.31E+03±5.67E+03**

Table 5 The Wilcoxon rank sum test for the mean of DBBA and comparison algorithms on the objective function

	win	*lose*	*tie*	*p-value*
OBA	23	6	0	0.003892
TBA	22	7	0	0.035008
EBA	22	7	0	0.033176
ODFA	28	1	0	0.000003

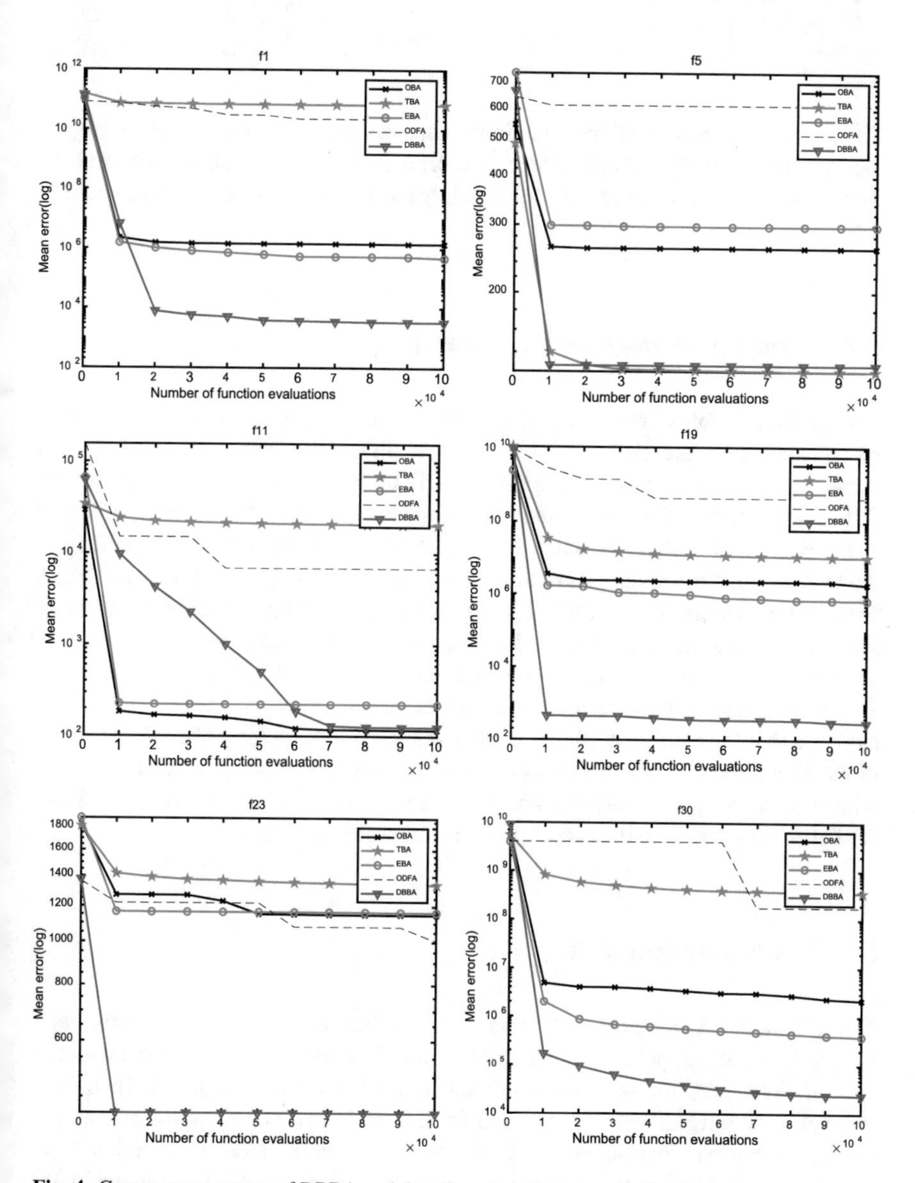

Fig. 4 Convergence curves of DBBA and the other algorithms on selected functions

Table 6 Mean ranks using the Friedman test for all algorithms

	OBA	TBA	EBA	ODFA	DBBA
Mean rank	3.28	2.84	2.67	4.48	**1.72**

Table 7 The Wilcoxon rank sum test for the mean of DBBA and the ordering variants

	win	*lose*	*tie*	*p-value*
DBBA_{rev}	16	13	0	0.888218
DBBA_{ran}	15	14	0	0.913896

the average rankings on all test functions, where the best average ranking among all algorithms is highlighted in boldface. It is clear in Table 6 that DBBA stands at the first position with a score of 1.72 followed by EBA ranked second with the score of 2.67.

4.4 Analysis on the Ordering Sensitivity

One of the possible concerns mentioned earlier in the chapter is that results could be influenced by the order in which the bat dimensions are considered. In this subsection, we explore the sensitivity of DBBA to this ordering. We denote the DBBA algorithm with reverse order of the dimensions when constructing the DBB as DBBA_{rev} and with a random order as DBBA_{ran}. We tested these alternatives on the test suite and the Wilcoxon rank sum test between DBBA, and these two DBBA variants are summarized in Table 7, showing by function where DBBA wins (win), loses ($lose$), and ties (tie). The statistical p-values are recorded as well.

It can be seen that both of the p-values are 0.88218 and 0.913896, respectively. Both of them are very much larger than the selected significance level of 0.05, which suggests that the results of DBBA are not significantly different when the ordering of the dimension components changes. That is to say, the ordering of the dimensions when constructing the DBB will not significantly affect the performance of DBBA. DBBA is very robust to the ordering of the dimensions.

5 Concluding Remarks

The canonical BA and most of its variants rely on the bat with the current global best fitness among the population to guide its search direction. However, the bat with the current global best fitness may not offer the best value in each dimension. Therefore, we propose a dimension-based best bat for BA, which is formed by integrating the current global best bat and more favorable search information of the other bats. Using each population member in turn, each dimension is considered individually using the information of the best bat for all other dimensional values. The best dimensional value within the population is then inserted to the best bat. Then, this dimension-based best bat is used by the BA to guide the bats to fly toward better directions and to conduct local search. We have shown the computational efficiency of DBBA and its improvement over the ordinary BA. To further evaluate the performance of DBBA, experiments have been carried out on the IEEE CEC 2017 benchmark suite. We compare DBBA against several other popular BA variants along with a firefly algorithm variant that uses a related idea to guide search. DBBA clearly shows superiority to the other algorithms in terms of both mean value and variance of the results. Statistical tests support this assertion. DBBA is also computationally efficient. A potential concern of the approach is the dependence on the ordering of considering the dimensions one by one. We performed a sensitivity analysis of this and show that the ordering does not influence the performance of the algorithm.

For future work, we could consider extending DBBA to multi-objective problems. We could also consider intelligent ways to choose the ordering of the consideration of the dimensional variables even though DDBA is robust to this aspect. We might also use a stochastic version to reduce computational effort where not all population members are considered for each dimension—instead, a subset could be selected based on the objective function value or some other criterion.

Acknowledgments The authors thank the Chinese National Natural Science Foundation (No. 61972424), the fund of the China Scholarship Council in 2019, and the Fundamental Research Funds for the Central Universities, South-Central University for Nationalities (No. CZY18012), for partial financial support for this work.

References

1. M.A. Al-Betar, M.A. Awadallah, Island bat algorithm for optimization. Expert Syst. Appl. **107**, 126–145 (2018)
2. M.A. Al-Betar, M.A. Awadallah, H. Faris, X.-S. Yang, A.T. Khader, O.A. Alomari, Bat-inspired algorithms with natural selection mechanisms for global optimization. Neurocomputing **273**, 448–465 (2018)
3. U. Arora, M.E.A. Lodhi, PID parameter tuning using modified bat algorithm. J. Autom. Control Eng. **4**(5), 347–352 (2016)
4. N.H. Awad, M.Z. Ali, J.J. Liang, B.Y. Qu, P.N. Suganthan, P. Definitions, Evaluation criteria for the CEC 2017 special session and competition on single objective real-parameter numerical optimization. Technical report, 2016. https://www.ntu.edu.sg/home/EPNSugan/index_files/CEC2017/CEC2017.htm
5. A. Chakri, R. Khelif, M. Benouaret, X.-S. Yang, New directional bat algorithm for continuous optimization problems. Expert Syst. Appl. **69**, 159–175 (2017)
6. M. Chawla, M. Duhan, Bat algorithm: a survey of the state-of-the-art. Appl. Artif. Intell. **29**(6), 617–634 (2015)
7. M. Dorigo, V. Maniezzo, A. Colorni, Ant system: optimization by a colony of cooperating agents. IEEE Trans. Syst. Man Cybern. B (Cybern.) **26**(1), 29–41 (1996)
8. Z. Haruna, M.B. Muazu, K.A. Abubilal, S.A. Tijani, Development of a modified bat algorithm using elite opposition-based learning, in *2017 IEEE 3rd International Conference on Electro-Technology for National Development (NIGERCON)* (IEEE, 2017), pp. 144–151

9. T. Jayabarathi, T. Raghunathan, A.H. Gandomi, The bat algorithm, variants and some practical engineering applications: A review, in *Nature-Inspired Algorithms and Applied Optimization* (Springer, 2018), pp. 313–330
10. K. Kaced, C. Larbes, N. Ramzan, M. Bounabi, Z. Elabadine Dahmane, Bat algorithm based maximum power point tracking for photovoltaic system under partial shading conditions. Solar Energy **158**, 490–503 (2017)
11. J. Kennedy, R. Eberhart, Particle swarm optimization, in *Proceedings of ICNN'95-International Conference on Neural Networks*, vol. 4 (IEEE, 1995), pp. 1942–1948
12. G. Ram, D. Mandal, R. Kar, S.P. Ghoshal, Opposition-based bat algorithm for optimal design of circular and concentric circular arrays with improved far-field radiation characteristics. Int. J. Numer. Modell. Electron. Networks Devices Fields **30**(3–4), e2087 (2017)
13. M.R. Ramli, Z. A. Abas, M.I. Desa, Z. Z. Abidin, M.B. Alazzam, Enhanced convergence of bat algorithm based on dimensional and inertia weight factor. J. King Saudi Univ. Comput. Inf. Sci. **31**(4), 452–458 (2019)
14. O.P. Verma, D. Aggarwal, T. Patodi, Opposition and dimensional based modified firefly algorithm. Expert Syst. Appl. **44**, 168–176 (2016)
15. Y. Wang, P. Wang, J. Zhang, Z. Cui, X. Cai, W. Zhang, J. Chen, A novel bat algorithm with multiple strategies coupling for numerical optimization. Mathematics **7**(2), 1–17 (2019)
16. X.-S. Yang, Nature-inspired metaheuristic algorithms, in *Nature-inspired Metaheuristic Algorithms* (Luniver Press, London, 2008), pp. 242–246
17. X.-S. Yang, Harmony search as a metaheuristic algorithm, in *Music-Inspired Harmony Search Algorithm* (Springer, 2009), pp. 1–14
18. X.-S. Yang, A new metaheuristic bat-inspired algorithm, in *Nature inspired Cooperative Strategies for Optimization (NICSO 2010)* (Springer, 2010), pp. 65–74
19. X.-S. Yang, X. He, Bat algorithm: literature review and applications. Int. J. Bio Inspired Comput. **5**(3), 141–149 (2013)
20. G. Yildizdan, Ö.K. Baykan, A novel modified bat algorithm hybridizing by differential evolution algorithm. Expert Syst. Appl. **141**, 112–949 (2020)

Lingyun Zhou is currently a lecturer at the College of Computer Science South-Central University for Nationalities, China. She received the Ph.D. degree (2018) from Computer Software and Theory, College of Computer Science, Wuhan University, China. She was sponsored by the Chinese Scholarship Council as a visiting scholar and conducted research in the department of Industrial and Systems Engineering at Auburn University (2019–2020).

Her main research interests are in the area of swarm intelligence and their applications in engineering. She is a member of China Computer Federation.

Message:
It is my honor to participate in this book project. I entered the research work when I started my doctoral career. This field is very interesting and I was attracted to it very soon. I think women researchers can give full play to their advantages and innovative potential in scientific research and balance family and research work. I hope every female researcher can pursue their dreams in her research areas.

Alice E Smith is the Joe W. Forehand/Accenture Distinguished Professor of the Industrial and Systems Engineering Department at Auburn University, where she served as Department Chair from 1999 to 2011. She also has a joint appointment with the Department of Computer Science and Software Engineering. Previously, she was on the faculty of the Department of Industrial Engineering at the University of Pittsburgh from 1991 to 1999, which she joined after industrial experience with Southwestern Bell Corporation. Dr. Smith has degrees from Rice University, Saint Louis University, and Missouri University of Science and Technology.

Dr. Smith's research focus is analysis, modeling, and optimization of complex systems with emphasis on computation inspired by natural systems. She holds one U.S. patent and several international patents and has authored more than 200 publications which have garnered over nearly 13,000 citations and an H Index of 47 (Google Scholar). She is the editor of the recent book *Women in Industrial and Systems Engineering: Key Advances and Perspectives on Emerging Topics* (https://www.springer.com/us/book/9783030118655#aboutBook). Several of her papers are among the most highly cited in their respective journals including the most cited paper of *Reliability Engineering & System Safety* and the 2nd most cited paper of *IEEE Transactions on Reliability*. She won the E. L. Grant Best Paper Awards in 1999 and in 2006 and the William A. J. Golomski Best Paper Award in 2002. Dr. Smith is the Editor in Chief of *INFORMS Journal on Computing* and an *Area Editor of Computers & Operations Research.*

Dr. Smith has been a principal investigator on over $10 million of sponsored research with funding by NASA, U.S. Department of Defense, Missile Defense Agency, National Security Agency, NIST, U.S. Department of Transportation, Lockheed Martin, Adtranz (now Bombardier Transportation), the Ben Franklin Technology Center of Western Pennsylvania, and U.S. National Science Foundation, from which she has been awarded 18 distinct grants including a CAREER grant in 1995 and an ADVANCE Leadership grant in 2001. Her industrial partners on sponsored research projects have included DaimlerChrysler Electronics, Toyota, Eljer, Frontier Technology Inc., Extrude Hone, Ford Motor, and Crucible Compaction Metals. International research collaborations have been sponsored by Germany, Mexico, Japan, Turkey, United Kingdom, The Netherlands, Egypt, South Korea, Iraq, China, Colombia, Chile, Algeria, and the USA and by the Institute of International Education. In 2013, she was a Fulbright Senior Scholar at Bilkent University in Ankara, Turkey, in 2016 a Fulbright Specialist at EAFIT in Medellin, Colombia, in 2017 a Senior Fulbright fellow at Pontifical Catholic University of Valparaiso, Chile, and in 2020 a Fulbright Specialist at University La Sabana in Bogota, Colombia.

For accomplishments in research, education, and service, she was named the Joe W. Forehand/Accenture Distinguished Professor in 2015. Previously, she was the H. Allen and Martha Reed Professor. In 2017, she received the inaugural Auburn University 100 Women Strong Leadership in Diversity Faculty Award. Dr. Smith was awarded the Wellington Award in 2016, the IIE Albert G. Holzman Distinguished Educator Award in 2012, and the INFORMS WORMS Award for the Advancement of Women in OR/MS in 2009. Dr. Smith was named the Philpott—WestPoint Stevens Professor in 2001, received the Senior Research Award of the College of Engineering at Auburn University in 2001, and the University of Pittsburgh School of Engineering Board of Visitors Faculty Award for Research and Scholarly Activity in 1996.

Dr. Smith is a fellow of the Institute of Electrical and Electronics Engineers (IEEE), a fellow of the Institute of Industrial and Systems Engineers (IISE), a fellow of Institute for Operations Research and Management Science (INFORMS), a senior member of the Society of Women Engineers, a member of Tau Beta Pi, and a Registered Professional Engineer in Alabama and Pennsylvania. She was elected to serve on the Administrative Committee of the IEEE Computational Intelligence Society from 2013 to 2018 and 2020 to 2022 and as IISE Senior Vice President—Publications from 2014 to 2017. She served as associate editor for two IEEE journals and is currently an IEEE Computational Intelligence Society Distinguished Lecturer and an INFORMS Official Speaker. She has served as Chair of the Council of Industrial Engineering Academic Department Heads and as President of the INFORMS Association of Chairs of Operations Research Departments. She was a keynote speaker at the International INFORMS Conference (2019) and at the IEEE World Congress on Computational Intelligence (2018). She was named a 2020 Yellowhammer Women of Impact (20 women are honored each year in the State of Alabama https://alabamawomen.org/#2020) and was named an INFORMS Diversity, Equity, and Inclusion Ambassador in 2021.

During her tenure as Chair, the Industrial and Systems Engineering Department at Auburn University witnessed unprecedented growth in student enrollments (+200%), research funding (+500%), and private donations (+400%). Facilities expanded significantly and the department became a leader of three federally funded research centers. Interdisciplinary educational programs were developed and diversity of student body and faculty flourished. Ranking (*U.S. News*) significantly surpassed all other Auburn University engineering departments.

Message:

I have always loved mathematics, problem solving, and art. I found that engineering, and computational intelligence specifically, brings all three together. As an undergraduate engineering student in the late 1970s (!) I was in the first cohort that included a group of women, so I did not feel so isolated or strange. Although many years have passed since that time, women are still underrepresented in our field. This does not need to be. Computational intelligence, whether you focus on its aspects of computer science, mathematics, or engineering, is a rich career choice full of possibilities and lifelong growth. I have also found the field very welcoming to everyone regardless of background, gender, nationality, ethnicity, etc. You can see by the many chapters of this book that women are leading important and creative research in a great variety of technical and application area all under the scope of computational intelligence. I encourage you to follow your intuition and strive towards your dreams. If engineering, and particularly computational intelligence, seems appealing to you, go for it. It will not be easy, but great personal and profession satisfactions await you.

Index

A. E. Smith (ed.), *Women in Computational Intelligence*, Women in Engineering and Science, https://doi.org/10.1007/978-3-030-79092-9

D

E

J

K

L

T

U

V

W

Printed in the United States
by Baker & Taylor Publisher Services